PRINCIPLES OF DIGITAL COMMUNICATION SYSTEMS AND COMPUTER NETWORKS

PRINCIPLES OF DIGITAL COMMUNICATION SYSTEMS AND COMPUTER NETWORKS

K. V. Prasad

CHARLES RIVER MEDIA, INC.
Hingham, Massachusetts

Acquisitions Editor: James Walsh
Cover Design: Sherry Stinson

CHARLES RIVER MEDIA, INC.
10 Downer Avenue
Hingham, Massachusetts 02043
781-740-0400
781-740-8816 (FAX)
info@charlesriver.com
www.charlesriver.com

This book is printed on acid-free paper.

K. V. K. K. Prasad. *Principles of Digital Communication Systems and Computer Networks*
ISBN: 1-58450-329-7

Library of Congress Cataloging-in-Publication Data
Prasad, K. V. K. K.
  Principles of digital communication systems and computer networks /
K.V.K.K. Prasad.— 1st ed.
    p. cm.
  ISBN 1-58450-329-7 (Paperback : alk. paper)
 1. Digital communications. 2. Computer networks. I. Title.
TK5103.7.P74 2003
004.6—dc22
                        2003021571

Printed in the United States of America
04 7 6 5 4 3 2 First Edition

CHARLES RIVER MEDIA titles are available for site license or bulk purchase by institutions, user groups,
corporations, etc. For additional information, please contact the Special Sales Department at 781-740-0400.

Contents

Preface

At the dawn of the twenty-first century, we are witnessing a revolution that will make the utopia of a "Global Village" a reality. Telecommunications technology is paving the way to a world in which distance is no longer a barrier. Telecommunication is enabling us to interact with one another irrespective of our physical locations, bringing people of different countries, cultures, and races closer to one another.

The revolution we are witnessing is due to the fast pace of the convergence of information, communications, and entertainment (ICE) technologies. These technologies help us to communicate and carry out business, to entertain and educate ourselves, and to share information and knowledge. However, we are at just the beginning of this revolution. To ensure that every individual on this planet is connected and is provided with quality communication services is a gigantic task. This task requires dedicated professionals with a commitment to technological excellence. These professionals will be the architects of the Global Village.

Prospective communications professionals need to have a strong foundation in digital communication and data communication protocols, as well as an exposure to the latest technological innovations. This book aims at providing that foundation to the students of computer science/communication engineering/information technology.

ABOUT THE BOOK

This book lays the foundation for a career in telecommunications and data communications. It can be used as a textbook for the following subjects:

- Digital Communication Systems
- Data Communication and Computer Networks
- Mobile Computing

Accordingly, the book is divided into three parts.

Part 1 focuses on digital communication systems. This part covers the basic building blocks of digital communications, such as transmission media, multiplexing, multiple access, source coding, error detecting and correcting codes, and modulation techniques. Representative telecommunication systems such as Public Switched Telephone Network, terrestrial radio systems, satellite radio systems, and optical communication systems are also discussed.

Part 2 focuses on data communication protocols and computer networking. The Open Systems Interconnection (OSI) protocol architecture and the Transmission Control Protocol/Internet Protocol (TCP/IP) architecture are covered in detail. Representative computer networks based on international standards are covered, including Signaling System No. 7 (SS7), Integrated Services Digital Network (ISDN), Frame Relay and Asynchronous Transfer Mode (ATM).

Part 3 focuses on mobile computing and convergence technologies. Radio paging, cellular mobile communications, global positioning system, wireless Internet, and wireless personal/home area networks are covered, highlighting the technology ingredients of mobile computing. Multimedia communication over IP networks and Computer Telephony Integration (CTI) are covered in detail, highlighting the technology ingredients of convergence. The architecture of Telecommunications Management Network, which is gaining importance to manage complex telecommunication networks, is presented. The last chapter gives a glimpse of futuristic technologies.

At the end of each chapter, references, questions, exercises, and projects are given. References are books and research papers published in professional journals as well as Web addresses. Needless to say, the Internet continues to be the best platform to obtain the latest information on a specific topic. Just use a search engine and enter the topic of your choice, and you will get millions of references! The Questions help the reader focus on specific topics while reading the book. The Exercise section is exploratory and the reader is urged to do these exercises to enhance his conceptual understanding. Final year students can choose some of the exercises as seminar topics. In the Projects section, brief problem definitions are given based on the requirements of the communications industry. Final year students can select some of these projects in their areas of interest. These projects are long-term projects (of one month to three months duration). Certainly, students need the help of their teachers in converting the project ideas into detailed requirement documents. Of course, the reader is welcome to contact the author at kvkk.prasad@acm.org for any additional information.

Appendix A gives a glossary of important terms. Appendix B gives a list of acronyms and abbreviations. Appendix C gives the solutions to selected exercises from various chapters of the book.

Planning a Career in Telecom/Datacom

Telecom/datacom is a very promising field, and those who choose this field will have very bright careers. This field requires software and hardware professionals of almost all skill sets. Software development in communication protocols is generally carried out in C or C++, though Java is picking up fast. Prospective telecom/datacom software professionals need to gain expertise in at least two of the following skill sets:

- Unix/Solaris/Linux with C and C++
- Windows NT/2000/XP with C, C++, and C#
- Real-time operating systems (such as pSOS, VxWorks, OS/9, and RTLinux) and C/C++
- Unix/Solaris/Linux with Java (including RMI, JDBC, CORBA, EJB, Jini, and JMF)
- RDBMS with frontend tools such as VB, VC++, and Java
- Mobile operating systems (such as Win CE, Palm Operating System, and Symbian Operating System) and C/C++
- Markup languages such as XML, WML, XHTML, and VoiceXML and scripting languages such as JavaScript and WMLScript

For those who have an interest in pursuing hardware-cum-software development, the following skill sets are suggested:

- Real-time operating systems (such as pSOS, VxWorks, OS/9 and RTLinux) and C/C++
- VLSI design languages and tools (such as VHDL and Verilog) and C/C++
- Micro-controllers (such as Intel 8051, MC 68HC11 and ARM), associated assembly language, and C/C++
- Digital Signal Processors and C/C++

Along with expertise in the development environment (operating systems, programming languages, etc.), domain expertise in communications is important for a successful career. However, communications is a vast field, and it would be futile to try to gain expertise in every aspect. A very good understanding of the basics of telecommunication systems and communication protocols is mandatory for everyone. After that, one needs to have an exposure to the latest developments and then focus on some specialization. The specialization can be in any of the following areas:

- TCP/IP protocol stack
- Multimedia communication over IP networks
- Wireless Internet

- Telecommunication switching software development
- Telecommunication network management
- Voice and video coding
- Wireless local area networks
- Wireless personal area networks
- Terrestrial radio communication
- Optical communication
- Satellite communication
- Signaling systems
- Computer Telephony Integration

If one has a strong foundation in the basic concepts of telecommunications and data communications, it is not difficult to move from one domain to another during one's career. The sole aim of this book is to cover the fundamentals in detail and also to give an exposure to the latest developments in the application domains listed.

Let us now begin our journey into the exciting world of communications.

I Digital Communication Systems

The objective of any telecommunication system is to facilitate communication between people—who may be sitting in adjacent rooms or located in different corners of the world. Perhaps one of them is traveling. The information people may like to exchange can be in different forms—text, graphics, voice, or video. In broadcasting, information is sent from a central location, and people just receive the information; they are passive listeners (in the case of radio) or passive watchers (in the case of television). It is not just people; devices may have to communicate with each other—a PC to a printer, a digital camera to a PC, or two devices in a process control system.

The basic principles of all these types of communication are the same. In this part of the book, we will study the fundamentals of digital communication: the building blocks of a communication system and the mechanisms of coding various types of information. We will study Shannon's information theory which laid the foundation of digital communications. We also will cover the characteristics of various transmission media and how to utilize a medium effectively through multiplexing and multiple access techniques. The issues involved in designing telecommunication systems are also discussed. Finally, we will study representative telecommunication systems using cable, terrestrial radio, satellite radio, and optical fiber as the transmission media.

This part contains 14 chapters that cover all the fundamental aspects of telecommunications and representative telecommunication networks.

1 | Basics of Communication Systems

In This Chapter

- The Building Blocks of a Communication System
- Types of Communication
- Analog Communication and Digital Communication
- Transmission Impairments
- Terminology Used in Communication Systems

We begin the journey into the exciting field of telecommunications by studying the basic building blocks of a telecommunication system. We will study the various types of communication and how the electrical signal is impaired as it travels through the transmission medium. With the advances in digital electronics, digital communication systems slowly are replacing analog systems. We will discuss the differences between analog communication and digital communication.

1.1 BASIC TELECOMMUNICATION SYSTEM

A very simple telecom system is shown in Figure 1.1. At the transmitting end, there will be a source that generates the data and a transducer that converts the data into an electrical signal. The signal is sent over a transmission medium and, at the receiving end, the transducer again converts the electrical signal into data and is given to the destination (sink). For example, if two people want to talk to each other using this system, the transducer is the microphone that converts the sound waves into equivalent electrical signals. At the receiving end, the speakers convert the electrical signal into acoustic waves. Similarly, if video is to be transmitted, the transducers required are a video camera at the

3

transmitting side and a monitor at the receiving side. The medium can be copper wire. The public address system used in an auditorium is an example of such a simple communication system.

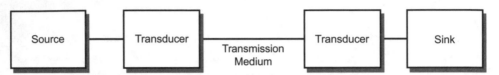

FIGURE 1.1 Basic telecommunication system.

In an electrical communication system, at the transmitting side, a transducer converts the real-life information into an electrical signal. At the receiving side, a transducer converts the electrical signal back into real-life information.

What is the problem with this system? As the electrical signal passes through the medium, the signal gets attenuated. The attenuated signal may not be able to drive the transducer at the receiving end at all if the distance between the sender and the receiver is large. We can, to some extent, overcome this problem by using amplifiers between. The amplifier will ensure that the electrical signals are of sufficient strength to drive the transducer.

But we still have a problem. The transmission medium introduces noise. The noise cannot be eliminated at all. So, in the above case, we amplify the signal, but at the same time, we also amplify the noise that is added to the actual signal containing the information. Amplification alone does not solve the problem, particularly when the system has to cover large distances.

As the electrical signal passes through the transmission medium, the signal gets attenuated. In addition, the transmission medium introduces noise and, as a result, the signal gets distorted.

The objective of designing a communication system is to reproduce the electrical signal at the receiving end with minimal distortion.

The objective of designing a communication system is for the electrical signal at the transmitting end to be reproduced at the receiving end with minimal distortion. To achieve this, different techniques are used, depending on issues such as type of data, type of communication medium, distance to be covered, and so forth.

Figure 1.2 shows a communication system used to interconnect two computers. The computers output electrical signals directly (through the serial port, for example), and hence there is no need for a transducer. The data can be passed directly through the communication medium to the other computer if the distance is small (less than 100 meters).

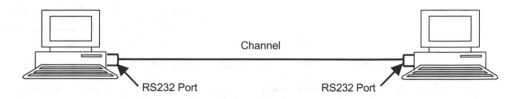

FIGURE 1.2 **PC-to-PC communication.**

The serial ports of two computers can be connected directly using a copper cable. However, due to the signal attenuation, the distance cannot be more than 100 meters.

Figure 1.3 shows a communication system in which two PCs communicate with each other over a telephone network. In this system, we introduced a new device called a *modem* (modulator-demodulator) at both ends. The PCs send digital signals, which the modem converts into analog signals and transmits through the medium (copper wires). At the receiving end, the modem converts the incoming analog signal into digital form and passes it on to the PC.

FIGURE 1.3 PC-to-PC communication over telephone network.

Two computers can communicate with each other through the telephone network, using a modem at each end. The modem converts the digital signals generated by the computer into analog form for transmission over the medium at the transmitting end and the reverse at the receiving end.

Figure 1.4 shows a generic communication system. In this figure, a block "medium access processing" is introduced. This block has various functions, depending on the requirement. In some communication systems, the transmission medium needs to be shared by a number of users. Sometimes the user is allowed to transmit only during certain time periods. Sometimes the user may need to send the same data to multiple users. Additional processing needs to be done to cater to all these requirements. At the transmitting side, the source generates information that is converted into an electrical signal. This signal, called the *baseband signal*, is processed and transmitted only when it is allowed. The signal is sent on to the transmission medium through a transmitter. At the receiving end, the receiver amplifies the signal and does the necessary operations to present the baseband signal to the user. Any telecommunication system is a special form of this system. Consider the following examples:

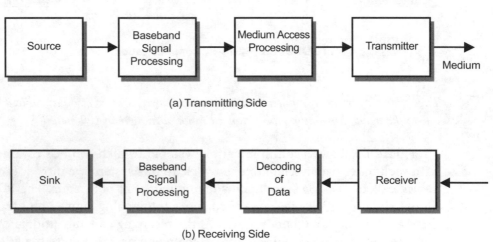

(a) Transmitting Side

(b) Receiving Side

FIGURE 1.4 Generic communication system.

In a radio communication system, the electrical signal is transformed into a high-frequency signal and sent over the air.

In the case of a radio communication system for broadcasting audio programs, the electrical signal is transformed into a high-frequency signal and sent through the air (free space). A radio transmitter is used to do this. A reverse of this transformation—converting the high-frequency signal into an audio signal—is performed at the receiving station. Since it is a broadcasting system, many receivers can receive the information.

In a communication system on which two persons communicate with two other persons located somewhere else, but only on one communication link, the voice signals need to be combined. We cannot mix the two voice signals directly because it will not be possible to separate them at the receiving end. We need to "multiplex" the two signals, using special techniques.

In a mobile communication system, a radio channel has to be shared by a number of users. Each user has to use the radio channel for a short time during

When multiple users have to share the same transmission medium, multiple access techniques are used. These techniques are required in both radio communication and cable communication.

which he has to transmit his data and then wait for his next turn. This mechanism of sharing the channel is known as *multiple access*.

Hence, depending on the type of communication, the distance to be covered, etc., a communication system will consist of a number of elements, each element carrying out a specific function. Some important elements are:

- **Multiplexer:** Combines the signals from different sources to transmit on the channel. At the receiving end, a demultiplexer is used to separate the signals.

- **Multiple access:** When two or more users share the same channel, each user has to transmit his signal only at a specified time or using a specific frequency band.

- **Error detection and correction:** If the channel is noisy, the received data will have errors. Detection, and if possible correction, of the errors has to be done at the receiving end. This is done through a mechanism called *channel coding*.

- **Source coding:** If the channel has a lower bandwidth than the input signal bandwidth, the input signal has to be processed to reduce its bandwidth so that it can be accommodated on the channel.

- **Switching:** If a large number of users has to be provided with communication facilities, as in a telephone network, the users are to be connected based on the numbers dialed. This is done through a mechanism called *switching*.

- **Signaling:** In a telephone network, when you dial a particular telephone number, you are telling the network whom you want to call. This is called *signaling information*. The telephone switch (or exchange) will process the signaling information to carry out the necessary operations for connecting to the called party.

The various functions to be carried out in a communication system are: multiplexing, multiple access, error detection and correction, source coding, switching and signaling.

Two voice signals cannot be mixed directly because it will not be possible to separate them at the receiving end. The two voice signals can be transformed into different frequencies to combine them and send over the medium.

1.2 TYPES OF COMMUNICATION

Based on the requirements, the communications can be of different types:

Point-to-point communication: In this type, communication takes place between two end points. For instance, in the case of voice communication using telephones, there is one calling party and one called party. Hence the communication is point-to-point.

Point-to-multipoint communication: In this type of communication, there is one sender and multiple recipients. For example, in voice conferencing, one person will be talking but many others can listen. The message from the sender has to be *multicast* to many others.

Broadcasting: In a broadcasting system, there is a central location from which information is sent to many recipients, as in the case of audio or video broadcasting. In a broadcasting system, the listeners are passive, and there is no reverse communication path.

Simplex communication: In simplex communication, communication is possible only in one direction. There is one sender and one receiver; the sender and receiver cannot change roles.

Half-duplex communication: Half-duplex communication is possible in both directions between two entities (computers or persons), but one at a time. A walkie-talkie uses this approach. The person who wants to talk presses a talk button on his handset to start talking, and the other person's handset will be in receive mode. When the sender finishes, he terminates it with an over message. The other person can press the talk button and start talking. These types of systems require limited channel bandwidth, so they are low cost systems.

Full-duplex communication: In a full-duplex communication system, the two parties—the caller and the called—can communicate simultaneously, as in a telephone system. However, note that the communication system allows simultaneous transmission of data, but when two persons talk simultaneously, there is no effective communication! The ability of the communication system to transport data in both directions defines the system as full-duplex.

In simplex communication, the communication is one-way only. In half-duplex communication, communication is both ways, but only in one direction at a time. In full-duplex communication, communication is in both directions simultaneously.

Depending on the type of information transmitted, we have voice communication, data communication, fax communication, and video communication systems. When various types of information are clubbed together, we talk of multimedia communications. Even a few years ago, different information media such as voice, data, video, etc. were transmitted separately by using their own respective methods of transmission. With the advent of digital communication and "convergence technologies," this distinction is slowly disappearing, and multimedia communication is becoming the order of the day.

1.3 TRANSMISSION IMPAIRMENTS

While the electrical signal is traversing over the medium, the signal will be impaired due to various factors. These transmission impairments can be classified into three types:

> The transmission impairments can be classified into: (a) attenuation distortion; (b) delay distortion; and (c) noise.

(a) Attenuation distortion

(b) Delay distortion

(c) Noise

The amplitude of the signal wave decreases as the signal travels through the medium. This effect is known as *attenuation distortion*. Delay distortion occurs as a result of different frequency components arriving at different times in the guided media such as copper wire or coaxial cable. The third type of impairment—noise—can be divided into the following categories:

- Thermal noise
- Intermodulation
- Crosstalk
- Impulse noise

Thermal noise: Thermal noise occurs due to the thermal agitation of electrons in a conductor. This is distributed uniformly across the spectrum and hence called *white noise*. This noise cannot be eliminated and hence, when designing telecom systems, we need to introduce some method to overcome the ill effects of thermal noise. Thermal noise for a bandwidth of 1 Hz is obtained from the formula:

$$No = kT$$

where   No is noise power density, watts per Hz

k is Boltzmann's constant. 1.3803×10^{-23} J/K

T is temperature, K.

Thermal noise for a bandwidth of B Hz is given by

$$N = kTB \text{ (watts)}$$

If N is expressed in dB (decibels)

$$N = 10 \log k + 10 \log T + 10 \log B \text{ dB watts}$$
$$= -228.6 + 10 \log T + 10 \log B$$

Using this formula, thermal noise for a given bandwidth is calculated.

Thermal noise for a bandwidth of B Hz is given by N = kTB (watts) where k is Boltzmann's constant and T is temperature. N is generally expressed in decibels.

Intermodulation noise: When two signals of different frequencies are sent through the medium, due to nonlinearity of the transmitters, frequency components such as f1 + f2 and f1 – f2 are produced, which are unwanted components and need to be filtered out.

Crosstalk: Unwanted coupling between signal paths is known as crosstalk. In the telephone network, this coupling is quite common. As a result of this, we hear other conversations. Crosstalk needs to be eliminated by using appropriate design techniques.

Impulse noise: This is caused by external electromagnetic disturbances such as lightning. This noise is unpredictable. When the signal is traversing the medium, impulse noise may cause sudden bursts of errors. This may cause a temporary disturbance in voice communication. For data communication, appropriate methods need to be devised whereby the lost data is retransmitted.

Impulse noise occurs due to external electromagnetic disturbances such as lightning. Impulse noise causes burst of errors.

Noise can be divided into four categories: (a) thermal noise, (b) intermodulation noise, (c) crosstalk and (d) impulse noise.

Noise is the source of bread and butter for telecom engineers! If there were no noise, there would be no need for telecom engineers—for we can then design perfect communication systems. Telecom engineering is all about overcoming the effects of noise.

1.4 ANALOG VERSUS DIGITAL TRANSMISSION

In analog communication, the signal, whose amplitude varies continuously, is transmitted over the medium. Reproducing the analog signal at the receiving end is very difficult due to transmission impairments. Hence, analog communication systems are badly affected by noise.

The electrical signal output from a transducer such as microphone or a video camera is an analog signal; that is, the amplitude of the signal varies continuously with time. Transmitting this signal (with necessary transformations) to the receiving end results in analog transmission. However, at the receiving end, it has to be ensured that the signal does not get distorted at all due to transmission impairments, which is very difficult.

The output of a computer is a digital signal. The digital signal has a fixed number of amplitude levels. For instance, binary 1 can be represented by one voltage level (say, 5 volts) and binary 0 can be represented by another level (say, 0 volt). If this signal is transmitted through the medium (of course with necessary transformations), the receiving end needs only to detect these levels. Even if the signal is slightly impaired due to noise, still there is no problem. For example, we can say that if the signal is above 2.5 volts, it is 1 and if it is below 2.5 volts, it is zero. Unless the signal is badly damaged, we can easily find out whether the transmitted bit is a 1 or a 0.

In a digital communication system, 1s and 0s are transmitted as voltage pulses. So, even if the pulse is distorted due to noise, it is not very difficult to detect the pulses at the receiving end. Hence, digital communication is much more immune to noise as compared to analog communication.

The voice and video signals (output of the transducer) are always analog. Then how do we take advantage of the digital transmission? Simple. Convert the analog signal into the digital format. This is achieved through analog-to-digital conversion. At this point, let us assume only that it is possible to convert an analog signal into its equivalent digital signal. We will study the details of this conversion process in later chapters.

Digital transmission is much more advantageous than analog transmission because digital systems are comparatively immune to noise. Due to advances in digital electronics, digital systems have become cheaper, as well. The advantages of digital systems are:

- More reliable transmission because only discrimination between ones and zeros is required.
- Less costly implementation because of the advances in digital logic chips.

- Ease of combining various types of signals (voice, video, etc.).
- Ease of developing secure communication systems.

The advantages of digital communication are more reliable transmission, less costly implementation, ease of multiplexing different types of signals, and secure communication.

Though a large number of analog communication systems are still in use, digital communication systems are now being deployed. Also, the old analog systems are being replaced by digital systems. In this book, we focus mainly on digital communication systems.

 All the newly developed communication systems are digital systems. Only in broadcasting applications, is analog communication used extensively.

Summary

This chapter has presented the basic building blocks of a communication system. The information source produces the data that is converted into electrical signal and sent through the transmission medium. Since the transmission medium introduces noise, additional processing is required to transmit the signal over large distances. Also, additional processing is required if the medium is shared by a number of users.

Communication systems are of various types. Point-to-point systems provide communication between two end points. Point-to-multipoint systems facilitate sending information simultaneously to a number of points. Broadcasting systems facilitate sending information to a large number of points from a central location. Simplex systems allow communication only in one direction. Half-duplex systems allow communication in both directions but in one direction at a time. Full-duplex systems allow simultaneous communication in both directions.

Communication systems can be broadly divided into analog communication systems and digital communication systems. In analog communication systems, the analog signal is transmitted. In digital communication system, even though the input signal is in analog form, it is converted into digital format and then sent through the medium. For a noise condition, the digital communication system gives better performance than the analog system.

The concepts of multiplexing and multiple access also are introduced in this chapter. The details will be discussed in later chapters.

References

G. Kennedy and B. Davis. *Electronic Communication Systems*. Tata McGraw-Hill Publishing Company Limited, 1993.

R. Horak. *Communication Systems and Networks*. Wiley-Dreamtech India Pvt. Ltd., 2002.

Questions

1. What are the advantages of digital communication over analog communication?
2. Explain the different types of communication systems.
3. What are the different types of transmission impairments?
4. What is multiplexing?
5. What is multiple access?
6. What is signaling?

Exercises

1. Write a program to generate a bit stream of ones and zeros.
2. Write a program to generate noise. You can use the random number generation function rand() to generate the random numbers. The conversion of the random numbers to binary form produces a pseudo-random noise.
3. Write a program that simulates a transmission medium. The bits at random places in the bit stream generated (in Exercise #1) have to be modified to create the errors—1 has to be changed to 0 and 0 has to be changed to 1.
4. Chips (integrated circuits) are available for generation of noise. Identify a noise generator chip.

Project

Write a report on the history of telecommunication, listing the important milestones in the development of telecommunication technology.

2 Information Theory

In This Chapter

- The Requirements of a Communication System
- The Building Blocks of a Communication System as Proposed by Shannon
- Entropy and Channel Capacity
- Shannon's Source Coding Theorem and Channel Coding Theorem

Claude Shannon laid the foundation of information theory in 1948. His paper "A Mathematical Theory of Communication" published in *Bell System Technical Journal* is the basis for the entire telecommunications developments that have taken place during the last five decades. A good understanding of the concepts proposed by Shannon is a must for every budding telecommunication professional. We study Shannon's contributions to the field of modern communications in this chapter.

2.1 REQUIREMENTS OF A COMMUNICATION SYSTEM

The requirement of a communication system is to transmit the information from the source to the sink without errors, in spite of the fact that noise is always introduced in the communication medium.

In any communication system, there will be an information source that produces information in some form, and an information sink absorbs the information. The communication medium connects the source and the sink. The purpose of a communication system is to transmit the information from the source to the sink without errors. However, the communication medium always introduces some errors because of noise. The fundamental requirement of a communication system is to transmit the information without errors in spite of the noise.

15

2.1.1 The Communication System

The block diagram of a generic communication system is shown in Figure 2.1. The information source produces symbols (such as English letters, speech, video, etc.) that are sent through the transmission medium by the transmitter. The communication medium introduces noise, and so errors are introduced in the transmitted data. At the receiving end, the receiver decodes the data and gives it to the information sink.

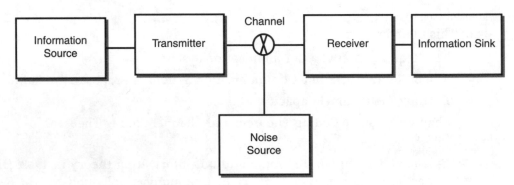

FIGURE 2.1 Generic communication system

As an example, consider an information source that produces two symbols A and B. The transmitter codes the data into a bit stream. For example, A can be coded as 1 and B as 0. The stream of 1's and 0's is transmitted through the medium. Because of noise, 1 may become 0 or 0 may become 1 at random places, as illustrated below:

Symbols produced:	A	B	B	A	A	A	B	A	B	A
Bit stream produced:	1	0	0	1	1	1	0	1	0	1
Bit stream received:	1	0	0	1	1	1	1	1	0	1

In a digital communication system, due to the effect of noise, errors are introduced. As a result, 1 may become a 0 and 0 may become a 1.

At the receiver, one bit is received in error. How to ensure that the received data can be made error free? Shannon provides the answer. The communication system given in Figure 2.1 can be expanded, as shown in Figure 2.2.

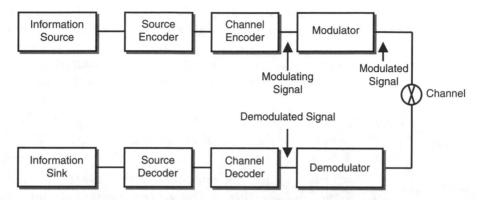

FIGURE 2.2 Generic communication system as proposed by Shannon.

As proposed by Shannon, the communication system consists of source encoder, channel encoder and modulator at the transmitting end, and demodulator, channel decoder and source decoder at the receiving end.

In this block diagram, the information source produces the symbols that are coded using two types of coding—*source encoding* and *channel encoding*—and then modulated and sent over the medium. At the receiving end, the modulated signal is demodulated, and the inverse operations of channel encoding and source encoding (channel decoding and source decoding) are performed. Then the information is presented to the information sink. Each block is explained below.

Information source: The information source produces the symbols. If the information source is, for example, a microphone, the signal is in analog form. If the source is a computer, the signal is in digital form (a set of symbols).

Source encoder: The source encoder converts the signal produced by the information source into a data stream. If the input signal is analog, it can be converted into digital form using an analog-to-digital converter. If the input to the source encoder is a stream of symbols, it can be converted into a stream of 1s and 0s using some type of coding mechanism. For instance, if the source produces the symbols A and B, A can be coded as 1 and B as 0. Shannon's source coding theorem tells us how to do this coding efficiently.

Source encoding is done to reduce the redundancy in the signal. Source coding techniques can be divided into lossless encoding techniques and lossy encoding techniques. In lossy encoding techniques, some information is lost.

In source coding, there are two types of coding—*lossless* coding and *lossy* coding. In lossless coding, no information is lost. When we compress our computer files using a compression technique (for instance, WinZip), there is no loss of information. Such coding techniques are called lossless coding techniques. In lossy coding, some information is lost while doing the source coding. As long as the loss is not significant, we can tolerate it. When an image is converted into JPEG format, the coding is lossy coding because some information is lost. Most of the techniques used for voice, image, and video coding are lossy coding techniques.

 The compression utilities we use to compress data files use lossless encoding techniques. JPEG image compression is a lossy technique because some information is lost.

In channel encoding, redundancy is introduced so that at the receiving end, the redundant bits can be used for error detection or error correction.

Channel encoder: If we have to decode the information correctly, even if errors are introduced in the medium, we need to put some additional bits in the source-encoded data so that the additional information can be used to detect and correct the errors. This process of adding bits is done by the channel encoder. Shannon's channel coding theorem tells us how to achieve this.

In channel encoding, redundancy is introduced so that at the receiving end, the redundant bits can be used for error detection or error correction.

Modulation: Modulation is a process of transforming the signal so that the signal can be transmitted through the medium. We will discuss the details of modulation in a later chapter.

Demodulator: The demodulator performs the inverse operation of the modulator.

Channel decoder: The channel decoder analyzes the received bit stream and detects and corrects the errors, if any, using the additional data introduced by the channel encoder.

Source decoder: The source decoder converts the bit stream into the actual information. If analog-to-digital conversion is done at the source encoder, digital-to-analog conversion is done at the source decoder. If the symbols are coded into 1s and 0s at the source encoder, the bit stream is converted back to the symbols by the source decoder.

Information sink: The information sink absorbs the information.

The block diagram given in Figure 2.2 is the most important diagram for all communication engineers. We will devote separate chapters to each of the blocks in this diagram.

2.2 ENTROPY OF AN INFORMATION SOURCE

What is information? How do we measure information? These are fundamental issues for which Shannon provided the answers. We can say that we received some information if there is "decrease in uncertainty." Consider an information source that produces two symbols A and B. The source has sent A, B, B, A, and now we are waiting for the next symbol. Which symbol will it produce? If it produces A, the uncertainty that was there in the waiting period is gone, and we say that "information" is produced. Note that we are using the term "information" from a communication theory point of view; it has nothing to do with the "usefulness" of the information.

Shannon proposed a formula to measure information. The information measure is called the entropy of the source. If a source produces N symbols, and if all the symbols are equally likely to occur, the entropy of the source is given by

$$H = \log_2 N \text{ bits/symbol}$$

For example, assume that a source produces the English letters (in this chapter, we will refer to the English letters A to Z and space, totaling 27, as symbols), and all these symbols will be produced with equal probability. In such a case, the entropy is

$$H = \log_2 27 = 4.75 \text{ bits/symbol}$$

The information source may not produce all the symbols with equal probability. For instance, in English the letter "E" has the highest frequency (and hence highest probability of occurrence), and the other letters occur with different probabilities. In general, if a source produces (i)th symbol with a probability of P(i), the entropy of the source is given by

$$H = -\sum_i P(i) \log_2 P(i) \text{ s}$$

If a large text of English is analyzed and the probabilities of all symbols (or letters) are obtained and substituted in the formula, then the entropy is

$$H = 4.07 \text{ bits/symbol}.$$

Consider the following sentence: "I do not knw wheter this is undrstandble." In spite of the fact that a number of letters are missing in this sentence, you can make out what the sentence is. In other words, there is a lot of redundancy in the English text.

This is called the *first-order approximation* for calculation of the entropy of the information source. In English, there is a dependence of one letter on the previous letter. For instance, the letter 'U' always occurs after the letter 'Q'. If we consider the probabilities of two symbols together (aa, ab, ac, ad,..ba, bb, and so on), then it is called the *second-order approximation*. So, in second-order approximation, we have to consider the conditional probabilities of digrams (or two symbols together). The second-order entropy of a source producing English letters can be worked out to be

If a source produces (i)th symbol with a probability of P(i), the entropy of the source is given by $H = -\sum P(i)\ \log_2 P(i)$ bits/symbol.

$$H = 3.36 \text{ bits/symbol}$$

The third-order entropy of a source producing English letters can be worked out to be

$$H = 2.77 \text{ bits/symbol}$$

As you consider the higher orders, the entropy goes down.

As another example, consider a source that produces four symbols with probabilities of 1/2, 1/4, 1/8, and 1/8, and all symbols are independent of each other. The entropy of the source is 7/4 bits/symbol.

As you consider the higher-order probabilities, the entropy of the source goes down. For example, the third-order entropy of a source producing English letters is 2.77 bits/symbol—each combination of three letters can be represented by 2.77 bits.

2.3 CHANNEL CAPACITY

Shannon introduced the concept of *channel capacity*, the limit at which data can be transmitted through a medium. The errors in the transmission medium depend on the energy of the signal, the energy of the noise, and the bandwidth of the channel. Conceptually, if the bandwidth is high, we can pump more data in the channel. If the signal energy is high, the effect of noise is reduced. According to Shannon, the bandwidth of the channel and signal energy and noise energy are related by the formula

$$C = W \log_2(1 + S/N)$$

where

C is channel capacity in bits per second (bps)

W is bandwidth of the channel in Hz

S/N is the signal-to-noise power ratio (SNR). SNR generally is measured in dB using the formula

(S/N) dB = 10 log(Signal power / Noise power)

The value of the channel capacity obtained using this formula is the theoretical maximum. As an example, consider a voice-grade line for which W = 3100Hz, SNR = 30dB (i.e., the signal-to-noise ratio is 1000:1)

$$C = 3100 \log_2(1 + 1000) = 30,894 \text{ bps}$$

So, we cannot transmit data at a rate faster than this value in a voice-grade line.

An important point to be noted is that in the above formula, Shannon assumes only thermal noise.

The bandwidth of the channel, signal energy, and noise energy are related by the formula $C = W \log_2(1 + S/N)$ bps where C is the channel capacity, W is the bandwidth, and S/N is the signal-to-noise ratio.

To increase C, can we increase W? No, because increasing W increases noise as well, and SNR will be reduced. To increase C, can we increase SNR? No, that results in more noise, called intermodulation noise.

The entropy of information source and channel capacity are two important concepts, based on which Shannon proposed his theorems.

2.4 SHANNON'S THEOREMS

In a digital communication system, the aim of the designer is to convert any information into a digital signal, pass it through the transmission medium and, at the receiving end, reproduce the digital signal exactly. To achieve this objective, two important requirements are:

1. To code any type of information into digital format. Note that the world is analog—voice signals are analog, images are analog. We need to devise mechanisms to convert analog signals into digital format. If the source produces symbols (such as A, B), we also need to convert these symbols into a bit stream. This coding has to be done efficiently so that the smallest number of bits is required for coding.

2. To ensure that the data sent over the channel is not corrupted. We cannot eliminate the noise introduced on the channels, and hence we need to introduce special coding techniques to overcome the effect of noise.

Shannon's source coding theorem addresses how the symbols produced by a source have to be encoded efficiently. Shannon's channel coding theorem addresses how to encode the data to overcome the effect of noise.

These two aspects have been addressed by Claude Shannon in his classical paper "A Mathematical Theory of Communication" published in 1948 in *Bell System Technical Journal*, which gave the foundation to information theory. Shannon addressed these two aspects through his source coding theorem and channel coding theorem.

2.4.1 Source Coding Theorem

The source coding theorem states that "the number of bits required to uniquely describe an information source can be approximated to the information content as closely as desired."

Again consider the source that produces the English letters. The information content or entropy is 4.07 bits/symbol. According to Shannon's source coding theorem, the symbols can be coded in such a way that for each symbol, 4.07 bits are required. But what should be the coding technique? Shannon does not tell us! Shannon's theory puts only a limit on the minimum number of bits required. This is a very important limit; all communication engineers have struggled to achieve the limit all these 50 years.

Consider a source that produces two symbols A and B with equal probability.

Symbol	Probability	Code Word
A	0.5	1
B	0.5	0

The two symbols can be coded as above, A is represented by 1 and B by 0. We require 1 bit/symbol.

Now consider a source that produces these same two symbols. But instead of coding A and B directly, we can code AA, AB, BA, BB. The probabilities of these symbols and associated code words are shown here:

Symbol	Probability	Code Word
AA	0.45	0
AB	0.45	10
BA	0.05	110
BB	0.05	111

Here the strategy in assigning the code words is that the symbols with high probability are given short code words and symbols with low probability are given long code words.

Assigning short code words to high-probability symbols and long code words to low-probability symbols results in efficient coding.

In this case, the average number of bits required per symbol can be calculated using the formula

$$L = \sum_i P(i) \, L(i)$$

where P(i) is the probability and L(i) is the length of the code word. For this example, the value is (1 * 0.45 + 2 * 0.45 + 3 * 0.05 + 3 * 0.05) = 1.65 bits/symbol. The entropy of the source is 1.469 bits/symbol.

So, if the source produces the symbols in the following sequence:

A A B A B A A B B B

then source coding gives the bit stream

0 110 110 10 111

This encoding scheme on an average, requires 1.65 bits/symbol. If we code the symbols directly without taking into consideration the probabilities, the coding scheme would be

AA 00
AB 01
BA 10
BB 11

Hence, we require 2 bits/symbol. The encoding mechanism taking the probabilities into consideration is a better coding technique. The theoretical limit of the number of bits/symbol is the entropy, which is 1.469 bits/symbol. The entropy of the source also determines the channel capacity.

As we keep considering the higher-order entropies, we can reduce the bits/symbol further and perhaps achieve the limit set by Shannon.

Based on this theory, it is estimated that English text cannot be compressed to less than 1.5 bits/symbol even if you use sophisticated coders and decoders.

The source coding theorem states "the number of bits required to uniquely describe an information source can be approximated to the information content as closely as desired."

This theorem provides the basis for coding information (text, voice, video) into the minimum possible bits for transmission over a channel. We will study the details of source coding in Chapter 4, "Coding of Text, Voice, Image, and Video Signals."

2.4.2 Channel Coding Theorem

Shannon's channel coding theorem states that "the error rate of data transmitted over a bandwidth limited noisy channel can be reduced to an arbitrary small amount if the information rate is lower than the channel capacity."

This theorem is the basis for error correcting codes using which we can achieve error-free transmission. Again, Shannon only specified that using 'good' coding mechanisms, we can achieve error-free transmission, but he did not specify what the coding mechanism should be! According to Shannon, channel coding may introduce additional delay in transmission but, using appropriate coding techniques, we can overcome the effect of channel noise.

Consider the example of a source producing the symbols A and B. A is coded as 1 and B as 0.

Symbols produced: A B B A B
Bit stream: 1 0 0 1 0

Now, instead of transmitting this bit stream directly, we can transmit the bit stream

111000000111000

that is, we repeat each bit three times. Now, let us assume that the received bit stream is

101000010111000

Two errors are introduced in the channel. But still, we can decode the data correctly at the receiver because we know that the second bit should be 1 and the eighth bit should be 0 because the receiver also knows that each bit is transmitted thrice. This is error correction. This coding is called Rate 1/3 error correcting code. Such codes that can correct the errors are called Forward Error Correcting (FEC) codes.

Ever since Shannon published his historical paper, there has been a tremendous amount of research in the error correcting codes. We will discuss error detection and correction in Chapter 5 "Error Detection and Correction".

All these 50 years, communication engineers have struggled to achieve the theoretical limits set by Shannon. They have made considerable progress. Take the case of line modems that we use for transmission of data over telephone lines. The evolution of line modems from V.26 (2400bps data rate, 1200Hz bandwidth), V.27 modems (4800bps data rate, 1600Hz bandwidth), V.32 modems (9600bps data rate, 2400Hz bandwidth), and V.34 modems (28,800bps data rate, 3400Hz bandwidth) indicates the progress in source coding and channel coding techniques using Shannon's theory as the foundation.

Shannon's channel coding theorem states that "the error rate of data transmitted over a bandwidth limited noisy channel can be reduced to an arbitrary small amount if the information rate is lower than the channel capacity."

Source coding is used mainly to reduce the redundancy in the signal, whereas channel coding is used to introduce redundancy to overcome the effect of noise.

Summary

In this chapter, we studied Shannon's theory of communication. Shannon introduced the concept of entropy of an information source to measure the number of bits required to represent the symbols produced by the source. He also defined channel capacity, which is related to the bandwidth and signal-to-noise ratio. Based on these two measures, he formulated the source coding theorem and channel coding theorem. Source coding theorem states that "the number of bits required to uniquely describe an information source can be approximated to the information content as closely as desired." Channel coding theorem states that "the error rate of data transmitted over a bandwidth limited noisy channel can be reduced to an arbitrary small amount if the information rate is lower than the channel capacity." A good conceptual understanding of these two theorems is important for every communication engineer.

References

C.E. Shannon. "A Mathematical Theory of Communication." *Bell System Technical Journal*, Vol. 27, 1948.

Every communications engineer must read this paper. Shannon is considered the father of modern communications. You have to be very good at mathematics to understand this paper.

W. Gappmair, Claude E. Shannon. "The 50th Anniversary of Information Theory." *IEEE Communications Magazine*, Vol. 37, No. 4, April 1999.

This paper gives a brief biography of Shannon and a brief overview of the importance of Shannon's theory.

cm.bell-labs.com/cm/ms/what/shannonday/paper.html You can download Shannon's original paper from this link.

Questions

1. Draw the block diagram of a communication system and explain the function of each block.
2. What is entropy of an information source? Illustrate with examples.
3. What is source coding? What is the difference between lossless coding and lossy coding?
4. Explain the concept of channel capacity with an example.
5. What is channel coding? Explain the concept of error correcting codes.

Exercises

1. A source produces 42 symbols with equal probability. Calculate the entropy of the source.
2. A source produces two symbols A and B with probabilities of 0.6 and 0.4, respectively. Calculate the entropy of the source.
3. The ASCII code is used to represent characters in the computer. Is it an efficient coding technique from Shannon's point of view? If not, why?
4. An information source produces English symbols (letters A to Z and space). Using the first- order model, calculate the entropy of the information source. You need to enter a large English text with the 27 symbols, calculate the frequency of occurrence of each symbol, and then calculate the entropy. The answer should be close to 4.07 bits/symbol.

5. In the above example, using the second-order model, calculate the entropy. You need to calculate the frequencies taking two symbols at a time such as aa, ab, ac, etc. The entropy should be close to 3.36 bits/symbol.

Projects

1. Simulate a digital communication system in C language. (a) Write a program to generate a continuous bit stream of 1s and 0s. (b) Simulate the medium by changing a 1 to a 0 and a 0 to a 1 at random places in the bit stream using a random number generator. (c) Calculate the bit error rate (= number of errors/number of bits transmitted).

2. Develop a file compression utility using the second-order approximation described in this chapter. The coding should be done by taking the frequencies of occurrence of combinations of two characters, as in Exercise 2.

3 | Transmission Media

In This Chapter

- The Characteristics of Various Transmission Media
- Radio as the Transmission Medium
- Radio Spectrum Management

To exchange information between people separated by a distance has been a necessity throughout the history of mankind. Long ago, people used fire for communicating over a limited distance; they used birds and messengers for long distance communication. The postal system still provides an excellent means of sending written communication across the globe.

Communication over long distances, not just through written text but using other media such as voice and video, has been achieved through electrical communication. This discipline deals with conversion of information into electrical signals and transmitting them over a distance through a transmission medium. In this chapter, we will study the various transmission media used for electrical communication, such as twisted pair, coaxial cable, optical fiber, and the radio. We will study the characteristics and the advantages and disadvantages of each for practical communication systems. Free space (or radio) communication particularly provides the unique advantage of support for mobility—the user can communicate while on the move (in a car or an airplane). However, the radio spectrum is a precious natural resource and has to be used efficiently. We will also discuss the issue of radio spectrum management briefly in this chapter.

3.1 TWISTED PAIR

Twisted pair gets its name because a pair of copper wires is twisted to form the transmission medium. This is the least expensive transmission medium and hence the most widely used. This medium is used extensively in the local underground telephone network, in Private Branch Exchanges (PBX's), and also in local area networks (LANs).

As the electrical signal traverses the medium, it becomes attenuated, that is, the signal level will go down. Hence, a small electronic gadget called a *repeater* is used every 2 to 10 kilometers. A repeater amplifies the signal to the required level and retransmits on the medium.

The twisted pair of copper wires is a low-cost transmission medium used extensively in telephone networks and local area networks.

The data rate supported by the twisted pair depends on the distance to be covered and the quality of the copper. Category 5 twisted pair supports data rates in the range of 10Mbps to 100Mbps up to a distance of 100 meters.

3.2 COAXIAL CABLE

Coaxial cable is used extensively for cable TV distribution, long-distance telephone trunks, and LANs. The cross-section of a coaxial cable used in an Ethernet local area network is shown in Figure 3.1.

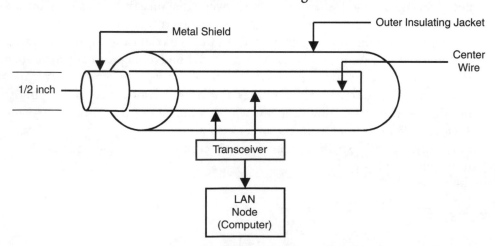

FIGURE 3.1 Coaxial cable used in Ethernet LAN.

Coaxial cable is used in long distance telephone networks, local area networks, and cable TV distribution networks.

Coaxial cable can support a maximum data rate of 500Mbps for a distance of about 500 meters. Repeaters are required every 1 to 10 kilometers.

The speed of transmission in copper cable is 2.3×10^8 meters/second. Note that this speed is less than the speed of light in a vacuum (3×10^8 meters/second).

Based on the speed of transmission and the distance (length of the cable), the propagation delay can be calculated using the formula

Delay = distance/speed

For example, if the distance is 10 kilometers, the propagation delay is

Delay = $10,000/(2.3 \times 10^8)$ seconds = 43.48 microseconds.

3.3 OPTICAL FIBER

Optical fiber is now being deployed extensively and is the most preferred medium for all types of networks because of the high data rates that can be supported. Light in a glass medium can carry more information over large distances, as compared to electrical signals in a copper cable or a coaxial cable.

Optical fiber is the most attractive transmission medium because of its support for very high data rates and low attenuation.

The challenge in the initial days of research on fiber was to develop glass so pure that at least 1% of the light would be retained at the end of 1 kilometer. This feat was achieved in 1970. The recent advances in fiber have been phenomenal; light can traverse 100km without any amplification, thanks to research in making purer glass. With the state of the art, the loss will be about 0.35dB/km for 1310 nanometers and 0.25dB/km for 1550nm.

Light transmission in the fiber works on the principle that the light waves are reflected within the core and guided to the end of the fiber, provided the angle at which the light waves are transmitted is controlled. Note that if the angle is not proper, the light is refracted and not reflected. The fiber medium has a core and cladding, both pure solid glass and protected by acrylate coating that surrounds the cladding.

As shown in Figure 3.2, there are two types of fiber: single mode and multimode. Single mode fiber has a small core and allows only one ray (or mode) of light to propagate at a time. Multimode fiber, the first to be commercialized, has a much larger core than single mode fiber and allows hundreds of rays of light to be transmitted through the fiber simultaneously. The larger core diameter allows low-cost optical transmitters and connectors and hence is cheaper.

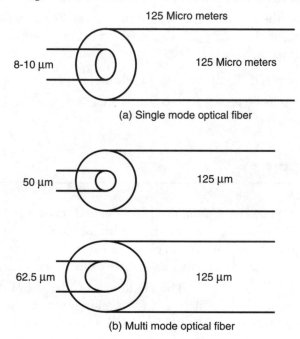

FIGURE 3.2 Optical fiber.

Gigabits and even terabits of data can be transmitted through the fibers, and the future lies in optical fiber networks. Currently twisted copper wire is being used for providing telephones to our homes, but soon fiber to the home will be a reality.

The two types of optical fiber are single mode and multimode fiber. Single-mode fiber allows only one ray (or mode) of light to propagate at a time whereas multimode fiber allows multiple rays (or modes).

The speed of transmission is 2×10^8 meters/second in optical fiber.

We will study optical communication systems in detail in Chapter 14, "Optical Fibre Communication Systems."

3.4 TERRESTRIAL RADIO

Free space as the medium has the main advantage that the receiver can be fixed or mobile. Free space is called an unguided medium because the electromagnetic waves can travel freely in all directions. Depending on the frequency of the radio waves, the propagation characteristics vary, and different frequencies are used for different applications, based on the required propagation characteristics. Radio is used for broadcasting extensively because a central station can transmit the program to be received by a large number of receivers spread over a large geographical area. In this case, the transmitter transmits at a specific frequency, and all the receivers tune to that frequency to receive the program.

Radio as a transmission medium has the main advantage that it supports mobility. In addition, installation and maintenance of radio systems are very easy.

In two-way communication systems such as for voice, data, or video, there is a base station located at a fixed place in the area of operation and a number of terminals. As shown in Figure 3.3, a pair of frequencies is used for communication—one frequency for transmitting from the base station to the terminals (the downlink) and one frequency from the terminal to the base station (the uplink). This frequency pair is called the *radio channel*.

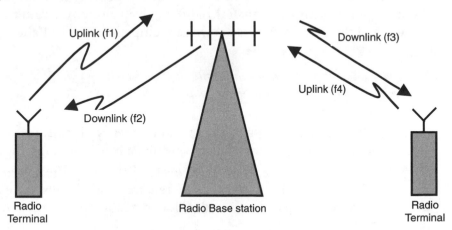

FIGURE 3.3 Two-way communication using radio.

A radio channel consists of a pair of frequencies—one frequency is used for uplink and one frequency is used for downlink. However, in some radio systems, a single frequency is used in both directions.

Radio as the transmission medium has special characteristics that also pose special problems.

Path loss: As the distance between the base station and the terminal increases, the received signal becomes weaker and weaker, even if there are no obstacles between the base station and the terminal. The higher the frequency, the higher the path loss. Many models are available (such as Egli's model and Okomura-Hata model) to estimate path loss. To compensate for path loss, we need to use high-gain antennas and also develop receivers of high sensitivity.

Path loss causes a heavy attenuation of the radio signal. Hence, the radio receiver should be capable of receiving very weak signals. In other words, the receiver should have high sensitivity.

Fading: Where there are obstacles between the base station and the terminal (hills, buildings, etc.), the signal strength goes down further, which is known as fading. In densely populated urban areas, the signal can take more than one path—one signal path can be directly from the base station to the terminal and another path can be from the base station to a building and the signal reflected from the building and then received at the terminal. Sometimes, there may not be a line of sight between the base station and terminal antennas, and hence the signals received at the terminals are from different paths. The received signal is the sum of many identical signals that differ only in phase. As a result, there will be fading of the signal, which is known as multipath fading or Raliegh fading.

Multipath fading is predominant in mobile communication systems. The mobile phone receives the signals that traverse different paths.

Rain attenuation: The rain affects radio frequency signals. Particularly in some frequency bands, rain attenuation is greater. When designing radio systems, the effect of rain (and hence the path loss) needs to be taken into consideration.

The radio spectrum is divided into different frequency bands, and each band is used for a specific application. The details of the radio spectrum are discussed in the next section.

Radio wave propagation is very complex, and a number of mathematical models have been developed to study the propagation in free space.

3.4.1 Radio Spectrum

Electrical communication is achieved by using electromagnetic waves, that is, oscillations of electric and magnetic fields in free space. The electromagnetic waves have two main parts: radio waves and light waves. Distinguishing between radio waves and light waves reflects the technology used to detect them. The radio waves are measured in frequency (Hz), and the other types of waves in terms of wavelength (meters) or energy (electron volts).

The electromagnetic spectrum consists of the following:

Radio waves	:	300GHz and lower (frequency)
Sub-millimeter waves	:	100 micrometers to 1 millimeter (wavelength)
Infrared	:	780 nanometers to 100 micrometers (wavelength)
Visible light	:	380 nanometers to 780 nanometers (wavelength)
Ultraviolet	:	10 nanometers to 380 nanometers (wavelength)
X-ray	:	120eV to 120keV (energy)
Gamma rays	:	120 keV and up (energy)

The radio spectrum spans from 3kHz to 300GHz. This spectrum is divided into different bands. Because of the differences in propagation characteristics of the waves with different frequencies, and also the effect of atmosphere and rain on these waves, different bands are used for different applications. Table 3.1 gives the various frequency bands, the corresponding frequency ranges, and some application areas in each band.

The radio spectrum is divided into a number of bands, and each band is used for specific applications.

TABLE 3.1 The radio frequency spectrum and typical applications

Frequency band	Frequency range	Application areas
Very Low Frequency (VLF)	3kHz to 30kHz	Radio navigation, maritime mobile (communication on ships)
Low Frequency (LF)	30kHz to 300kHz	Radio navigation, maritime mobile
Medium Frequency (MF)	300kHz to 3MHz	AM radio broadcast, aeronautical mobile
High Frequency (HF)	3MHz to 30MHz	Maritime mobile, aeronautical mobile
Very High Frequency (VHF)	30MHz to 300MHz	Land mobile, FM broadcast, TV broadcast, aeronautical mobile, radio paging, trunked radio
Ultra-High Frequency (UHF)	300MHz to 1GHz	TV broadcast, mobile satellite, land mobile, radio astronomy
L band	1GHz to 2GHz	Aeronautical radio navigation, radio astronomy, earth exploration satellites
S band	2GHz to 4GHz	Space research, fixed satellite communication
C band	4GHz to 8GHz	Fixed satellite communication, meteorological satellite communication
X band	8GHz to 12GHz	Fixed satellite broadcast, space research
Ku band	12GHz to 18GHz	Mobile and fixed satellite communication, satellite broadcast
K band	18GHz to 27GHz	Mobile and fixed satellite communication
Ka band	27GHz to 40GHz	Inter-satellite communication, mobile satellite communication
Millimeter	40GHz to 300GHz	Space research, Inter-satellite communications

International Telecommunications Union (ITU) assigns specific frequency bands for each application. Every country's telecommunications authorities in turn make policies on the use of these frequency bands. The specific frequency bands for some typical applications are listed here:

AM radio	535 to 1605 MHz
Citizen band radio	27MHz
Cordless telephone devices	43.69 to 50 MHz
VHF TV	54 to 72 MHz, 76 to 88 MHz, 174 to 216 MHz
Aviation	118 to 137 MHz
Ham radio	144 to 148 MHz 420 to 450 MHz
UHF TV	470 to 608 MHz 614 to 806 MHz
Cellular phones	824 to 849 MHz, 869 to 894 MHz
Personal communication services	901–902 MHz, 930–931 MHz, 940–941 MHz
Search for extra-terrestrial intelligence	1420 to 1660 MHz
Inmarsat satellite phones	1525 to 1559 MHz, 1626.5 to 1660.5 MHz

The representative terrestrial radio systems are discussed in Chapter 12, "Terrestrial Radio Communication Systems."

Some frequency bands such as ham radio band and the Industrial, Scientific, and Medical (ISM) band are free bands—no prior government approvals are required to operate radio systems in those bands.

3.5 SATELLITE RADIO

Arthur C. Clarke proposed the concept of communication satellites. A communication satellite is a relay in the sky. If the satellite is placed at a distance of about 36,000 km above the surface of the earth, then it appears stationary with respect to the earth because it has an orbital period of 24 hours. This orbit is called a geostationary orbit, and the satellites are called geostationary satellites. As shown in Figure 3.4, three geostationary communication satellites can cover the entire earth.

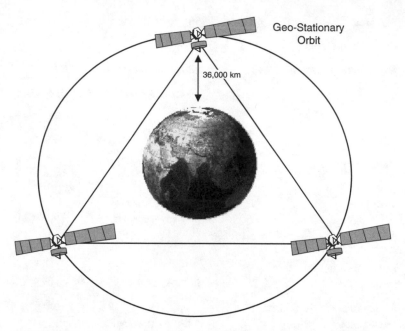

FIGURE 3.4 Three geostationary satellites covering the entire earth.

On Earth, we need satellite antennas (which are a part of the Earth stations) that point toward the satellite for communication. A pair of frequencies is used for communication with the satellite—the frequency used from Earth station to the satellite is called the uplink frequency, and the frequency from the satellite to the Earth station is called the downlink frequency. The signals transmitted by an Earth station to the satellite are amplified and then relayed back to the receiving Earth stations.

The main attraction of communication satellites is distance insensitivity. To provide communication facilities across the continents and also to rural and remote areas where laying cables is difficult, satellite communication will be very attractive. However, satellite communication has a disadvantage—delay. The propagation time for the signal to travel all the way to the satellite and back is nearly 240 msec. Also, because the signal has to travel long distances, there will be signal attenuation, and high-sensitivity receivers are required at both the satellite and the Earth stations.

To develop networks using satellite communications, there are two types of configurations—mesh and star, which are shown in Figure 3.5.

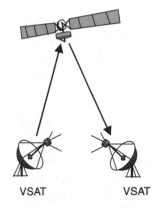

VSAT   VSAT

(a) Mesh Configuration

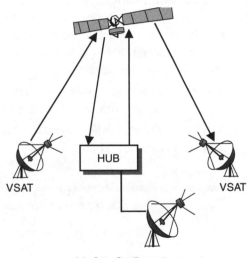

HUB

VSAT   VSAT

(b) Star Configuration

FIGURE 3.5 Satellite communication network configurations.

In mesh configuration, two Earth stations can communicate via the satellite. In this configuration, the size of the antennas at the Earth stations is large (starting from 4.5 meters diameter).

In star configuration, the size of the antenna will be very small, so the cost of an Earth station will be low. However, the disadvantage of star configuration is that the propagation delay is high.

Satellite communication systems work in two configurations: mesh and star. In mesh configuration, Earth stations communicate with each other directly; in star configuration, the Earth stations communicate via a central station.

In star configuration, there will be a central station (called the *hub*) and a number of Earth stations each with a Very Small Aperture Terminal (VSAT). When a VSAT has to communicate with another VSAT, the signal will be sent to the satellite, the satellite will relay the signal to the hub, the hub will amplify the signal and resend it to the satellite, and then the satellite will relay it to the other VSAT. In this configuration, the roundtrip delay will be double that of the mesh configuration. However the advantage is that smaller Earth stations can be used. VSAT communication is now used extensively for communication networks because of the low cost.

ITU allocated specific frequency bands for satellite communications. Some bands used for satellite communications are 6/4GHz, 14/12GHz, and 17/12GHz bands. As the frequency goes up, the size of the antenna goes down, and use of higher frequencies results in lower cost of equipment. However, the effect of rain on signals of different frequencies varies. Rain attenuation is more in 14/12GHz bands as compared to 6/4GHz bands.

In addition to the geostationary satellites, low Earth orbiting satellites also are being deployed for providing communication facilities.

In Chapter 13, "Satellite Communcation Systems," we will discuss satellite communication systems in greater detail.

The size of the satellite Earth station antenna decreases as the frequency of operation increases. Hence, the higher the frequency of operation, the smaller the size of antenna.

3.6 RADIO SPECTRUM MANAGEMENT

In the electromagnetic spectrum that spans from 0Hz to 10^{25}GHz (cosmic rays), the radio spectrum spans from 3kHz to 300GHz in VLF, LF, MF, HF, VHF, UHF, SHF, and EHF bands. The only portion of the radio spectrum not allocated to anyone is 3 to 9 kHz (rather, it is allocated to every individual for freedom of speech).

The radio spectrum is a limited natural resource, and hence international and national authorities regulate its use.

The International Telecommunications Union (ITU), through the World Administrative Radio Conferences (WARC), allots frequency bands for different application areas. For administrative convenience, the world is divided into three regions. The allocations for these regions differ to some extent. All nations are bound by these regulations for frequency use. The WARC regulations are only broad guidelines, because a centralized authority in every country manages the radio spectrum. In the United States, the frequency spectrum is managed by the FCC (Federal Communications Commission). In India, the agency is WPC (Wireless Planning and Coordination Cell) under the Government of India.

The complexity of radio spectrum management results from several factors:

- There are some frequency bands for the exclusive use of governmental agencies and some for nongovernmental agencies. Some frequency bands are shared. When the same band is shared by different agencies for the same application or for different applications, it is necessary to ensure that there is no interference between various systems.

- When a frequency band is allocated for a particular application (e.g., cellular mobile communication), the service can be provided by different operators in the same area. Without a coordinated effort in band allocation, the spectrum cannot be used efficiently to support a large number of users for that service in the same frequency band.

- Because of higher user demands, the frequency band allocated for a particular service may become congested and new bands need to be allocated. For example, in the case of mobile communications, the 900MHz band got congested, and the 1800MHz band was allocated. This type of new allocation of bands calls for long-term planning of spectrum use.

- As new application areas emerge, accommodating these applications along with the existing applications in the required frequency bands is another challenge in spectrum management.

- New technologies (better techniques for reducing bandwidth requirements, new frequency reuse technologies, antenna technologies, etc.) lead to better utilization of spectrum. Ensuring that the new technologies are incorporated is important in spectrum management.

Radio spectrum management ensures that the allotted spectrum is being used efficiently, to ensure that there is no interference between different radio systems and to allocate new frequency bands for new services.

- Agencies will be allocated fixed frequencies for use. For various reasons, however, these frequencies may not be used at all or may not be used efficiently. A periodic review of the use of the spectrum is also required. A review process needs to be followed by which an application for frequency allotment will be processed, frequencies allocated, and use monitored.

All these aspects make spectrum management a difficult task, and the need for an efficient spectrum management methodology cannot be overemphasized.

3.6.1 Spectrum Management Activities

Radio spectrum management involves three major activities:

(1) Spectrum assignment and selection involves recommending a specific frequency band of operation for use in a given location. For this, extensive databases containing all the information regarding the present uses of radio spectrum have to be developed and maintained so that the effect of the proposed frequency band uses on existing systems can be studied. Depending on the interference considerations, specific frequency bands can be allocated.

(2) Spectrum engineering and analysis involves computations for installations of radio equipment at specific locations and for predicting the system performance in the radio environment.

(3) Spectrum planning involves long-term/emergency planning, keeping in view, among other things, the demands for new services and technological changes.

Because all these activities involve huge computation, various national authorities are deploying computerized spectrum management systems. Expert systems are also being developed to manage the spectrum efficiently.

> Radio spectrum management involves spectrum management and selection, spectrum engineering and analysis, and long-term/emergency spectrum planning for new services.

3.6.2 Cost of Spectrum

Radio spectrum is a limited natural resource, and its optimal utilization must be ensured. Government agencies charge users for the spectrum. The charges are generally on an annual basis.

Nowdays, the government agencies are also using innovative methods for making money out of the spectrum. The present trend is to auction the spectrum. The highest bidder will be given the spectrum for a specific application. For the

> Operators that obtain specific frequency bands for radio services need to pay for the cost of the spectrum.

3rd Generation (3G) wireless systems, this approach has been followed, and it turned out that in most countries the spectrum cost is much higher than the infrastructure (equipment) cost.

Summary

The details of various transmission media used in telecommunications systems are presented in this chapter. Based on the considerations of cost, data rates required, and distance to be covered, the transmission medium has to be chosen. The transmission media options are twisted copper pair, coaxial cable, optical fiber, and radio. Twisted pair is of low cost, but the attenuation is very high and the data rates supported are low. Because of the low cost, it is used extensively in the telephone network, in PBX, and for LANs. Coaxial cable supports higher data rates compared to twisted pair and is used in cable TV, LANs, and the telephone network. Optical fiber supports very large data rates and is now the preferred medium for LANs and the telephone networks. The main attraction of radio as the transmission medium is its support for mobility. Furthermore, installation and maintenance of radio systems is easy because there is no need to dig below ground. Terrestrial radio systems are used extensively in the telephone network as well as for mobile communications. Wireless LANs also are becoming predominant nowadays. Satellite radio has the main advantage that remote and rural areas can be connected easily. Satellite radio is also used extensively for broadcasting.

Because radio spectrum is a precious natural resource, the spectrum has to be used effectively. ITU allocates the frequency bands for different applications. In each country, there is a government organization that coordinates the allocation and use of the spectrum for different applications. We have studied the intricacies of spectrum management, which include planning, allocation, and monitoring the use of the spectrum.

References

R. Horak. *Communications Systems and Networks*, Third Edition. Wiley-Dreamtech India Pvt. Ltd., 2002. This book gives comprehensive coverage of many topics in telecommunication and details of the transmission media.

www.inmarsat.com Inmarsat operates a worldwide mobile satellite network. You can get the details of the satellites and the services offered from this URL.

www.iec.org/tutorials The Web site of International Engineering Consortium. This link provides many tutorials on telecommunications topics.

Questions

1. List the different transmission media and their applications.
2. List the frequency bands of the radio spectrum and the applications in each frequency band.
3. What are the issues involved in radio spectrum management?
4. Some frequency bands such as for ham radio and the Industrial, Scientific and Medical (ISM) band are not regulated, and anyone can use those bands. Debate the pros and cons of such unregulated bands.

Exercises

1. A terrestrial radio link is 30 kilometers long. Find out the propagation delay, assuming that the speed of light is 3×10^8 meters/second.
2. The coaxial cable laid between two telephone switches is 40 kilometers long. Find out the propagation delay if the speed of transmission in the coaxial cable is 2.3×10^8 meters/second.
3. Calculate the propagation delay in an optical fiber of 100 kilometers. The speed is 2×10^8 meters/second.
4. Compile the list of frequency bands used for satellites for the following applications: (a) broadcasting, (b) telephone communications, (c) weather monitoring, (d) military applications.
5. Calculate the propagation delay from one Earth station to another Earth station in a satellite communication network for (a) mesh configuration and (b) star configuration.

Projects

1. Develop a database of frequency bands used for different applications given in Table 3.I. Develop a graphical user interface to facilitate the display of the frequency bands for a given application.
2. Study the frequency bands of operation for Inmarsat satellites. Inmarsat operates a worldwide mobile satellite network.

4 Coding of Text, Voice, Image, and Video Signals

In This Chapter

- Coding of Different Types of Signals
- How Text is Represented in Computers
- Waveform Coding and Vocoding of Voice Signals
- Image and Video Coding
- Standards for Coding of Text, Voice, Image, and Video Signals

The information that has to be exchanged between two entities (persons or machines) in a communication system can be in one of the following formats:

- Text
- Voice
- Image
- Video

In an electrical communication system, the information is first converted into an electrical signal. For instance, a microphone is the transducer that converts the human voice into an analog signal. Similarly, the video camera converts the real-life scenery into an analog signal. In a digital communication system, the first step is to convert the analog signal into digital format using analog-to-digital conversion techniques. This digital signal representation for various types of information is the topic of this chapter.

4.1 TEXT MESSAGES

Text messages are generally represented in ASCII (American Standard Code for Information Interchange), in which a 7-bit code is used to represent each character. Another code form called EBCDIC (Extended Binary Coded Decimal Interchange Code) is also used. To transmit text messages, first the text is converted into one of these formats, and then the bit stream is converted into an electrical signal.

ASCII is the most widely used coding scheme for representation of text in computers. ISCII is used to represent text of Indian languages.

Using ASCII, the number of characters that can be represented is limited to 128 because only 7-bit code is used. The ASCII code is used for representing many European languages as well. To represent Indian languages, a standard known as Indian Standard Code for Information Interchange (ISCII) has been developed. ISCII has both 7-bit and 8-bit representations.

In extended ASCII, each character is represented by 8 bits. Using 8 bits, a number of graphic characters and control characters can be represented.

Unicode is used to represent any world language in computers. Unicode uses 16 bits to represent each character. Java and XML support Unicode.

Unicode has been developed to represent all the world languages. Unicode uses 16 bits to represent each character and can be used to encode the characters of any recognized language in the world. Modern programming languages such as Java and markup languages such as XML support Unicode.

It is important to note that the ASCII/Unicode coding mechanism is not the best way, according to Shannon. If we consider the frequency of occurrence of the letters of a language and use small codewords for frequently occurring letters, the coding will be more efficient. However, more processing will be required, and more delay will result.

The best coding mechanism for text messages was developed by Morse. The Morse code was used extensively for communication in the old days. Many ships used the Morse code until May 2000. In Morse code, characters are represented by dots and dashes. Morse code is no longer used in standard communication systems.

Morse code uses dots and dashes to represent various English characters. It is an efficient code because short codes are used to represent high-frequency letters and long codes are used to represent low-frequency letters. The letter E is represented by just one dot and the letter Q is represented by dash dash dot dash.

4.2 VOICE

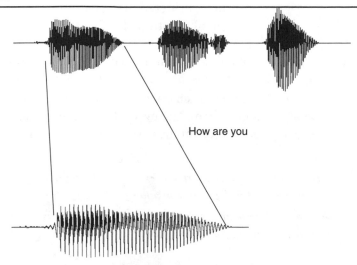

How are you

FIGURE 4.1 Speech waveform.

To transmit voice from one place to another, the speech (acoustic signal) is first converted into an electrical signal using a transducer, the microphone. This electrical signal is an analog signal. The voice signal corresponding to the speech "how are you" is shown in Figure 4.1. The important characteristics of the voice signal are given here:

- The voice signal occupies a bandwidth of 4kHz i.e., the highest frequency component in the voice signal is 4kHz. Though higher frequency components are present, they are not significant, so a filter is used to remove all the high-frequency components above 4kHz. In telephone networks, the bandwidth is limited to only 3.4kHz.

- The pitch varies from person to person. Pitch is the fundamental frequency in the voice signal. In a male voice, the pitch is in the range of 50–250 Hz. In a female voice, the pitch is in the range of 200–400 Hz.

- The speech sounds can be classified broadly as voiced sounds and unvoiced sounds. Signals corresponding to voiced sounds (such as the vowels a, e, i, o, u) will be periodic signals and will have high amplitude. Signals corresponding to unvoiced sounds (such as th, s, z, etc.) will look like noise signals and will have low amplitude.

- Voice signal is considered a nonstationary signal, i.e., the characteristics of the signal (such as pitch and energy) vary. However, if we take small portions

The voice signal occupies a bandwidth of 4KHz. The voice signal can be broken down into a fundamental frequency and its harmonics. The fundamental frequency or pitch is low for a male voice and high for a female voice.

of the voice signals of about 20msec duration, the signal can be considered stationary. In other words, during this small duration, the characteristics of the signal do not change much. Therefore, the pitch value can be calculated using the voice signal of 20msec. However, if we take the next 20msec, the pitch may be different.

These characteristics are used while converting the analog voice signal into digital form. Analog-to-digital conversion of voice signals can be done using one of two techniques: waveform coding and vocoding.

The characteristics of speech signals described here are used extensively for speech processing applications such as text-to-speech conversion and speech recognition.

Music signals have a bandwidth of 20kHz. The techniques used for converting music signals into digital form are the same as for voice signals.

4.2.1 Waveform Coding

Waveform coding is done in such a way that the analog electrical signal can be reproduced at the receiving end with minimum distortion. Hundreds of waveform coding techniques have been proposed by many researchers. We will study two important waveform coding techniques: pulse code modulation (PCM) and adaptive differential pulse code modulation (ADPCM).

Pulse Code Modulation

Pulse Code Modulation (PCM) is the first and the most widely used waveform coding technique. The ITU-T Recommendation G.711 specifies the algorithm for coding speech in PCM format.

PCM coding technique is based on Nyquist's theorem, which states that if a signal is sampled uniformly at least at the rate of twice the highest frequency component, it can be reconstructed without any distortion. The highest frequency component in voice signal is 4kHz, so we need to sample the waveform at 8000 samples per second—every 1/8000th of a second (125 microseconds). We have to find out the amplitude of the waveform for every 125 microseconds and transmit that value instead of transmitting the analog signal as it is. The sample values are still analog values, and we can "quantize" these values into a fixed number of levels. As shown in Figure 4.2, if the number of quantization levels is

ITU-T standard G.711 specifies the mechanism for coding of voice signals. The voice signal is band limited to 4kHz, sampled at 8000 samples per second, and each sample is represented by 8 bits. Hence, using PCM, voice signals can be coded at 64kbps.

256, we can represent each sample by 8 bits. So, 1 second of voice signal can be represented by 8000 × 8 bits, 64kbits. Hence, for transmitting voice using PCM, we require 64 kbps data rate. However, note that since we are approximating the sample values through quantization, there will be a distortion in the reconstructed signal; this distortion is known as *quantization noise*.

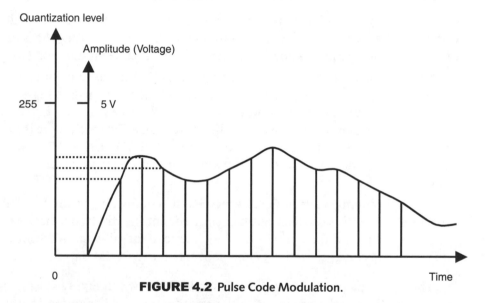

FIGURE 4.2 Pulse Code Modulation.

In the PCM coding technique standardized by ITU in the G.711 recommendation, the nonlinear characteristic of human hearing is exploited— the ear is more sensitive to the quantization noise in the lower amplitude signal than to noise in the large amplitude signal. In G.711, a logarithmic (non-linear) quantization function is applied to the speech signal, and so the small signals are quantized with higher precision. Two quantization functions, called A-law and m-law, have been defined in G.711. m-law is used in the U.S. and Japan. A-law is used in Europe and the countries that follow European standards. The speech quality produced by the PCM coding technique is called *toll quality speech* and is taken as the reference to compare the quality of other speech coding techniques.

50   Principles of Digital Communication Systems and Computer Networks

For CD-quality audio, the sampling rate is 44.1kHz (one sample every 23 microseconds), and each sample is coded with 16 bits. For two-channel stereo audio stream, the bit rate required is $2 \times 44.1 \times 1000 \times 16 = 1.41$Mbps.

The quality of speech obtained using the PCM coding technique is called toll quality. To compare the quality of different coding techniques, toll quality speech is taken as the reference.

Adaptive Differential Pulse Code Modulation

One simple modification that can be mode to PCM is that we can code the difference between two successive samples rather than coding the samples directly. This technique is known as differential pulse code modulation (DPCM).

Another characteristic of the voice signal that can be used is that a sample value can be predicted from past sample values. At the transmitting side, we predict the sample value and find the difference between the predicted value and the actual value and then send the difference value. This technique is known as adaptive differential pulse code modulation (ADPCM). Using ADPCM, voice signals can be coded at 32kbps without any degradation of quality as compared to PCM.

ITU-T Recommendation G.721 specifies the coding algorithm. In ADPCM, the value of speech sample is not transmitted, but the difference between the predicted value and the actual sample value is. Generally, the ADPCM coder takes the PCM coded speech data and converts it to ADPCM data.

The block diagram of an ADPCM encoder is shown in Figure 4.3(a). Eight-bit μ-law PCM samples are input to the encoder and are converted into linear format. Each sample value is predicted using a prediction algorithm, and then the predicted value of the linear sample is subtracted from the actual value to generate the difference signal. Adaptive quantization is performed on this difference value to produce a 4-bit ADPCM sample value, which is transmitted. Instead of representing each sample by 8 bits, in ADPCM only 4 bits are used. At the receiving end, the decoder, shown in Figure 4.3(b), obtains the dequantized version of the digital signal. This value is added to the value generated by the adaptive predictor to produce the linear PCM coded speech, which is adjusted to reconstruct m-law-based PCM coded speech.

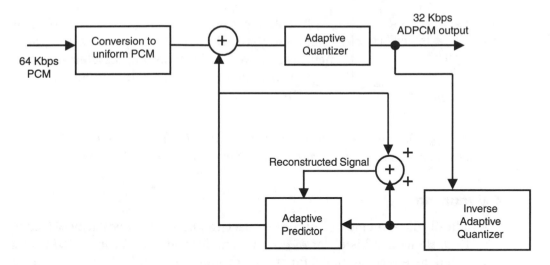

FIGURE 4.3 (a) ADPCM Encoder.

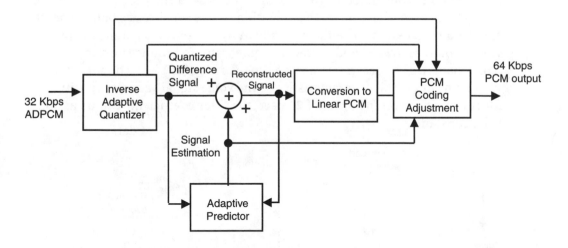

FIGURE 4.3 (b) ADPCM Decoder.

There are many waveform coding techniques such as delta modulation (DM) and continuously variable slope delta modulation (CVSD). Using these, the coding rate can be reduced to 16kbps, 9.8kbps, and so on. As the coding rate reduces, the quality of the speech is also going down. There are coding techniques using good quality speech which can be produced at low coding rates.

In ADPCM, each sample is represented by 4 bits, and hence the data rate required is 32kbps. ADPCM is used in telephone networks as well as radio systems such as DECT.

The PCM coding technique is used extensively in telephone networks. ADPCM is used in telephone networks as well as in many radio systems such as digital enhanced cordless telecommunications (DECT).

NOTE

Over the past 50 years, hundreds of waveform coding techniques have been developed with which data rates can be reduced to as low as 9.8kbps to get good quality speech.

4.2.2 Vocoding

A radically different method of coding speech signals was proposed by H. Dudley in 1939. He named his coder *vocoder*, a term derived from VOice CODER. In a vocoder, the electrical model for speech production seen in Figure 4.4 is used. This model is called the *source–filter model* because the speech production mechanism is considered as two distinct entities—a filter to model the vocal tract and an excitation source. The excitation source consists of a pulse generator and a noise generator. The filter is excited by the pulse generator to produce voiced sounds (vowels) and by the noise generator to produce unvoiced sounds (consonants). The vocal tract filter is a time-varying filter—the filter coefficients vary with time. As the characteristics of the voice signal vary slowly with time, for time periods on the order of 20msec, the filter coefficients can be assumed to be constant.

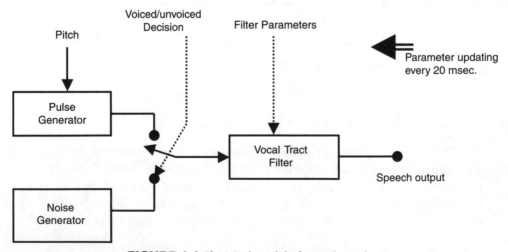

FIGURE 4.4 Electrical model of speech production.

In vocoding techniques, at the transmitter, the speech signal is divided into frames of 20msec in duration. Each frame contains 160 samples. Each frame is analyzed to check whether it is a voiced frame or unvoiced frame by using parameters such as energy, amplitude levels, etc. For voiced frames, the pitch is determined. For each frame, the filter coefficients are also determined. These parameters—voiced/unvoiced classification, filter coefficients, and pitch for voiced frames—are transmitted to the receiver. At the receiving end, the speech signal is reconstructed using the electrical model of speech production. Using this approach, the data rate can be reduced as low as 1.2kbps. However, compared to voice coding techniques, the quality of speech will not be very good. A number of techniques are used for calculating the filter coefficients. Linear prediction is the most widely used of these techniques.

In vocoding techniques, the electrical model of speech production is used. In this model, the vocal tract is represented as a filter. The filter is excited by a pulse generator to produce voiced sounds and by a noise generator to produce unvoiced sounds.

The voice generated using the vocoding techniques sounds very mechanical or robotic. Such a voice is called synthesized voice. Many speech synthesizers, which are integrated into robots, cameras, and such, use the vocoding techniques.

Linear Prediction

The basic concept of linear prediction is that the sample of a voice signal can be approximated as a linear combination of the past samples of the signal.

If S_n is the nth speech sample, then

$$S_n = \Sigma a_k S_{n-k} + G U_n$$

where a_k ($k = 1,\ldots,P$) are the linear prediction coefficients, G is the gain of the vocal tract filter, and U_n is the excitation to the filter. Linear prediction coefficients (generally 8 to 12) represent the vocal tract filter coefficients. Calculating the linear prediction coefficients involves solving P linear equations. One of the most widely used methods for solving these equations is through the Durbin and Levinson algorithm.

Coding of the voice signal using linear prediction analysis involves the following steps:

In linear prediction technique, a voice sample is approximated as a linear combination of the past *n* samples. The linear prediction coefficients are calculated every 20 milliseconds and sent to the receiver, which reconstructs the speech samples using these coefficients. Using this approach, voice signals can be compressed to as low as 1.2kbps.

- At the transmitting end, divide the voice signal into frames, each frame of 20msec duration. For each frame, calculate the linear prediction coefficients and pitch and find out whether the frame is voiced or unvoiced. Convert these values into code words and send them to the receiving end.

- At the receiver, using these parameters and the speech production model, reconstruct the voice signal.

Using linear prediction vocoder, voice signals can be compressed to as low as 1.2kbps. Quality of speech will be very good for data rates down to 9.6kbps, but the voice sounds synthetic for further lower data rates. Slight variations of this technique are used extensively in many practical systems such as mobile communication systems, speech synthesizers, etc.

Variations of LPC technique are used in many commercial systems, such as mobile communication systems and Internet telephony.

4.3 IMAGE

In image coding, the image is divided into small grids called pixels, and each pixel is quantized. The higher the number of pixels, the higher will be the quality of the reconstructed image.

To transmit an image, the image is divided into grids called *pixels* (or *picture elements*). The higher the number of grids, the higher the resolution. Grid sizes such as 768 × 1024 and 400 × 600 are generally used in computer graphics. For black-and-white pictures, each pixel is given a certain grayscale value. If there are 256 grayscale levels, each pixel is represented by 8 bits. So, to represent a picture with a grid size of 400 × 600 pixels with each pixel of 8 bits, 240kbytes of storage is required. To represent color, the levels of the three fundamental colors—red, blue, and green—are combined together. The shades of the colors will be higher if more levels of each color are used.

For example, if an image is coded with a resolution of 352 × 240 pixels, and each pixel is represented by 24 bits, the size of the image is 352 × 240 × 24/8 = 247.5 kilobytes.

To store the images as well as to send them through a communication medium, the image needs to be compressed. A compressed image occupies less storage space if stored on a medium such as hard disk or CD-ROM. If the image is sent through a communication medium, the compressed image can be transmitted fast.

One of the most widely used image coding formats is JPEG format. Joint Photograph Experts Group (JPEG) proposed this standard for coding of images. The block diagram of JPEG image compression is shown in Figure 4.5.

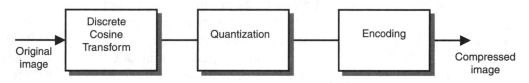

FIGURE 4.5 JPEG compression.

For compressing the image using the JPEG compression technique, the image is divided into blocks of 8 by 8 pixels and each block is processed using the following steps:

JPEG compression of an image is done in three steps: (a) division of the image into 8 × 8 matrix and applying discrete cosine transform (DCT) on each matrix, (b) quantization of the frequency coefficients obtained in step (a), and (c) conversion of the quantization levels into bits. Compression ratios of 30:1 can be achieved using this technique.

1. Apply discrete cosine transform (DCT), which takes the 8 × 8 matrix and produces an 8 × 8 matrix that contains the frequency coefficients. This is similar to the Fast Fourier Transform (FFT) used in Digital Signal Processing. The output matrix represents the image in spatial frequency domain.

2. Quantize the frequency coefficients obtained in Step 1. This is just rounding off the values to the nearest quantization level. As a result, the quality of the image will slightly degrade.

3. Convert the quantization levels into bits. Since there will be little change in the consecutive frequency coefficients, the differences in the frequency coefficients are encoded instead of directly encoding the coefficients.

Compression ratios of 30:1 can be achieved using JPEG compression. In other words, a 300kB image can be reduced to about 10kB.

JPEG image compression is used extensively in Web page development. As compared to the bit mapped files (which have a .bmp extension), the JPEG images (which have a .jpg extension) occupy less space and hence can be downloaded fast when we access a Web site.

4.4 VIDEO

For video coding, the video is considered a series of frames. At least 16 frames per second are required to get the perception of moving video. Each frame is compressed using the image compression techniques and transmitted. Using this technique, video can be compressed to 64kbps, though the quality will not be very good.

A video signal occupies a bandwidth of 5MHz. Using the Nyquist sampling theorem, we need to sample the video signal at 10 samples/msec. If we use 8-bit PCM, video signal requires a bandwidth of 80Mbps. This is a very high data rate, and this coding technique is not suitable for digital transmission of video. A number of video coding techniques have been proposed to reduce the data rate.

Video encoding is an extension of image encoding. As shown in Figure 4.6, a series of images or frames, typically 16 to 30 frames, is transmitted per second. Due to the persistence of the eye, these discrete images appear as though it is a moving video. Accordingly, the data rate for transmission of video will be the number of frames multiplied by the data rate for one frame. The data rate is reduced to about 64kbps in desktop video conferencing systems where the resolution of the image and the number of frames are reduced considerably. The resulting video is generally acceptable for conducting business meetings over the Internet or corporate intranets, but not for transmission of, say, dance programs, because the video will have many jerks.

Moving Picture Experts Group (MPEG) released a number of standards for video coding. The following standards are used presently:

A variety of video compression standards have been developed. Notable among them is MPEG-2, which is used for video broadcasting. MPEG-4 is used in video conferencing applications and HDTV for high-definition television broadcasting.

MPEG-2: This standard is for digital video broadcasting. The data rates are 3 and 7.5Mbps. The picture quality will be much better than analog TV. This standard is used in broadcasting through direct broadcast satellites.

MPEG-4: This standard is used extensively for coding, creation, and distribution of audio-visual content for many applications because it supports a wide range of data rates. The MPEG-4 standard addresses the following aspects:

- Representing audio-visual content, called media objects.
- Describing the composition of these objects to create compound media objects.
- Multiplexing and synchronizing the data.

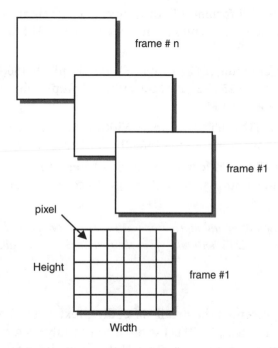

FIGURE 4.6 Video coding through frames and pixels.

The primitive objects can be still images, audio, text, graphics, video, or synthesized speech. Video coding between 5kbps and 10Mbps, speech coding from 1.2kbps to 24kbps, audio (music) coding at 128kbps, etc. are possible. MP3 (MPEG Layer–3) is the standard for distribution of music at 128kbps data rate, which is a part of the MPEG-4 standards.

For video conferencing, 384kbps and 2.048Mbps data rates are very commonly used to obtain better quality as compared to 64kbps. Video conferencing equipment that supports these data rates is commercially available.

MPEG-4 is used in mobile communication systems for supporting video conferencing while on the move. It also is used in video conferencing over the Internet.

In spite of the many developments in digital communication, video broadcasting continues to be analog in most countries. Many standards have been developed for digital video applications. When optical fiber is used extensively as the transmission medium, perhaps then digital video will gain popularity. The important European digital formats for video are given here:

Multimedia CIF format: Width in pixels 360; height in pixels 288; frames/ second 6.25 to 25; bit rate without compression 7.8 to 31 Mbps; with compression 1 to 3 Mbps.

Video conferencing (QCIF format): Width in pixels 180; height in pixels 144; frames per second 6.25 to 25; bit rate without compression 1.9 to 7.8 Mbps; with compression 0.064 to 1 Mbps.

Digital TV, ITU-R BT.601 format: Width 720; height 526; frames per second 25; bit rate without compression 166 Mbps; with compression 5 to 10 Mbps.

HDTV, ITU-R BT.109 format: Width 1920; height 1250; frames per second 25; bit rate without compression 960 Mbps; with compression 20 to 40 Mbps.

Commercialization of digital video broadcasting has not happened very fast. It is expected that ultilization of HDTV will take off in the first decade of the twenty-first century.

Summary

This chapter presented the details of coding text, voice, image, and video into digital format. For text, ASCII is the most commonly used representation. Seven bits are used to represent characters. Unicode, which uses 16 bits is now being used for text representation. Characters of any world language can be represented using Unicode.

For audio, Pulse Code Modulation (PCM) is the most widely used coding technique. In PCM, voice is coded at 64kbps data rate by sampling the voice signal at 8000 samples per second and representing each sample by 8 bits. Using Adaptive Differential Pulse Code Modulation (ADPCM), the coding rate can be reduced to 32kbps without any reduction in quality. Another technique used for voice coding is Linear Prediction Coding (LPC), with which the data rate can be reduced to as low as 1.2kbps. However, as the bit rate goes down, quality goes down. Variants of LPC are used in many applications such as mobile communications, Internet telephony, etc.

For image compression, the Joint Photograph Experts Group (JPEG) standard is used, through which compression ratios up to 30:1 can be achieved. For video coding, the most widely used standard was developed by Moving Picture Experts Group (MPEG). MPEG-2 is used for broadcasting. MPEG-4 defines standards for video encoding from 5kbps to 10Mbps. MPEG-4 is used in mobile communications as well as in multimedia communication over the Internet.

References

J. Campbell. *C Programmer's reference guide to Serial communication*. Prentice-Hall, Inc., 1997.

G. Karlsson. "Asynchronous Transfer of Video". *IEEE Communications Magazine*, Vol. 34, No. 8, August 1996.

G. K. Wallace. "The JPEG Still Picture Compression Standard". *Communications of the ACM*, Vol. 34, No. 1, April 1991, pp. 30-44.

D. LeGall. "MPEG: A Video Compression Standard for Multimedia Applications". *Communications of the ACM*, Vol. 34, No. 1, April 1994.

www.cdacindia.com Web site of Center for Development of Advanced Computing. You can obtain the details of the ISCII standard from this site.

Questions

1. What are the different standards for coding of text messages?
2. What is waveform coding? Explain the PCM and ADPCM coding techniques.
3. What is a vocoder? Describe the speech production model.
4. Explain the LPC coding technique.
5. Explain the JPEG compression technique.
6. What are the salient features of the MPEG-4 standard?

Exercises

1. On your multimedia PC, record your voice and observe the speech waveform. Store the speech data in a file and check the file size. Vary the sampling rate

and quantization levels (bits per sample), store the speech data, and observe the file sizes.

2. Install a desktop video camera on your PC and, using a software package such as Microsoft's NetMeeting, participate in a video conference over the LAN. Observe the video quality.

3. Calculate the bit rate required for video transmission if the video is transmitted at the rate of 30 frames per second, with each frame divided into 640 × 480 pixels, and coding is done at 3 bits per pixel.

4. Describe the Indian Standard Code for Information Interchange.

5. Download freely available MP3 software and find out the compression achieved in MP3 software by converting WAV files into MP3 files.

6. Calculate the memory required to store 100 hours of voice conversation if the coding is done using (a) PCM at 64kbps (b) ADPCM at 32kbps and (c) LPC at 2.4kbps.

7. If the music signal is band-limited to 15 kHz, what is the minimum sampling rate required? If 12 bits are used to represent each sample, what is the data rate?

8. An image is of size 640 × 480 pixels. Each pixel is coded using 4 bits. How much is the storage requirement to store the image?

Projects

1. Develop a program to generate Morse code. The output of the Morse code (sounds for dot and dash) should be heard through the sound card. The duration of a dash is three times that of the dot.

2. Study the Durbin–Levinson algorithm for calculation of linear prediction coefficients. Implement the algorithm in software.

3. Develop software for compression of images using the JPEG standard.

5 Error Detection and Correction

In This Chapter

- Error Detection and Correction
- Error Detection Techniques
- Using the C Language to Calculate CRC
- The Mechanism of Error Correction

In a digital communication system, totally error-free transmission is not possible due to transmission impairments. At the receiving end, there should be a mechanism for detection of the errors and if possible for their correction. In this chapter, we will study the various techniques used for error detection and correction.

5.1 NEED FOR ERROR DETECTION AND CORRECTION

In a digital communication system, some bits are likely to be received in error due to the noise in the communication channel. As a result, 1 may become 0 or 0 may become 1. The Bit Error Rate (BER) is a parameter used to characterize communication systems.

Consider a communication system in which the transmitted bit stream is

1 0 11 0 111 0

The transmitted electrical signal corresponding to this bit stream and the received waveform are shown in Figure 5.1. Due to the noise introduced in the transmission medium, the electrical signal is distorted. By using a threshold, the receiver determines whether a 1 is transmitted or a 0 is transmitted. In this case, the receiver decodes the bit stream as

1 0 1 0 0 1 0 1 0

At two places, the received bit is in error—1 has become 0 in both places.

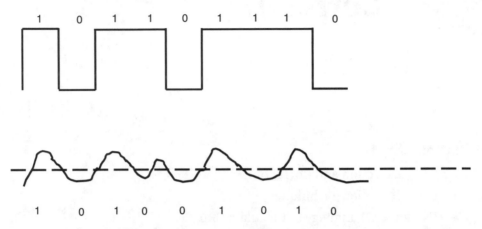

FIGURE 5.1 Errors introduced by transmission medium.

How many errors can be tolerated by a communication system? It depends on the application. For instance, if English text is transmitted, and a few letters are received in error, it is tolerable. Studies indicate that even if 20% of the letters are missing, human beings can understand the text.

Suppose the communication system is used to transmit digitized voice from one place to another. Studies indicate that even if the Bit Error Rate is 10^{-3}, the listener will be able to understand the speech. In other words, a voice communication system can tolerate one error for every 1000 bits transmitted.

NOTE

The performance of a communication system can be characterized by the Bit Error Rate (BER). If BER is 10^{-3}, there is one error per 1000 bits.

Errors can be classified as random errors and burst errors. Random errors, as the name implies, appear at random intervals. Burst errors are caused by sudden disturbances such as lightning. Such disturbances cause many consecutive bits to be in error.

Now consider the case of a banking application. Suppose I need to transfer $100 from my account to a friend's account through a data communication network. If the digit 1 becomes 3 due to one bit error during transmission, then instead of $100, $300 will be deducted from my account! So, for such applications, not even a single error can be tolerated. Hence, it is very important to detect the errors for data applications.

Errors can be classified as

- Random errors
- Burst errors

Random errors occur at random places in the bit stream. Burst errors occur due to sudden disturbances in the medium, caused by lightning, sudden interference with the nearby devices, etc. Such disturbances result in a sequence of bits giving errors.

Detection and correction of errors is done through channel coding. In channel coding, additional bits are added at the transmitter end, and these additional bits are used at the receiving end to check whether the transmitted data is received correctly or not and, if possible, to correct the errors.

5.2 ERROR DETECTION

The three widely used techniques for error detection are parity, checksum, and cyclic redundancy check (CRC). These techniques are discussed in the following sections.

5.2.1 Parity

Parity is used in serial communication protocols whereby we transmit one character at a time. For example, if the information bits are

1 0 1 1 0 1 0

then an additional bit is added, which is called a parity bit. The parity bit can be added in such a way that the total number of ones becomes even. In such a case, it is called *even parity*. In the above bit stream, already there are four ones, and hence a 0 is added as the parity bit. The bit stream transmitted is

1 0 1 1 0 1 0 0

In case of odd parity, the additional bit added will make the total number of ones odd. For odd parity, the additional bit added in the above case is 1 and the transmitted bit stream is

1 0 1 1 0 1 0 1

At the receiving end, from the first 7 bits, the receiver will calculate the expected parity bit. If the received parity and the calculated parity match, it is assumed that the character received is OK.

Parity bit is the additional bit added to a character for error checking. In even parity, the additional bit will make the total number of ones even. In case of odd parity, the additional bit will make the total number of ones odd. Parity bit is used in serial communication.

The various parities can be even, odd or none. In the case of none parity, the parity bit is not used and is ignored.

It is very easy to verify that parity can detect errors only if there is an odd number of errors; if the number of errors is 1, 3, or 5, the error can be detected. If the number of errors is even, parity bit cannot detect the error.

5.2.2 Block Codes

In block coders, a block of information bits is taken and additional bits are generated. These additional bits are called checksum or cyclic redundancy check (CRC). Checksum or CRC is used for error detection.

The procedure used in block coding is shown in Figure 5.2. The block coder takes a block of information bits (say 8000 bits) and generates additional bits (say, 16). The output of the block coder is the original data with the additional 16 bits. The additional bits are called checksum or cyclic redundancy check (CRC). Block codes can detect errors but cannot correct errors.

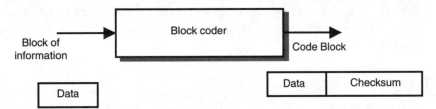

FIGURE 5.2 Block coder.

Checksum

Suppose you want to send two characters, C and U.

The 7-bit ASCII values for these characters are

    C   1 0 0 0 0 1 1
    U   1 0 1 0 1 0 1

In addition to transmitting these bit streams, the binary representation of the sum of these two characters is also sent. The value of C is 67 and the value of U is 85. The sum is 152. The binary representation of 152 is 1 0 0 1 1 0 0 0. This bit stream is also attached to the original binary stream, corresponding to C and U, while transmitting the data.

Checksum of information bits is calculated using simple binary arithmetic. Checksum is used extensively because its computation is very easy. However, checksum cannot detect all errors.

So, the transmitted bit stream is

1 0 0 0 0 1 1 1 0 1 0 1 0 1 1 0 0 1 1 0 0 0

At the receiving end, the checksum is again calculated. If the received checksum matches this calculated checksum, then the receiver assumes that the received data is OK. The checksum cannot detect all the errors. Also, if the characters are sent in a different order, i.e., if the sequence is changed, the checksum will be the same and hence the receiver assumes that the data is correct.

However, checksum is used mainly because its computation is very easy, and it provides a reasonably good error detection capability.

 Checksum is used for error detection in TCP/IP protocols to check whether packets are received correctly. Different algorithms are used for calculation of checksum.

Cyclic Redundancy Check

CRC is a very powerful technique for detecting errors. Hence, it is extensively used in all data communication systems. Additional bits added to the information bits are called the CRC bits. These bits can be 16 or 32. If the additional bits are 16, the CRC is represented as CRC-16. CRC-32 uses 32 additional bits. There are international standards for calculation of CRC-16 and CRC-32. Since CRC calculation is very important, the C programs to calculate the CRC are in Listings 5.1 and 5.2. When these programs are executed, the information bits and the CRC in hexadecimal notation will be displayed.

CRC-16 and CRC-32 are the two standard algorithms used for calculation of cyclic redundancy check. The additional CRC bits (16 and 32) are appended to the information bits at the transmitting side. At the receiving side, the received CRC is compared with the calculated CRC. If the two match, the information bits are considered as received correctly. If the two do not match, it indicates that there are errors in the information bits.

Error detection using CRC is very simple. At the transmitting side, CRC is appended to the information bits. At the receiving end, the receiver calculates CRC from the information bits and, if the calculated CRC matches the received CRC, then the receiver knows that the information bits are OK.

Program for calculation of CRC-16

Listing 5.1 Program for calculation of CRC-16.

```
#include <stdio.h>
#include <stdlib.h>
```

```c
#include <string.h>

long CRC = 0x0000;
long GenPolynomial = 0x8005; //Divisor for CRC-16 Polynomial

void bitBybit(int bit);

int main()
{
    unsigned int MsgLength;
    int  i=0,j=0;
    char SampleMsg[] = "Hello World";
    char tempBuffer[100];

    MsgLength = sizeof(SampleMsg)-1;

    printf("\nActual Message: %s\n",SampleMsg);

    strcpy(tempBuffer,SampleMsg);
    tempBuffer[MsgLength] = 0x00;
    tempBuffer[MsgLength+1] = 0x00;
    tempBuffer[MsgLength+2] = '\0';

    printf("\nAfter padding 16 0-bits to the Message:");

    for(i=0;i<MsgLength+2;++i)
    {
        unsigned char ch = tempBuffer[i];
        unsigned char mask = 0x80;
        for(j=0;j<8;++j)
        {
            bitBybit(ch&mask);
            mask>>=1;
        }
        printf(" ");
    }
    printf("\n\nCalculated CRC:0x%x\n\n",CRC);
    return 0;
}

void bitBybit(int bit)
{
```

```
long firstBit = (CRC & 0x8000);

CRC = (CRC << 1);
if(bit)
{
   CRC = CRC ^ 1;
   printf("1");
}
else
{
   CRC = CRC ^ 0;
   printf("0");
}

if(firstBit)
{
   CRC = (CRC^GenPolynomial);
}

}
```

In this listing, the actual message to be transmitted is "Hello World." The message is padded with sixteen 0 bits, and the message bit stream is

01001000 01100101 01101100 01101100 01101111 00100000 01010111 01101111 01110010 01101100 01100100 00000000 00000000

The calculated CRC value in hexadecimal notation is 0x303f70c3.

Program for calculation of CRC-32

Listing 5.2 Program for calculation of CRC-32.

```
#include <stdio.h>
#include <stdlib.h>
#include <string.h>

long CRC = 0x00000000L;
long GenPolynomial = 0x04c11db7L; //Divisor for CRC-32 Polynomial

void bitBybit(int bit);

int main()
```

```
{
    unsigned int MsgLength;
    int i=0,j=0;
    char SampleMsg[] = "Hello World";
    char tempBuffer[100];

    MsgLength = sizeof(SampleMsg)-1;

    printf("\nActual Message: %s\n",SampleMsg);
    strcpy(tempBuffer,SampleMsg);
    tempBuffer[MsgLength]   = 0x00;
    tempBuffer[MsgLength+1] = 0x00;
    tempBuffer[MsgLength+2] = 0x00;
    tempBuffer[MsgLength+3] = 0x00;
    tempBuffer[MsgLength+4] = '\0';

    printf("\nAfter padding 32 0-bits to the Message:");

    for(i=0;i<MsgLength+4;++i)
    {
        unsigned char ch = tempBuffer[i];
        unsigned char mask = 0x80;
        for(j=0;j<8;++j)
        {
                bitBybit(ch&mask);
                mask>>=1;
        }
        printf(" ");
    }
    printf("\n\nCalculated CRC:0x%x\n\n",CRC);
        return 0;
}

void bitBybit(int bit)
{
    long firstBit = (CRC & 0x80000000L);

    CRC = (CRC << 1);
    if(bit)
    {
        CRC = CRC ^ 1;
        printf("1");
    }
    else
    {
        CRC = CRC ^ 0;
```

```
        printf("0");
    }

    if(firstBit)
    {
        CRC = (CRC^GenPolynomial);
    }

}
```

Listing 5.2 gives the C program to calculate CRC-32. In this program the message for which CRC has to be calculated is "Hello World." The message bit stream is

01001000 01100101 01101100 01101100 01101111 00100000 01010111 01101111 01110010 01101100 01100100 00000000 00000000 00000000 00000000

The calculated CRC is 0x31d1680c.

In CRC calculation, a standard polynomial is used. This polynomial is different for CRC-16 and CRC-32. The bit stream is divided by this polynomial to calculate the CRC bits.

Using error detection techniques, the receiver can detect the presence of errors. In a practical communication system, just detection of errors does not serve much purpose, so the receiver has to use another mechanism such as asking the transmitter to resend the data. Communication protocols carry out this task.

5.3 ERROR CORRECTION

If the error rate is high in transmission media such as satellite channels, error-correcting codes are used that have the capability to correct errors. Error correcting codes introduce additional bits, resulting in higher data rate and bandwidth requirements. However, the advantage is that retransmissions can be reduced. Error correcting codes are also called forward-acting error correction (FEC) codes.

Convolutional codes are widely used as error correction codes. The procedure for convolutional coding is shown in Figure 5.3. The convolutional coder takes a block of information (of n bits) and generates some additional bits (k bits). The additional k bits are derived from the information bits. The output is (n + k) bits. The additional k bits can be used to correct the errors that occurred in the original n bits. n/(n+k) is called rate of the code. For instance, if 2 bits are sent for every 1 information bit, the rate is 1/2. Then the coding technique is called Rate 1/2 FEC. If 3 bits are sent for every 2 information bits, the coding technique is called Rate 2/3 FEC. The additional bits are derived from the information bits, and hence redundancy is introduced in the error correcting codes.

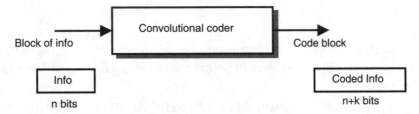

FIGURE 5.3 Convolutional coder.

In error correcting codes such as convolutional codes, in addition to the information bits, additional redundant bits are transmitted that can be used for error correction at the receiving end. Error correcting codes increase the bandwidth requirement, but they are useful in noisy channels.

For example, in many radio systems, error rate is very high, and so FEC is used. In Bluetooth radio systems, Rate 1/3 FEC is used. In this scheme, each bit is transmitted three times. To transmit bits b0b1b2b3, the actual bits transmitted using Rate 1/3 FEC are

b0b0b0b1b1b1b2b2b2b3b3b3

At the receiver, error correction is possible. If the received bit stream is

101000111000111

it is very easy for the receiver to know that the second bit is received in error, and it can be corrected.

A number of FEC coding schemes have been proposed that increase the delay in processing and also the bandwidth requirement but help in error correction. Shannon laid the foundation for channel coding, and during the last five decades, hundreds of error-correcting codes have been developed.

It needs to be noted that in source coding techniques, removing the redundancy in the signal reduces the data rate. For instance, in voice coding, low-bit rate coding techniques reduce the redundancy. In contrast, in error correcting codes, redundancy is introduced to facilitate error correction at the receiver.

Summary

In a communication system, transmission impairments cause errors. In many data applications, errors cannot be tolerated, and error detection and correction are required. Error detection techniques use parity, checksum, or cyclic redundancy check (CRC). Additional bits are added to the information bit stream at the transmitting end. At the receiving end, the additional bits are used to check whether the information bits are received correctly or not. CRC is the most effective way of detecting errors and is used extensively in data communication. If the receiver detects that there are errors in the received bit stream, the receiver will ask the sender to retransmit the data.

Error correction techniques add additional redundancy bits to the information bits so that at the receiver, the errors can be corrected. Error correcting codes increase the data rate and hence the bandwidth requirement, but they are needed if the channel is noisy and if retransmissions have to be avoided.

References

Dreamtech Software Team. *Programming for Embedded Systems*. Wiley-Dreamtech India Pvt. Ltd., 2002. This book contains the source code for calculation of CRC as well as serial programming.

J. Campbell. *C Programmer's Guide to Serial Communications*. Prentice-Hall Inc., 1994.

Questions

1. Explain the need for error detection and correction.
2. What is parity? Explain with examples.
3. What is checksum? Explain with an example.
4. What is CRC? What are the different standards for CRC calculation?
5. Explain the principles of error correcting codes with an example.

Exercises

1. Connect two PCs using an RS232 cable. Using serial communication software, experiment with parity bits by setting even parity and odd parity.
2. "Parity can detect errors only if there is an odd number of errors." Prove this statement with an example.

3. Calculate the even parity bit and the odd parity bit if the information bits are 1101101.

4. If the information bits are 110110110, what will be the bit stream when Rate 1/3 FEC is used for error correction? Explain how errors are corrected using the received bit stream. What will be the impact of bit errors introduced in the channel?

Projects

1. Write C programs to calculate CRC for the standard CRC algorithms, CRC-16 and CRC-32.

2. Write a program for error correction using Rate 1/3 FEC.

3. Studies indicate that even if 20% of the characters are received in error in a communication system for transmitting English text messages, the message can be understood. Develop software to prove this study. You need to simulate errors using a random number generation program.

6 ▪ Digital Encoding

In This Chapter

- Digital Encoding Schemes
- Categories of Encoding Schemes

In a digital communication system, the first step is to convert the information into a bit stream of ones and zeros. Then the bit stream has to be represented as an electrical signal. In this chapter, we will study the various representations of the bit stream as an electrical signal.

6.1 REQUIREMENTS FOR DIGITAL ENCODING

The bit stream is encoded into an equivalent electrical signal using digital encoding schemes. The encoding scheme should be chosen keeping in view the bandwidth requirement, clocking, error detection capability, noise immunity, and complexity of the decoder.

Once the information is converted into a bit stream of ones and zeros, the next step is to convert the bit stream into its electrical representation. The electrical signal representation has to be chosen carefully for the following reasons:

- The electrical representation decides the bandwidth requirement.
- The electrical representation helps in clocking—the beginning and ending of each bit.
- Error detection can be built into the signal representation.
- Noise immunity can be increased by a good electrical representation.
- The complexity of the decoder can be decreased.

A variety of encoding schemes have been proposed that address all these issues. In all communication systems, the standards specify which encoding technique has to be used. In this chapter, we discuss the most widely used encoding schemes. We will refer to these encoding schemes throughout the book, and hence a good understanding of these is most important.

6.2 CATEGORIES OF ENCODING SCHEMES

Encoding schemes can be divided into the following categories:

- Unipolar encoding
- Polar encoding
- Bipolar encoding

The encoding schemes are divided into three categories: (a) unipolar encoding; (b) polar encoding; and (c) bipolar encoding. In unipolar encoding, only one voltage level is used. In polar encoding, two voltage levels are used. In bipolar encoding, three voltage levels are used. Both polar encoding and bipolar encoding schemes are used in practical communication systems.

Unipolar encoding: In the unipolar encoding scheme, only one voltage level is used. Binary 1 is represented by positive voltage and binary 0 by an idle line. Because the signal will have a DC component, this scheme cannot be used if the transmission medium is radio. This encoding scheme does not work well in noisy conditions.

Polar encoding: In polar encoding, two voltage levels are used: a positive voltage level and a negative voltage level. NRZ-I, NRZ-L, and Manchester encoding schemes, which we discuss in the following sections, are examples of this encoding scheme.

Bipolar encoding: In bipolar encoding, three levels are used: a positive voltage, a negative voltage, and 0 voltage. AMI and HDB3 encoding schemes are examples of this encoding scheme.

The encoding scheme to be used in a particular communication system is generally standardized. You need to follow these standards when designing your system to achieve interoperability with the systems designed by other manufacturers.

6.3 NON-RETURN TO ZERO INVERTIVE (NRZ-I)

In NRZ-I, 0 is represented by 0 volt and 1 by 0 volt or +V volts, based on the previous voltage level.

In NRZ-I, binary 0 is represented by 0 volt and binary 1 by 0 volt or +V volts, according to the previous voltage. If the previous voltage was 0 volt, the current binary 1 is +V volts. If the previous voltage was +V volts, then the current binary 1 is 0 volt.

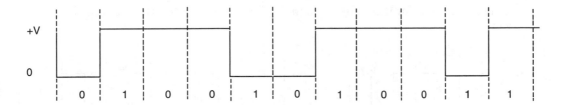

6.4 NON-RETURN TO ZERO LEVEL (NRZ-L)

In NRZ-L, 1 is represented by positive voltage and 0 by negative voltage. Synchronization is a problem in this encoding scheme.

In NRZ-L, binary 1 is represented by positive voltage and 0 by negative voltage. This scheme, though simple, creates problems: if there is a synchronization problem, it is difficult for the receiver to synchronize, and many bits are lost.

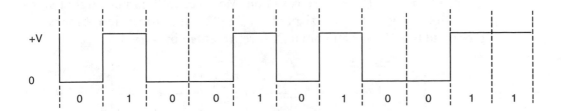

6.5 MANCHESTER

In Manchester encoding, 1 is represented by low-to-high voltage transition and 0 is represented by high-to-low voltage transition. This scheme is useful for deriving the clock, and errors can be detected.

In Manchester encoding, the voltage transition is in the middle of the bit period. A low-to-high transition represents 1 and high-to- low transition represents 0. The advantage of this scheme is that the transition serves as a clocking mechanism, and errors can be detected if there is no transition. However, bandwidth requirement for this scheme is higher as compared to other schemes.

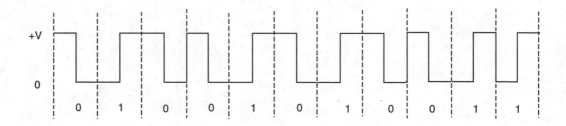

6.6 RS232 STANDARD

In RS232, 0 is represented by +V volts and 1 is represented by –V volts.

The RS232 interface (commonly known as a *serial port*) is used in embedded systems as well as to connect PC modems. The RS232 standard specifies a data rate of 19.2kbps, but higher data rates are supported in most PCs, speeds up to 115kbps are common. Most controllers and digital signal processors support the serial port. In RS232 communication, binary 0 is represented by +V volts and binary 1 is represented by –V volts.

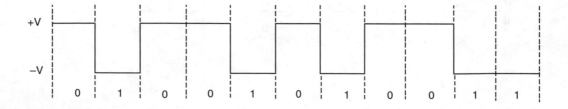

6.7 BIPOLAR ALTERNATE MARK INVERSION (BIPOLAR AMI)

In the bipolar AMI encoding scheme, 0 is represented by no signal and 1 by either positive or negative voltage. Binary 1 bits alternate in polarity. Ease of synchronization is the main advantage of this scheme.

In the bipolar alternate mark inversion (Bipolar AMI) encoding scheme, 0 is represented by no signal and 1 by positive or negative voltage. Binary 1 bits must alternate in polarity. The advantage of this coding scheme is that if a long string of ones occurs, there will be no loss of synchronization. If synchronization is lost, it is easy to resynchronize at the transition.

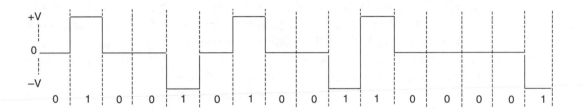

6.8 HIGH-DENSITY BIPOLAR 3 (HDB3)

HDB3 encoding is similar to AMI, except that two pulses called violation pulse (V) and balancing pulse (B) are used when consecutively four or more zeros occur in the bit stream. These pulses eliminate the DC component.

The HDB3 encoding scheme is almost the same as AMI, except for a small change: Two pulses, called violation pulse and the balancing pulse, are used when consecutively four or more zeros occur in the bit stream. When there are four consecutive binary 0 bits, the pulses will be 000V, where V will be same as the previous non-zero voltage. However, this V pulse creates a DC component. To overcome this problem, the B bit is introduced. If there are four consecutive binary 0 bits, the pulse train will be B00V. B is positive or negative to make alternate Vs of opposite polarity.

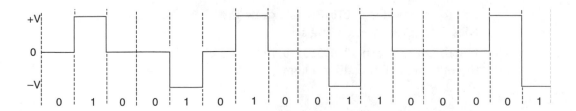

The HDB3 encoding scheme is a standard adopted in Europe and Japan.

Summary

This chapter has presented the various schemes used for digital encoding. The digital encoding scheme should take into consideration the bandwidth requirement, immunity to noise, and synchronization. Encoding schemes can be broadly divided into unipolar encoding, polar encoding, and bipolar encoding. In unipolar encoding, there will be only one level—binary 1 is represented by positive voltage and binary 0 by keeping the line idle. Because the signal will have a DC component, it cannot be used in radio systems. Polar encoding uses two levels: a positive voltage level and a negative voltage level. Non-return to zero level (NRZ-L) and non-return to zero invertive (NRZ-I) are examples of this type of encoding. In biploar encoding, three voltage levels are used: positive, negative, and zero levels. Alternate mark inversion (AMI) and high-density bipolar 3 (HDB3) are examples of this type of encoding. All these types of encoding schemes are used in communication systems.

References

W. Stallings. *Data and Computer Communication*. Fifth Edition. Prentice Hall Inc., 1999.

J. Campbell. *C Programmer's Guide to Serial Communications*. Prentice Hall Inc., 1997.

Questions

1. What are the requirements of digital encoding?
2. What are the three categories of digital encoding?
3. Explain polar encoding techniques.
4. Explain bipolar encoding techniques.

Exercises

1. For the bit pattern 10101010100, draw the electrical signal waveforms for NRZ-I and NRZ-L encoding schemes.
2. For the bit pattern 10101010110, draw the electrical signal waveforms for the Manchester encoding scheme.

3. If the bit pattern is 101010101110, how does the waveform appear on an RS232 port?

4. For the bit pattern 10101010100, draw the waveforms for Bipolar AMI and Manchester encoding schemes.

5. Study the UART (universal asynchronous receive-transmit) chips used for RS232 communication.

6. Is it possible to generate 5-bit codes from an RS232 interface of your PC? If not, what is the alternative?

Projects

1. Write a C program that takes a bit stream and a type of encoding scheme as inputs and produces the electrical signal waveforms for all the codes explained in this chapter.

2. A variety of codes such as CCITT2, CCITT3, CCITT5, TORFEC, etc. have been developed to represent the alphabet (A to Z). Compile a list of such codes. Write a program that takes English text as input and displays the bit stream corresponding to various codes.

7

Multiplexing

In This Chapter

- The Need for Multiplexing
- Different Types of Multiplexing Techniques
- Frequency, Time, and Wave Division Multiplexing

In a communication system, the costliest element is the transmission medium. To make the best use of the medium, we have to ensure that the bandwidth of the channel is utilized to its fullest capacity. *Multiplexing* is the technique used to combine a number of channels and send them over the medium to make the best use of the transmission medium. We will discuss the various multiplexing techniques in this chapter.

7.1 MULTIPLEXING AND DEMULTIPLEXING

Use of multiplexing technique is possible if the capacity of the channel is higher than the data rates of the individual data sources. Consider the example of a communication system in which there are three data sources. As shown in Figure 7.1, the signals from these three sources can be combined together (multiplexed) and sent through a single transmission channel. At the receiving end, the signals are separated (demultiplexed).

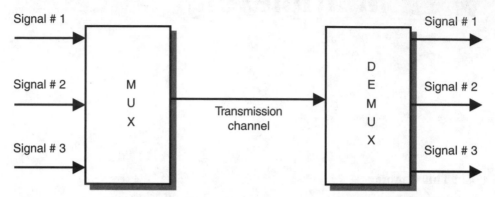

FIGURE 7.1 Multiplexing and demultiplexing.

A multiplexer (MUX) combines the data of different sources and sends it over the channel. At the receiving end, the demultiplexer (DEMUX) separates the data of the different sources.

Multiplexing is done when the capacity of the channel is higher than the data rates of the individual data sources.

At the transmitting end, equipment known as a multiplexer (abbreviated to MUX) is required. At the receiving end, equipment known as a demultiplexer (abbreviated to DEMUX) is required. Conceptually, multiplexing is a very simple operation that facilitates good utilization of the channel bandwidth. The various multiplexing techniques are described in the following sections.

7.2 FREQUENCY DIVISION MULTIPLEXING

In frequency division multiplexing (FDM), the signals are translated into different frequency bands and sent over the medium. The communication channel is divided into different frequency bands, and each band carries the signal corresponding to one source.

Consider three data sources that produce three signals as shown in Figure 7.2. Signal #1 is translated to frequency band #1, signal #2 is translated into frequency band #2, and so on. At the receiving end, the signals can be demultiplexed using filters. Signal #1 can be obtained by passing the multiplexed signal through a filter that passes only frequency band #1.

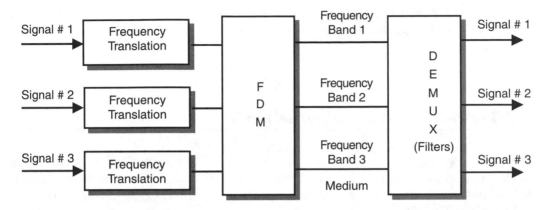

FIGURE 7.2 Frequency division multiplexing.

FDM is used in cable TV transmission, where signals corresponding to different TV channels are multiplexed and sent through the cable. At the TV receiver, by applying the filter, a particular channel's signal can be viewed. Radio and TV transmission are also done using FDM, where each broadcasting station is given a small band in the frequency spectrum. The center frequency of this band is known as the *carrier frequency*.

Figure 7.3 shows how multiple voice channels can be combined using FDM. Each voice channel occupies a bandwidth of 3.4kHz. However, each channel is assigned a bandwidth of 4kHz. The second voice channel is frequency translated to the band 4–8kHz. Similarly, the third voice channel is translated to 8–12 kHz and so on. Slightly higher bandwidth is assigned (4kHz instead of 3.4kHz) mainly because it is very difficult to design filters of high accuracy. Hence an additional bandwidth, known as *guard band*, separates two successive channels.

In FDM, the signals from different sources are translated into different frequency bands at the transmitting side and sent over the transmission medium. In cable TV, FDM is used to distribute programs of different channels on different frequency bands. FDM is also used in audio/video broadcasting.

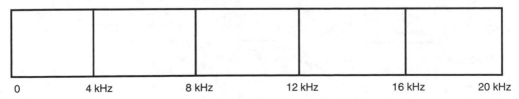

| 0 | 4 kHz | 8 kHz | 12 kHz | 16 kHz | 20 kHz |

FIGURE 7.3 FDM of voice channels.

FDM systems are used extensively in analog communication systems. The telecommunication systems used in telephone networks, broadcasting systems, etc. are based on FDM.

7.3 TIME DIVISION MULTIPLEXING

In time division multiplexing, the digital data corresponding to different sources is combined and transmitted over the medium. The MUX collects the data from each source, and the combined bit stream is sent over the medium. The DEMUX separates the data corresponding to the individual sources.

In synchronous time division multiplexing (TDM), the digitized signals are combined and sent over the communication channel. Consider the case of a communication system shown in Figure 7.4. Three data sources produce data at 64kbps using pulse code modulation (PCM). Each sample will be 8 bits, and the time gap between two successive samples is 125 microseconds. The job of the MUX is to take the 8-bit sample value of the first channel and the 8 bits of the second channel and then the 8 bits of the third channel. Again, go back to the first channel. Since no sample should be lost, the job of the MUX is to complete scanning all the channels and obtain the 8-bit sample values within 125 microseconds. This combined bit stream is sent over the communication medium. The MUX does a scanning operation to collect the data from each data source and also ensures that no data is lost. This is known as time division multiplexing. The output of the MUX is a continuous bit stream, the first 8 bits corresponding to Channel 1, the next 8 bits corresponding to Channel 2, and so on.

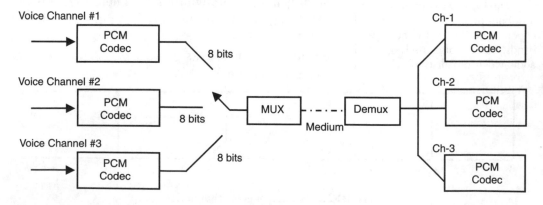

FIGURE 7.4 Time division multiplexing.

In a telephone network, switches (or exchanges) are interconnected through trunks. These trunks use TDM for multiplexing 32 channels. This is shown in Figure 7.4. The 32 channels are, by convention, numbered as 0 to 31. Each channel produces data at the rate of 64kbps. The MUX takes the 8 bits of each channel and produces a bit stream at the rate of 2048kbps (64kbps × 32). At the receiving end, the DEMUX separates the data corresponding to each channel. The TDM frame is also shown in Figure 7.5. The TDM frame depicts the number of bits in each channel. Out of the 32 slots, 30 slots are used to carry voice and two slots (slot 0 and slot 16) are used to carry synchronization and signaling information.

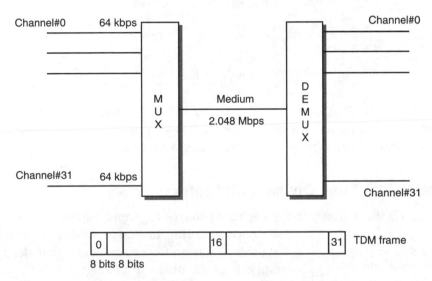

FIGURE 7.5 Time division multiplexing of voice channels.

In the Public Switched Telephone Network (PSTN), the switches are interconnected through trunks that use TDM. Trunks in which 30 voice channels are multiplexed are called E1 trunks.

Though TDM appears very simple, it has to be ensured that the MUX does not lose any data, and hence it has to maintain perfect timings. At the DEMUX also, the data corresponding to each channel has to be separated based on the timing of the bit stream. Hence, synchronization of the data is very important. Synchronization helps in separating the bits corresponding to each channel. This TDM technique is also known as *synchronous TDM*.

The trunks used in the telephone network using this TDM mechanism are known as T1 trunks or T1 carriers. The multiplexing of 24 channels is known as

Level 1 multiplexing. Four such T1 carriers are multiplexed to form T2 carrier. Seven T2 carriers are multiplexed to form T3 carrier and six T3 carriers are multiplex to form T4 carriers. The various levels of multiplexing, the number of voice channels and the data rates are given in Table 7.1. Note that at each level, additional bits are added for framing and synchronization.

Table 7.1 Digital Hierarchy of T-carriers

Level	Number of Voice Channels	Data Rates (Mbps)
1	24	1.544
2	96	6.312
3	672	44.736
4	4032	274.176

In a T1 carrier, the total number of voice channels is 24—there are 24 voice slots in the TDM frame. In Europe, a different digital hierarchy is followed. At the lowest level, 30 voice channels are multiplexed and the trunk is called E1 trunk.

7.3.1 Statistical Time Division Multiplexing

In the TDM discussed above, each data source is given a time slot in which the data corresponding to that source is carried. If the data source has no data to transmit, that slot will be empty. To make best use of the time slots, the data source can be assigned a time slot only if it has data to transmit. A centralized station can assign the time slots based on the need. This mechanism is known as statistical time division multiplexing (STDM).

In STDM, the data source is assigned a time slot only if it has data to transmit. As compared to synchronous TDM, this is a more efficient technique because time slots are not wasted.

7.4 WAVE DIVISION MULTIPLEXING

Wave division multiplexing (WDM) is used in optical fibers. In optical fiber communication, the signal corresponding to a channel is translated to an optical frequency (generally expressed in wavelength) and transmitted. This optical frequency is expressed by its equivalent wavelength and denoted by λ (lambda).

Instead of transmitting only one signal on the fiber, if two (or more) signals are sent on the same fiber at different frequencies (or wavelengths), it is called WDM. In 1994, this was demonstrated—signal frequencies had to be separated widely, typically 1310 nm and 1550 nm. Therefore, just using the two wavelengths can double the fiber capacity.

Wave Division Multiplexing is used in optical fibers. Data of different sources is sent through the fiber using different wavelengths. The advantage of WDM is that the full capacity of an already laid optical fiber can be used.

As shown in Figure 7.6, the wave division multiplexer takes signals from different channels, translates them to different wavelengths, and sends them through the optical fiber. Conceptually, it is the same as FDM.

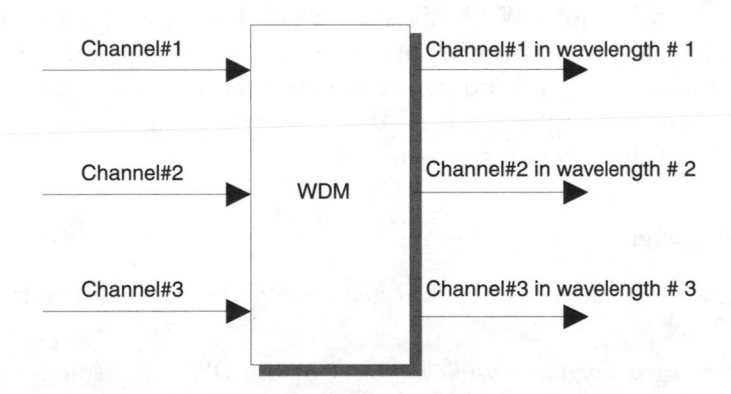

FIGURE 7.6 Wave division multiplexing.

Note that in WDM, single-mode fiber carries the data of different channels in different wavelengths. The advantage of WDM is that the full capacity of an already laid optical fiber can be increased by 16 to 32 times by sending different channels in different wavelengths. Because each wavelength corresponds to a different color, WDM effectively sends data corresponding to different channels in different colors.

7.4.1 Dense Wave Division Multiplexing

As the name implies, dense wave division multiplexing (DWDM) is the same as WDM except that the number of wavelengths or ones can be much higher than 32. Multiple optical signals are combined on the same fiber to increase the

DWDM is an extension of WDM—a large number of wavelengths are used to transmit data of different sources. Data rates up to the terabit level per second can be achieved using DWDM.

capacity of the fiber. One hundred twenty-eight and 256 wavelengths can be used to transmit different channels at different rates. DWDM is paving the way for making high bandwidths available without the need for laying extra fiber cables. Data rates up to terabits per second can be achieved on a single fiber.

Summary

Various multiplexing techniques have been discussed in this chapter. In frequency division multiplexing (FDM), different channels are translated into different frequency bands and transmitted through the medium. In time division multiplexing (TDM), the data corresponding to different channels is given separate time slots. Wave division multiplexing (WDM) and dense WDM (DWDM) facilitate transmission of different wavelengths in the same fiber. All these multiplexing techniques are used to combine data from different sources and to send it over the medium. At the receiving end, demultiplexing separates the data of the different sources.

References

R. Horak. *Communications Systems and Networks*. Wiley-Dreamtech India Pvt. Ltd., 2002.

www.iec.org/online/tutorials/dwdm Link to the DWDM resources of IEC online tutorials.

www.cisco.com Web site of Cisco Corporation. Gives a wealth of information on different multiple access techniques and commercial products.

Questions

1. Explain the different multiplexing techniques.
2. What is FDM? Give examples of practical systems in which FDM is used.
3. What is the difference between synchronous TDM and statistical TDM?
4. Explain the digital hierarchy of E-carriers.
5. Explain the importance of WDM and DWDM in enhancing the capacity of optical fibers.

Exercises

1. The digital hierarchy given in Table 7.1, calculate the overhead data bits added to the T2 carrier.
2. Carry out a survey on the use of the multiplexing techniques used in satellite communications.
3. Study the multiplexing hierarchy used in optical fiber systems.
4. Explore the various bands used in optical fiber for WDM and DWDM.

Projects

1. Implement a time division multiplexer with two PCM inputs using digital hardware.
2. Prepare a technical report on the WDM and DWDM giving the details of various commercial products.

8 | Multiple Access

In This Chapter

- What Multiple Access is
- Multiple Access Techniques
- Multiple Access Techniques Used in Practical Communication Systems

Multiple access is a technique with which multiple terminals share the bandwidth of the transmission medium. Multiple access techniques are of paramount importance in radio systems where the channel bandwidth is very limited.

Multiple access is a technique used to make best use of the transmission medium. In multiple access, multiple terminals or users share the bandwidth of the transmission medium. In this chapter, we study the various multiple access techniques. Multiple access attains much importance in radio communications because the radio spectrum is a precious natural resource, and to make best use of the radio bandwidth is important. We discuss the various multiple access techniques with illustrative examples.

8.1 FREQUENCY DIVISION MULTIPLE ACCESS

In radio communication systems, a certain frequency band is allocated for a specific use. This band is divided into smaller bands, and this pool of bands is available for all the terminals to share. Depending on the need, the base station allocates a frequency (the center frequency of a band) for the terminal to transmit. This is known as frequency division multiple access (FDMA). In FDMA systems, there is a central station that allocates the band (or frequency) to different terminals, based on their needs.

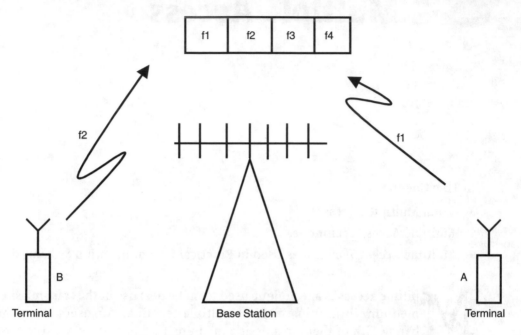

FIGURE 8.1 Frequency division multiple access.

In FDMA, the frequency band is divided into smaller bands, and the pool of these bands is available for a number of radio terminals to share, based on the need. A central station allocates the bands to different terminals based on the request.

Consider the example of a mobile communication system. The system consists of a number of base stations. As shown in Figure 8.1, a base station has a pool of frequencies (here, f1 to f4). All four frequencies are available for sharing by the mobile terminals located around the base station. The frequency will be allocated based on need. Mobile terminal A will be allocated the frequency f1, and then mobile terminal B will be allocated frequency f2.

To ensure that there is no overlap in transmission by different terminals using adjacent frequencies, a guard band is provided. A small separation between adjacent bands is called the *guard band*.

FDMA is used in many radio systems. Mobile communication systems and satellite communication systems use this access technique.

8.2 SPACE DIVISION MULTIPLE ACCESS

Because radio spectrum is a natural resource, we have to make best use of it. When a set of frequencies is reused in another location, it is called space division multiple access (SDMA). Again, consider a mobile communication system. A base station will be allocated a few frequencies. The same set of frequencies can be allocated to another region, provided there is sufficient distance between the two regions.

In SDMA, a service area is divided into small regions called cells and each cell is allocated certain frequencies. Two cells can make use of the same set of frequencies, provided these two cells are separated by a distance called *reuse distance*. SDMA is used in mobile communication systems.

As shown in Figure 8.2, the area covered by a mobile communication system can be divided into small regions called *cells*. A cell is represented as a hexagon. In each cell, there will be a base station that will be given a pool of frequencies. The frequencies of cell A (1) can be reused in cell A(2). We will study the detail of SDMA in more detail when we discuss cellular mobile communications in the chapter "Cellular Mobile Communication Systems."

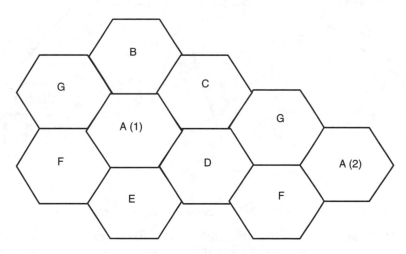

FIGURE 8.2 Space division multiple access.

The main attraction of SDMA is frequency reuse. Frequencies allocated to one cell can be allocated to another cell, provided there is sufficient distance between the two cells to avoid interference. This minimum distance is called reuse distance.

8.3 TIME DIVISION MULTIPLE ACCESS

In time division multiple access (TDMA), one frequency is shared by a number of terminals. Each station is given a time slot in which it can transmit. The number of time slots is fixed, and the terminals transmit in the time slots allocated to them. For example, in mobile communication systems, each frequency is divided into eight time slots and hence eight mobile terminals use the same frequency, communicating in different time slots.

Figure 8.3 illustrates the concept of TDMA. Terminal A transmits in time slot 1 using frequency f1. Terminal B also transmits using frequency f1 but in a different time slot. Which terminal gets which time slot is decided by the base station.

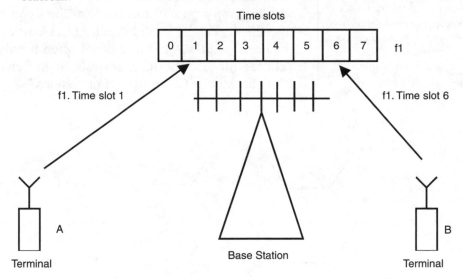

FIGURE 8.3 Time division multiple access.

In TDMA systems, the time slots can be fixed or dynamic. In fixed TDMA, each station is given a fixed time slot (say, time slot 1 for station A, time slot 2 for station B, and so on). This results in a simple system design, but the disadvantage is that if a station has no data to transmit, that time slot is wasted. In dynamic TDMA, a station is assigned a time slot only when it makes a request for it. This leads to a more complex system, but the channel is used effectively.

In TDMA, a single frequency is shared by a number of terminals. Each terminal is assigned a small time slot during which it can transmit. The time slot assignment can be fixed, or the assignment can be dynamic—a slot is assigned only when a terminal has data to transmit.

An important issue in TDMA systems is synchronization: each station should know precisely when it can start transmission in its time slot. It is easy to tell terminal A that it should transmit in time slot 1, but how does terminal A know when exactly time slot 1 starts? If it starts transmission slightly early, the data will collide with the data in time slot 0. If it starts transmission slightly late, the data will collide with the data in time slot 2. The complexity of TDMA lies in the synchronization. Synchronization is achieved by the central station sending a bit pattern (101010101...pattern), and all the stations use this bit pattern to synchronize their clocks.

Synchronization is a major problem in TDMA systems. To ensure that each terminal transmits only during its time slot, very strict timings have to be followed. The central station sends a bit pattern to all the terminals with which all the terminals synchronize their clocks.

In fixed TDMA, the time slots are assigned to each terminal permanently. This results in a simple implementation, but time slots are wasted if the terminal has no data to transmit. On the other hand, in dynamic TDMA, time slots are assigned by a central station based on the request of a terminal. Hence, a separate signaling slot is required to transmit requests for slots. In mobile communication systems, dynamic TDMA is used.

8.3.1 Time Division Multiple Access–Frequency Division Duplex (TDMA-FDD)

As mentioned earlier, in radio systems, a pair of frequencies are used for communication between two stations—one uplink frequency and one downlink frequency. We can use TDMA in both directions. Alternatively, the base station can multiplex the data in TDM mode and broadcast the data to all the terminals; the terminals can use TDMA for communication with the base station. When two frequencies are used for achieving full-duplex communication along with TDMA, the multiple access is referred to as TDMA-FDD.

In TDMA-FDD, two frequencies are used for communication between the base station and the terminals—one frequency for uplink and one frequency for downlink.

TDMA-FDD is illustrated in Figure 8.4. The uplink frequency is f1, and the downlink frequency is f2. Terminal-to-base station communication is achieved in time slot 4 using f1. Base station-to-terminal communication is achieved in time slot 4 using f2.

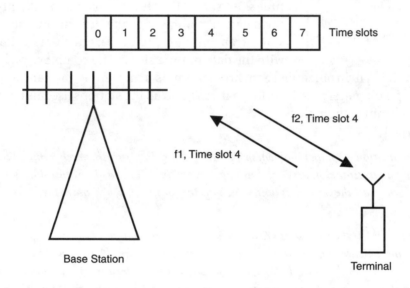

FIGURE 8.4 TDMA-FDD.

8.3.2 Time Division Multiple Access–Time Division Duplex (TDMA-TDD)

Instead of using two frequencies for full-duplex communication, it is possible to use a single frequency for communication in both directions between the base station and the terminal. When two-way communication is achieved using the single frequency but in different time slots, it is known as TDMA-TDD.

As shown in Figure 8.5, the time slots are divided into two portions. The time slots in the first portion are for communication from the base station to the terminals (downlink), and the time slots in the second portion are for communication from the terminals to the base station (uplink). The base station will transmit the data in time slot 2 using frequency f1. The terminal will receive the data and then switch over to transmit mode and transmit its data in time slot 13 using frequency f1.

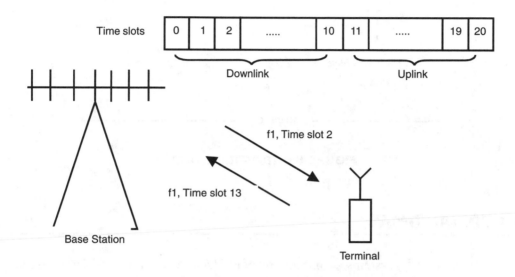

f1: Uplink & Downlink Frequency

FIGURE 8.5 TDMA-TDD.

In TDMA-FDD, two frequencies are used for communication between the base station and the terminals—one frequency for uplink and one frequency for downlink.

The advantage of TDMA-TDD is that a single frequency can be used for communication in both directions. However, it requires complex electronic circuitry because the terminals and the base station have to switch to transmit and receive modes very quickly. For illustration of the timings, the TDMA-TDD scheme used in digital enhanced cordless telecommunications (DECT) system is shown in Figure 8.6. Here the TDMA frame is of 10 milliseconds duration, which is divided into 24 time slots—12 for downlink and 12 for uplink.

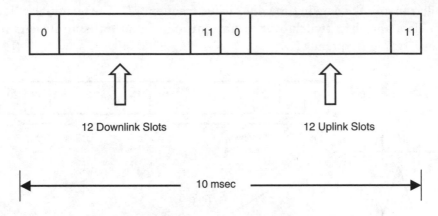

FIGURE 8.6 TDMA-TDD in DECT.

8.4 FDMA/TDMA

In many radio systems, a combination of FDMA and TDMA is used. For example, in cellular mobile communication systems, each base station is given a set of frequencies, and each frequency is divided into time slots to be shared by different terminals.

The concept of FDMA/TDMA is illustrated in Figure 8.7. The base station is assigned four frequencies: f1 to f4. Each frequency is in turn shared in TDMA mode with eight time slots. Terminal A is assigned time slot 1 of frequency f1. The base station keeps allocating the time slots in f1 until all are exhausted. Then it allocates time slots in f2 and so on. After some time, terminal B wants to make a call. At that time, time slot 2 in frequency f3 is free and is allocated to terminal B.

In FDMA/TDMA, a pool of frequencies is shared by a number of terminals. In addition, each frequency is shared in different time slots by the terminals. FDMA/TDMA is used in mobile communication systems and satellite communication systems.

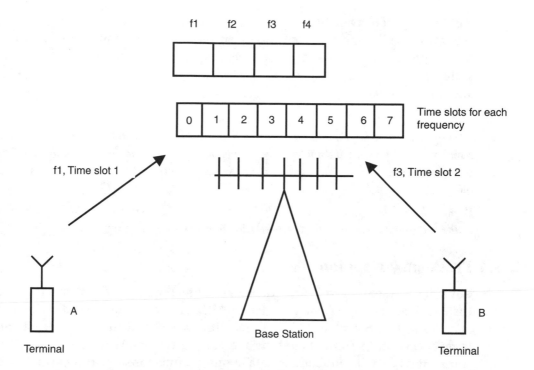

f1   f2   f3   f4

| 0 | 1 | 2 | 3 | 4 | 5 | 6 | 7 |

Time slots for each frequency

f1, Time slot 1

f3, Time slot 2

A

Base Station

B

Terminal

Terminal

FIGURE 8.7 FDMA/TDMA

The combination of FDMA and TDMA increases the capacity of the system. In the above example, 32 subscribers can make calls simultaneously using the eight time slots of four frequencies.

8.5 CODE DIVISION MULTIPLE ACCESS

In spread spectrum multiple access, a wide bandwidth channel is used. Frequency hopping and direct sequence CDMA are the two types of SSMA techniques.

Code division multiple access (CDMA) technology has been developed for defense applications where secure communication is very important. In CDMA systems, a very large bandwidth channel is required, many times more than the bandwidth occupied by the information to be transmitted. For instance, if the actual bandwidth required is 1MHz, in CDMA systems, perhaps 80MHz is allocated. Such large bandwidths were available only with defense organizations, and hence CDMA was used initially only for defense applications. Because the spectrum is spread, these systems are also known as spread spectrum multiple access (SSMA) systems. In this category, there are two types of techniques: frequency hopping and direct sequence.

CDMA requires a large radio bandwidth. Becasue radio spectrum is a precious natural resource, CDMA systems did not become commercially popular and were used only in defense communication systems. However, in recent years, commercial CDMA systems are being widely deployed.

Wireless local loops are the wireless links between subscriber terminals and the base stations connected to the telephone switches. CDMA is widely used in wireless local loops.

8.5.1 Frequency Hopping (FH)

Consider a system in which 1MHz bandwidth is required to transmit the data. Instead of allocating a radio channel of 1MHz only, a number of radio channels (say 79) will be allocated, each channel with 1MHz bandwidth. We need a very large spectrum, 79 times that of the actual requirement. When a station has to transmit its data, it will send the data in one channel for some time, switch over to another channel and transmit some more data, and again switch over to another channel and so on. This is known as frequency hopping (FH). When the transmitting station hops its frequency of transmission, only those stations that know the hopping sequence can receive the data. This will be a secure communication system if the hopping sequence is kept a secret between the transmitting and the receiving stations.

In frequency hopping (FH) systems, each packet of data is transmitted using a different frequency. A pseudo-random sequence generation algorithm decides the sequence of hopping.

Frequency hopping, as used in Bluetooth radio system, is illustrated in Figure 8.8. Here the frequency hopping is done at the rate of 1600 hops per second. Every 0.625 milliseconds, the frequency of operation will change. The terminal will receive the data for 0.625 msec in frequency f1, for 0.625 msec in f20, for 0.625 msec in f32, and so on. The hopping sequence (f1, f20, f32, f41...) is decided between the transmitting and receiving stations and is kept secret.

Frequency hopping is used in Global System for Mobile Communications (GSM) and Bluetooth radio systems.

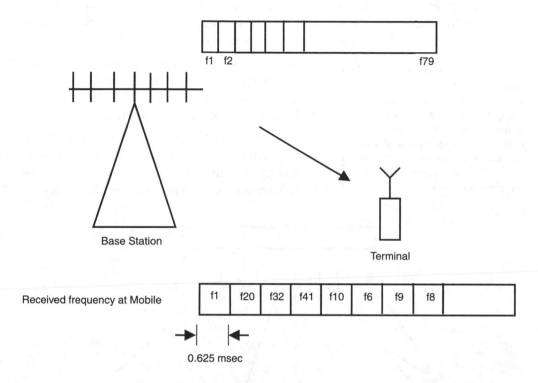

FIGURE 8.8 Frequency hopping.

  *Bluetooth radio system, which interconnects devices such as desktop, laptop, mobile phone,
headphones, modems, and so forth within a range of 10 meters, uses the frequency hopping
technique.*

NOTE

8.5.2 Direct Sequence CDMA

In direct sequence CDMA (DS-CDMA), each bit to be transmitted is represented
by multiple bits. For instance, instead of transmitting a 1, a pattern of say 16
ones and zeros is transmitted, and instead of transmitting a 0, another pattern
of 16 ones and zeros is transmitted. Effectively, we are increasing the data rate
and hence the bandwidth requirement by 16 times. The number of bits to be
transmitted in place of 1 or 0 is known as *chipping rate*. If the chipping code is
kept a secret, only those stations that have the chipping code can decode the
information. When multiple stations have to transmit, the chipping codes will

be different for each station. If they are chosen in such a way that they are orthogonal to each other, then the data from different stations can be pushed on to the channel simultaneously without interference.

In DS-CDMA, multiple terminals transmit on the same channel simultaneously, with different chipping codes. If the chipping code length is say 11 bits, both 1 and 0 are replaced by the 11-bit sequence of ones and zeros. This sequence is unique for each terminal.

As shown in Figure 8.9, in DS-CDMA, multiple terminals transmit on to the channel simultaneously. Because these terminals will have different chipping codes, there will be no interference.

CDMA systems are now being widely deployed for cellular communications as well as 3G systems for accessing the Internet through wireless networks. CDMA systems are used in wireless local loops.

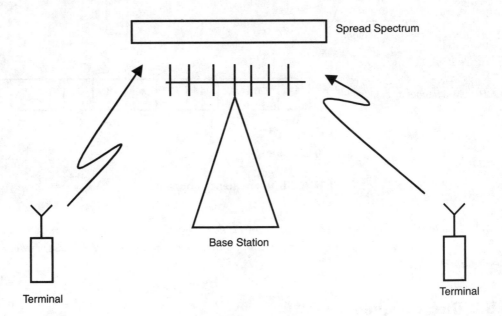

FIGURE 8.9 DS-CDMA.

In IEEE 802.11 wireless local area network standard, 11-bit chipping code is used.

8.6 ORTHOGONAL FREQUENCY DIVISION MULTIPLEXING (OFDM)

In CDMA systems, a single carrier is used to spread the spectrum. In orthogonal frequency division multiplexing (OFDM), the transmission by individual terminals is split over a number of carriers. It is closer to frequency hopping except that the transmission is done on the designated channels simultaneously.

In OFDM, the data is transmitted simultaneously on a number of carriers. This technique is used to overcome multipath fading. OFDM is used in wireless local area networks.

The advantage of this scheme is that because multiple carriers are used to carry the data, multipath fading is less compared to CDMA. OFDM is considered a big competitor to CDMA in wireless networks, though OFDM systems are yet to be widely deployed.

8.7 CARRIER SENSE MULTIPLE ACCESS (CSMA)

In local area networks, the medium is either twisted copper pair or coaxial cable. This medium is shared by a number of computers (nodes). While sharing the medium, only one node can transmit at a time. If multiple nodes transmit simultaneously, the data is garbled. Sharing the medium is achieved through a protocol known as Medium Access Control (MAC) protocol. The MAC protocol works as follows. When a node has to transmit data, first it will sense the medium. If some other node is transmitting, it will keep sensing the medium until the medium is free, and then it will start the transmission. This process of checking whether the medium is free or not is known as carrier sense. Since multiple nodes access the medium through carrier sensing, the multiple access is known as carrier sense multiple access.

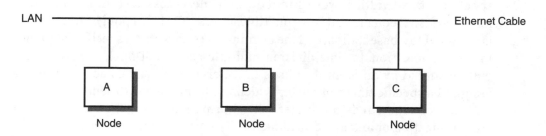

FIGURE 8.10 CSMA/CD.

In local area networks, the medium (cable) is shared by a number of nodes using CSMA/CD. Before transmitting its data, a node senses the carrier to check whether the medium is free. If it is free, the node sends its data. If the medium is not free, the node waits for a random amount of time and again senses the carrier. If two or more nodes send their data simultaneously, it causes collision, and the nodes would need to retransmit their data.

Now consider the two nodes A and C in Figure 8.10. Node A sensed the carrier, found it free, and started the transmission. For the data to reach the point where node C is connected, it will take some finite time. Before the data on the medium reaches that point, node C senses the carrier, finds the carrier to be free, and then it also sends its data. As a result, the data of A and C collide and get garbled. The garbling of data can be detected because there will be a sudden rise in the voltage level on the cable. It is not enough if the nodes find the carrier to be free, they also need to detect collisions. When a collision is detected by a node, that node has to retransmit its data, waiting for some more time after the carrier is free. If more collisions are detected, the waiting time has to be increased. This mechanism is known as carrier sense multiple access/collision detection (CSMA/CD). CSMA/CD is the protocol used in Ethernet LANs. A variation of CSMA/CD is CSMA/CA, in which CA stands for collision avoidance. In CSMA/CA, collisions are avoided by reserving the medium for a specific duration. We will study the details of these protocols in Chapter 17, "Local Area Networks," where we will study wired and wireless LANs.

CSMA with collision avoidance is used in wireless local area networks. To avoid collisions, time slots have to be reserved for the nodes.

Summary

Multiple access techniques provide the mechanism to share the bandwidth efficiently by multiple terminals. Particularly in radio systems where the spectrum bandwidth is very limited, multiple access techniques are very important. In frequency division multiple access (FDMA), a pool of frequencies is assigned to a base station, and the terminals are allocated a specific frequency for communication. In time division multiple access (TDMA), one frequency will be shared by different terminals in different time slots. Each terminal is assigned a specific time slot during which the terminal has to send the data. In TDMA frequency division duplex (FDD), a pair of frequencies is used for communication uplink and downlink. In TDMA time division duplex (TDD), a single frequency is used for communication in both directions—the time slots will be divided into two portions—downlink slots and uplink slots. The code division multiple access technique requires large bandwidth, but the advantage is that it provides secure communication. In CDMA, there are two mechanisms: frequency hopping (FH) and direct sequence. In FH, the frequency of transmission changes very fast in a pseudo-random fashion. Only those terminals that know the sequence of hopping will be able to decode the data. In direct sequence CDMA, instead of sending the bit stream directly, chipping codes are used to transmit ones and zeros. In local area networks (LANs), the medium is shared by multiple nodes using carrier sense multiple access (CSMA) protocols.

References

R.O. LaMarive et al. "Wireless LANs and Mobile Networking: Standards and Future Directions." *IEEE Communications Magazine*, Vol. 34, No. 8, August 1996.

J. Karaoguz. "High Rate Personal Area Networks." *IEEE Communications Magazine*, Vol. 39, No. 11, December 2001.

www.3gpp.org The official site of the 3G partnership program. You can get the details of the CDMA- and TDMA-based wireless networks from this site.

www.cdg.org Web site of CDMA development group.

www.iec.org IEC web site, which has excellent online tutorials.

www.qualcomm.com Web site of Qualcomm Corporation. You can get information on CDMA technologies and products from this site.

www.standards.ieee.org The IEEE Standards web site. The details of various multiple access techniques used in local area networks can be obtained from this site.

Questions

1. Explain FDMA with examples.
2. What is the difference between TDMA-FDD and TDMA-TDD?
3. Explain how SDMA is used in mobile communication systems.
4. Explain frequency hopping with an example.
5. What is direct sequence CDMA?
6. Explain the variations of CSMA protocol.

Exercises

1. Explore the details of the multiple access techniques used in the following systems: (a) mobile communication systems based on GSM standards, (b) Bluetooth radio system, (c) 802.11 wireless local area network, (d) HiperLAN, and (e) digital enhanced cordless telecommunications (DECT) system.

2. Prepare a technical paper on the various multiple access techniques used in satellite communication systems.

3. Explain the multiple access technique used in mobile communication system based on the Global System for Mobile Communications (GSM) standard.

4. In TDMA systems, the channel data rate is higher than the data rate of the information source. Explain with an example.

Projects

1. Make a paper design of a radio communication system that has one base station and 32 remote stations. The base station communicates with the remotes using broadcast mode. All the remotes listen to the base station transmission and, if the address transmitted matches with its own address, the data is decoded, otherwise discarded. The remotes transmit using TDMA. Design the TDMA time frame. Hint: Access is TDMA/FDD. Because there are 32 remotes, the address length should be minimum 5 bits. After carrying out the design, look for details in Chapter 12.

2. Simulate a direct sequence CDMA system. Replace 1 with 11-bit code and replace 0 with another 11-bit code. Study how these codes have to be chosen.

9 ▪ Carrier Modulation

In This Chapter

- The Need for Modulation
- Analog Modulation and Digital Modulation
- Analog Modulation Techniques
- Digital Modulation Techniques

When a signal is to be transmitted over a transmission medium, the signal is superimposed on a carrier, which is a high-frequency sine wave. This is known as *carrier modulation*. In this chapter, we will study the various analog and digital carrier modulation techniques. We also will discuss the various criteria based on which a particular modulation technique has to be chosen.

9.1 WHAT IS MODULATION?

Modulation can be defined as superimposition of the signal containing the information on a high-frequency carrier. If we have to transmit voice that contains frequency components up to 4kHz, we superimpose the voice signal on a carrier of, say, 140MHz. The input voice signal is called the *modulating signal*. The transformation of superimposition is called the *modulation*. The hardware that carries out this transformation is called the *modulator*. The output of the modulator is called the *modulated signal*. The modulated carrier is sent through the transmission medium, carrying out any other operations required on the modulated signal such as filtering. At the receiving end, the modulated signal is passed through a demodulator, which does the reverse operation of the modulator and gives out the modulating signal, which contains the original information. This process is depicted in Figure 9.1. The modulating signal is

107

Modulation is the superimposition of a signal containing the information on a high-frequency carrier. The signal carrying the information is called the modulating signal, and the output of the modulator is called the modulated signal.

also called the *baseband signal*. In a communication system, both ends should have the capability to transmit and receive, and therefore the modulator and the demodulator should be present at both ends. The modulator and demodulator together are called the modem.

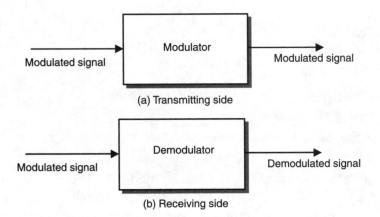

FIGURE 9.1 Modulation and demodulation.

9.2 WHY MODULATION?

Suppose we want to transmit two voice channels from one place to another. If we combine the two voice signals and transmit them on the medium, it is impossible to separate the voice conversations at the receiving end. This is because both voice channels occupy the same frequency band, 300Hz to about 4kHz. A better way of transmitting the two voice channels is to put them in different frequency bands and then send them, translating the voice channels into different frequency bands.

Low-frequency signals have poor radiation capability and so low-frequency signals such as voice signals are translated into high frequencies. We need to superimpose the voice signal onto a high-frequency signal to transmit over large distances. This high-frequency signal is called the carrier, and the modulation is called *carrier modulation*.

When different voice signals are modulated to different frequencies, we can transmit all these modulated signals together. There will be no interference.

If radio is used as the transmission medium, the radio signal has to be sent through an antenna. The size of the antenna decreases as the frequency of the signal goes up. If the voice is transmitted without superimposing it on a high-frequency carrier, the antenna size should be 5,000 (yes, five thousand) meters!

For these reasons, modulation is an important transformation of the signal that is used in every communication system.

To summarize, modulation allows:

- Transmitting signals over large distances, because low-frequency signals have poor radiation characteristics.

> Modulation allows transmission of signals over a large distance, because low-frequency signals have poor radiation characteristics. It is also possible to combine a number of baseband signals and send them through the medium.

- It is possible to combine a number of baseband signals and send them through the medium, provided different carrier frequencies are used for different baseband signals.

- Small antennas can be used if radio is the transmission medium.

In transmission systems using radio as the medium, the higher the frequency of operation, the smaller the antenna size. So, using high-frequency carrier reduces the antenna size considerably.

NOTE

9.3 TYPES OF MODULATION

In general, "modulation" can be defined as transformation of a signal. In Chapter 4, "Coding of Text, Voice, Image, and Video Signals", we studied pulse code modulation (PCM) and its derivatives. It is important to note that these are source coding techniques that convert the analog signal to digital form. In this chapter, we will study carrier modulation, which transforms a carrier in such a way that the transformed carrier contains the information of the modulating signal.

Many carrier modulation techniques have been proposed in the literature. Here, we will study the most fundamental carrier modulation techniques that are used extensively in both analog and digital communication systems.

Carrier modulation can be broadly divided into two categories:

- Analog modulation
- Digital modulation

The various analog modulation techniques are:

- Amplitude modulation (AM)
- Frequency modulation (FM)
- Phase modulation (PM)

FM and PM together are known as angle modulation techniques.

The various digital modulation techniques are:

- Amplitude shift keying (ASK)
- Frequency shift keying (FSK)
- Phase shift keying (PSK)

> Analog modulation techniques can be broadly divided into amplitude modulation (AM), frequency modulation (FM), and phase modulation (PM). FM and PM together are known as angle modulation techniques.

> The three digital modulation techniques are (a) amplitude shift keying (ASK); (b) frequency shift keying (FSK); and (c) phase shift keying (PSK).

9.4 COMPARISON OF DIFFERENT MODULATION TECHNIQUES

Before we discuss the details of different modulation techniques, it is important to understand why so many modulation schemes are available. The reason for having different modulation schemes is that the performance of each modulation scheme is different. The performance criteria on which modulation techniques can be compared are:

Bandwidth: What is the bandwidth of the modulated wave?

Noise immunity: Even if noise is added to the modulated signal on the transmission medium, can the original modulating signal be obtained by the demodulator without much distortion?

> The performance of a modulation scheme can be characterized by the bandwidth of the modulated signal, immunity to noise and the complexity of the modulator/demodulator hardware.

Complexity: What is the complexity involved in implementing the modulator and demodulator? Generally, the modulator and demodulator are implemented as hardware, though nowdays, digital signal processors are used for implementation, and hence a lot of software is also used.

Based on these performance criteria, a modulation technique has to be chosen for a given application.

 In the past, both modulator and demodulator were implemented completely in hardware. With the advent of digital signal processors, modulator and demodulator implementations are now software oriented.

 How well a modulation scheme performs on a noisy channel is characterized by the Bit Error Rate (BER). The BER is related to the signal-to-noise ratio (SNR). For a given BER, say 10^{-3}, the modulation technique that requires the least SNR is the best.

9.5 ANALOG MODULATION TECHNIQUES

Analog modulation is used extensively in broadcasting of audio and video programs, and many old telecommunication systems are also based on analog modulation. All newly developed systems use digital modulation. For broadcasting, analog modulation continues to play a vital role.

9.5.1 Amplitude Modulation

In AM, the amplitude of the carrier is proportional to the instantaneous amplitude of the modulating signal. The frequency of the carrier is not changed.

In Figure 9.2(a), the modulating signal (a sine wave) is shown. Figure 9.2(b) shows the amplitude-modulated signal. It is evident from this figure that the carrier's amplitude contains the information of the modulating signal. It can also be seen that both the upper portion and the lower portion of the carrier amplitude contain the information of the modulating signal.

FIGURE 9.2 (a) Modulating signal.

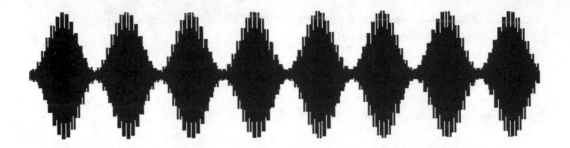

FIGURE 9.2 (b) Amplitude modulated signal.

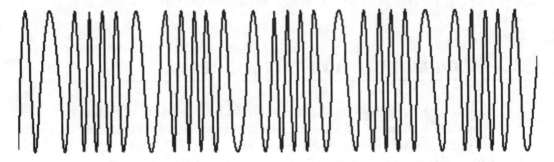

FIGURE 9.2 (c) Frequency modulated signal.

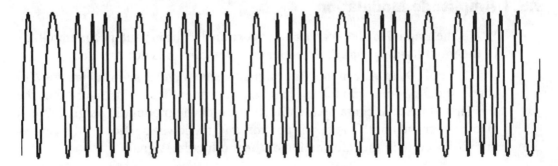

FIGURE 9.2 (d) Phase modulated signal.

If f_c is the frequency of the carrier, the carrier can be represented mathematically as

A $\sin(2\pi f_c t)$

If f_m is the frequency of the modulating signal, the modulating signal is represented by

B $\cos(2\pi f_m t)$

The amplitude-modulated wave is represented by

$(A + B \cos 2\pi f_m t) \sin(2\pi f_c t)$ or

$A\{1 + (B/A) \cos(2\pi f_m t)\} \sin(2\pi f_c t)$

The value of B/A denoted by m is called the *modulation index*. The value of (m × 100) is the modulation index, represented as a percentage.

On expanding the above equation, we get three terms: one term with f_c, the second term with $(f_c + f_m)$, and the third term with $(f_c - f_m)$.

The terms with $(f_c + f_m)$ and $(f_c - f_m)$ represent sidebands. Both the sidebands contain the information of the modulating signal. The frequency component $(f_c + f_m)$ is called the upper sideband, and the frequency component $(f_c + f_m)$ is called the lower sideband. Only the upper sideband or lower sideband can be transmitted and demodulated at the receiving end.

The bandwidth required for AM is twice the highest modulating frequency. If the modulating frequency has a bandwidth of 15kHz (the bandwidth used in audio broadcasting), the amplitude-modulated signal requires 30kHz.

It is very easy to implement the modulator and demodulator for AM. However, as the amplitude-modulated signal travels on the medium, noise gets added, and the amplitude changes; hence, the demodulated signal is not an exact replica of the modulating signal; i.e., AM is not immune to noise. Another problem with AM is that when the AM waves are transmitted, the carrier will take up most of the transmitted power, though the carrier does not contain any information; the information is present only in the sidebands.

The bandwidth required for AM is twice the highest modulating frequency. In AM radio broadcasting, the modulating signal has bandwidth of 15kHz, and hence the bandwidth of an amplitude-modulated signal is 30kHz.

Amplitude modulation is used in audio broadcasting. Different audio programs are amplitude modulated, frequency division multiplexed, and transmitted over the radio medium. In video broadcasting, a single sideband is transmitted.

9.5.2 Frequency Modulation

In frequency modulation (FM), the frequency of the carrier is varied according to the amplitude of the modulating signal. The frequency deviation is proportional to the amplitude of the modulating signal. The amplitude of the carrier is kept constant.

Figure 9.2(c) shows the frequency-modulated signal when the modulating signal is a sine wave as shown in Figure 9.2(a).

If the carrier is represented by

$$A \sin 2 \pi f_c t$$

and the modulating signal is represented by

$$B \cos (2 \pi f_m t),$$

the instantaneous frequency of the frequency-modulated carrier is given by

$$f = f_c + k B \cos 2 \pi f_m t$$

where k is a proportionality constant.

According to Carlson's rule, the bandwidth of an FM signal is twice the sum of the modulating signal frequency and the frequency deviation. If the frequency deviation is 75kHz and the modulating signal frequency is 15kHz, the bandwidth required is 180kHz.

As compared to AM, FM implementation is complicated and occupies more bandwidth. Because the amplitude remains constant, FM is immune to noise.

Many audio broadcasting stations now use FM. In FM radio, the peak frequency deviation is 75kHz, and the peak modulating frequency is 15kHz. Hence the FM bandwidth requirement is 180kHz. The channels in FM band are separated by 200kHz. If we compare the quality of audio in AM radio and FM radio, we can easily make out that FM radio gives much better quality. This is because FM is more immune to noise as compared to AM. FM also is used to modulate the audio signal in TV broadcasting.

In FM radio broadcasting, the peak frequency deviation is 75kHz, and the peak modulating signal frequency is 15kHz. Hence the bandwidth of the modulated signal is 180kHz.

NOTE

9.5.3 Phase Modulation

In phase modulation, the phase deviation of the carrier is proportional to the instantaneous amplitude of the modulating signal. It is possible to obtain frequency modulation from phase modulation. The phase modulated signal for the modulating signal shown in

Figure 9.2(d) looks exactly same as Figure 9.2(c).

No practical systems use phase modulation.

9.6 DIGITAL MODULATION TECHNIQUES

The three important digital modulation techniques are

- Amplitude shift keying (ASK)
- Frequency shift keying (FSK)
- Phase shift keying (PSK)

For a bit stream of ones and zeros, the modulating signal is shown in Figure 9.3(a). Figures 9.3(b), 9.3(c), and 9.3(d) show modulated signals using ASK, FSK, and binary PSK, respectively.

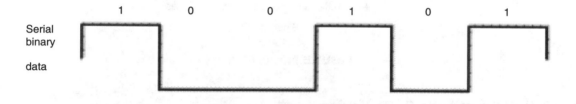

FIGURE 9.3 (a) Modulating signal.

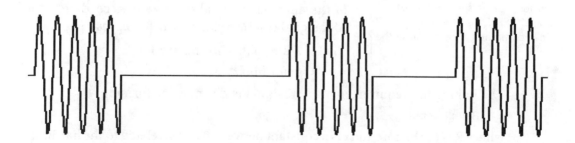

FIGURE 9.3 (b) ASK.

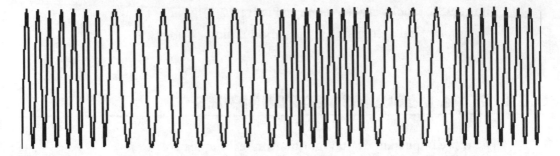

FIGURE 9.3 (c) FSK.

BPSK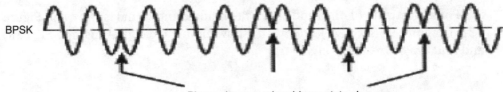

Phase changes when binary state changes

Binary 1 = 0°
Binary 0 = 180°

FIGURE 9.3 (d) BPSK.

9.6.1 Amplitude Shift Keying

In amplitude shift keying (ASK), 1 and 0 are represented by two different amplitudes of the carrier. ASK is susceptible to noise. ASK is used in optical fiber communication because the noise is less.

Amplitude shift keying (ASK) is also known as on off keying (OOK). In ASK, two amplitudes of the carrier represent the binary values (1 and 0). Generally, one of the amplitudes is taken as zero. Accordingly, the ASK signal can be mathematically represented by

$$S(t) = A \sin(2\Pi f_c t) \text{ for binary 1}$$
$$= 0 \text{ for binary 0}$$

The bandwidth requirement of ASK signal is given by the formula

$$B = (1 + r)R$$

where R is the bit rate and r is a constant between 0 and 1, related to the hardware implementation.

ASK is susceptible to noise and is not used on cable. It is used in optical fiber communication.

9.6.2 Frequency Shift Keying

In FSK, the binary values are represented by two different frequencies close to the carrier frequency. An FSK signal is mathematically represented by

$$S(t) = A \sin(2\pi f_1 t) \text{ for binary 1}$$

$$= A \sin(2\pi f_2 t) \text{ for binary 0}$$

f_1 can be $f_c + f_m$ and f_2 can be $f_c - f_m$, where f_c is the carrier frequency and $2f_m$ is the frequency deviation.

The bandwidth requirement of FSK signal is given by

$$B = 2 \times f_m + (1 + r)R$$

where R is the data rate and r is a constant between 0 and 1.

In frequency shift keying (FSK), 1 and 0 are represented by two different frequencies of the carrier. FSK is used widely in cable and radio communication systems.

FSK is used widely in cable communication and also in radio communication.

9.6.3 Phase Shift Keying

The two commonly used PSK techniques are binary PSK (BPSK) and quadrature PSK (QPSK).

In PSK, the phase of the carrier represents a binary 1 or 0. In BPSK, two phases are used to represent 1 and 0. Mathematically, a PSK signal is represented by

In BPSK, binary 1 and 0 are represented by two phases of the carrier.

$$S(t) = A \sin(2\pi f_c t + p) \text{ for binary 0}$$

$$= A \sin(2\pi f_c t) \text{ for binary 1}$$

The phase is measured relative to the previous bit interval. The bandwidth occupied by BPSK is the same as that of ASK.

In quadrature PSK (QPSK), two bits in the bit stream are taken, and four phases of the carrier frequency are used to represent the four combinations of the two bits.

In quadrature phase shift keying (QPSK), different phases of the carrier are used to represent the four possible combinations of two bits: 00, 01, 10, and 11. QPSK is used widely in radio communication systems.

$$S(t) = A \sin(2\pi f_c t + 45°) \text{ for } 11$$
$$A \sin(2\pi f_c t + 135°) \text{ for } 10$$
$$A \sin(2\pi f_c t + 225°) \text{ for } 00$$
$$A \sin(2\pi f_c t + 315°) \text{ for } 01$$

The bandwidth required for a QPSK modulated signal is half that of the BPSK modulated signal. Phase shift keying (BPSK and QPSK) is used extensively in radio communication systems. In mobile communication systems also, different PSK techniques are used.

Summary

Various carrier modulation techniques are reviewed in this chapter. Carrier modulation is the technique used to transform the signal such that many baseband signals can be multiplexed and sent over the medium for transmitting over large distances without interference. Modulation techniques can be broadly divided into analog modulation techniques and digital modulation techniques. Amplitude modulation (AM) and frequency modulation (FM) are the widely used analog modulation techniques. In AM, the information is contained in the amplitude of the carrier. In FM, the frequency deviation of the carrier contains the information. AM and FM are used extensively in broadcasting audio and video. The important digital modulation techniques are amplitude shift keying (ASK), frequency shift keying (FSK), and phase shift keying (PSK). In ASK, binary digits are represented by the presence or absence of the carrier. In FSK, the binary digits are represented by two frequencies of the carrier. In PSK, the binary values are represented by different values of the phase of the carrier. ASK is used in optical fiber communication. FSK and PSK are used when the transmission medium is cable or radio. When designing a communication system, the modulation scheme is chosen, keeping in mind the bandwidth of the modulated signal, ease of implementation of the modulator/demodulator, and noise immunity.

References

G. Kennedy and B. Davis. *Electronic Communication Systems*. Tata McGraw Hill Publishing Company Limited, 1993.

S. Haykin. *Communication Systems*, Third Edition, 1994.

The Web sites of digital signal processor (DSP) chip manufacturers such as Analog Devices, Motorola, Lucent Technologies, Texas Instruments, and others provide a wealth of information on modulations and development of modems using DSP.

Questions

1. Explain the need for modulation.
2. List the various analog and digital modulation techniques.
3. What criteria are used for comparing different modulation schemes?
4. Explain the various analog modulation schemes.
5. Explain the various digital modulation schemes.

Exercises

1. For the bit pattern 1 0 1 1 0 1 1 1 0 0 1, draw the modulated waveform signals if the modulation used is (a) ASK, (b) FSK, (c) BPSK.
2. Write a C program to generate a carrier of different frequencies and to generate amplitude modulated waves if the modulating signal is a sine wave of 1kHz. Give a provision to change the modulation index.
3. Make a comparative statement on different digital modulation techniques.
4. Find out the modulation techniques used in (a) Global System for Mobile Communication (GSM); (b) Bluetooth; (c) IEEE 802.11 local area networks; and (d) digital subscriber lines (DSL).
5. If the bandwidth of a modulating signal is 20kHz, what is the bandwidth of the amplitude modulated signal?
6. If the bandwidth of a modulating signal is 20kHz and the frequency deviation used in frequency modulation is 75kHz, what is the bandwidth of the frequency-modulated signal?

Projects

1. Develop a software package that can be used for teaching different modulation schemes. The graphical user interface (GUI) should (a) facilitate giving a bit stream (1010100...) or a sine wave as input; (b) enable the user to select a modulation technique (AM, FM, ASK, FSK, and so on); and (c) select the modulation parameters such as carrier frequency for AM and FM, modulation index for AM, frequency deviation for FM, and so forth. The bit pattern/modulating signal and the modulated signal should be displayed as output.

2.   Digital signal processors are used extensively for modulation. Using a DSP evaluation board (that of Analog Devices, Texas Instruments, Motorola, or Lucent Technologies), develop DSP software to generate various modulations.

3.   Using MATLAB, generate various modulations. You can obtain an evaluation copy of MATLAB from *www.mathworks.com*.

4.   Study the various modulation schemes used in line modems. Make a list of various ITU-T standards for line modems, such as V.24, V.32, and so forth.

10 Issues in Communication System Design

In This Chapter

- Design of Communication Systems
- Design Parameters
- Problems in Radio System Design
- Telecommunication Standardization Bodies

To design a communication system according to user requirements is a challenging task because the requirements differ from user to user and there are many design trade-offs to be considered. The design of a communication system has to be carried out keeping in view the following factors:

- What are the information sources? Data, voice, fax, video, or all of them? Depending on the requirements, the bandwidth requirements will be different.

- What is the coverage area? The coverage area decides which transmission medium has to be chosen, and sometimes a combination of media may be required (for instance, a combination of twisted pair and satellite).

- Is a secure network required? For defense and corporate networks, a highly secure system is required to ensure that the information is kept confidential. Special security features have to be incorporated in such cases.

- What are the performance criteria? For data applications, particularly such as in banking and other financial transactions, performance requirements are very stringent. For instance, even a one-bit error in a million bits transmitted is not acceptable. Such criteria call for efficient modulation techniques and error detection/correction mechanisms. On the other hand, for voice and video applications, delay should be minimum.

- What are the signaling requirements? Is a separate signaling channel/network required?

- Which national/international standards have to be followed? The days of providing proprietary solutions are gone; every communication system has to be designed according to the available international standards. Before designing a communication system, the relevant standards need to be studied, and the design has to be carried out.

- Given an unlimited budget, we can design a world-class communication system for every user segment. But the user is always constrained by budget. For a given budget, to design the optimal system that meets the requirements of the user is of course the biggest challenge, and as usual, we need to consider the various trade-offs.

When designing a communication system, the following requirements need to be considered: coverage area, information sources, security issues, performance issues, signaling requirements, international/national standards to be followed, and the cost.

In this chapter, we discuss the issues involved in designing communication systems. We also study the important aspects in designing radio communication systems. The special attraction of the radio systems is that they provide mobility to users, but every attraction comes with a premium—radio systems pose special design challenges.

10.1 DATA RATES

In a digital system, the information to be transmitted is converted into binary data (ones and zeros). In the case of text, characters are converted into ASCII format and transmitted. In audio or video, the analog signal is converted into a digital format and then transmitted.

To make best use of a communication channel, the data rate has to be reduced to the extent possible without compromising quality. All information (text, graphics, voice, or video) contains redundancy, and this redundancy can be removed using compression techniques. Use of low bit rate coding techniques (also called source coding techniques) is very important to use the bandwidth efficiently. Particularly in radio systems, where radio bandwidth has a price, low-bit rate coding is used extensively.

When designing a communication system, the designer has to consider the following issues related to information data rates:

The services to be supported by the communication system—data, voice, fax, and video services—decide the data rate requirement. Based on the available communication bandwidth, the designer has to consider low bit rate coding of the various information sources.

- What are the information sources to the communication system: data, voice, fax, video, or a combination of these?
- How many information sources are there, and is there a need for multiplexing them before transmitting on the channel?
- How many information sources need to use the communication channel simultaneously? This determines whether the channel bandwidth is enough, whether multiple access needs to be used, etc.

- If the channel bandwidth is not sufficient to cater to the user requirements, the designer has to consider using data compression techniques. A trade-off is possible between quality and bandwidth requirement. For instance, voice signals can be coded at 4.8kbps, which allows many more voice channels to be accommodated on a given communication channel, but quality would not be as good as 64kbps PCM-coded voice.

NOTE

Compression techniques can be divided into two categories: (a) lossless compression techniques; and (b) lossy compression techniques. The file compression techniques such as Winzip are lossless because the original data is obtained by unzipping the file. Compression techniques for voice, image, and video are lossy techniques because compression causes degradation of the quality.

10.2 ERROR DETECTION AND CORRECTION

Because the communication channel introduces errors in the bit stream, error detection techniques need to be incorporated. If errors are detected, the receiver can ask for retransmission. If retransmissions have to be reduced, error correction techniques need to be incorporated.

The source coding techniques are used to reduce the redundancy in the signal. Because the transmission medium introduces errors, we need to devise methods so that the receiver can either detect the errors or correct the errors. To achieve this, error detection techniques and error correction techniques are used. These techniques increase the bandwidth requirement, but they provide reliable communication. Error detection is done through CRC and error correction through FEC.

NOTE
In voice and video communications, a higher bit error rate can be tolerated. Even if there is one error in 10,000 bits, there will not be perceptible difference in voice/video quality. However, for data applications, a very reliable data transfer is a must.

10.3 MODULATION TECHNIQUES

For designing an analog communication system, analog modulation techniques such as AM and FM are used. If you are designing a digital communication system, the choices are digital modulation techniques such as ASK, FSK, and PSK. Many variations of these basic schemes are available that give slightly different performance characteristics. Choice of a modulation scheme needs to take into consideration the performance requirements as well as the availability of hardware for implementing the modulator and demodulator.

In a communication system, an important design parameter is Bit Error Rate (BER). To achieve a good BER (to reduce the bit errors as much as possible), the Signal-to-Noise Ratio (SNR) should be high. SNR and E_b/N_o are related by the formula

$$E_b/N_o = S/(N_o R)$$

where N_o is the noise power density in watts/Hertz. The total noise in a signal with bandwidth of B is given by $N = N_o \times B$. Hence,

$$E_b/N_o = (S/N)(B/R).$$

The BER can be reduced by increasing the E_b/N_o value, which can be achieved either by increasing the bandwidth or by decreasing the data rate.

The curves shown in Figure 10.1 give the performance of the modulation schemes. These curves are known as the *waterfall curves*. In designing a communication system, based on the required BER, the E_b/N_o value is obtained for a given modulation scheme. PSK and QPSK perform better compared to ASK and FSK because for a given BER, the value of E_b/N_o is less, and hence with less energy of the signal, we can achieve good performance. Another criterion to be considered is the ease of implementing the modulator/demodulator. It is much easier to implement ASK and FSK modulators and demodulators as compared to PSK and its variations.

For digital communication systems, the BER is an important design parameter. BER can be reduced by increasing the value of E_b/N_o (where E_b is the energy per bit and N_o is the noise power density).

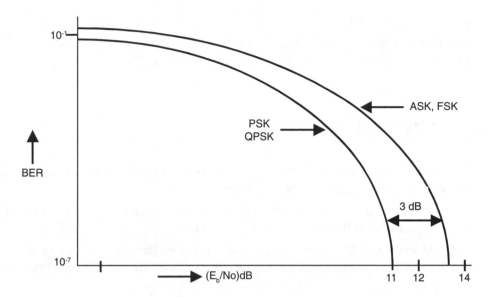

FIGURE 10.1 Performance curves for Digital Modulation Systems.

Choosing a specific modulation technique depends on two important factors—BER performance and complexity of the modulator/demodulator circuitry.

10.4 PERFORMANCE CRITERIA

The performance of a digital communication system is measured in terms of the Bit Error Rate (BER). BER is measured in terms of the ratio of the number of bits in error as a percentage of the total number of bits received.

BER = number of bits received in error / total number of bits received

Depending on the application, the BER requirements vary. For applications such as banking, a very high BER is required, of the order of 10^{-12} i.e., out of 10^{12} bits only one bit can go wrong (even that has to be corrected or a retransmission requested). For applications such as voice, a BER of 10^{-4} is acceptable.

As we discussed in the previous section, BER is dependent on the modulation technique used. The performance in terms of the BER also is dependent on the transmission medium. A satellite channel, for example, is characterized by a high bit error rate

Because it is not possible to achieve a completely error-free transmission, errors should be detected or corrected using error-detection or error-correcting codes. After detection of errors, using automatic repeat request (ARQ) protocols, the receiver has to request retransmission of data.

(generally, around 10^{-6}; in such a case, a higher-layer protocol (data link layer) has to implement error detection techniques and automatic repeat request (ARQ) protocols such as stop-and-wait and sliding window.

10.5 SECURITY ISSUES

Security is of paramount importance because systems are prone to attacks. The various security threats can be the following:

Interruption: The intended recipient is not allowed to receive the data—this is an attack on the availability of the system.

Interception: The intended recipient receives the data, but unauthorized persons also receive the data—this is an attack on the confidentiality of the data.

Modification: An unauthorized person receives the data, modifies it, and then sends it to the intended recipient—this is an attack on the integrity of the data.

Fabrication: An unauthorized person generates the data and sends it to a person—this is an attack on the authenticity of the system.

To overcome these security threats, the data has to be encrypted. Encryption is a mechanism wherein the user data is transformed using an encryption key. Only those who have the encryption key can decrypt the data.

There are two possibilities: link encryption and end-to-end encryption. In link encryption, at the transmitting end, the data is encrypted and sent over the communication link. At the receiving end of the link, the data is decrypted. In end-to-end encryption, the user encrypts the data and sends it over the communication link, and the recipient decrypts the data. To provide high security, both types of encryption can be employed. Note that encryption does not increase the data rate (or bandwidth). Length of the encryption key decides how safe the encryption mechanism is. Though 56- and 64-bit keys were used in earlier days, now 512- and 1024-bit keys are being used for highly secure communication systems.

The major security threats are: interruption, interception, modification, and fabrication. The data is encrypted at the transmitting end to overcome these security threats. At the receiving end, the data is decrypted.

We will study security issues in detail in Chapter 38, "Information Security."

For encryption, there will be an encryption algorithm and an encryption key. The encryption algorithm specifies the procedure for modifying the data using an encryption key. The algorithm can be made public (known to everyone), but the encryption key is kept secret.

10.6 RADIO SYSTEM DESIGN ISSUES

Radio system design poses special problems because of the special nature of the radio signal propagation. The important design issues are:

Frequency of operation: Radio systems cannot be operated in any frequency of our choice. Frequency allocation needs to be obtained from the centralized authority of the government. Only certain bands such as ham radio band and Industrial, Scientific, and Medical (ISM) band are unlicensed, and anyone can use these bands without getting a license from the government authorities.

Radio survey: Radio frequency propagation characteristics depend on many factors, such as natural terrain (presence of hills and valleys, lakes) and artificial terrain (presence of high-rise buildings). A radio survey must be carried out to decide where to keep the antennas to achieve the maximum possible coverage. Multipath fading causes signal degradation. Measures have to be taken to reduce the effect of multipath fading.

The propagation characteristics differ for different frequency bands. A number of mathematical models are available to analyze radio propagation. The natural terrain (presence of hills, lakes, greenery) and artificial terrain (presence of tall buildings) also affect radio propagation.

Line of sight communication: Some radio systems are line of sight systems, that is, there should not be any obstructions such as tall buildings/hills between the transmitting station and the receiving station. In the case of broadcasting applications, the transmitting antennas have to be located at the right places to obtain the maximum coverage. Systems such as AM broadcast systems do not have this limitation because the radio waves are reflected by the ionosphere, and hence the range is very high.

Path loss calculations: When a signal is transmitted with a particular signal strength, the signal traverses a large distance and becomes attenuated. The loss of signal strength due to the propagation in the atmosphere and attenuation in the communication subsystems (such as filters and the cable connecting the radio equipment to the antenna) is called *path loss*. The path loss calculations have to be done to ensure that the minimum required signal strength is available to the receiver to decode the information content. The receiver should be sensitive enough to decode the signals. The required BER, SNR, gain of the antenna, modulation technique used, rain attenuation, and gain of the amplifiers used are some of the parameters considered during the path loss calculations.

For all radio systems, path loss calculations are very important. Based on the path loss calculations, the receiver sensitivity, antenna gain, amplifier gains, etc. are calculated when designing radio systems.

Rain attenuation: The attenuation of the radio signals due to rain varies, depending on the frequency band. For instance, in satellite communication, at 17/12GHz the rain attenuation is very high as compared to 6/4GHz. This aspect has to be taken into consideration in the path loss calculations.

Radio bandwidth: Radio spectrum being a limited natural resource, the bandwidth of a radio channel has to be fully utilized. To achieve this, efficient source coding techniques have to be used. For example, to transmit voice over a radio channel, it is not advisable to use 64kbps PCM (though many systems still use it). A better approach would be to use low bit rate coding techniques (such as ADPCM, LPC, or its variations) so that in a given radio bandwidth, more voice channels can be pumped in.

As radio spectrum is a limited natural resource, the radio channel has to be fully utilized. Using low bit rate coding of voice/video signals and choosing an efficient modulation technique are very important in radio system design.

Radio channels: A radio channel consists of a pair of frequencies—one frequency from base station to the end station and one frequency from end station to the base station. A minimum separation is required between the uplink and downlink frequencies.

Design of radio systems involves special issues to be addressed. These include frequency of operation, radio propagation characteristics, path loss calculations, rain attenuation, efficient usage of radio spectrum through low-bit rate coding of voice and video signals, and usage of multiple access techniques.

Multiple access: Radio systems use multiple access techniques to make efficient use of bandwidth. FDMA, TDMA, and CDMA systems, as discussed earlier, have different spectrum requirements and different complexities.

All these issues need to be kept in mind when designing radio systems.

10.7 TELECOMMUNICATION STANDARDS

In telecommunication system design, the standards play a very important role. The days when organizations used to develop proprietary interfaces and protocols are gone. Before embarking on a system design, the designer has to look at international/national standards for the interfaces and protocols. The various standards bodies for telecommunication/data communications are:

- American National Standards Institute (ANSI)
- Electronics Industries Association (EIA)
- European Telecommunications Standards Institute (ETSI)
- Internet Engineering Task Force (IETF)
- International Organization for Standardization (ISO)
- International Telecommunications Union Telecommunications Services Sector (ITU-T), earlier known as CCITT
- Institute of Electrical and Electronics Engineers (IEEE)

Nowadays, communication system design is driven by international standards. The standards formulated by standardization bodies such as ANSI, EIA, ETSI, IETF, ISO, ITU-T and IEEE need to be followed while designing communication systems and protocols.

Throughout this book, we will mention a number of relevant standards related to telecommunication systems and interfaces as well as communication protocols. Referring to the standards documents is very important to get an in-depth knowledge of the specifications, particularly during implementation.

10.8 COST

Cost is the most important design parameter while designing communication systems. To design a system that meets all the performance requirements at the lowest cost is the callenge for all communcation engineers.

The cost is the most important of the design parameters. To design a communication system that meets all the performance criteria at minimum cost is the major challenge to communication engineers. The choice of the transmission medium (twisted pair, coaxial cable, radio or fiber, etc.) needs to keep in view the cost. In communication system design, engineers do not develop each and every subsystem. The various subsystems are procured from different vendors and integrated. In such a case, the engineer has to choose the subsystem that meets all the performance requirements and is cost effective. Experience is the best teacher for choosing the right subsystems.

Summary

In designing a communication system, a number of issues need to be taken into consideration. These include services to be supported such as data, voice, video, etc., which decide the data rate requirements; error detection and correction mechanisms; type of modulation; performance criteria such as bit error rates; security aspects; standards to be followed for the implementation; and the cost. These issues are discussed in detail in this chapter.

References

K. Krechmer. "Standards Make the GIH Possible." *IEEE Communications Magazine*, Vol. 34, No. 8, August 1996. This paper gives information about the history of standardization and the various standards authorities.

IEEE Communications Magazine, Vol. 34, No. 8, August 1996. This is a special issue on "Emerging Data Communications Standards."

www.ansi.org The Web site of American National Standards Institute.

www.iso.org The Web site of International Organization for Standardization.

www.itu.int The Web site of International Telecommunications Union.

www.etsi.org The Web site of European Telecommunications Standards Institute.

www.amdah.com The Web site of Fiber Channel Association.

www.ietf.org The Web site of Internet Engineering Task Force.

Questions

1. List the various issues involved in telecommunication systems design.
2. What are the special issues to be considered for radio system design?
3. List some important international standardization bodies.
4. What criteria are used to compare different modulation techniques?
5. What are the common security threats? Explain the concept of encryption.

Exercises

1. Conduct a survey of the various international and national standardization bodies and prepare a report.

2. Conduct a survey of commercial products available for planning and designing a cellular mobile communication system.

3. Study the performance of different digital modulation schemes using Bit Error Rate (BER) and bandwidth requirement as the performance criteria.

4. Design a local area network for your college campus. Design it based on the available international standards. List the various options available.

Projects

1. Prepare the paper design of a communication system to provide Internet access to 50 communities that are located within a radius of about 30 km from the district headquarters. Consider the various design options: data rate support, transmission medium, multiple access technique, performance, delay, and of course the cost of the system.

2. Simulate a digital communication system. The software has to (a) generate a continuous stream of bits, (b) introduce errors by converting 1 to 0 or 0 to 1 at random places in the bit stream, and (c) modulate the bit stream using different modulations.

3. Prepare a paper design of a communication system interconnecting all the colleges providing technical education in your state. The system is to be used for providing distance education. From a central location (state capital), lectures have to be broadcast with a video data rate of 384kbps, audio and data at 64kbps. At each college, there should be a provision to send data or voice to interact with the professor at the central location.

4. Design a communication system that links two branch offices of an organization. The two branches are separated by a distance of 10km. The employees of the two branches should be able to talk to each other, exchange computer files, and also do video-conferencing using 64kbps desktop video conferencing systems. Make appropriate assumptions while preparing the design.

11 Public Switched Telephone Network

In This Chapter

- Architecture of the Telephone Network
- Network Elements of the PSTN
- Switching Concepts
- Signaling Used in the PSTN

Ever since Alexander Graham Bell made the first telephone call, the telephone network has expanded by leaps and bounds. The technical term for the telephone network is Public Switched Telephone Network (PSTN). PSTN now interconnects the whole world to enable people to communicate using the most convenient and effective means of communication—voice.

In this chapter, we will study the architecture of the PSTN and also discuss the latest technologies being introduced in the oldest network to provide better services to telephone subscribers. The PSTN is becoming more and more digital, and hence our description of the PSTN is oriented toward digital technology.

11.1 PSTN NETWORK ELEMENTS

The elements of PSTN are shown in Figure 11.1. The PSTN consists of:

- Subscriber terminals
- Local loops
- Switches (or exchanges)
- Trunks

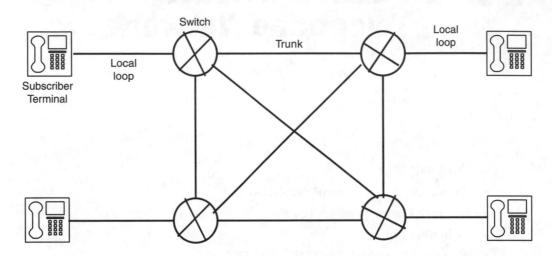

FIGURE 11.1 Elements of PSTN.

11.1.1 The Subscriber Terminal

The Public Switch Telephone Network (PSTN) consists of subscriber terminals, local loops, switches or exchanges, and trunks.

In its simplest form, the subscriber terminal is the ordinary telephone with a keypad to dial the numbers. There are two types of dialing: (a) pulse dialing; and (b) DTMF dialing.

Pulse dialing: In pulse dialing, when a digit is dialed, a series of pulses is sent out. When the user dials 1, 1 pulse is transmitted to the exchange, when 2 is dialed, 2 pulses are sent, and so on; when 0 is dialed, 10 pulses are sent. The exchange uses a pulse counter to recognize the digits. Since pulses are likely to be distorted over the medium due to attenuation, pulse recognition accuracy is not very high. Many old switches and telephones support only pulse dialing, though slowly pulse dialing is becoming outdated.

The two types of dialing supported by the subscriber terminals are (a) pulse dialing and (b) tone dialing or DTMF dialing. In pulse dialing, for each digit, a series of pulses is sent to the switch. In tone dialing, for each digit, a combination of two sine waves is sent.

DTMF dialing: DTMF stands for Dual Tone Multi Frequency. DTMF dialing is also known as tone dialing or speed dialing. When a digit is dialed, a combination of two sine waves is sent. The various combinations of tones are shown in Figure 11.2. When 1 is dialed, a combination of 697Hz and 1209Hz is sent from the terminal to the exchange. A DTMF recognition chip is used at the exchange to decode the digits. DTMF recognition is highly accurate and is becoming predominant. Most present-day telephones support DTMF.

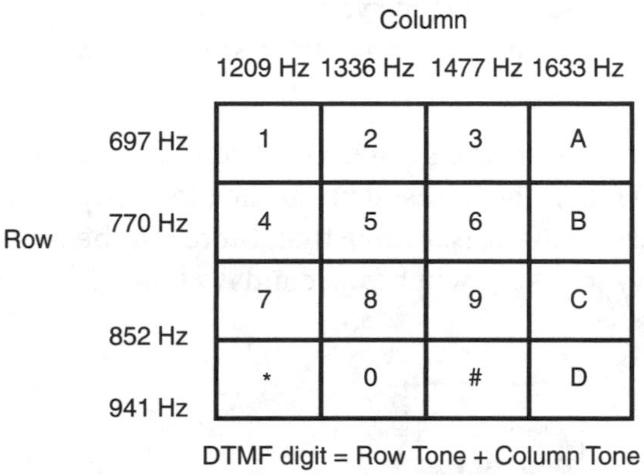

Column

	1209 Hz	1336 Hz	1477 Hz	1633 Hz
697 Hz	1	2	3	A
770 Hz	4	5	6	B
852 Hz	7	8	9	C
941 Hz	*	0	#	D

Row

DTMF digit = Row Tone + Column Tone

FIGURE 11.2 DTMF digits.

DTMF dialing is more reliable as compared to pulse dialing. Pulses are likely to get distorted due to transmission impairments, and so pulse dialing is not always reliable. On the other hand, tone detection is very reliable, and hence DTMF dialing is now extensively used.

11.1.2 Local Loop

The local loop is a dedicated link between a subscriber terminal and the switch. Present local loop uses twisted-pair copper wire as the local loop. In the future, fiber is being planned to provide high bandwidth services to subscribers. In remote and rural areas, where laying the cable is costly or infeasible (due to terrains such as hills etc.), radio is used. This wireless local loop (WLL) has many advantages: fast installation, low maintenance costs. Moreover, it obviates the need for digging below ground. Hence WLL deployment is also catching up, even in urban areas.

Local loop is the dedicated link between the subscriber terminal and the switch. Twisted pair copper wire is the most widely used medium for local loop. Nowdays, wireless local loop is gaining popularity. In the future, optical fiber will be used as the local loop to support very high data rates.

11.1.3 Switch

In earlier days, mechanical and electromechanical switches (strowger and crossbar switches) were used extensively. Present switches use digital technology. These digital switches have the capacity to support several thousand to a few million telephones.

To cater to large areas, the switching system is organized as a hierarchy as shown in Figure 11.3. At the lowest level of the hierarchy, the switches are called end offices or local exchanges. Above that, there will be toll exchanges (class 4 switches), primary (class 3) switches, secondary (class 2) switches, and regional (class 1) switches.

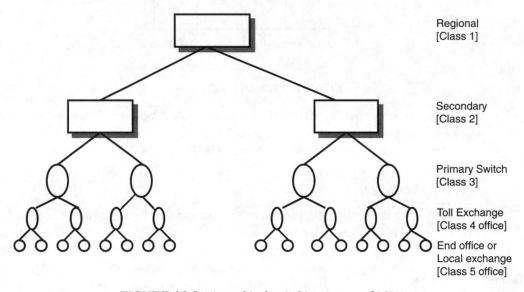

Regional
[Class 1]

Secondary
[Class 2]

Primary Switch
[Class 3]

Toll Exchange
[Class 4 office]

End office or
Local exchange
[Class 5 office]

FIGURE 11.3 Hierarchical switching system of PSTN.

In PSTN, the switching system is organized as a hierarchical system. The local switches are connected to toll exchanges. Toll exchanges are connected to primary switches which are in turn connected to secondary and regional switches.

In a city, an exchange is designated as a toll exchange and acts as the gateway for all long distance calls. Similarly, a few gateway switches carry calls from one nation to another. However, the billing for subscribers is always done by the parent exchange (the exchange to which the subscriber is connected).

The switch to which a subscriber is connected is called the parent switch. Billing is always done by the parent switch.

11.1.4 Trunks

Trunks interconnect the switches. Based on traffic considerations as well as administrative considerations, the interconnection between the switches through trunks is decided. Nowdays, trunks are mostly digital: speech is converted to PCM format, multiplexed, and transmitted through the trunks. The trunks can be T1 or E1 links if the switches are of small capacity (say, 512 ports). Depending on which switches are connected, the trunks are categorized as intracity trunks and intercity trunks.

The switches are interconnected through trunks. Most of the trunks are digital and use PCM format for carrying the voice traffic. E1 trunks carry 30 voice channels.

In the following sections, we will study the important technical aspects of local loops, switches, and trunks.

11.2 LOCAL LOOP

The local loop, which is a dedicated connection between the terminal equipment (the telephone) and the switch, is the costliest element in the telephone network. Generally, from the switch, a cable is laid up to a distribution box (also called a distribution point) from which individual cable pairs are taken to the individual telephone instruments.

To reduce the cable laying work, particularly to provide telephone connections to dense areas such as high-rise residential complexes, digital loop carrier (DLC) systems are being introduced. The DLC system is shown in Figure 11.4. The telephone cables are distributed from the DLC system, and the DLC is connected to the digital switch using a single high-bandwidth cable.

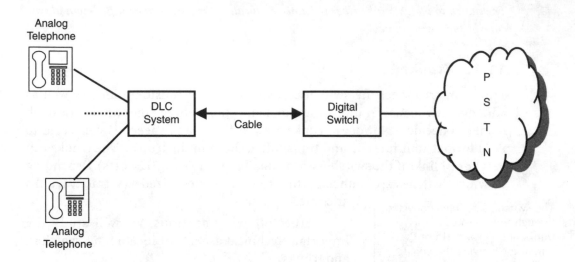

FIGURE 11.4 Digital loop carrier (DLC) in PSTN.

To reduce the installation and maintenance effort (finding out where the cable fault is), wireless local loops are now being introduced. Wireless local loops (WLLs) using CDMA technology are becoming widespread. The advantages of WLL are (a) low maintenance costs because no digging is required; (b) low maintenance costs because equipment will be only at two ends (either the switch or the distribution point and the terminal equipment; (c) fast installation; and (d) possibility of limited mobility.

Most of the present local loops using copper can support very limited data rates. To access the Internet using the telephone network, the speed is generally limited to about 56kbps. Nowdays, user demands are increasing for voice and video services that cannot be supported by the present local loops. Hence, fiber will be the best choice so that very high bandwidth can be available to subscribers, supporting services such as video conferencing, graphics, etc. In the future, optical fiber would be the choice for the local loop. Experiments are going on to develop plastic optical fibers that can take slight bends and support high data rates. Plastic optical fiber would be the ideal choice for fiber to the home.

Because the data rate supported by twisted pair copper cable is limited, optical fiber will be used as the preferred medium for local loop in the future to provide high-bandwidth services. For remote/rural areas, wireless local loop is the preferred choice because of fast installation and low maintenance costs.

11.3 SWITCHING CONCEPTS

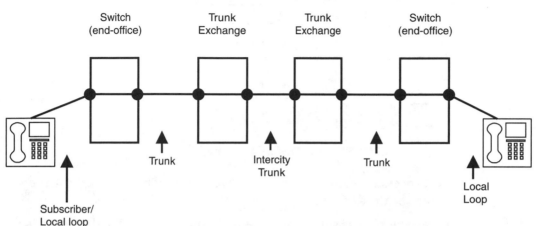

FIGURE 11.5 Circuit switching.

PSTN operates on circuit switching. For two subscribers to converse, a circuit is established between the two subscribers, and after conversation, the circuit is disconnected. The circuit is a concatenation of various trunks between the switches.

To establish a call and subsequently disconnect the call, information needs to be exchanged between the subscriber terminal and the switch and also between the switches. For billing the subscriber and for network management, information is exchanged between the switches. This information is called signaling information and is carried on a signaling trunk.

The PSTN operates on the circuit switching principle illustrated in Figure 11.5. When a subscriber calls another subscriber, a circuit is established that is a concatenation of various channels on the trunks between the switch connected to the calling subscriber and the switch connected to the called subscriber. Circuit switching operation involves the following steps:

1. Call establishment
2. Data transfer (conversation)
3. Call disconnection

To establish and disconnect the calls, information needs to be passed from the subscriber to the switch and also between the switches. This information is known as signaling information. In PSTN, the signaling is carried by the same physical channel that is used to transmit voice. The signaling between two switches is carried on the trunks. As shown in Figure 11.6, some trunks are assigned as signaling trunks and some as traffic (or voice) trunks. This is known as channel associated signaling (CAS).

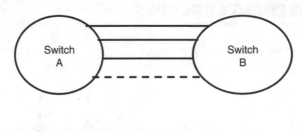

——— Voice Trunk

— — — Signaling link

FIGURE 11.6 Channel associated signaling.

Call processing software resides in the switch. The functions of the call processing software are:

- To keep track of the subscriber terminals and to feed the dial tone when the subscriber goes off-hook (lifts the telephone).
- To collect the digits dialed by the subscriber. Note that the subscriber may dial a few digits and then pause—the software should be capable of handling such cases as well.
- Analyze the digits and switch the call to the right destination by seizing the trunk.
- Feed various tones to the subscriber terminal (such as hunting, busy, call hold, etc.).
- When the subscriber goes on-hook, free the trunk.
- Keep track of the call records (known as CDRs or call details records) that contain call information such as date and time when the call is made, the called party number, whether the call is local/long distance, and duration of the call.
- Based on the CDRs, do an offline analysis to generate billing information.

The call details record (CDR) is generated by the switch. The CDR contains the details of all the calls made. These details include calling subscriber number, called subscriber number, date and time when the call was initiated, duration of the call, and so on. Billing information is generated by processing the CDR.

The switch also contains the diagnostic software to carry out tests on the subscriber loop and the subscriber terminals. This is done through special "line testing software" which feeds signals on the subscriber loop and measures various parameters to check whether the local loop is OK or faulty.

The call processing software that resides on the switch carries out the following functions: collection of the digits dialed by the subscriber, to switch the call to the called subscriber by seizing the trunks, to feed the various tones to the subscribers, to free the trunks after the call is completed and to collect statistics related to the calls.

Though circuit switching has been used extensively for many years, its disadvantage is that the communication channels are not used efficiently. Particularly when voice is transmitted, nearly 40% of the time, the channel is idle because of the gaps in the speech signal. Another drawback is that the signaling information is carried using the same channels resulting in inefficient usage of the channel. In Chapter 26, we will discuss how signaling can be carried by a separate signaling network called Signaling System No. 7 which is being introduced on a large scale in PSTN.

Circuit switching is not an efficient switching mechanism because a lot of time is wasted for establishing and disconnecting the circuit. An alternative switching mechanism is packet switching used in computer networks.

11.4 TRUNKING SYSTEMS

Two switches are connected together through trunks. The trunks are of different types:

1. Two-wire analog trunks, which are used to interconnect small switches.
2. Four-wire trunks, which are also used to interconnect small switches.
3. T1 carriers which are digital trunks. Each T1 carrier carries 24 voice channels. In Europe, the equivalent standard is referred to as E1 trunk. Each E1 trunk supports 30 voice channels.

Small switches are interconnected using 2-wire or 4-wire analog trunks or digital T1 carriers. T1 carrier supports 24 voice channels using Time Division Multiplexing (TDM) technique.

Generally, the switches are connected through T1 carriers. Data corresponding to 24 voice channels is multiplexed to form the T1 carrier. For every 125 microseconds, the bit stream from each voice channel consists of 8 bits out of which 7 bits are data and one bit is control information. Hence, for each voice channel, the total data consist of $7 \times 8000 = 56,000$ bps of voice and 1×8000 bps $= 8000$ bps of control information.

In T1 carrier, a frame consists of 193 bits—192 bits corresponding to 24 voice channels' data and one additional bit for framing. The frame duration is 125 microseconds. Hence, the gross data rate of T1 carrier is 1.544 Mbps.

Frame duration: 125 microseconds, number of bits: 193

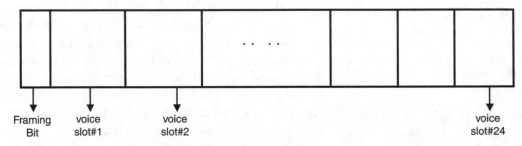

Framing Bit   voice slot#1   voice slot#2   voice slot#24

FIGURE 11.7 Carrier Frame Format.

In major towns and cities, because of high traffic, T1 carriers will not suffice. In such a case, T2, T3, and T4 carriers are used. Higher capacity trunks are obtained by multiplexing T1 carriers. The standard for this digital hierarchy is shown in Figure 11.8. Four T1 carriers are multiplexed to obtain T2 carrier. Seven T2 carriers are multiplexed to obtain T3 carrier. Six T3 carriers are multiplexed to obtain T4 carrier.

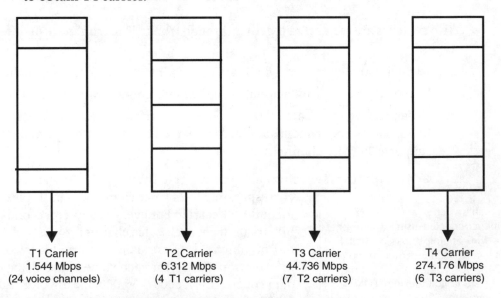

T1 Carrier
1.544 Mbps
(24 voice channels)

T2 Carrier
6.312 Mbps
(4 T1 carriers)

T3 Carrier
44.736 Mbps
(7 T2 carriers)

T4 Carrier
274.176 Mbps
(6 T3 carriers)

FIGURE 11.8 Trunks with Higher Capacities.

11.5 SIGNALING

Four T1 carriers are multiplexed to obtain T2 carrier. Note that the data rate of T2 carrier is 6.312 Mbps and not 4 X 1.544 Mbps (=6.176 Mbps). The extra bits are added for synchronization. Similarly, extra bits are added to form T3 and T4 carriers. The digital hierarchy is standardized by ITU-T to provide high capacity trunks between large switches.

In a telephone network, before the conversation takes place, a circuit has to be established. Lot of information is to be exchanged between the subscriber terminal and the switch, and between the switches for the call to materialize and later get disconnected. This exchange of information is known as signaling.

Signaling is used to indicate/exchange the following information:

- Calling and called party numbers
- Availability/non-availability of network resources such as trunks
- Availability of the called party
- Billing information
- Network information such as busy trunks, faulty telephones etc.
- Routing information as regards how the call has to be routed.
- To provide special services such as calling card facility, toll-free numbers, called party, paying the telephone bill etc.

In the telephone network, there are three types of signaling.

The signaling information exchanged between the subscriber and the switch consists of dialed digits and the various tones such as dial tone, busy tone, ring back tone, etc. This signaling information is carried on the local loop using the in-band signaling mechanism.

In-band signaling: When a subscriber lifts his telephone, he gets a dial tone which is fed by the switch. The subscriber dials the called number and the switch interprets the number and finds out to which switch the called subscriber is connected. The switch establishes a connection to the other switch and the other switch checks whether the called subscriber is available. If she is available, the path is established and the conversation takes place. When the caller puts back the telephone, the circuit is freed. Prior to the conversation and after the conversation, the information exchanged is the signaling information. Normally, this signaling information is exchanged in the same communication link in which the conversation takes place. This signaling is known as in-band signaling. In-band signaling is simple, but creates problems because the tones corresponding to the signaling fall in the voice band and cause disturbances to the speech.

Channel-Associated Signaling (CAS): Between two switches, separate channels are used for signaling and information transfer as shown in Figure 11.6. For instance, when two switches are connected using an E1 link, one time slot is used for signaling. This results in substantial savings as traffic channels are not used for transferring the signaling information.

Common channel signaling (CCS): Another mechanism for signaling is to have a separate communication network for exchanging signaling information. When two network elements have to exchange signaling information, they use this independent network, and the actual voice conversation takes place using the voice trunks. This mechanism (though it appears complex and calls for additional infrastructure) is extremely efficient and is now being widely used.

Two switches in the PSTN exchange signaling information using dedicated time slots in the trunks. This is known as channel associated signaling.

An ITU-T standard called Signaling System No. 7 (SS7) is used for common channel signaling. SS7 uses concepts of data communications, and we will discuss the details of SS7 in a later portion of the book (in Chapter 26, "Signaling System No. 7").

In common channel signaling (CCS), a separate data communication network is used to exchange signaling information. CCS is much more efficient and reliable compared to Channel-Associated Signaling. An ITU-T standard called Signaling System No. 7 (SS7) is used for CCS. SS7 is now used in PSTN, ISDN, and mobile communication systems.

Summary

In this chapter, the architecture of PSTN is presented. The PSTN consists of subscriber terminals, local loops, switches, and trunks. The local loop is a dedicated link between the subscriber terminal and the switch. Currently, twisted copper pair is used as the local loop, but optical fiber-based local loops are likely to be installed in the future. The switches are interconnected through trunks. Normally digital trunks are used. The basic digital trunk supports 30 voice channels and is called an E1 trunk. The PSTN operates in circuit switching mode: a connection is established between two subscribers and, after the conversation, the circuit is disconnected. The signaling information is exchanged between the subscriber and the switch as well as between switches. The signaling used in PSTN is in-band signaling and channel associated signaling.

References

Ray Horak. *Communications Systems and Networks*, Third Edition. Wiley-Dreamtech India Pvt. Ltd., 2002. Chapter 5 of this book is on PSTN.

50th Anniversary Commemorative Issue of IEEE Communications Magazine. May 2002. This issue traces the major developments in communications.

Questions

1. What are the network elements of PSTN? Explain the function of each network element.
2. What is the difference between pulse dialing and DTMF dialing?
3. Explain circuit-switching operation.
4. What are the different types of signaling used in PSTN?

Exercises

1. Two switches need to be interconnected to carry 60 voice channels. Calculate the number of T1 carriers required.
2. Design a circuit to generate/decode DTMF tones. You can use a chip such as MITEL 8880.
3. Study the characteristics of an integrated circuit (IC) that does PCM coding and decoding.
4. The present switches are hardware driven. Nowdays, "soft switches" are being developed. Prepare a technical paper on soft switches.

Projects

1. Simulate a DTMF generator on a PC. The telephone keypad has to be simulated through software. When you click on a digit, the corresponding tones have to be generated (the tone frequencies are given in Figure 11.2) and played through the sound card of your PC.
2. Develop a telephone tapping software package. You can connect a voice/data modem in parallel to the telephone line. When someone calls your telephone, the complete conversation has to be recorded on the PC to which the modem is connected. You can use the Telephony API (TAPI) to develop the software.

12

Terrestrial Radio Communication Systems

In This Chapter

- The Architecture of a Radio Communication System
- Broadcasting Systems
- Wireless Local Loop Technologies
- Cordless Systems and Trunking Systems

Terrestrial radio as a transmission medium is attractive when the terrain does not permit laying of cables. To cover hilly areas, areas separated by lakes, rivers, etc., terrestrial radio is used for local loops and trunks. For providing telecommunication facilities to remote/rural areas, again radio is the best option. To avoid digging up the roads, even in urban areas, radio systems are being used. Radio has another advantage—it provides the capability of mobility to the users. For broadcasting applications, radio is the best choice because radio waves travel long distances. In this chapter, we will study some representative terrestrial radio systems—broadcasting systems, wireless local loops, cordless telecommunication systems, and trunked radio systems.

12.1 ADVANTAGES OF TERRESTRIAL RADIO SYSTEMS

As compared to the guided media such as twisted pair, coaxial cable, and optical fiber, terrestrial radio has many advantages:

- Installation of radio systems is easier compared to cable because digging can be avoided. The radio equipment has to be installed only at the two endpoints.

147

- Maintenance of the radio systems is also much easier as compared to cable systems. If the cable becomes faulty, it is difficult to locate the fault and more difficult to rectify the fault.

- Radio provides the most attractive feature—mobility of the user. Even if the user is moving at a fast pace in a car or even in an airplane, communication is possible.

- Radio waves can propagate over large distances. The coverage area depends on the frequency band. HF waves can travel hundreds of kilometers, VHF and UHF systems can cover up to 40 kilometers, and microwave systems can cover a few kilometers. With the use of repeaters, the distance can be increased.

> Terrestrial radio as the transmission medium has the following advantages: easy installation, easy maintenance, and ability to cover large distances. Another main attraction of radio is that it provides mobility to users.

However, while designing a radio system, the following need to be taken into consideration:

- Radio wave propagation is affected by many factors—the natural terrain (hills, mountains, valleys, lakes, seashores, etc.), artificial terrain (multistoried buildings), and weather conditions (rainfall, snow, fog).

- Radio waves are subject to interference with other radio systems operating in the same vicinity. Radio waves are also affected by power generation equipment, aircraft engine noise, etc.

- Radio waves are attenuated as they travel in the atmosphere. This loss of signal strength is known as path loss. To overcome the effect of path loss, the radio receiver should have high sensitivity—it should be capable of receiving weak signals and amplifying the signal for later decoding.

> In designing radio systems, the following aspects need to be taken into consideration: propagation characteristics, which vary for different frequencies, based on the terrain; interference with other radio systems, and path loss.

The path loss in a radio system is the cumulative loss due to the attenuation of the signal while traveling in the free space and the attenuation in the various subsystems such as the filters, cable connecting the radio equipment to the antenna, etc.

12.2 RADIO COMMUNICATION SYSTEM

The block diagram of a radio communication system is shown in Figure 12.1. The transmission section consists of a baseband processor, which does the necessary filtering of the input signal by limiting the input signal bandwidth to the required value and digitizes the signal using analog-to-digital conversion. If the input signal is voice, the filter limits the bandwidth to 4kHz. If the input signal is video, the bandwidth will be limited to 5MHz. If the radio communication system is a digital system, necessary source coding is also done by the baseband processor. This signal is modulated using analog or digital modulation techniques.

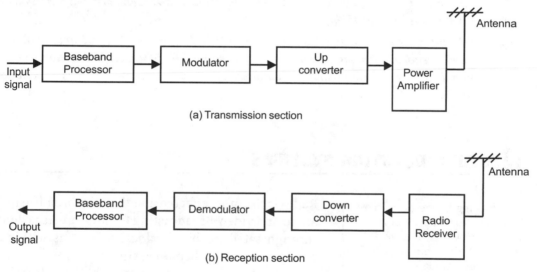

(a) Transmission section

(b) Reception section

FIGURE 12.1 Simplified block diagram of a radio communication system.

Suppose the radio communication system operates in the VHF band with a carrier frequency of 140MHz. The baseband signal is converted into the radio frequency in two stages. In the first stage, called the intermediate frequency (IF) stage, the signal is translated to an intermediate frequency. The most widely used standard IFs are 455kHz, 10.7MHz, and 70MHz. In the second stage, the signal is translated to the required radio frequency using an up-converter as shown in Figure 12.1(a). The up-converted signal is given to a power amplifier that pumps out the modulated radio waves with the desired power level through the antenna. The antenna can be an omnidirectional antenna, a sectoral antenna, or a directional antenna. An omnidirectional antenna radiates equally in all directions. A sectoral antenna radiates at a fixed area, such as a 60° arc, a 120° arc, and so on. A directional antenna transmits in a specific direction. Omnidirectional and sectoral antennas are used at base stations, and directional antennas are used at remote stations.

In a radio system, the baseband signal is first up-converted into an intermediate frequency (IF) and then to the desired radio frequency. Sometimes the up-conversion is done in two or more stages.

At the receiving end, the signal is received by the antenna, down-converted to the IF frequency, demodulated, and filtered, and the original signal is obtained. The baseband processor in the receiving section carries out the necessary decoding.

Some standard IF frequencies are: 455kHz, 10.7MHz, and 70MHz. HF and VHF systems generally use an IF of 455kHz.

The antennas are broadly classified as directional antennas, sectoral antennas, and omnidirectional antennas. A directional antenna radiates in a specific direction. and an omnidirectional antenna radiates in all directions. Sectoral antennas radiate in a sector of 60/120°.

12.3 BROADCASTING SYSTEMS

Broadcasting is one of the most important applications of radio systems. However, the present audio and video broadcasting systems are analog systems.

Radio is the most effective medium for broadcasting audio and video. Broadcasting for large areas is done through satellites, but for local programs within a country or state, broadcasting is done through terrestrial radio. Unfortunately, the broadcasting systems have not evolved as fast as other communication systems. Even now, the audio and video broadcasting systems are analog systems. Though a number of digital broadcasting standards have been developed, they are yet to take off on a large scale. Digital broadcasting systems are now being commercialized.

12.3.1 Audio Broadcasting

In audio broadcasting systems, amplitude modulation (AM) is used. The analog audio signal with 15kHz bandwidth is modulated using AM and then up-converted to the desired frequency band. Many such audio programs of different broadcast stations are frequency division multiplexed (FDM) and sent over the air. The AM radio frequency band is 550–1610 kHz. The receiving stations consist of radio receivers that can be tuned to the desired frequency band. The received signal in that frequency band is down-converted and demodulated, and baseband audio signal is played through the speakers. Presently, frequency modulated (FM) audio broadcasting is becoming predominant. As compared to AM, FM gives a better performance, so the quality of the audio from FM stations is much better. FM stations operate in the frequency band 88–108 MHz.

> In AM broadcasting systems, the modulating signal of bandwidth 15kHz is amplitude modulated and up-converted to the frequency band 550–1610 kHz. A number of audio programs are multiplexed using frequency division multiplexing.

FM broadcasting is now becoming popular. In FM radio, the frequency band of operation is 88–108 MHz. FM gives better quality audio because of its noise immunity.

NOTE

12.3.2 TV Broadcasting

The TV broadcasting system is similar to the audio broadcasting system. The TV signal requires a bandwidth of 5MHz. The video signal is modulated and up-converted to the desired band and transmitted over the radio. At the receiving end, the receiver filters out the desired signal and demodulates the signal, and the video is displayed. VHF and UHF bands are used for TV broadcasting.

To provide very high quality video broadcasting using digital communication, efforts have been going on for the past 20 years. In digital video broadcasting, the moving picture is divided into frames, and each frame is divided into pixels. The number of frames and the pixels in each frame decide the resolution and hence the quality of the video. ETSI has developed two standards: digital TV and HDTV (high-definition TV).

Digital TV standard ITU-R BT.601 uses 25 frames/second, with each frame divided into 720 pixels width and 526 pixels height. For this format, the uncompressed data rate is 166Mbps. Using compression techniques, the data rate can be brought down to 5–10 Mbps.

> Present TV broadcasting systems operating in the VHF and UHF bands are analog systems. For digital TV transmission, two standards, Digital TV standard and High Definition TV (HDTV) standard, have been developed.

HDTV standard ITU-R BT.709 uses 25 frames/second, with each frame divided into 1920 pixels width and 1250 pixels height. For this format, the uncompressed data rate is 960Mbps. Using compression, the data rate can be brought down to 20–40 Mbps.

For digital TV transmission, the moving video is considered a series of frames, and each pixel in the frame is quantized. Using compression techniques, the data rate is reduced to a few Mbps.

12.4 WIRELESS LOCAL LOOPS

In the telephone network, the local loop is the costliest network element. To provide telephone services to remote and rural areas, wireless local loop is the most cost-effective alternative.

For providing telecommunication facilities, the network elements required are the switch, the trunks, the local loops, and the subscriber terminals. The local loop is the dedicated link between the subscriber terminal and the switch. In cities and towns, the local loop uses twisted pair as the transmission medium because the distance between the switch and the subscriber terminal generally will be less than 5 km. Because the subscriber density is high in cities and towns, the cost of installing a switch for subscribers within a radius of 5 km is justified. In remote and rural areas, the subscriber density will be less, the number of calls made by the subscribers will not be very high, and the areas are separated by long distances from the nearby towns. As a result, laying a cable from one town to another is not cost effective. Installing a switch to cater to a small number of subscribers is also prohibitively costly.

Wireless local loops can be in two configurations. Figure 12.2 shows configuration 1. A radio base station will be connected to the switch. The base

Wireless local loop can have two configurations. In one configuration, the subscriber telephone is connected to the switch using radio as the medium. In the other configuration, wireless connectivity is provided between the subscriber terminal and the distribution point, and the connectivity between the switch and the distribution point is through a wired medium.

station is generally located in a town at the same premises as the switch. A number of remote stations communicate with the base station through radio. Each remote station can be installed in an area, and it can support anywhere between 1 and 32 telephones. The distance between the base station and each remote generally can be up to 30 km. A base station can provide telephone facilities to subscribers in a radius of 30 km. This configuration is used extensively for providing telephone facilities in rural and remote areas.

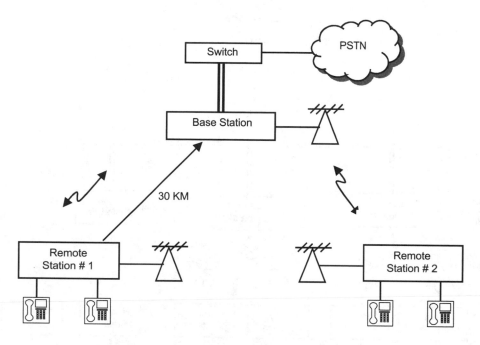

FIGURE 12.2 Wireless local loop configuration 1.

Figure 12.3 shows configuration 2 of wireless local loops. In this configuration, a number of base stations are connected to the switch using cable. Each base station in turn communicates with a number of remote stations. Each remote station can support a number of telephones. In this configuration, the local loop is a combination of wired and wireless media. This configuration is used extensively in urban areas. TDMA and CDMA technologies are used in this configuration. The number of subscribers supported by the base station/remote station depends on the access technology. In the following sections, some representative wireless local loop systems are described.

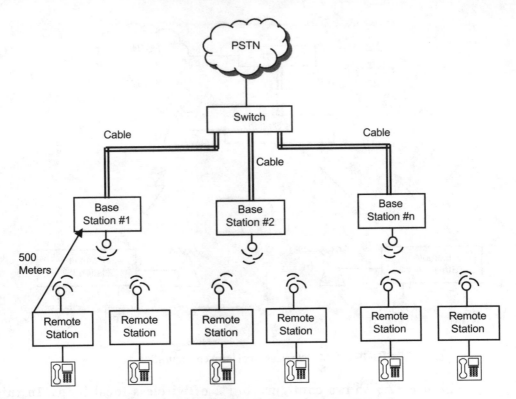

FIGURE 12.3 Wireless local loops configuration 2.

 Wireless local loop also is gaining popularity in urban areas because of reduced installation and maintenance costs.

NOTE

12.4.1 Shared Radio Systems

The block diagram of a shared radio system is shown in Figure 12.4. This system can cater to a number of communities within a radius of about 30 km from the base station. The system consists of a base station and a number of remote stations (say, 15). Each remote station will provide one telephone connection. This remote telephone can be used as a public call office (PCO) to be shared by a number of people. The base station consists of a base station controller and a radio. The base station controller is connected to the PSTN through the switch located at a town. The base station radio will have transmitter /receivers for two radio channels. Both the channels can be used for either incoming or outgoing calls. So, the system works in the FDMA mode. When a remote user wants to make a call, one of the free channels will be assigned to him, with one uplink frequency and one downlink frequency. Since the base station can support only two radio channels at a time, only two remote telephones can use the system. This is not a major problem because the traffic in rural areas is not very high. This system is called a 2/15 shared radio system, 2 indicating the number of radio channels, and 15 indicating the number of remotes supported by the system. For each channel, 25kHz of bandwidth is allocated. These systems operate in the VHF (150MHz and 170MHz) and UHF (400MHz) bands. Shared radio systems are analog systems.

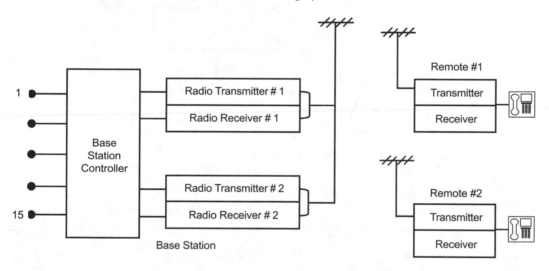

FIGURE 12.4 Shared radio system as wireless local loop.

In a shared radio system (SRS), a number of remote stations communicate via a central station. A few radio channels are shared by the remotes using Frequency Division Multiple Access. SRS provides low-cost analog wireless local loops.

This concept can be extended to develop higher order systems such as a 4/32 shared radio system, which will have 4 radio channels and 32 remotes, or an 8/64 shared radio system, which will have 8 radio channels and 64 remotes.

These systems also can support data services up to a data rate of 9.6kbps. Using normal line modems, the data can be sent in the voice channels.

12.4.2 Digital Local Loops

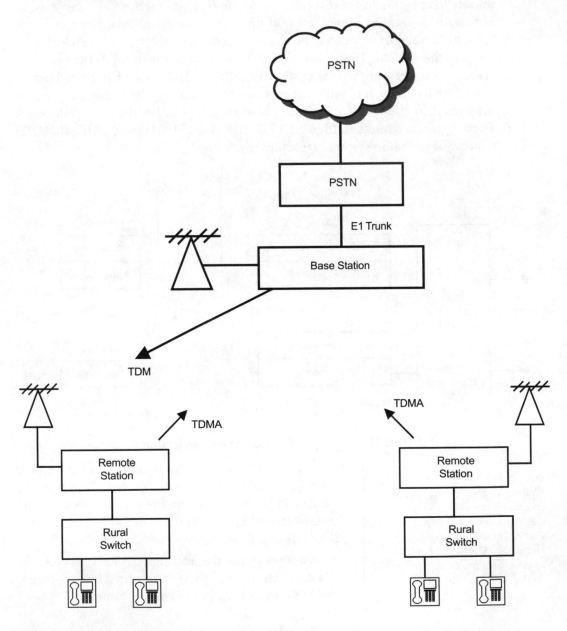

FIGURE 12.5 Digital wireless local loop using TDM/TDMA.

The digital wireless local loop using TDM/TDMA is shown in Figure 12.5. Unlike the shared radio systems, this system uses digital communication. The system consists of a base station and number of remotes (up to 64). The base station is connected to the switch using a T1 trunk (to support 24 voice channels). Each remote station can be connected to a small rural switch. The rural switch can handle up to 32 subscribers. The advantage of having a small switch is that the local calls within a community can be switched locally without the need for a radio. All the remotes will be within a radius of 30 km from the base station. Communication from the base to the remote will be in TDM mode, and from remote to the base will be in TDMA mode. The salient features of the system are as follows:

Frequency of operation: The system operates in the 800MHz frequency band. The base station transmits in the 840–867.5 MHz band (downlink frequency). The remote stations transmit in the frequency band 885–912.5 MHz band. The number of channels will be 12 with a channel spacing of 2.5MHz. The modulation in both directions can be QPSK. Alternatively, the base station can use FSK for transmission, and remote stations can use QPSK—this scheme has the advantage that the remote station electronics will be less complex because demodulation of FSK is much easier than demodulation of QPSK. Base station and the remote stations transmit a power of 2 watts. The base station uses an omni-directional antenna, and the remote stations use directional antennas.

Voice coding: To conserve the spectrum, ADPCM can be used for voice coding. In ADPCM, the voice is coded at 32kbps.

Access scheme: The base station transmits the data in TDM mode. The data for all the remote stations will be multiplexed, and the multiplexed data will be broadcasted. All the remote stations will receive the data, and each remote will decode the data only if the data is meant for it by matching the received address with its own address.

The remote stations will use the TDMA for accessing the base station. Each remote will be assigned a time slot and transmit the data in that time slot.

Assignment of TDMA slots: Since there is less traffic in rural areas, fixed assignment of time slots results in wasted time slots because even if there is no call, the slot will be assigned to that remote. In dynamic TDMA, each remote will get a time slot periodically that is exclusively for signaling. When a subscriber at a remote picks up the telephone and dials a telephone number to make a call, a request for a voice slot is sent to the base station along with the called number. The base station checks for the availability of the called party. If available, a time slot is allocated in which the remote can send the digitized voice.

Frame and slot sizes: There is a trade-off between buffer storage required and synchronization overhead. If the frame size is small, the synchronization overhead will be high; if the frame size is large, buffer storage requirement will be high. As an optimal choice, 8 msec can be used as the frame size.

The TDM/TDMA frame formats are shown in Figure 12.6. From base to remote, the 8 msec frame consists of a signaling slot of 416 bits and 27 voice slots of 288 bits. From the remote to the base, the signaling slot is also 418 bits and voice slot 288 bits. The voice slot contains 12 bits of guard time, 20 bits of preamble and sync, and 256 bits of 32 kbps ADPCM voice data. The guard time is used to take care of the propagation delay. The preamble and sync bits are for synchronization. Since the frame has 27 time slots, 27 subscribers connected to different remote stations can make calls simultaneously.

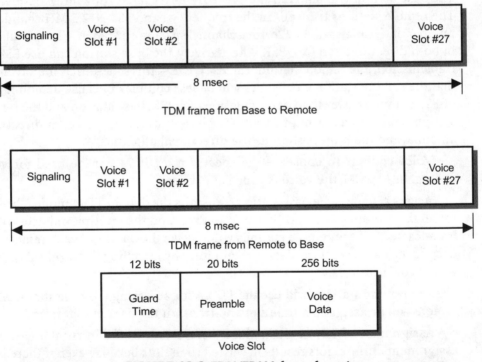

FIGURE 12.6 TDM/TDMA frame formats.

Because the base transmits the data in broadcast mode, all the remotes can receive the data. Each remote synchronizes itself using the synchronization pattern and processes only the signaling slot data and the voice slots allocated

Digital local loops using the TDM/TDMA technique provide wireless connectivity between a number of remote stations and a central station. The central station transmits the data in TDM broadcast mode, and the remote stations use TDMA slots to send their data.

to it. The remaining slots are ignored. The remote communicates with the base in the channel allocated to it for signaling. This type of signaling is called common channel signaling because a separate slot is used for signaling, and the voice slots contain only voice data and no signaling information.

The digital wireless local loop systems based on TDM/TDMA provide very low cost solutions for rural/ sub-urban areas. Hughes Network Systems, USA, SR Telecom, Canada, and Japan Radio Company, Japan are the major organizations that developed such systems.

In the TDM/TDMA time slots, a separate slot is dedicated for carrying signaling information. This is known as common channel signaling.

NOTE

12.5 CORDLESS TELEPHONY

Cordless telephone systems provide limited mobility and a short range. To provide cordless telephone services in residential and office environments, Cordless Telephony standards have been developed that have a maximum range of 200 meters.

At home and the office, we use cordless telephones that provide limited mobility. ETSI developed a series of Cordless Telephony (CT) standards to provide cordless telephony services in residential and office areas. The first generation CT systems were analog systems, and hence the performance was poor. A second generation Cordless Telephony (CT-2) system can be used as a two-way cordless telephone in an office environment or as a two way cordless in the home. CT-2 is an FDMA-based system that operates in the 800MHz band. CT-3 is a digital version of CT-2 and uses TDMA-TDD as the access scheme. CT-3 operates in the 800–1000 MHz band with an overall data rate of 640kbps. The speech data rate is 32kbps. In CT-2 and CT-3, the range is 5 meters in built-up areas and 200 meters for line of sight. To enhance the capabilities of cordless telephone systems, ETSI developed the DECT standards.

12.5.1 Digital Enhanced Cordless Telecommunications (DECT)

The DECT standard developed by ETSI has three configurations: (a) for residential operation; (b) for access to telephone networks from public places; and (c) for use in office complexes.

European Telecommunications Standards Institute (ETSI) developed the DECT standard for cordless telephony. DECT-based handsets can be used in homes, at offices, or at public places such as shopping malls and airports.

DECT at Home

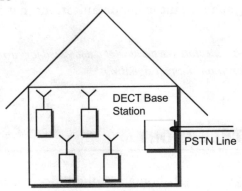

FIGURE 12.7 DECT at home.

In the home configuration, a DECT system operates as a cordless PBX with one base station and up to four extensions.

In this configuration, in the home, an incoming telephone line can be shared by up to four extensions, as shown in Figure 12.7. There will be a DECT base station connected to the incoming telephone line that can communicate with any of the four extensions using radio. This configuration is a cordless PBX with four extensions.

DECT Telepoint Access System

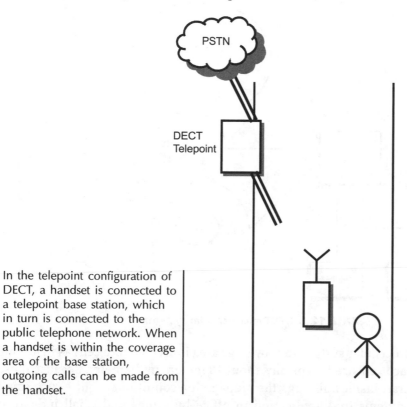

In the telepoint configuration of DECT, a handset is connected to a telepoint base station, which in turn is connected to the public telephone network. When a handset is within the coverage area of the base station, outgoing calls can be made from the handset.

FIGURE 12.8 Telepoint for public access.

In places such as airports, railway stations, and large shopping complexes, a base station (called telepoint) can be installed. Using the DECT handset, the PSTN can be accessed through the telepoint. In such a case, it is expected that the user will not move considerably, so the handset should be within the coverage area of the base station. This service has the following disadvantages:

- Only outgoing calls can be supported. If incoming calls are to be supported, a radio pager has to be integrated with the handset.

- The environment in which the telepoint base stations are installed being noisy, special enclosures need to be provided (called telepoint booths).

DECT Micro-Cellular System

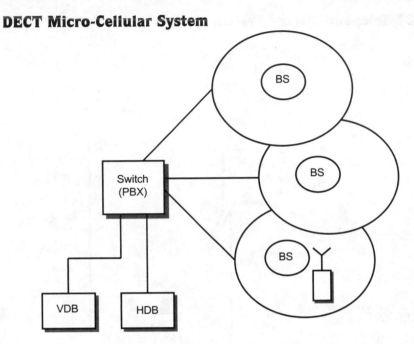

FIGURE 12.9 DECT micro-cellular system.

In a microcellular system, the coverage area is divided into cells as in mobile networks; each cell size is typically about 10 meters in diameter. Because these systems operate inside buildings, the propagation conditions require such small cells. Such systems find application in office buildings and small industrial complexes. The cordless system using microcellular concept is also known as cordless PBX. Microcellular systems are also useful for installing in temporary locations such as construction sites, big exhibitions, etc. Unlike the telepoint system, microcellular systems do not provide public service, they serve only closed user groups. Each cell will have a base station. All the base stations will be connected to the switch (PBX). The PBX will have two databases: the home database (HDB) and the visitor database (VDB). These databases contain the information about the subscribers and their present locations. The DECT handset is connected to the base station using radio as the medium. When the handset moves from one cell to another, the radio connection automatically is transferred to the other base station.

In a DECT microcellular system, the service area is divided into small cells of about 10 meters diameter, and each cell has a base station. As a person moves from one cell to another cell while talking, the handset is connected to the nearest base station.

For all three configurations, the DECT standard specifies the frequency of operation, access mechanism, coding schemes, and such, which are described in the following sections.

DECT Standard

The broad specifications of DECT are as follows:

Range: The range of a DECT system is about 300 meters for a maximum peak power of 250 milliwatts. Average power is 10 milliwatts. However, in the home and telepoint configurations, the range can be about 10 to 50 meters so that the power consumption can be less. It is also possible to extend the range up to 5 kilometers by increasing the transmit power.

Frequency band: The DECT system operates in the frequency band 1880–1900 MHz. A total of 10 carriers can be assigned to different cells.

Voice coding: Voice is coded at 32kbps data rate using adaptive differential pulse code modulation (ADPCM).

Multiple access: The multiple access used is TDMA-TDD. The frame structure is shown in Figure 12.10. The frame duration is 10 milliseconds. The first 12 slots are for base to remote communication, and the next 12 slots are for remote to base communication. The same frequency is used for both base to remote and remote to base communication. The transmission data rate is 1152kbps. The channel assignment is dynamic: when a user has to make a call, the frequency and time slot will be allocated.

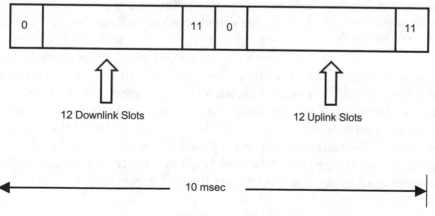

FIGURE 12.10 DECT TDMA-TDD frame structure.

Since there are 10 carriers, and each carrier will have 12 time slots, total number of voice channels supported is 120. If a telepoint base station is installed in a shopping complex, simultaneously 120 persons can make calls through the base station. In the home configuration, this is not required. However, the advantage of a DECT system is that the same handset used as a cordless telephone at home can be carried to the shopping complex or to the office to make calls.

DECT operates in the 1880–1900 MHz band in TDMA/TDD mode. For voice communication, 32kbps ADPCM is used.

A DECT system can be used as a wireless local loop using the telepoint configuration. In India, a number of villages are provided with telephone facilities using the DECT system developed at IIT, Madras. The only limitation of DECT is that mobility should not be more than 20 km/hour.

DECT can be used to provide wireless local loops for rural areas by increasing the radio transmit power. The handsets can be mobile, but with a restriction that the speed should not be more than 20 km/hour.

12.6 TRUNKED RADIO SYSTEMS

Trunked radio systems are used to provide low-cost mobile communication services to closed user groups such as taxi operators, ambulance service operators, and police.

Radio systems used in cities and towns to provide mobile communication for closed user groups such as taxi operators, ambulance service operators, police, etc. are called trunked radio systems. Trunked radio systems can also be installed by organizations that need communication among their employees in their area of operation such as construction sites. The trunked radio system consists of a base station at a central location in a city/town. The user terminals are mobile devices that communicate with the operator at the base station or with another mobile device using FDMA/TDMA. The operator of the trunked radio system will be assigned a set of frequencies that are shared by all the mobile devices. As compared to cellular mobile communication systems, trunked radio systems are low-cost systems because the entire service area is just a single cell.

In trunked radio systems, a city is generally covered by a single base station. The base station and the mobile terminals communicate using FDMA/TDMA.

12.6.1 TETRA

ETSI developed a standard for trunked radio system called TETRA (Terrestrial Trunked Radio). The configuration of TETRA is shown in Figure 12.11.

A TETRA system consists of TETRA nodes, which are radio base stations, to provide mobile communication facilities for users in a service area. A trunked radio operator can interconnect a number of TETRA nodes using cable or microwave radio. Each node can cater to a portion of a large city.

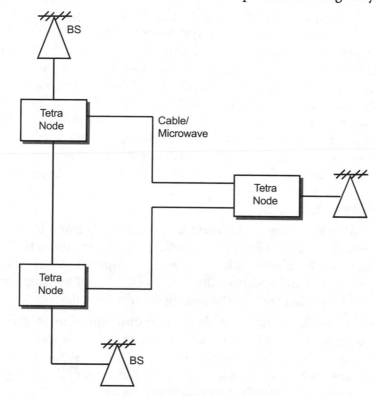

FIGURE 12.11 TETRA.

The TETRA standard has been formulated by ETSI. TETRA uses TDMA/FDD mechanism with 380–390 MHz uplink frequency and 390–400 MHz downlink frequency. Both circuit switching and packet switching are supported.

The salient features of TETRA are as follows:

Frequency of operation: The uplink frequency is 380 to 390 MHz, and the downlink frequency is 390 to 400 MHz. Each channel has a bandwidth of 25kHz.

Multiple access: The system operates in TDMA/FDD. The TDMA frame is divided into four time slots that are dynamically assigned.

Types of services: Two types of services are supported. Voice and data services use circuit switched operation. This service uses the TDMA scheme with four slots per carrier. Packet data optimized (PDO) services use packet switching for data communication. Data rates up to 36kbps can be achieved.

TETRA-based trunked systems are used extensively in Europe.

Summary

In this chapter, we discussed the details of representative terrestrial radio communication systems. Radio systems provide many advantages—easy installation and maintenance and support for mobility. The radio as the transmission medium is used in audio and video broadcasting, in wireless local loops, in cordless telephony applications, and for trunked radio systems. The radio broadcasting systems are mostly analog, though in the next few years, digital broadcasting systems are likely to increase. Wireless local loops are now being deployed extensively in both urban and rural areas. Analog wireless local loops are shared radio systems in which a few radio channels (2 or 4 or 8) are shared by a number of remote stations (16 or 32 or 64). Digital wireless local loops use TDMA and CDMA technologies. Low-cost local loops can be provided using these technologies.

The digital enhanced cordless telecommunications (DECT) standard developed by ETSI can be used for cordless telephony applications at home or office or at public places such as shopping complexes. DECT operates in the 1880–1900 MHz band and uses the TDMA-TDD scheme to support up to 120 voice channels using the ADPCM coding scheme for voice.

Trunked radio systems provide mobile communication facilities for closed user groups such as police, taxi operators, and ambulances. The Terrestrial Trunked Radio (TETRA) standard developed by ETSI supports both voice and data services. TETRA offers a low-cost solution for mobile communication as compared to cellular mobile communication systems.

References

G. Karlsosson. "Asynchronous Transfer of Video," Vol. 34, No. 8, August 1996. Proceedings of the IEE Conference on Rural Telecommunications, May 1988, London.

www.etsi.org The Web site of European Telecommunications Standards Institute. You can obtain the ETSI standards referred to in this chapter from this site.

www.motorola.com Motorola is a leading manufacturer of trunked radio systems. You can obtain the details from this site.

Questions

1. Draw the block diagram of a radio communication system and explain the various blocks.

2. What is a wireless local loop? Explain the architecture of analog wireless local loop systems.

3. Explain the architecture of digital wireless local loop systems based on TDMA.

4. Explain the various configurations of a DECT system. List the salient features of the DECT standard.

5. Describe the operation of a trunked radio system. What are the salient features of the TETRA standard?

Exercises

1. Prepare a technical report on digital broadcasting using HDTV.

2. Prepare a technical report on a wireless local loop using CDMA technology.

3. Prepare a technical report on the various commercial products available for trunked radio.

4. Study the details of path loss calculations for terrestrial radio systems.

5. Instead of air, water is used as the transmission medium for underwater communication. What are the frequency bands used for this type of communication? As compared to terrestrial radio systems, what are the major differences in underwater communication systems?

Projects

1. Study the various propagation models for modeling the radio wave propagation in free space (such as the Okumura-Hata model). Develop software to calculate the path loss in a radio communication system and to carry out link analysis.

2. You are asked to design a cordless PBX for your office. Work out the details of location of base stations, if a DECT-based system has to be installed.

13 ▪ Satellite Communication Systems

In This Chapter

- Satellite Communication Systems
- The Problems Associated with Satellite Communication
- Multiple Access Techniques Used in Satellite Communication
- A System That Supports Multimedia Applications

E ver since the first communication satellite was launched in 1962 by the United States, satellites have been used extensively for communications. In this chapter, we will study the various applications of satellites, frequency bands in which the satellite communication systems operate, and the multiple access techniques used in satellite communication systems. We also will study the architecture of a representative communication system that uses satellite as the transmission medium.

13.1 APPLICATIONS OF SATELLITES

Satellites are used for a variety of applications such as these:

- Astronomy
- Atmospheric studies
- Communication
- Navigation
- Remote sensing
- Search and rescue operations
- Space exploration

- Surveillance
- Weather monitoring

In communications, satellites are used for broadcasting, providing trunks between switches of telephone networks, providing telephone facilities for remote and rural areas, land mobile communications, marine communication, and many other uses. Many corporate networks use satellite communication for their interoffice communication.

Satellites also are used to send location information to people on the move (on the earth, in aircraft, or underwater). Global Positioning System (GPS) uses 24 satellites that continuously broadcast their positional parameters. The users have GPS receivers. The GPS receiver calculates its own positional parameters (longitude, latitude, and altitude) based on the data received from the satellites. We will discuss the details of GPS in Chapter 32, "Global Positioning System."

Satellites are used for a variety of applications such as communication, broadcasting, surveillance, navigation, weather monitoring, atmospheric studies, remote sensing, and space exploration.

Surveillance satellites are fitted with video cameras. These satellites continuously monitor the enemy's territory and send video data to a ground station. Surveillance satellites are used by many countries to keep track of the activities of other countries.

13.2 ARCHITECTURE OF A SATELLITE COMMUNICATION SYSTEM

Communication satellites operate in two configurations: (a) mesh; and (b) star. In mesh configuration, a remote station can communicate directly with another remote station. In star configuration, two remote stations communicate via a central station or hub.

As discussed in Chapter 3, "Transmission Media," satellite communication systems operate in two configurations: (a) mesh; and (b) star. In mesh configuration, two satellite terminals communicate directly with each other. In star configuration, there will be a central station (called a hub), and remote stations communicate via the hub. The star configuration is the most widely used configuration because of its cost-effectiveness, and we will study the details of satellite communication systems based on star configuration in this chapter.

The architecture of a satellite communication system is shown in Figure 13.1. The system consists of two segments:

- Space segment
- Ground segment

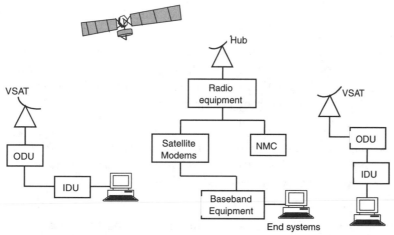

FIGURE 13.1 Architecture of a satellite communication system.

The space segment consists of the satellite, which has three systems: fuel system, telemetry control system, and transponders.

Space segment: The space segment consists of the satellite, which has three main systems: (a) fuel system; (b) satellite and telemetry control system; and (c) transponders. The fuel system is responsible for making the satellite run for years. It has solar panels, which generate the necessary energy for the operation of the satellite. The satellite and telemetry control system is used for sending commands to the satellite as well as for sending the status of onboard systems to the ground stations. The transponder is the communication system, which acts as a relay in the sky. The transponder receives the signals from the ground stations, amplifies them, and then sends them back to the ground stations. The reception and transmission are done at two different frequencies. The transponder needs to do the necessary frequency translation.

Ground segment: The ground segment consists of a number of Earth stations. In a star configuration network, there will be a central station called the hub and a number of remote stations. Each remote station will have a very small aperture terminal (VSAT), an antenna of about 0.5 meter to 1.5 meters. Along with the antenna there will an outdoor unit (ODU), which contains the radio hardware to receive the signal and amplify it. The radio signal is sent to an indoor unit (IDU), which demodulates the signal and carries out the necessary baseband processing. IDU is connected to an end systems, such as a PC, LAN, or PBX.

The central station consists of a large antenna (4.5 meters to 11 meters) along with all associated electronics to handle a large number of VSATs. The central station also will have a Network Control Center (NCC) that does all the management functions, such as configuring the remote stations, keeping a database of the remote stations, monitoring the health of the remotes, traffic analysis, etc. The NCC's main responsibility is to assign the necessary channels to various remotes based on the requirement.

The central station or the hub consists of a large antenna and associated electronics to handle a large number of VSATs. The network control center (NCC) at the hub is responsible for all management functions to control the satellite network.

Communication satellites are stationed at 36,000 km above the surface of the earth, in geostationary orbit. Two separate frequencies are used for uplink and downlink.

The communication path from a ground station to the satellite is called the uplink. The communication link from the satellite to the ground station is called the downlink. Separate frequencies are used for uplink and downlink. When a remote transmits data using an uplink frequency, the satellite transponder receives the signal, amplifies it, converts the signal to the downlink frequency, and retransmits it. Because the signal has to travel nearly 36,000 km in each direction, the signal received by the satellite as well as the remote is very weak. As soon as the signal is received, it has to be amplified before further processing.

Due to the large distance to be traversed by the signals, the attenuation is very high in satellite systems. Hence, the sensitivity of the radio receivers at the Earth stations should be very high.

13.2.1 Frequencies of Operation

The three widely used frequency bands in satellite communication systems are C band, Ku band, and Ka band. The higher the frequency, the smaller will be the antenna size. However, the effect of rain is greater at higher frequencies.

The various bands of operation are:

C band: Uplink frequency band 6GHz (5.925 to 6.425 GHz)

Downlink frequency band: 4GHz (3.7 to 4.2 GHz)

Ku band: Uplink frequency band: 14GHz (13.95 to 14.5 GHz)

Downlink frequency band: 11/12GHz (10.7-11.25 GHz, 12.2-12.75 GHz)

Ka band, with uplink frequency band of 30GHz and downlink frequency band of 20GHz, is used for broadcasting applications. Direct broadcast satellites, which broadcast video programs directly to homes (without the need for distribution through cable TV networks) operate in the frequency band 17/12GHz, with uplink frequency band being 17.3 to 18.1 GHz and downlink frequency band being 11.7 to 12.2 GHz.

Because the frequency of operation is higher in the Ku band, the antenna size will be much smaller as compared to C band antennas. However, the effect of rain is greater in Ku band than in C band. For many years, only C band was used for satellite communication. With advances in radio components such as amplifiers, filters, modems, and so on, the effect of rain on Ku band can be nullified by necessary amplification. Presently, Ku band is used extensively for communication.

> Communication satellites operate in different frequency bands: C band (6/4GHz), Ku band (14/12GHz), and Ka band (30/20GHz) band. The higher the frequency, the smaller the antenna.

13.3 PROBLEMS IN SATELLITE COMMUNICATION

The main attraction of satellite communication is that it provides communication facilities to any part on the earth—satellites are insensitive to the distance. However, the problems associated with satellites are:

Propagation delay: In a star network, the total delay from one VSAT to another VSAT is nearly 0.5 seconds if the VSAT has to communicate via the hub. This type of delay is not acceptable particularly for voice communication, because it results in echo and talker overlap. Propagation delay also causes problems for many data communication protocols such as TCP/IP. Special protocols need to be designed for data communication networks that use satellites.

If the VSAT communicates directly with another VSAT, the propagation delay is nearly 0.25 seconds. We will discuss multiple access techniques that facilitate direct communication from one VSAT to another VSAT.

Low bandwidth: As compared to the terrestrial media, particularly the optical fiber, the bandwidth supported by satellites is much less. Though present satellites provide much more bandwidth than the satellites of the 1970s and 1980s, the bandwidth is nowhere comparable to the optical fiber bandwidth.

The problems associated with satellite communication are: high propagation delay, low bandwidths as compared to terrestrial media, and noise due to the effect of rain and atmospheric disturbances.

Noise: Satellite channels are affected by rain, atmospheric disturbances, etc. As a result, the performance of satellite links is generally poor as compared to terrestrial links. If data is received with errors, the data has to be retransmitted by the sender. To reduce retransmissions, forward error correcting (FEC) codes are implemented.

The large propagation delay in satellite networks poses problems for voice communication. High delay causes echo and talker overlap. Echo cancellers need to be used to overcome this problem.

The TCP/IP protocol stack used in computer communication will not perform well on satellite networks. The stack is suitably modified to overcome the problems due to propagation delay.

13.4 MULTIPLE ACCESS TECHNIQUES

A number of multiple access techniques are used in satellite communication. Based on the type of application and the cost of the equipment, a particular multiple access technique has to be chosen. A VSAT network operating in the star configuration can use the TDM/TDMA access scheme. The central station will multiplex the data of all the remotes and broadcast it. All the remotes will receive the data, and the remote for which the data is meant will decode the data; the rest of the remotes will ignore the data. Each remote will transmit in TDMA mode in the time slot allocated to it. A signaling slot is available to each

VSAT networks operating in star architecture use the TDM/TDMA access mechanism. The hub multiplexes the data of all remotes and broadcasts it. The remote stations use the TDMA slots to send their data.

of the remotes for making a request for a TDMA traffic slot. This mechanism is useful if the network has a large number of remotes and traffic flow is mostly from the central station to the remotes. If direct communication from one remote to another remote is required, the multiple access techniques discussed in the following sections are used.

13.4.1 DAMA-SCPC

In demand assigned multiple access–single channel per carrier (DAMA-SCPC), a channel is assigned to a remote only when the remote has data to transmit. The channel assignment is done by one station that acts as the network control center (NCC). Once the channel is assigned, the remote can directly transmit data to another remote (as in a mesh configuration). Because the channel is assigned based on demand, the access mechanism is DAMA. Because the data corresponding to one channel (say, voice) can be transmitted on one carrier assigned to the remote, the transmission mechanism is SCPC. SCPC got its name due to the earlier analog systems that used one channel per carrier. Now, it is possible to multiplex different channels and send the data on one carrier, which is known as multi-channel per carrier (MCPC).

The configuration of the DAMA-SCPC network is shown in Figure 13.2. At both the NCC and the remote, there will be a control channel modem and a number of modems (modem bank) to modulate and demodulate many carriers. When a remote has to communicate with another remote, it will send a request in a TDMA request channel. The NCC will send the control information to both the remotes, indicating the carrier assigned to each remote using TDM broadcast mode. The control channel modem is used exclusively for requests and commands. Once a carrier is assigned to the remote, the modem corresponding to that carrier is used to transmit the data.

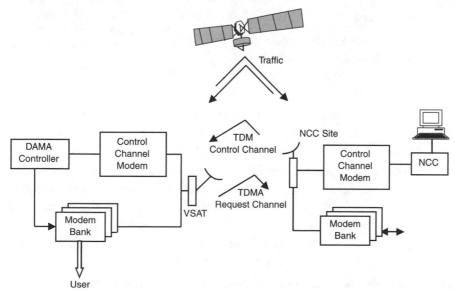

FIGURE 13.2 DAMA-SCPC mesh architecture.

In DAMA-SCPC-based networks, each remote should have a control channel modem and a modem bank for different carrier frequencies.

In a satellite network using DAMA-SCPC, a remote sends a request to the network control center (NCC), and the NCC will assign different carriers to the two remotes that need to communicate with each other.

The TDM control frame and TDMA request frame formats are shown in Figure 13.3. The TDM control frame consists of unique word (UW) and control fields that are used for framing, synchronization, and timing. These fields are followed by a number of data slots, each slot for one remote. The data slot consists of:

- Preamble to indicate the beginning of the slot
- Header that contains the remote address and configuration information.
- Data indicating the information for call setup and disconnection (carrier assigned).
- Frame check sequence (FCS) containing the checksum.
- Postamble to indicate the end of the slot.

The TDMA request channel contains a series of slots for each remote. A remote has to send its requests for call setup using its slot. The slot structure is the same as the structure of the data slot in the TDM frame.

The procedure for call setup is as follows:

The TDM control frame consists of a unique word, control word, and a number of time slots. Each slot consists of the following fields: preamble, header, data, frame check sequence, and postamble. The TDMA request channel has the same format as the slot of the TDM control frame.

- The remote sends a request in its slot of the TDMA request channel indicating the address of the called remote. The control channel modem is used for sending the request.
- The network control center sends the control information in the TDM slot assigned to the remote indicating the carrier assigned. The control channel modem is used for sending the command.
- Using the modem for the assigned carrier, the remote sends its data to the other remote. It needs to be noted that the data sent by a remote goes to the satellite, and the satellite broadcasts the data so that the other remotes can receive the data.
- Once the data transfer is complete, the remote sends the request for disconnection in the TDMA request channel.
- The network control center sends the command to the remote to free the modem corresponding to the carrier assigned earlier.

- The carrier assigned to the remote is now available in the pool of carriers that can be assigned to the other remotes based on demand.

This configuration is very useful if remotes need to communicate with each other directly. However, one of the remotes has to act as a network control center.

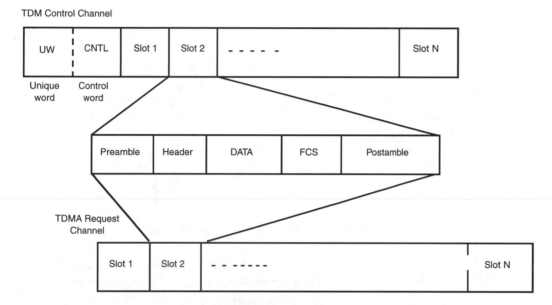

FIGURE 13.3 DAMA-SCPC control and request channels.

13.4.2 TDM-SCPC

In time division multiplex–single channel per carrier (TDM-SCPC), every remote broadcasts its data in TDM mode. Each remote is assigned a carrier frequency permanently, and so each remote will have one modulator. However, each remote will have a bank of demodulators to demodulate the data received from other remotes. Every remote will listen to transmissions from other remotes and decode the data meant for it based on the address.

The attractive feature of this configuration is that there is no need for a network control center. Also, there is no need for call setup.

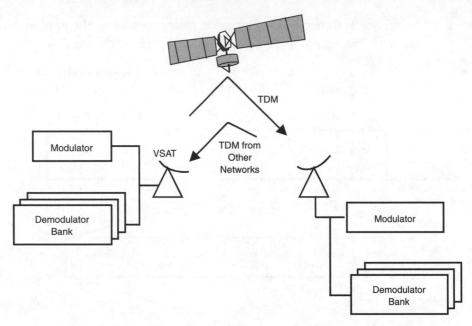

FIGURE 13.4 TDM-SCPC mesh architecture.

In TDM-SCPC, each remote is assigned a carrier frequency permanently, and the remote sends its data in TDM mode using broadcasting. Every remote listens to the broadcast data and decodes the data meant for it.

The TDM frame format is shown in Figure 13.5. The format of each slot is the same as that discussed in the previous section.

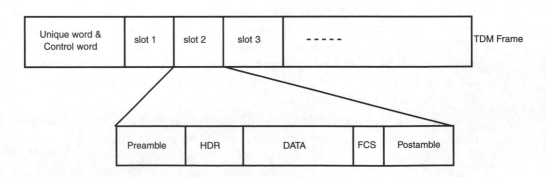

FIGURE 13.5 TDM-SCPC frame.

In TDM-SCPC-based networks, each remote will have one modulator to transmit using the carrier frequency assigned to it and also a bank of demodulators to demodulate the signals received from different remote stations.

NOTE

13.4.3 TDMA

In TDMA, all the remotes use the same frequency for transmission. As shown in Figure 13.6, at each remote there will be a burst modem. Each remote will transmit its data as a burst in the TDMA time slot assigned to it. The time slot allocation is done by the network control center.

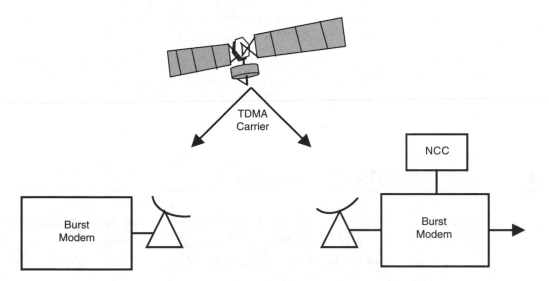

FIGURE 13.6 TDMA mesh architecture.

In TDMA-based networks, all remotes use the same frequency for transmission. Each remote will be assigned a time slot by the network control center. Using a burst modem, the remote sends its data in the time slot assigned to it.

The TDMA frame format is shown in Figure 13.7. Each time slot will have the same format as discussed earlier. Signaling from the NCC is sent in the control word. The remote obtains information about the slot allocation by analyzing the control field. Signaling from the remotes is sent in the header field. The NCC may also allocate a time slot to a free remote so that the remote can send its signaling information in the data portion of its slot.

TDM Frame

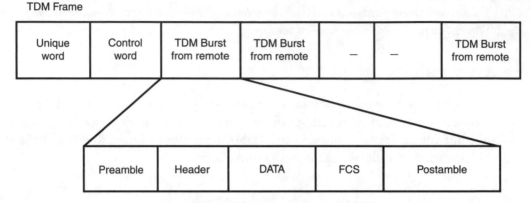

FIGURE 13.7 TDMA frame format.

All the access techniques we have mentioned are used in commercial systems. Because the hardware requirements vary for each type of access technique, the total system cost also varies. Based on cost considerations, a particular type of access mechanism is chosen.

13.5 A REPRESENTATIVE NETWORK

Using any of the multiple access techniques discussed, a variety of applications can be supported on the satellite network. Some typical applications are:

- Interconnecting the local area networks of different branch offices of an organization spread over large distances.

- Interconnecting the PBXs of different branch offices of an organization.

- Providing audio/video broadcasting from a central station with provision for audio/text interaction with users at the remote site.

A satellite-based network is the best choice to broadcast multimedia content to a large number of remote stations which are distributed geographically. From the hub, the information can be broadcast in TDM mode, and all the remotes will be in receive-mode. Low-speed channels are used from the remotes to the hub for interaction with the hub location.

Figure 13.8 shows the equipment configuration for providing distance education through satellite from a central location to a number of remotes located at different colleges/universities in a country. At the hub, there will be video transmission equipment. Additional infrastructure such as local area network, PBX, etc. also can be connected to the baseband interface equipment of the satellite hub. The baseband equipment provides the necessary interfaces to the video equipment, router connected to the LAN, PBX, etc. At each remote, there will be video reception equipment and also PBX and LAN. A lecture can be broadcast from the hub in the TDM mode. All the remotes can receive the lecture containing video, audio, and text. Whenever a user at a remote needs to interact with the professor at the hub, he can make a request, and a carrier is assigned to the remote in which he can send either text or voice.

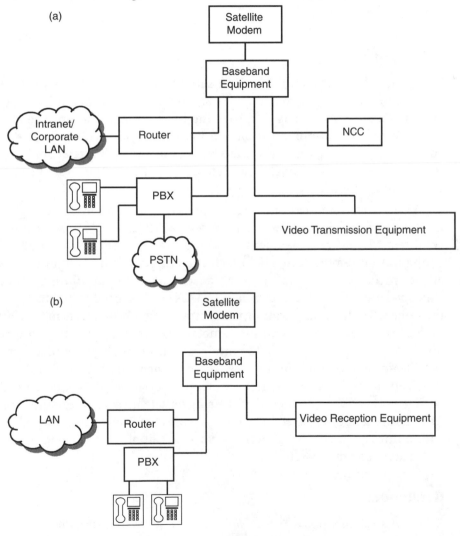

FIGURE 13.8 (a) Hub configuration for video conferencing and (b) remote configuration.

Systems with similar architecture also are used to provide telemedicine service to rural areas. The central location is connected to a hospital, and the remotes are located at remote locations. The patients can consult the doctor at the central location using video conferencing.

To conserve radio spectrum, multimedia communication over satellite networks use Voice/ Video over IP protocols in which voice and video are encoded using low-bit rate coding techniques.

Summary

Ever since the first communication satellite was launched in 1962, satellite communication systems have been used for broadcasting and tele-communications as well as for providing location information. The attractive feature of satellite communication is its insensitivity to distance. Hence, satellite communication is a very cost-effective way of providing telecommunication facilities to rural and remote areas. Satellite communication systems operate in C, Ku, and Ka bands.

Satellite communication systems operate in star and mesh configurations. In star configuration, there will be a central station (hub) and a number of remotes. All the remotes communicate via the central station. In mesh configuration, remotes can talk to each other directly. Star configuration is more attractive than mesh configuration because in star configuration, small antennas called very small aperture terminals (VSATs) can be used to reduce the cost of the remote. In star configuration, the central station broadcasts in time division multiplex (TDM) mode to all the remotes, and remotes can transmit in time division multiple access (TDMA) mode. The three multiple access techniques with which mesh configuration can be obtained are: DAMA-SCPC (demand assigned multiple access-single channel per carrier), TDM-SCPC, and TDMA. With the availability of onboard processing on satellites, satellite communication systems are now being used to provide mobile communications as well. Satellite communications can be effectively used for applications such as distance education and telemedicine.

References

R. Horak. *Communications Systems and Networks*. Wiley-Dreamtech India Pvt. Ltd., 2002.

www.intelsat.int Web site of Intelsat, which provides satellite services.
www.isro.org Web site of Indian Space Research Organization.

Questionss

1. List the various applications of satellites.
2. What are the problems associated with satellite communication?
3. What is the impact of high propagation delay on voice communication?
4. What is the impact of high propagation delay on data communication?
5. What are the frequency bands of operation for communication satellites?
6. Explain the various multiple access techniques used in satellite communication.

Exercises

1. Prepare a technical report on direct broadcast satellites. Direct broadcast satellites are used for transmitting TV programs directly to homes. These satellites operate in the 17/12GHz band.
2. Prepare a technical report on remote sensing satellites.
3. For supporting voice services on satellite-based networks, study the various voice encoding techniques used.
4. Carry out a paper design to develop a satellite-based video surveillance system. The system has to capture video data of a specific location and send it to an Earth station.
5. In a satellite network, the roundtrip delay is about 0.5 second. If the stop-and-wait protocol is used on such a network for data communication, study how effective the satellite channel utilization is.

Projects

1. Work out the details of the commercial equipment required to implement a satellite-based distance education system for which the architecture is explained in this chapter. You can obtain the details of the commercial products from the Web sites of equipment vendors such as Hughes Network Systems, Paragea, Scientific Atlanta, etc.

2. Design and develop a communication system used in surveillance satellites. The video signal has to be encoded at 384kbps data rate using a commercially available video codec and transmitted to the Earth station.

14 Optical Fiber Communication Systems

In This Chapter

- An Optical Fiber Communication System
- Wave Division Multiplexing Systems
- The Architecture of Optical Communication Networks

During the last two decades, there has been an exponential growth in optical fiber communication systems. Optical fiber has many attractive features: support for very high data rates, low transmission loss, and immunity to interference. Presently, optical fiber is used in the backbone network, but in the future, it can be extended to subscriber premises—offices and homes. This chapter gives an overview of optical fiber communication systems, with an emphasis on various network elements that make up the communication system.

14.1 EVOLUTION OF OPTICAL FIBER COMMUNICATION

The feasibility of transmitting information coded into light signals was demonstrated in the 1960s. However, development of pure glass to carry light signals was successfully achieved only in the 1970s.

In the 1960s, K.C. Kao and G.A. Hockham demonstrated the feasibility of transmitting information coded into light signals through a glass fiber. However, fabrication of pure glass that can carry the light signals without loss was successful only in the 1970s in Bell Laboratories. Subsequently, single-mode fiber was developed, which has less loss and support for higher data rates. The next milestone was the development of wave division multiplexing (WDM), which increased the capacity of a fiber significantly. Dense wave division multiplexing (DWDM) is the next evolutionary step which increased the capacity of fiber to terabits.

14.1.1 Multimode Optical Fiber Communication

In a multimode fiber optic system, there will be a light emitting diode or a laser at the transmitter and a photodetector at the receiver.

A communication system using the multimode optical fiber is shown in Figure 14.1. A light emitting diode (LED) or a semiconductor laser is the light source at the transmitter. LED is a low-power device, whereas laser is a high-power device. Laser is used for larger distances. To transmit 1, the LED is switched on for the duration of the pulse period; to transmit 0, the LED is switched off for the duration of the pulse period. The photodetector at the receiving end converts the light signal to an electrical signal. The 800 and 1300 nm bands are used for the transmission. (Note that in optical fiber communication, we will represent the bands in wavelength and not in Hertz). Data rates up to 140Mbps can be supported by multimode fiber systems.

Multimode optical fiber has a loss of about 0.5 dB/km. Hence, a regenerator is required every 10 km. The main attraction of multimode fiber is that it is of low cost, and hence it is used as the medium for communication over small distances.

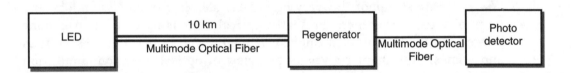

FIGURE 14.1 Multimode optical fiber communication system.

The multimode fiber has a loss of about 0.5 dB/km. If the distance between the transmitter and the receiver is more than 10 km, a regenerator is required. The regenerator converts the light signal into an electrical signal and then back to a light signal for pushing it on to the fiber. Regenerators are very costly, and proved to be very expensive if used for very large distance communication. In a multimode fiber, light travels in multiple propagation modes, and each mode travels at a different speed. The pulse received at the receiving end is distorted. This is called multimodal dispersion.

The main attraction of multimode fiber is that it is cheaper and hence a cost-effective way of providing communication for small distances.

14.1.2 Single-Mode Optical Fiber Communication

In single-mode optical fiber communication systems, transmission is done at 1300 and 1550 nm wavelengths. Single mode fiber is of low loss and hence regenerators are required only for every 40 km.

Single-mode optical fiber communication system is shown in Figure 14.2. The transmission is done at 1300 and 1550 nm wavelengths. Data rates up to 1 Gbps can be achieved using this system. The main advantage of the single-mode fiber is that the loss is less and hence regenerators are required only for every 40 km. A laser is used at the transmitting end and a photodetector at the receiving end. Developments in lasers and photo detectors as well as manufacturing of the pure glass are the main reason for the dramatic increase in data rate and decrease in loss.

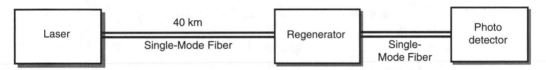

FIGURE 14.2 Single-mode optical fiber communication system.

14.1.3 Wave Division Multiplexing Systems

In wave division multiplexing, signals corresponding to different sources are transmitted at different wavelengths. Hence, the capacity of an already laid fiber can be increased significantly.

Wave division multiplexing (WDM) increases the capacity of an already laid optical fiber. As shown in Figure 14.3, signals corresponding to different sources are transmitted over the fiber at different wavelengths. At the receiving end, the demultiplexer separates these different wavelengths, and the original signals are obtained. Up to 16 or 32 wavelengths can be multiplexed together. The next development was dense wave division multiplexing in which 256 wavelengths can be multiplexed and sent on a single fiber. Systems that can support terabits per second data rates have been demonstrated.

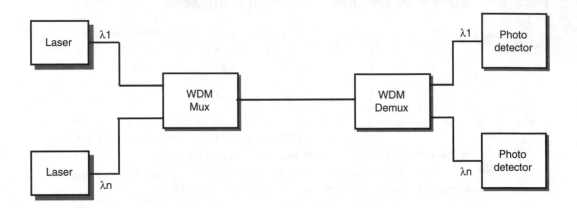

FIGURE 14.3 Wave Division Multiplexing.

The only problem with single-mode optical fiber is that regenerators are required every 40 km. The development of an optical amplifier eliminated the need for regenerators. The optical amplifier is a specially made fiber of about 10 meters. These amplifiers replaced the costly regenerators, resulting in tremendous cost savings for using optical fiber for very long distance communication.

Optical amplifiers eliminate the need for regenerators. The optical amplifier is a specially made fiber of about 10 meters.

Using these systems (Figure 14.4), 10Tbps data rates can be transmitted over a distance of a few hundred kilometers.

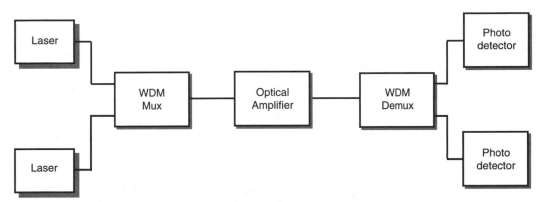

FIGURE 14.4 Optical communication with optical amplifier.

The various bands used in single-mode optical fiber are given in Table 14.1.

TABLE 14.1 Communication Bands Used in Single-Mode Fiber

Band	Wavelength Range (nm)
O band	1260 to 1360
E band	1360 to 1460
S band	1460 to 1530
C band	1530 to 1565
L band	1565 to 1625
U band	1625 to 1675

C band (1530 to 1565 nm) and L band (1565 to 1625 nm) are presently being used in single-mode optical fiber communication due to the availability of optical amplifiers.

Presently, C and L bands are being used due to the availability of the optical amplifiers. Developments in optical fiber manufacturing, Raman amplification, and so forth will lead to use of the other bands in the future.

14.2 OPTICAL NETWORKS

So far, we have discussed optical communication systems for point-to-point links. These systems are transmission systems that can carry optical signals at very high data rates over very large distances. As the optical communication technology matured, a number of proprietary networking solutions were developed. Subsequently, standardization activities resulted in a number of international standards to develop optical networks.

14.2.1 SONET/SDH

Synchronous optical network (SONET) is a standard for optical networking. SONET was developed by ANSI for North America. A slightly different standard used in Europe is called the synchronous digital hierarchy (SDH).

SONET (synchronous optical network) is a standard developed by American National Standards Institute (ANSI) for optical networking in North America. International Telecommunications Union Telecommunications Sector (ITU-T) developed a slightly different standard that is called synchronous digital hierarchy (SDH). SONET/SDH standards specify how to access single-mode optical fiber using standard interfaces and how to multiplex the digital signals using synchronous TDM. These standards specify the rate hierarchy and interfaces for data rates from 51.84Mbps to 39.813Gbps.

The signal hierarchy in SONET/SDH is shown in Table 14.2.

TABLE 14.2 SONET/SDH Signal Hierarchy

Optical Carrier (OC) Level (STM) Level	SDH Synchronous Transfer Model	Data Rate	No. of 64kbps Channels
OC-1		51.84Mbps	672
OC-2		103.68Mbps	1,344
OC-3	STM-1	155.52Mbps	2,016
OC-4	STM-3	207.36Mbps	2,688
OC-9	STM-3	466.56Mbps	6,048
OC-12	STM-4	622.08Mbps	8,064
OC-18	STM-6	933.12Mbps	12,096
OC-24	STM-8	1.24416Gbps	16,128
OC-36	STM-12	1.86624Gbps	24,192
OC-48	STM-16	2.48832Gbps	32,256
OC-96	STM-32	4.976Gbps	64,512
OC-192	STM-64	9.953Gbps	129,024
OC-768	STM-256	39.813Gbps	516,096

A SONET/SDH network operates in dual-ring topology. One fiber is used for transmitting in one direction and the other fiber for transmission in the opposite direction. This topology facilitates development of survivable networks—the network can survive even if one link fails.

A typical network based on SONET/SDH standards is shown in Figure 14.5. Though the network can operate in any topology such as star or mesh, dual-ring topology is the preferred choice. In dual-ring topology, there will be two fibers. One fiber will transmit in one direction and the other fiber in the opposite direction. The advantage of this topology is that even if one link fails, the communication does not fail. Such a topology leads to survivable networks—networks that can survive even if some links fail. As shown in the figure, there can be a backbone ring that operates say at OC-192/STM-64. The backbone network may interconnect major cities in a country. In each city, there can be a ring operating at different speeds from OC-3 to OC-12. The add-drop multiplexer (ADM) is used to insert the traffic channels into the SONET/SDH transmission pipe as well as to take out traffic channels from the pipe. DXCs (digital cross connects) connect two rings and also do the multiplexing/demultiplexing and switching functions. The line terminating equipment (LTE) provides user access to the network.

The ADM and the DXC are the two important network elements that provide the networking capability. The functioning of ADM and DXC can be understood through the analogy of a train. When a train stops at a railway station, some people get off and some people get on; the traffic is dropped and added. ADM does a similar function. The traffic for a particular place (in the form of TDM slots) can get dropped and added. The input to the ADM can be an E1 link with 30 voice channels. Some channels, say 10 voice channels, meant for a particular station can be dropped and another 10 voice channels can be added.

Continuing with our train analogy, a train may stop at a railway junction, and some coaches can be detached and some coaches can be attached to it. The detached coaches are connected to another train. The attached coaches would have come from some other train of a different route. The DXC performs a similar function. Traffic from different stations is multiplexed and demultiplexed, and also switching is done at a DXC.

The add-drop multiplexer (ADM) and the digital cross connect (DXC) are the two important network elements in an optical fiber network.

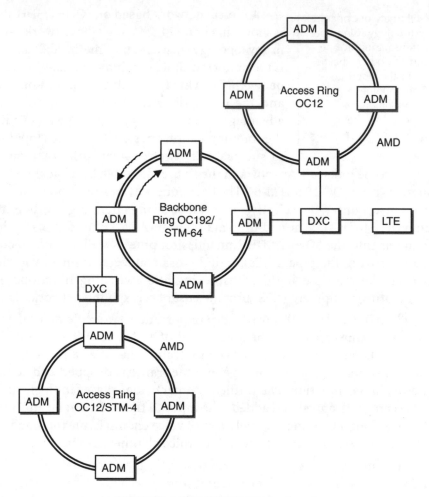

FIGURE 14.5 SONET/SDH network.

14.2.2 Optical Transport Networks

An optical transport network consists of a backbone WDM ring to which SONET/SDH ring or passive optical network (PON) can be connected. The network elements in this configuration are optical ADM and WDM cross connect.

Similar to a SONET/SDH network, a WDM network can be developed as shown in Figure 14.6. A WDM ring can act as a backbone network. The backbone ring can be connected to a SONET/SDH ring or to passive optical network (PON). The network elements in the WDM network will be OADM (optical ADM) and WXC (WDM cross connect), which do similar functions as ADM and DXC, but in the optical wavelengths. At each WXC, a SONET/SDH ring can be connected that serves different cities. Compared to the SONET/SDH network, this network offers much better capacity and flexibility. The PON can be any of the present optical networks based on star or ring topology.

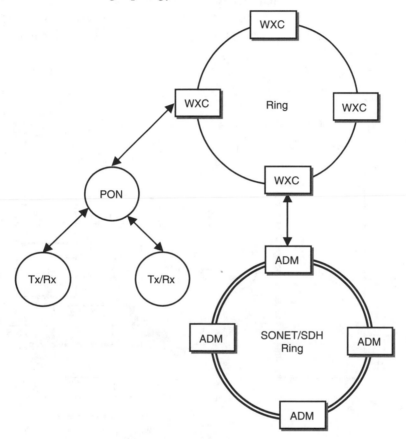

FIGURE 14.6 WDM network.

The network shown in Figure 14.6 has one problem that needs to be addressed. The data needs to be converted from electrical domain to optical domain as well as from optical domain to electrical domain. This calls for additional hardware. To avoid these conversions, all-optical networks are being developed. Research in this direction is very promising, and commercial products are in the works.

14.3 BROADBAND SERVICES TO THE HOME/OFFICE

To prsovide broadband services to the home, the office, and so on, the Full Services Access Networks (FSAN) group has been established by network operators to evolve optical access for providing a variety of voice/data/video services. The conceptual architecture for such an access network is shown in Figure 14.7. Work is in progress to define the various standards for deploying such a network.

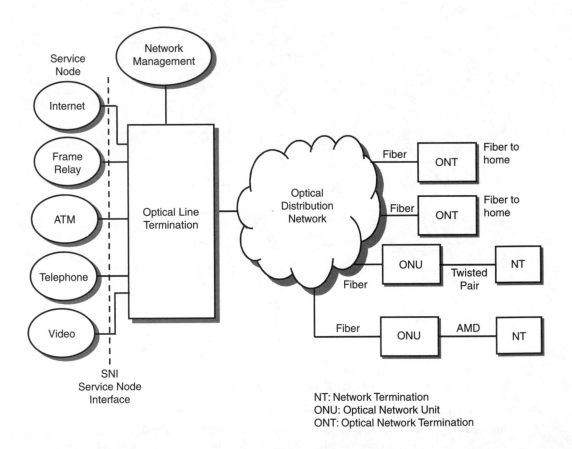

FIGURE 14.7 Fiber to home/office.

The architecture consists of an optical distribution system that provides the connectivity to homes/offices, etc. through standard interfaces. The distribution network is a collection of optical fibers and associated interface equipment. In the case of fiber to home or fiber to office, an optical network termination (ONT) will be used to which the end systems can be connected. In places where it is not feasible/economical to use optical fiber at the end, twisted pair or very high speed digital subscriber line (VDSL) can be used. The optical network unit (ONU) provides the interface between the network termination (NT) and the distribution network. The services supported can be based on a variety of networks such as Asynchronous Transfer Mode (ATM), Frame Relay, Internet, PSTN, and so on. The objective of FSAN is very clear: to develop a single worldwide broadband access system.

The Full Services Access Networks (FSAN) group developed a conceptual architecture to provide broadband multimedia services to the home/office using optical fiber as the main transmission medium.

 The objective of the FSAN architecture is to develop a single worldwide broadband access system. In addition to the optical network elements, even the non-optical fiber-based systems can be integrated into this architecture.

Summary

This chapter presented advances in optical communication systems. Initially, multimode optical fibers were used for transmission, which required regenerators every 10 kilometers. Because regenerators are very costly, multimode optical fiber is used for small distances. With the development of single-mode optical fiber, the data rates increased tremendously. The loss in single-mode fiber is also very low, and hence regenerators are required only for every 40 km. Wave division multiplexing (WDM) and dense WDM (DWDM) facilitate sending multiple wavelengths on the same fiber. Development of optical amplifiers resulted in a lot of cost savings. As a result of all these developments, optical fiber communication at terabits per second data rates is achieved. Standardization activities for optical networks resulted in synchronous optical network (SONET)/synchronous digital hierarchy (SDH) standards. These standards specify the multiplexing hierarchy for transmitting data at rates up to 39.813Gbps, referred to as OC-768. In SONET/SDH, a dual-ring topology is used that provides a very reliable communication infrastructure. A backbone ring can interconnect the major cities in a country, and smaller rings can cover each city. These rings can be interconnected through digital cross connect (DXC) equipment. Now work is in progress to develop standard interfaces to achieve broadband access to homes/offices using optical fiber communication.

References

R.Ramaswami. "Optical fiber communication: From Transmission to Networking." *IEEE Communications Magazine*, 50th Anniversary Issue, May 2002.

Y. Maeda et al. "FSAN OAN-WG and Future Issues for Broadband Optical Access Networks." *IEEE Communications Magazine*, Vol. 31, No. 12, December 2001.

www.fsanet.net Web site of Full Services Access Networks Group.

www.ansi.org Web site of American National Standards Institute.

Questions

1. What are the advantages and disadvantages of multimode optical fiber communication systems?
2. What is wave division multiplexing?
3. Explain the architecture of a SONET/SDH ring.
4. What is the function of add-drop multiplexer (ADM)?
5. Explain the conceptual architecture proposed by FSAN.

Exercises

1. Study the specifications of commercially available optical components such as lasers, LEDs, and photodetectors.
2. Study the recent developments in DWDM and the data rates achieved.
3. Prepare a technical report on Raman amplification.
4. Prepare a technical report on the various wavelength bands used in single-mode fiber and the issues involved in making use of these bands (availability of optical components, amplifiers, loss in the cable, etc.).

Projects

1. Carry out the paper design of a communication system that provides connectivity between all the county seats and state capital of your state. From the county seat to all the major towns in the county, connectivity should be provided. Suggest the design alternatives for mesh and star architecture. High-bandwidth services such as video conferencing, high-speed Internet access, and so forth should be provided.

2. An organization has five branches in a city. Each branch has to be provided with high-speed connectivity for interbranch communication. Propose a solution using fiber optic communication. Study the alternatives in using (a) ring, star, mesh topology; (b) single-mode fiber, multimode fiber; and (c) SONET, WDM. Choose the best alternative based on the following criteria: (a) cost; (b) future expandability in case more branches are opened later; and (c) reliability and availability of the network.

II Data Communication Protocols and Computer Networking

The twentieth century's two great gifts to humankind are the PC and the Internet. The PC has now become ubiquitous and is an integral part of the daily lives of most of us. The Internet, the network of computer networks spreading across the globe, is now making distance irrelevant. The Internet is the platform to access information that is available anywhere in the world through a click of the mouse. Unlike PSTN, the Internet is a recent phenomenon, just about three decades old. But then, during the past three decades, the field of computer networking has seen developments at a breath-taking pace.

For entities (people, computers, telephones, or any appliances) to communicate with each other, established procedures are mandatory. These established procedures or protocols, fundamental to networking, are discussed in detail in this part of the book. We will study the OSI reference model, the TCP/IP architecture, and the various protocols for distributed applications. The technologies and standards for local and wide area networks are also covered. We will study the developments in network computing and the exciting new applications using this technology such as application service provisioning, dynamic distributed systems, etc. We will also study Signaling System No. 7, Integrated Services Digital Network (ISDN), Frame Relay, and Asynchronous Transfer Mode (ATM) systems.

Computer networks are being used extensively not just for data, but for voice, fax and video communication as well. For designing any communication network, a good understanding of the protocols described in this part of the book is most important for every budding telecommunications professional. This part of the book has 15 chapters, which cover the details of the data communication protocols and representative computer networks.

15

Issues in Computer Networking

In This Chapter

- Packet Switching
- Virtual Circuit and Datagram Services
- Designing Computer Networks
- A Layered Approach to Protocol Development

With the widespread use of computers in all walks of life, the need arose for making the computers communicate with one another to share data and information. Circuit switching, which works well for voice communication, is inefficient for computer communication, and a radically new approach, called packet switching, was developed. In this chapter, we will study the concept of packet switching, which is fundamental to computer networks. We will also study the various services and applications supported on computer networks.

To make two computers talk to each other is a tough task. The two computers may be running different operating systems, using different data formats, have different speeds, and so on. We need to establish very detailed procedures to make computers exchange information. To tackle this big problem, we divide the problem into small portions and tackle each portion. This philosophy has led to the layered approach to protocol architecture—we will discuss the importance of the layered approach and protocol standards.

15.1 THE BEGINNING OF THE INTERNET

The threat of a war between the Soviet Union and the United States in the 1960s led to the development of the Internet. The U.S. Department of Defense wanted to develop a computer network that would continue to work even if some portions of the communication network were destroyed by the enemy.

In circuit switching, a connection is established between the two users, data is transferred, and then the circuit is disconnected. The telephone network uses circuit switching to establish voice calls between subscribers.

Consider a network of four computers shown in Figure 15.1. If we use the circuit switching operation for communication between any two computers, the procedure is:

1. Establish the connection
2. Transfer the data
3. Disconnection

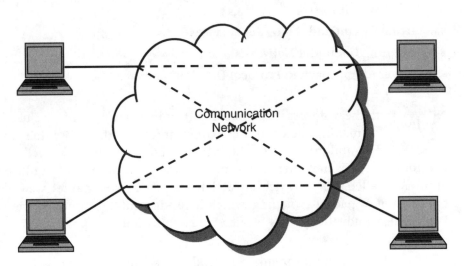

FIGURE 15.1 A computer network.

If the communication link is destroyed, then it is not possible to make the two computers exchange data. This is the inherent problem in a circuit switching operation. If there are alternate communication links, the computers can exchange data, provided the data is routed through the alternate paths, even without the user's knowledge. This led to the concept of packet switching, a revolutionary concept that is fundamental to data communications.

Instead of circuit switching, packet switching is used in computer communication. Packet switching has two main advantages: reliable communication even if some links are not working and call setup and disconnection can be eliminated.

Using the concept of packet switching, the TCP/IP (Transmission Control Protocol/Internet Protocol) suite was developed during 1969-70 when the first form of the Internet, known as DARPANET (Defense Advanced Research Projects Agency Network) was deployed, interconnecting a few sites in the U.S.

Packet switching has two main attractions—providing reliable communications even if some communication links fail and efficient usage of the communication links because no time is lost for call setup and disconnection. However, packet switching requires new protocols to be designed, and we will study these issues in detail in this chapter.

15.2 SERVICES AND APPLICATIONS

The need for a network of computers arises mainly to share the information present in different computers. The information can be transported electronically from one computer to another, obviating the need for physical transportation through floppies or CD-ROMs. In addition, resources can be shared—for example, a printer connected to a computer on the network can be accessed from other computers as well. The services that can be supported on the network are:

Electronic mail: This is the most popular service in computer networks. People can exchange mail and the mail can contain, in addition to the text, graphics, computer programs, and such.

File transfer: A file on one computer can be transferred to another computer electronically.

Remote login: A person at one computer can log in to another computer and access the programs or files on the other computer. For instance, a person using a PC can log in to a mainframe computer and execute the programs on the mainframe.

These are the basic services supported by computer networks. Using these services, many applications can be developed, such as bibliographic services to obtain literature on a specific topic, to search different computers on a network for specific information, to find out where the information is available and then obtain the information, etc.

The three basic services supported by computer networks are: email, file transfer and remote login.

15.3 PACKET SWITCHING CONCEPTS

Suppose you want to transfer a file from one computer to another. In packet switching, a file is divided into small units (for instance, 1,024 bytes) called packets, and each packet is sent over the transmission medium. At the receiving end, these packets will be put together, and the file is given to the user. The sender just gives a command to send a file, and the recipient receives the file—the underlying operation of packetization is transparent and is not known to the user.

In packet switching, the data is divided into small units called packets, and each packet is sent over the network. The packet switch analyzes the destination address in the packet and routes the packet toward the destination.

To transmit data using this concept, we need special equipment called *packet switches,* as shown in Figure 15.2. A packet switch has ports to receive packets from incoming lines and ports to send packets on outgoing lines. The packet switch receives each packet, analyzes the data fields in the packet to find the destination address, and puts the packet in the required outgoing line. This mechanism is known as *packet switching.* The packet switch should have buffers to hold the packets if they arrive at a rate faster than the rate at which the packets can be processed. When the packets arrive on the input ports, they are kept in buffers, each packet is analyzed by the packet switch (to find out the destination address), and based on the destination address, the packet will be sent through one of the outgoing ports. In other words, the packet is "routed" to the appropriate destination.

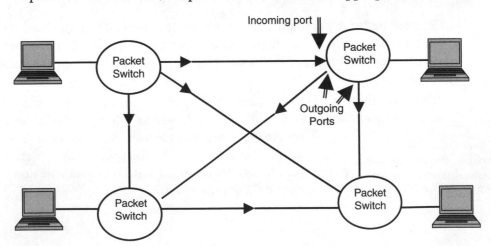

FIGURE 15.2 A packet switching network.

Good concept. The main advantage is that each packet can take a different route to reach the destination, and it is not necessary that all packets follow the same route. If one communication link fails, packets can take a different route; if one communication link has too much traffic, the packets can take a route with less traffic. What are the problems?

- Since each packet can travel independently, the packet has to carry the address of the destination.

- Since different packets can take different routes, they may not be received in sequence at the destination, and hence each packet has to carry a sequence number so that at the destination all the packets can be put in sequence.

- Due to transmission errors, some packets may be received with errors. Each packet must contain some error detection mechanism.

- Some packets may get lost due to intermittent problems in the medium or packet switches. There must be some mechanism for acknowledgements— the receiver has to inform the sender whether a particular packet is received correctly or request a retransmission of the packet. Each packet should also have the source address in addition to the destination address.

There are a few more problems, but these are the major problems. Is packet switching worth it, then? Well, yes, because it provides a very reliable mechanism for transfer of data provided we establish the necessary procedures. Also, we can avoid the call setup and call disconnection procedures, saving time.

For packet switching to work, we need a packet switch. The job of a packet switch is to get the packets from an incoming port, analyze the packet to see its destination address, and then put the packet in the outgoing port. In the packet switch, the incoming packet is kept in a buffer (in a queue). The software takes each packet, analyzes it, and keeps it in an outgoing buffer. This switching has to be done very fast. For fast packet switching, the size of the packet is an important issue.

A packet switch will have incoming ports and outgoing ports. The packet received on the incoming port will be analyzed by the packet switch for the destination address, and the packet is given to one of the outgoing ports towards the destination.

If the packet size is very large, the packet switch should have a large buffer. If the packet is small, for each packet there will be overhead (additional data to be inserted such as addresses, CRC, etc.). The size of the packet is a design parameter. Just to get an idea of the packet size used in practical networks, in Ethernet LANs, the maximum packet size is 1,526 bytes, and in X.25-based networks, it is 1,024 bytes.

Packet switching takes two forms: virtual circuit and datagram service.

The size of the packet is an important design parameter. If the packet is small, the overhead on each packet will be very high. On the other hand, if the packet is large, a high capacity buffer is required in the packet switch.

15.3.1 Virtual Circuit

When a computer (source node) has to transmit some information to another computer (destination node) over a packet network, a small packet, called the call setup packet, is sent by the source node to the destination node. The call setup packet will take a route to reach the destination node, and the destination can send a call accept packet. The route taken by these packets is subsequently used to transmit the data packets. A virtual circuit is established between the transmitting and receiving nodes to transfer the data. This is similar to circuit switching, and hence some time is lost in call setup and call disconnection. However, the advantage is that all the data packets are received at the destination in the same sequence, so reassembly is easier. The virtual circuit service (also known as *connection-oriented service*) concept is shown in Figure 15.3.

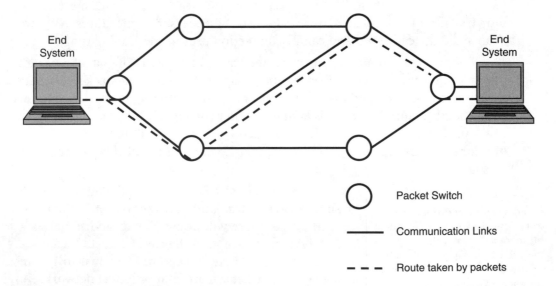

FIGURE 15.3 Virtual circuit (connection-oriented) service.

At the time of setting up the call, at each packet switch, the necessary buffers will be allocated, and the outgoing port will also be allocated. When a packet arrives, the packet switch knows what to do. If the packet switch cannot accept a call for any reason, it informs the sender so that the sender can find an alternative path.

A virtual circuit can be set up on a per-call basis or on a permanent basis. A virtual circuit set up on a per-call basis is called a switched virtual circuit (SVC), similar to the call set up in a PSTN when we call someone else's telephone. A virtual circuit set up on a permanent basis is called a permanent virtual circuit (PVC), which is similar to a leased line.

For creating a permanent virtual circuit (PVC), each packet switch on the route between the source and destination has to be programmed. The PVC is a concatenation of the routes between packet switches. A PVC is used when there is heavy traffic between two nodes so that no time is wasted in setting up a call for each data transfer session.

In virtual circuit service, also known as connection-oriented service, first a circuit is established between the source and destination, then data transfer takes place, and finally the circuit is disconnected. This is similar to circuit-switching operation.

X.25, Asynchronous Transfer Mode (ATM), and Frame Relay–based networks use the virtual circuit mechanism.

15.3.2 Datagram Service

In datagram service, there is no procedure for call setup (and hence for call disconnection as well). Each data packet (called a datagram) is handled independently: the data is divided into packets, and each packet can take its own route to reach the destination, as shown in Figure 15.4. However, in this scheme, the packets can reach the destination at different times and hence not in sequence. It is the responsibility of the destination to put the packets in order and check whether all the packets are received or whether some packets are lost. If packets are lost, this has to be communicated to the source with a request for retransmission. The advantage of datagram service is that there is no call setup.

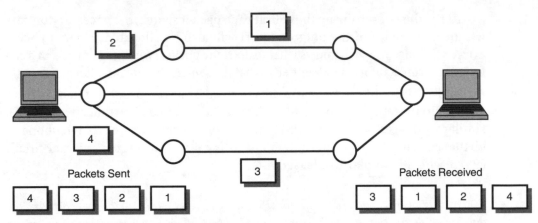

FIGURE 15.4 Datagram service.

In datagram service, each packet is handled independently by the packet switch. The advantage of datagram service is that there is no call setup and disconnection procedures, and hence it is much more efficient as compared to a virtual circuit.

For data communication (such as e-mail or file transfer), datagram service is quite efficient because the channel can be used efficiently.

15.3.3 Source Routing

A third mechanism used to send packets from the source to destination is source routing. In this mechanism, the source decides which route the packet has to take. This information is also included in the packet. Each packet switch receives the packet and then forwards it to the next packet switch based on the information available in the packet. For source routing to work, the source should know the complete topology of the network (how many packet switches are there, how each packet switch is connected to other packet switches, and so on). Hence, this mechanism is rarely used.

In source routing, the source specifies the route to be taken by the packets. However, this is difficult to implement because the topology of the network needs to be known by the source.

15.4 ISSUES IN COMPUTER NETWORKING

To design and develop computer networks is a challenging task because there are many issues to be considered. Here are some of the major issues:

Geographic area: The area to be covered by the network is of foremost importance. The area of coverage can be a single floor in a building, multiple floors in a building, a large campus with different buildings, an entire city with offices in different locations, a country with offices in different cities, or the world with offices in different countries. Based on the area of coverage, the communication media is chosen. A typical corporate network can use different media—fiber for LAN, copper wire or terrestrial radio for MAN (metropolitan area network), fiber or satellite for WAN (wide area network), and so on. In addition to the geographic coverage, what services are to be supported also determines the type of medium—voice and video communication require larger bandwidths.

A wide area network can use a combination of different transmission media—such as satellite radio, coaxial cable and optical fiber. Generally, the speeds of wide area networks are much lower compared to the speeds of local area networks.

Services to be supported: Nowadays, computer networks need to support not just data, but voice, fax, and video services as well. This requirement needs to keep in mind the available bandwidths and cost considerations. The higher the bandwidth requirement, the higher the cost. For different types of services to be supported, different application-level protocols are required.

Though in the initial days of computer networks, mostly data services were supported, nowadays voice and video services are also becoming predominant. Real-time communication of voice and video over computer networks requires additional protocols.

Security: To ensure that the network provides secure communication is of paramount importance. To ensure secrecy, necessary protocols need to be implemented. To support applications such as e-commerce, security is a must.

Different computing platforms: Computers have a wide variety of processors and operating systems. Because of these differences, filesystems will be different, data formats will be different, and filename conventions will be different (for instance, Unix versus Windows 9x). The protocols need to be designed in such a way that computers with different operating systems and data formats can communicate with one another.

Error control: The transmission media introduce errors during the transmission. Also, because of congestion in the networks, some packets may be lost. The protocols need to take care of the errors. Error detection is the first step: the receiver has to check whether the data that has been transmitted is received correctly using some additional bits called CRC (cyclic redundancy check). If the data is incorrect, the receiver has to ask for retransmission of the packet. Alternatively, error-correcting codes can be used to correct a few errors at the receiver without asking for retransmission.

If errors are detected at the receiving end, the receiver has to ask for retransmission of packets. If the transmission medium is not very reliable, this results in a lot of retransmissions, causing waste of bandwidth. To reduce retransmissions, error-correction is a good approach and is generally used in satellite communication systems.

Flow control: As computers on a network will be of different capacities in terms of processing power, memory, and so forth, it may happen that one computer may not be able to receive packets with the same speed with which the other computer is sending. Protocols should be designed to control the flow of packets—the receiver may have to inform the sender "hold on, do not send any more packets until I ask you to." This mechanism is handled through flow control protocol. Flow control poses special problems in high-delay networks, such as satellite networks where the roundtrip delay is quite high.

Addressing: When two computers in a network have to share information, the sender has to specify to whom the packets are addressed. The receiver has to know from where the packets have come. So, addressing needs to be handled by the protocols.

The global Internet solves the addressing problem by assigning a unique address to every machine connected to it. This address is known as the IP address.

Type of communication: Generally, two computers need to talk to each other. Cases also arise when a computer has to broadcast a packet to all the computers on a network. Also, there may be a packet being sent from one computer to a selected number of computers (for video conferencing between computers). This type of communication can be point-to-point, broadcasting, or multicasting. The protocols must have the capability to take care of these different types of communications.

Multicasting is a requirement in applications such as video conferencing. For instance, if five persons situated at five different locations want to participate in a video conference, the data from each location needs to be sent to the other four locations.

Signaling packets: Before the actual data transfer takes place, it may be necessary to set up the call (as in the case of virtual circuit service) and disconnect the call. Protocols need to be designed for this to happen through transmission of special packets.

Congestion control: The packet switch will have a limited buffer for queuing the incoming packets. If the queue is full and the packet switch cannot take any more packets, then it results in congestion. Protocols need to take care of congestion to increase the throughput and reduce the delay. Two alternatives can be implemented: (a) send a control packet from the congested node to the other nodes informing them to reduce the number of packets, or (b) inform the other nodes to follow different routes. The strategies followed for congestion control are similar to the strategies adopted by traffic police at traffic islands.

In computer networks, it is difficult to predict the traffic. Suddenly, there may be heavy traffic as a result of which some packet switches cannot handle incoming packets at all. In such cases, the packets might be discarded.

In developing computer networks using packet switching, a number of issues need to be considered. These include the geographical area to be covered, services to be supported, security, different computing platforms, mechanisms to control errors and the different speeds of the computers, addressing, support for real-time communication services such as voice/video, whether networking of networks is required, etc.

Internetworking: Based on the need, different networks need to be connected together, such as a LAN and a WAN. The protocols used in the two networks are likely to be different, and to inter-work, there must be some machines that do the protocol conversion. This protocol conversion is achieved by entities known as *routers* or *gateways*.

Segmentation and reassembly: When two networks are interconnected, there is a possibility that the packet sizes supported by the two networks are different. There must be some designated machines that will take care of the differences in packet size. The large packets have to be broken down into smaller packets (segmented) and later put together (reassembled).

Real-time communication: For data communication (such as e-mail and file transfer), there may not be a strict limit on the delay with which a packet has to be received at the receiver. However, for real-time communication such as fax, voice, or video over packet networks, the delay is an important parameter. For instance, once the speech starts being played at the receiving end, all the packets should be received with constant delay; otherwise there will be breaks in the speech. Special protocols need to be designed to handle real-time communication.

Network management: To ensure that the computer networks operate as per the user requirements, to detect faults, to analyze traffic on the network, to rectify problems if any, etc., we need to design protocols whereby the networks can be managed well. Network management protocols take care of these issues.

These are just a few issues in computer communication, we will discuss how these issues are resolved for developing user-friendly computer networks, throughout this book.

15.5 NEED FOR LAYERED APPROACH AND PROTOCOL STANDARDS

A protocol can be defined as a set of conventions between two entities for communication. As discussed in the previous section, many protocols are required in computer communication to tackle different issues. One way of achieving computer communication is to write monolithic software for all the protocols to be implemented. This approach, being not modular, leads to lots of problems in debugging while developing the software and also in maintenance. On the other hand, a "layered approach" leads to modularity of the software. In a layered approach, each layer is used only for some specific protocols. Layered approach has many advantages:

- Every layer will perform well-defined, specific functions.
- Due to changes in the standards or technology, if there are modifications in one layer's functionality or implementation, the other layers are not affected and hence changes are easier to handle.
- If necessary, a layer can be divided into sub-layers for handling different functions (as in the case of LAN).
- If necessary, a layer can be eliminated or by-passed.

- If the protocols for different layers are based on international standards, software or hardware can be procured from different vendors. This multi-vendor approach has a major advantage—because of the competition among the vendors, the prices will be competitive and in the bargain, the end user will be benefited.

In Local Area Networks, the datalink layer is divided into two sub-layers viz., logical link sub-layer and medium access control sub-layer. A LAN need not have the network layer at all. This type of flexibility is provided by the layered approach to protocol development.

However, while deciding on the number of layers, the following points need to be kept in mind:

- If the number of layers is high, there will be too much protocol overhead. Hence, the number of layers should be optimized.
- The interfaces between two adjacent layers should be minimal so that when a layer's software/hardware is modified, the impact on the adjacent layers is minimal.

For each layer, as shown in Figure 15.5, there will be: (a) service definition that specifies the functions of the layer (b) protocol specification that specifies the precise syntax and semantics for interoperability, and (c) addressing or service access point to interface with the higher layer. These three form the specifications of the protocol layer. International bodies such as International Organization for Standardization (ISO) and Internet Engineering Task Force (IETF) standardized these specifications so that any equipment vendor or software developer can develop networking hardware/software that will interoperate.

The layered approach to protocol development is a very important concept. Each layer does a specific job and interfaces with the layers above and below it. This results in modular development of protocols for computer communication.

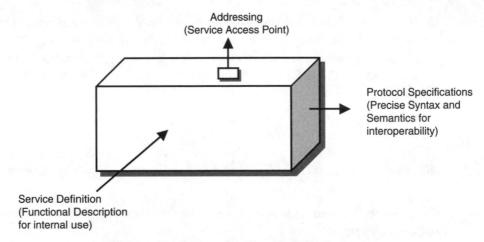

FIGURE 15.5 Specification of the protocol layer.

To make two computers talk to each other, we run the layered software on both computers. As shown in Figure 15.6, each layer interacts with the layer above it and the layer below it. It also communicates with the peer layer in the other machine. Each layers provides defined services to the layers above and below it.

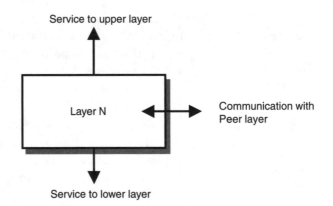

FIGURE 15.6 Layer's services and protocols.

The two important layered architectures for computer communication are: ISO/OSI architecture in which there are 7 layers and TCP/IP architecture in which there are 5 layers.

For computer communication, the layered approach has been well accepted, and the ISO/OSI protocol suite and the TCP/IP protocol suite follow this approach. The ISO/OSI protocol suite is a seven-layer architecture, whereas TCP/IP is a five-layer architecture. The beauty of the layered architecture will be evident when we study the ISO/OSI protocol architecture in the next chapter.

Summary

In this chapter, we studied the important concepts of packet switching. Packet switching is fundamental to computer networking. In packet switching, the data to be sent is divided into small packets, and each packet is transmitted over the network. In a virtual circuit, the route to be taken by each packet is determined before the data transfer takes place. In a virtual circuit, there will be three phases—call setup, data transfer, and call disconnection. ATM, Frame Relay, and X.25 based networks are based on a virtual circuit. In datagram service, each packet is handled independently, and hence, there is no call setup and call disconnection. The Internet uses datagram service.

We also studied the various issues involved in developing computer networks. Coverage area, services to be provided to end users, computing platforms (hardware and operating systems), error control, flow control, addressing, signaling, networking of networks, real-time communication, segmentation and reassembly, congestion control, and network management are the main issues to be handled. To address all these issues, protocols need to be developed for making computers talk to each other. We introduced the concept of the layered approach to computer networking. Instead of handling all the issues together, each layer can handle specific functions so that the software/hardware for making computers talk to each other can be developed in a modular fashion.

References

50th Anniversary Commemorative Issue of IEEE Communications Magazine, May 2002. This issue contains excellent articles on the history of communications as well as the papers that have influenced communications technology developments during the past 50 years.

Larry L. Peterson and Bruce S. Davie. *Computer Networks: A Systems Approach*. Morgan Kaufmann Publishers Inc., 2000. A systems approach rather than a layered approach to computer networking makes this book very interesting.

www.acm.org The Web site of the Association for Computing Machinery (ACM). If you are a member of ACM, you can access the online education portal, which gives excellent tutorials on different aspects of computing, including computer networks.

www.computer.org The Web site of the IEEE Computer Society. If you are a member of the IEEE Computer Society, you will have access to excellent online educational material on computer networking.

www.ieee.org The Web site of IEEE, the largest professional body of electrical and electronic engineers. This site provides access to many standards developed by IEEE on computer networks.

Questions

1. What is packet switching? As compared to circuit switching, what are its advantages and disadvantages?

2. Explain virtual circuit service and datagram service. Compare the two services in terms of quality of service provided, reliability of service provided, and implementation complexity.

3. If you connect two PCs using a point-to-point link, what are the issues involved in providing various applications such as file transfer, chat, and e-mail?

4. When you connect three or more PCs in a network (as a local area network), list the additional issues involved for providing the applications in Question 3.

5. What is the fundamental concept in a layered approach to computer networking? Discuss the pros and cons of developing monolithic software for each application (for file transfer, e-mail, etc.).

Exercises

1. On the local area network installed in your department/organization, find out the address of the computer on which you are working (it is called the IP address).

2. When you access a Web site through a browser, you get a message "Connecting to" followed by the IP address of the Web server. What is the use of this address to you? Is it necessary that this address be displayed at all? Discuss.

3. The Ethernet LAN uses a maximum packet size of 1526 bytes. An X.25 network uses a maximum packet size of 1024 bytes. In both cases, the packet size is variable. Discuss the pros and cons of having variable size packets as compared to fixed size packets.

4. ATM uses a fixed packet size of 53 bytes. This is a small packet as compared to Ethernet or X.25 packet sizes. Discuss the merits and disadvantages of having small fixed size packets.

Projects

1. Install Microsoft NetMeeting software on two systems connected over a LAN in your department/laboratory. Run the NetMeeting application and try different services provided by this application (audio conferencing, video conferencing, white board, etc.)

2. Search the Internet to find various sites that provide free networking software source code with which you can experiment. (You can try the open source sites that provide Linux-based networking software.)

16 ISO/OSI Protocol Architecture

In This Chapter

- The ISO/OSI Protocol Architecture
- The Functions of the Seven Layers
- Peer-to-Peer Communication
- The Layered Approach to Protocol Development

As we saw in the previous chapter, to make two computers talk to each other, protocols play an important role. How do we tackle the big problem of taking care of so many issues involved in networking computers— the speeds, formats, interfaces, applications, etc.? The famous riddle provides the solution: How do you eat an elephant? One bite at a time!

When the problem is big, divide it into smaller problems and solve each problem. The OSI protocol architecture does precisely that. A layered approach is followed whereby the problem is divided into seven layers. Each layer does a specific job. This seven-layer architecture, developed by the International Organization for Standardization (ISO), is the topic of this chapter.

16.1 OSI REFERENCE MODEL

The Open Systems Interconnection (OSI) protocol architecture developed by the International Organization for Standardization (ISO) is a seven-layer architecture. This architecture is defined in ITU-T Recommendation X.200.

The International Organization for Standardization (ISO) has developed the Open Systems Interconnection (OSI) protocol architecture, which is a seven-layer architecture for computer communications. This standard is specified in ISO 7498 and ITU-T X.200 recommendations.

During the late 1980s, a number of vendors started marketing software based on ISO/OSI protocol suite. However, by that time, there already was a large number of TCP/IP-based networks, and in the race between ISO/OSI and TCP/IP, TCP/IP won without much effort.

The ISO/OSI model (also referred to as the ISO/OSI protocol suite) still is an important model to study because it is considered as the reference model for computer communications. We can map any protocol suite on to the ISO/OSI model for studying the functionality of the layers.

In the battle between ISO/OSI protocol architecture and TCP/IP protocol architecture, TCP/IP won mainly because of the large installation base of the TCP/IP protocol software. Still, studying the ISO/OSI protocol architecture is important because it is acknowledged as the reference model for computer communication protocols.

The ISO/OSI protocol suite is shown in Figure 16.1. The seven layers are (from the bottom)

- Physical layer
- Datalink layer
- Network layer
- Transport layer
- Session layer
- Presentation layer
- Application layer

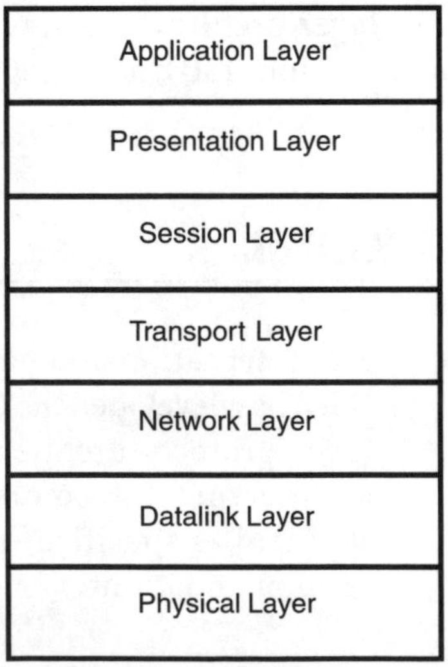

Application Layer
Presentation Layer
Session Layer
Transport Layer
Network Layer
Datalink Layer
Physical Layer

FIGURE 16.1 ISO/OSI Protocol Suite.

What is the rationale for seven layers? Why not six layers or eight layers? During the standardization process, there were two proposals—one proposal with six layers, and the other with eight layers. To achieve consensus, finally the seven-layer architecture was standardized—just the average of six and eight. That is how standardization is done!

Each protocol layer adds a header and passes the packet to the layer below. Because the header is interpreted only by the corresponding layer in the receiving system, the communication is called peer-to-peer communication. Peer means a layer at the same level.

Each layer performs a specific set of functions. If we consider two systems, the protocol stack has to run on each system to exchange useful information for a particular application. As shown in Figure 16.2, the two application programs on the two end systems communicate with each other via the protocol suite. The application program on one end system sends the data to the layer below (application layer), which in turn adds its header information and sends it to the layer below (presentation layer). Each layer adds the header and forwards to the layer below. Finally, the physical layer sends the data over the communication medium in the form of bits. This bit stream is received by the other system, and each layer strips off the header and passes the remaining data to the layer above it. The header information of one layer is interpreted by the corresponding layer in the other system. This is known as *peer-to-peer communication*. For instance, the header added by the transport layer is interpreted by the transport layer of the other system. So, though the two peer layers do not communicate with each other directly, the header can be interpreted only by the peer layer. We will study the functionality of each layer in the following sections.

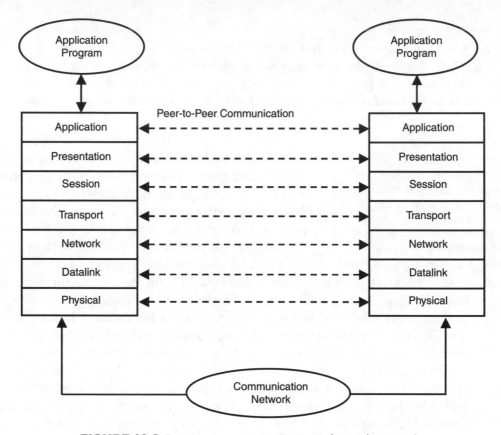

FIGURE 16.2 Peer-to-peer communication in layered approach.

16.2 PHYSICAL LAYER

The physical layer specifies the physical interface between devices. This layer describes the mechanical, electrical, functional, and procedural characteristics of the interface.

An example of the physical layer is Electronic Industries Association (EIA) RS232, which specifies the serial communication interface. We connect a modem to the PC through the RS232 interface. The modem is referred to as DCE (data circuit terminating equipment) and the PC as DTE (data terminal equipment). RS232 specifications are briefly described below to give an idea of the detail with which each layer in the protocol architecture is specified.

The physical layer specifies the mechanical, electrical, functional, and procedural characteristics of the interface. RS232 is an example of physical layer specifications.

Mechanical specifications of RS232 specify that a 25-pin connector should be used with details of pin assignments. Pin 1 is for shield, pin 2 is for transmit data, pin 3 for receive data, pin 4 for request to send, pin 5 for clear to send, pin 6 for DCE ready, pin 7 for signal ground, and so on.

Electrical specifications of RS232 give the details of voltage levels for transmitting binary one and zero, data rates, and so on. With respect to ground, a voltage more negative than –3 volts is interpreted as binary one and a voltage more positive than +3 volts is interpreted as binary zero. The data rate supported is less than 20kbps for a distance less than 15 meters. With good hardware design, higher data rate and higher distance can be supported; these are the minimum requirements given in the standard for data rate and distance.

Functional specifications of RS232 give the details of the data signals, control signals, and timing signals and how to carry out loop-back testing.

Procedural specifications of RS232 give the details of sequence of operations to be carried out for specific applications. For instance, if a PC is connected to the modem using RS232, the procedure is as follows:

- When modem (DCE) is ready, it gives the DCE ready signal.
- When PC (DTE) is ready to send data, it gives the request to send signal.
- DCE gives the clear to send signal indicating that data can be sent.
- DTE sends the data on the transmit data line.

A detailed procedure is specified to establish the connection, transfer the data, and then to disconnect. The procedural specifications also take care of the possible problems encountered during the operation such as modem failure, PC failure, link failure, and so forth.

When you connect two PCs using an RS232 cable, you need to specify the communication parameters. These parameters are the speed (300bps, 600bps, 19.2kbps, etc.), the number of data bits (seven or eight), the number of stop bits (one or two), and the parity (even or odd). Only when the same parameters are set on both the PCs can you establish the communication link and transfer the data.

Every computer has an RS232 interface. The modem is connected to the computer through an RS232 interface. The computer is the DTE, and the modem is the DCE. The protocol described here is used for communication between the computer and the modem.

16.3 DATALINK LAYER

The datalink layer's job is to activate the link, maintain the link for data transfer and deactivate the link after the data transfer is complete. Error detection and control, and flow control are also done by the datalink layer.

The physical layer's job is to push the bit stream through a defined interface such as RS232. Before pushing the data, the link has to be activated, and it has to be maintained during the data transfer. After data transfer is complete, the link has to be disconnected. It is possible that one computer may be sending the data very fast, and the receiving computer may not be able to absorb the data at that rate because of its slow processing power. In such a case, the receiving computer has to inform the sending computer to stop transmission for some time and then resume it again. This mechanism is known as *flow control*, which is the job of the *datalink* layer. The *datalink* layer's other function is to ensure error control—this may be in the form of sending acknowledgements or CRC. So, the functions of the data link layer are:

- Activate the link, maintain the link, and deactivate the link
- Error detection and control
- Flow control

Some standard datalink layer protocols are high level datalink control (HDLC), link access protocol balanced (LAPB) used in X.25 networks, and link access protocol D (LAPD) used in ISDN.

In some communication protocols, error detection and flow control are handled by the transport layer.

16.3.1 High-Level Datalink Control (HDLC)

HDLC is the most widely used datalink layer protocol. Many other datalink layer protocols such as LAPB and LAPD are derived from HDLC. We will briefly describe the HDLC protocol here.

Consider a simple case of two PC's that would like to communicate over an RS232 link. To exchange meaningful data, one of the PCs can take the responsibility of controlling the operations, and the other can operate under the control of the first one. One system can act as a primary node and issue commands; the other acts as a secondary node and gives responses to the commands. Alternatively, both machines can give commands and responses, in which case the node is called the combined node.

Depending on the type of node, the link can be of: (a) unbalanced configuration in which there will be one primary node and one or more secondary nodes; or (b) balanced configuration in which there will be two combined nodes.

The data transfer can be in one of the following three modes:

- Normal response mode (NRM): This mode is used in unbalanced configuration. The primary node will initiate the data transfer, but the secondary node can send data only on command from the primary node. NRM is used for communication between a host computer and the terminals connected to it.

- Asynchronous balanced mode (ABM): This mode is used with balanced configuration. A combined node can initiate transmission. ABM is used extensively for point-to-point full-duplex communication.

- Asynchronous response mode (ARM): This mode is used with unbalanced configuration. The primary node will have the responsibility to initiate the link, error recovery, and logical disconnection, but the secondary node may initiate data transmission without permission from the primary. ARM is rarely used.

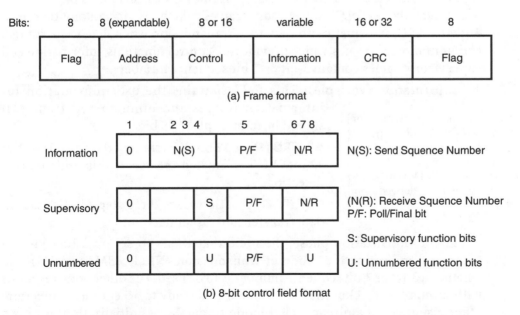

FIGURE 16.3 HDLC frame format: (a) information frame format, (b) 8-bit control field format

The HDLC frame structure is shown in Figure 16.3. The details of the frame structure are as follows:

Flag (8 bits): The flag will have the pattern 01111110. This is to indicate the beginning of the frame. The receiver continuously looks for this pattern to find the beginning of the frame. The frame also ends with the flag so that the receiver can detect the end of frame as well. The actual data may contain the same bit pattern, in which case the frame synchronization is lost. To overcome this, a procedure known as bit stuffing is used—when six ones appear continuously in the data, after five ones, a zero is inserted. At the receiver, after the starting flag is detected, if five ones are detected followed by a zero, that zero is removed. If six ones are detected followed by one zero, then it is taken as the ending flag, and the frame is processed further. However, note that bit stuffing is not foolproof if there are bit errors.

Address (8 bits): This field identifies the secondary node that has to receive the frame. Though an address is not necessary for point-to-point connections, it is included for uniformity. Normally, the address is of 8 bits in length, though higher lengths can be used. The address with all ones is used as the broadcast address.

Control (8 or 16 bits): The control field specifies the type of frame. Three types of frames are defined: information frames which carry data of the HDLC user (layer above HDLC); supervisory frames, which provide automatic repeat request (ARQ) information to request retransmission when data is corrupted; and unnumbered frames, which provide link control functions. Information and supervisory frames contain 3-bit or 7-bit sequence numbers.

Information (variable): This field contains the user information for information frames and unnumbered frames. It should be in multiples of 8 bits.

HDLC protocol is an example of datalink layer protocol. Many datalink layer protocols used in X.25, Frame Relay, and Integrated Services Digital Network (ISDN) are derived from HDLC.

CRC (16 or 32 bits): Excluding the flags, 16-bit or 32-bit CRC is calculated and kept in this field for error detection.

HDLC operation takes three steps: initialization, data transfer, and disconnection.

In the initialization phase, the node that wants to set up a link gives a command indicating the mode of transmission (NRM, ABM, or ARM) and whether a 3-bit or 7-bit sequence number is to be used. The other node responds with unnumbered acknowledged frame if it is ready to accept the connection; otherwise it sends a disconnected mode frame. If the initialization has been accepted, a logical connection is established between the two nodes.

In the data transfer mode, data is exchanged between the two nodes using the information frames starting with sequence number 0. To ensure proper flow control and error control, the N(S) and N(R) fields in the control field are used. N(S) is the sequence number and N(R) is the acknowledgement for the information frames received (refer to Figure 16.3b for the formats of the control field.)

Supervisory frames are also used for error control and flow control. The receive ready (RR) frame is used to indicate the last information frame received. Hence, the sender knows that up to that frame, all frames have been received intact. RR frame is used when there is no information frame containing the acknowledgement to be sent. Receive-not-ready (RNR) frame is sent to acknowledge the receipt of frames up to a particular sequence number and also to tell the sender to suspend further transmission. When the node that sent RNR is ready again, it sends an RR frame.

The disconnect phase is initiated by one of the nodes sending a disconnect (DISC) frame. The other node has to respond with an unnumbered acknowledged frame to complete the disconnection phase.

The HDLC protocol is the basis for a number of datalink layer protocols to be discussed in subsequent chapters. Note that in the datalink layer, a number of bytes are put together and a new entity is formed, which is referred to as a frame.

16.4 NETWORK LAYER

The functions of the network layer are switching and routing of packets. Additional functionality such as management of multiple links and assigning priority to packets is also done at this layer.

The important function of the network layer is to relieve the higher layers of the need to know anything about the underlying transmission and switching technologies. Network layer protocol provides for transfer of information between the end systems. The functions of the network layer are:

- Switching and routing of packets
- Management of multiple datalinks
- Negotiating with the network for priority and destination address

16.5 TRANSPORT LAYER

The transport layer provides end-to-end reliable data transfer between two end systems. This layer ensures that all packets are received without error and the packets are received in sequence.

The transport layer can provide either connection-oriented service or connectionless service. TCP provides connection-oriented service and UDP provides connectionless service. Note that only TCP provides reliable data transfer.

The transport layer can provide two types of services—connection-oriented and connectionless. In connection-oriented service, a connection is established between the two end systems before the transfer of data. The transport layer functionality is to ensure that data is received error-free, packets are received in sequence, and that there is no duplication of packets. The transport layer also has to ensure that the required quality of service is maintained. Quality of service can be specified in terms of bit error rate or delay. In connectionless service, the packets are transported without any guarantee of their receipt at the other end.

In TCP/IP networks, TCP is an example of a connection-oriented transport layer, and UDP is an example of a connectionless transport layer.

Quality of service parameters such as delay and throughput are important for some applications. For example, in voice communication, there should not be variation in delay in receipt of packets. The transport layer provides the facility to specify quality of service parameters.

16.6 SESSION LAYER

The functionality of session layer is to provide recovery mechanism in case of intermittent failure of links and to negotiate whether the communication should be full-duplex or half-duplex.

The session layer specifies the mechanism for controlling the dialogue in the end systems. Session layer functionality is as follows:

- Dialogue discipline: whether the communication should be full duplex or half duplex.
- Grouping: to group data into logical units.
- Recovery: a mechanism for recovery in case of intermittent failures of the links.

Because it does not have much work to do, in many networks, there is no session layer.

In many practical networks, the functionality of the session layer is taken by other layers, and the session layer is just a null layer.

16.7 PRESENTATION LAYER

The functions of the presentation layer are negotiation of data formats, data encryption and data compression.

In computer networks, computers may be running different operating systems having varying file systems and file storage formats. For example, the file formats of Windows and Unix operating systems are different. Some computers may be using ASCII for character representation, and some computers may be using EBCDIC. Before exchanging the data for a file transfer, for instance, the two systems have to negotiate and agree on the format to be used for exchanging the data. This is done at the presentation layer. To provide secure communication, the data may need to be encrypted, which is also a part of this layer's job. Compression of the data for faster data transfer is also done in this layer. Hence, the functions of the presentation layer are:

- Provide for selection of data formats to be exchanged between applications
- Encryption
- Data compression

16.8 APPLICATION LAYER

The application layer provides management functions to support distributed applications such as e-mail, file transfer, remote login, and the World Wide Web. With the widespread use of TCP/IP-based networks, applications using the ISO/OSI protocol suite are very limited. Some of the prominent application layer protocols are these:

- FTAM (file transfer, access and management) for file transfer applications.

The application layer provides the functionality required for specific applications such as e-mail, file transfer and remote login. A separate protocol is defined for each application. Examples are X.400 for e-mail, FTAM for file transfer and X.500 for directory services.

- ■ X.400 MHS (message handling systems) for electronic mail.
- ■ X.500 for directory services.

The ISO/OSI reference model is fundamental to computer communication because it serves as a reference model for all protocol suites.

The application layer protocols of OSI are not widely used in practical networks. However, some application layer protocols used on the Internet are derived from the OSI application layer protocols. For example, Lightweight Directory Access protocol (LDAP) is derived from X.500.

Summary

The ISO/OSI protocol architecture is presented in this chapter. This layered model has seven layers: physical layer, datalink layer, network layer, transport layer, session layer, presentation layer, and application layer. Each layer has a specific functionality. The physical layer specifies the electrical, mechanical, functional, and procedural characteristics for transmission of the bit stream. The datalink layer's job is to establish and maintain the link and disconnect after the data transfer is complete. Error control and flow control are also done in this layer. The network layer takes care of routing of the packets in the network. The transport layer provides an end-to-end reliable data transfer. Even if the packets are received out of sequence or with error, the transport layer ensures that the packets are put in sequence and all the packets are received without error by giving the information about the erroneously received packets to the sender and obtaining the packets again. The job of the session layer is to establish and maintain sessions. The presentation layer takes care of the presentation formats, including compression and encryption. The application layer provides the interface to different application programs such as e-mail, file transfer, remote login, and so on.

Two important concepts in a layered model are protocol encapsulation and peer-to-peer communication. A layer sends its data in the form of a protocol data unit (PDU) to the layer below. The lower layer adds its own header to the PDU (without making any changes to the PDU) and then sends it to the layer below. This mechanism is known as protocol encapsulation. Finally when the data stream is received at the other end, the data passes through the layers again.

Each layer strips off the corresponding header and takes appropriate action based on the information in the header. For instance, the header information of the transport layer is interpreted only by the transport layer at the other end. Even though the transport layers on two machines do not talk to each other directly, the header information is interpreted only by the peer layer (layer at the same level). This is known as peer-to-peer communication.

Though there are hardly any networks that run the ISO/OSI protocol, a good understanding of this protocol is important because it acts as a reference model for many protocol stacks.

References

Dreamtech Software Team. *Programming for Embedded Systems*. Wiley Dreamtech India Pvt. Ltd., 2002. This book contains chapters on serial communication programming and protocol conversion software development.

Joe Campbell. *C Programmer's Guide to Serial Communications* (Second Edition). Prentice Hall Inc., 1997. This is an excellent book for developing a variety of applications and systems software using serial communications. A must-read for getting good expertise in serial communication programming.

A.S. Tanenbaum. *Computer Networks*. Prentice Hall Inc., 1997.

W. Stallings. *Data and Computer Communications* (Fifth Edition). Prentice Hall Inc., 1999.

Questions

1. Explain the ISO/OSI protocol architecture.
2. Explain the serial communication protocol using the RS232 standard.
3. Explain the HDLC protocol.
4. List the application layer protocols of the OSI reference model.

Exercises

1. Standardization of various protocols by the international standards bodies is done through consensus. Everyone has to accept the proposal, then only the proposal becomes a standard. For standardizing the ISO/OSI architecture, there were two proposals—one proposal based on six layers and the other proposal based on eight layers. Finally, seven layers were accepted (just the average of six and eight, but no other reason!). Develop six layer and eight layer architectures and study the pros and cons of each.

2. Interconnect two PCs running the Windows 9x/2000/XP operating system through RS232 and experiment on the communication parameters.

3. Interconnect two PCs running the Linux operating system through RS232 and experiment on the communication parameters.

Projects

1. Interconnect two PCs using an RS232 cable. Write a program to transfer a file from one computer to the other through the RS232 link.

2. Interconnect two PCs using an RS232 cable and write a program for a chat application. Make the software modular to give a layered approach to software development.

3. If you have a processor-based board (such as 8051 or 8085/8086) in your microprocessor laboratory, interconnect this board with a PC using an RS232 link. You need to write the software on both the PC and the processor board to achieve communication.

17 Local Area Networks

In This Chapter

- The Ethernet Local Area Network
- The LAN Protocols
- The LAN Standards
- Wireless LANs

During the1980s, PCs became ubiquitous in all organizations. Every executive started having his own PC for automation of individual activities. The need arose for sharing information within the organization—sending messages, sharing files and databases, and so forth. Whether the organization is located in one building or spread over a large campus, the need for networking the computers cannot be overemphasized. Today, we hardly find a computer, that is not networked. A local area network (LAN) interconnects computers over small distances up to about 10 kilometers. In this chapter, we will study the various configurations, technologies, protocols, and standards of LANs.

17.1 THE ETHERNET LAN

The Ethernet local area network developed by Xerox, Intel, and DEC in 1978 became the most popular LAN standard. IEEE released a compatible standard called IEEE 802.3.

In 1978, Xerox Corporation, Intel Corporation, and Digital Equipment Corporation standardized Ethernet. This has become the most popular LAN standard. IEEE released a compatible standard, IEEE 802.3. A LAN is represented in Figure 17.1.

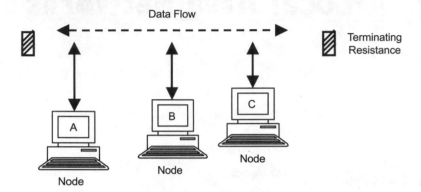

FIGURE 17.1 Ethernet LAN.

The cable, called the Ether, is a coaxial cable of ½ inch diameter and up to 500 meters long. A resistor is added between the center wire and the shield at each end of the cable to avoid reflection of the electrical signal. The cable is connected to a transceiver (transmitter and receiver) whose job is to transmit the signals onto the Ether and also to sense the Ether for the presence of the signals. The transceiver is connected through a transceiver cable (also called attachment unit interface cable) to a network card (also known as a network adapter card). The network card is a PC add-on card that is plugged into the motherboard of the computer. The coaxial cable used in Ethernet LAN is shown in Figure 17.2.

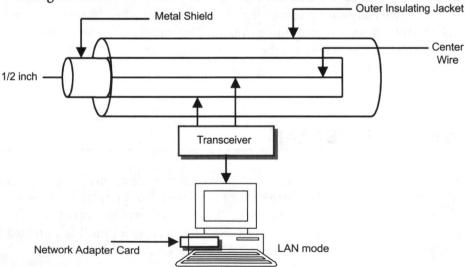

FIGURE 17.2 Coaxial cable used in an Ethernet LAN.

This architecture is known as *bus architecture* because all the nodes share the same communication channel. Ethernet operates at 10Mbps data rate. When a node has data to transmit, it divides the data into packets, and each packet is broadcast on the channel. Each node will receive the packet, and if the packet is meant for it (based on the destination address in the packet), the packet will be processed. All other nodes will discard the packet.

17.1.1 LAN Protocol Layers

In LAN protocol architecture, the bottom two layers—physical layer and datalink layer—are defined. The datalink layer is divided into two sublayers: medium access control sublayer and logical link control sublayer.

The protocol architecture of a LAN is shown in Figure 17.3. Every node on the LAN will have hardware/software corresponding to all these layers. The bottom two protocol layers are

- Physical layer
- Datalink layer, which is divided into
- Medium access control (MAC) sublayer
- Logical link control (LLC) sublayer

Above the datalink layer, the TCP/IP protocol suite is run to provide various applications on, the LAN. The network card and the associated software in the PC provide the first two layers' functionality, and the TCP/IP stack is integrated with every desktop operating system (Windows, Unix/Linux, etc.). While studying LANs, we will focus only on these two layers. All the standards for LANs, which we will study in subsequent sections, will also address only these two layers.

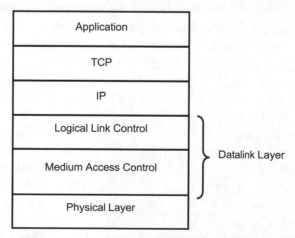

FIGURE 17.3 LAN protocol layers.

Physical layer: This layer specifies the encoding/decoding of signals, preamble generation, and removal for synchronization and bit transmission and reception.

MAC sublayer: The MAC sublayer governs the access to the LAN's transmission medium. A special protocol is required because a number of nodes share the same medium. This protocol is known as the medium access control protocol. In Ethernet, the medium access protocol is carrier sense multiple access/collision detection (CSMA/CD).

LLC sublayer: Above the MAC sublayer, the LLC layer runs. The data received from the higher layer (IP layer) is assembled into frames, and error detection and address fields are added at the transmitter and sent to the MAC layer. The MAC layer, when it gets a chance to send its data using CSMA/CD protocol, sends the frame via the physical layer. At the receiver, frames are disassembled, the address is recognized, and error detection is carried out. The LLC layer provides interface to the higher layers.

> The MAC sublayer defines the protocol using which multiple nodes share the medium. In Ethernet, the MAC protocol is carrier sense multiple access/collision detection (CSMA/CD).

17.1.2 CSMA/CD Protocol

In Ethernet LANs, all the nodes share the same medium (Ether). The protocol used to share the medium is known as carrier sense multiple access with collision detection (CSMA/CD).

When a node has to send a packet, it will broadcast it on the medium. All the nodes on the LAN will receive the packet and check the destination address in the packet. The node whose address matches the destination address in the packet will accept the packet. To ensure that more than one node does not broadcast its packet on the Ether, just before transmitting a packet, each node has to first monitor the Ether to determine if any signal is present; in other words, sense the carrier. However, there is still a problem—take the case of two nodes (A and C) on the LAN in Figure 17.1. If node A sent a packet, it takes finite time for the signal to reach node C. Meanwhile, node C will sense the carrier and find that there is no activity on the Ether, so it also will send a packet. These two packets will collide on the medium, resulting in garbling of data. To avoid these collisions, each node will wait for a random amount of time if a collision occurs. Binary exponential back-off policy is used to avoid collisions to the maximum possible extent—a node will wait for a random amount of time if there is a collision, twice that time if there is a second collision, thrice that time if there is third collision, and so on. This ensures that the probability of collisions is minimized. But a collisions cannot be eliminated altogether. That is why, even though on the Ethernet the data can be transmitted at the rate of 10Mbps, the effective data rate or throughput will be much lower.

When a collision of packets occurs on the medium, each node has to follow the binary exponential back-off policy wherein a node has to wait for a random amount of time if a collision occurs, twice that time if a collision occurs a second time, and so on.

17.1.3 Ethernet Addresses

To identify each node (computer) attached to the Ethernet, each node is given a 48-bit address. This address is also known as the hardware address or physical address because the network card on the computer carries this address, which is given by the hardware manufacturer. Therefore, each computer is uniquely identified.

The Ethernet address can be of three types:

1. Unicast address
2. Broadcast address
3. Multicast address

Each computer on an Ethernet LAN is identified by a 48-bit address. This address is contained in the network card on the computer.

When a node sends a packet with a unicast address, the packet is meant for only one node specified by that address. When a node sends a packet with a broadcast address (all ones), the packet is meant for all the nodes on the network. A set of nodes can be grouped together and given a multicast address. When a node sends a packet with a multicast address, the packet is meant for all the nodes in that group. Generally, network cards accept the unicast address and broadcast address. They can also be programmed to accept multicast addresses.

Ethernet address can be of three types: unicast, broadcast, and multicast. Generally, network cards accept unicast addresses and broadcast addresses. They need to be specially programmed to accept multicast addresses.

17.1.4 Ethernet Frame Format

The Ethernet frame format is shown in Figure 17.4.

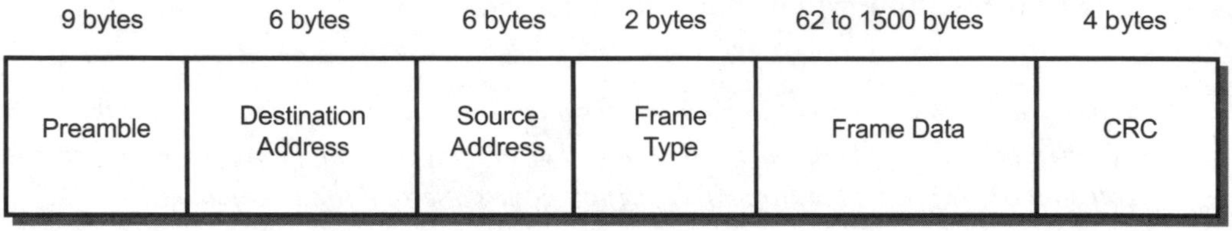

9 bytes	6 bytes	6 bytes	2 bytes	62 to 1500 bytes	4 bytes
Preamble	Destination Address	Source Address	Frame Type	Frame Data	CRC

FIGURE 17.4 Ethernet frame format.

Preamble (8 bytes): The beginning of the frame.

Destination address (6 bytes): Destination address of the packet.

Source address (6 bytes): Source address of the packet.

Frame type (2 bytes): Type of the frame.

Frame data (variable): User data, which can vary from 64 bytes to 1500 bytes.

CRC (4 bytes): 32-bit CRC computed for each frame.

The maximum allowed size of an Ethernet frame is 1526 bytes, out of which 26 bytes are used for the header and the CRC. Hence, 1500 bytes is the maximum allowed user data in an Ethernet frame.

Hence, the maximum allowed size of an Ethernet frame is 1526 bytes. This is the frame format of Ethernet developed by Digital, Intel, and Xerox.

Above the LLC, the IP layer runs, and each node on the LAN is assigned an IP address. Above the IP layer, the TCP layer and other application layer protocols run to provide applications such as e-mail, file transfer, and so on.

Ethernet LANs have become widely popular. Initially, 10Mbps Ethernet LANs were used, and now 100Mbps Ethernet LANs are common.

17.2 LAN TRANSMISSION MEDIA

For a LAN, the transmission medium can be twisted copper cable, coaxial cable, optical fiber, and radio. The topology, data rates, and medium access protocols will differ for the different media.

LANs can be broadly classified as baseband LANs and broadband LANs. In a baseband LAN, the baseband signals are transmitted over the medium. In broadband LANs, the signals are multiplexed using frequency division multiplexing.

LANs can be broadly categorized as baseband LANs and broadband LANs. Broadband LANs can span larger distances up to tens of kilometers. Broadband LAN uses frequency division multiplexing, in which multiple channels are used for data, voice, and video. RF modems are required for communication. They operate in unidirectional mode because it is difficult to design amplifiers that pass signals of one frequency in both directions. To achieve full connectivity, two data paths are required—one frequency to transmit and one to receive.

17.3 LAN TOPOLOGIES

A LAN can operate in the following topologies: bus, tree, ring, and star, as shown in Figure 17.5.

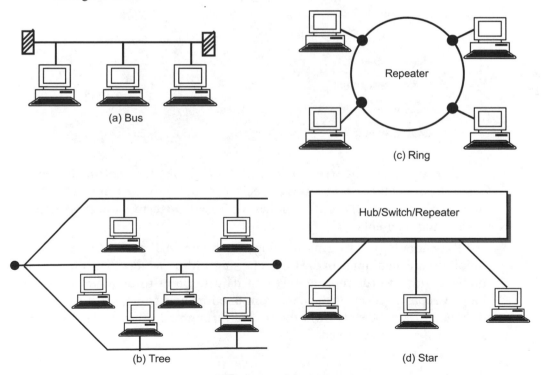

FIGURE 17.5 LAN topologies.

Bus topology: In bus topology, the transmission medium has terminating resistance at both ends. Data transmitted from a node will flow in both directions. Terminating resistance absorbs the data, removing it from the bus when it reaches the end points. When a node has data to transmit, it broadcasts it over the medium, and all the other nodes receive it.

In baseband LANs, binary data is inserted on the cable directly, and the entire bandwidth is consumed by the signal. The signal propagates in both directions. Distance up to only a few meters is possible due to attenuation of the signal.

In an IEEE 802.3 LAN, 50-ohm cable of 0.4 inch diameter is used. Maximum cable length is 500 meters per segment. This standard is called 10 BASE 5 to indicate 10Mbps data rate, baseband, and 500 meter cable length.

For low-cost PC LANs (called cheapernet), 10 BASE 2 standard is followed.

The differences between 10 BASE 5 and 10 BASE 2 LANs are listed below.

	10 BASE 5	10 BASE 2
Data rate	10Mbps	10Mbps
Maximum segment length	500 meters	185 meters
Network span	2500 meters	1000 meters
Nodes per segment	100	30
Node spacing	2.5 meters	0.5 meter
Cable diameter	0.4 inch	0.25 inch

Tree topology: The tree topology is shown in Figure 17.5(b). The tree is a general form of bus, or the bus is a special case of tree. Data transmitted from a node is received by all the other nodes. There will be terminating resistances at each of the segments.

Ring topology: The ring topology is shown in Figure 17.5(c). In a ring topology, the medium is in the form of a ring with repeaters that connect to the nodes. In other words, repeaters are joined by the point-to-point links in a closed loop. A repeater receives the data from one link and transmits it on the other link. Each frame travels from the source through all other nodes back to the source where it is removed.

NOTE

In ring topology, if the transmission medium fails (for instance, due to a cut in the cable), the network is down. To avoid such situations, dual-ring LANs are used in which two cables are used to connect the nodes.

The various LAN topologies are bus, tree, ring, and star. Choice of a particular topology is dependent on factors such as reliability, data rate, cost, and medium of transmission.

The repeater inserts data into the medium sequentially, bit by bit. Each repeater regenerates the data and retransmits each bit. The data packet contains the header with the destination address field. Packet removal is done either by the specified repeater or the source, after the ring is traversed.

The repeater can be in one of the following three states:

Listen state: Scan the data for the address, for permission to retransmit; pass the data to the attached station.

Transmit state: Retransmit the data to the next repeater and modify a bit to indicate that the packet has been copied (serves as acknowledgement).

Bypass state: Data is sent without delay.

Baseband coaxial cable, twisted pair, or fiber can be used for repeater-to-repeater links.

To recover data from a repeater and to transmit data to the next repeater, precise timing is required. The timing is recovered from the data signals. Deviation of timing recovery is called *timing jitter*. Timing jitter places a limitation on the number of repeaters in a ring.

Ring topology has the following problems:

- Failure of a repeater results in the breakdown of the network.
- New repeater installation is cumbersome.
- Timing jitter problems have to be solved (extra hardware is required).

Star topology: In star topology (Figure 17.5(d)), there will be a central hub to which all the nodes are connected. The central node can operate in broadcast mode or as a switch. In star networks, addition and deletion of nodes is very easy; however, if the central hub fails, the entire network is down. The hub acts as a repeater—data transmitted by a station is received by all stations. Hence logically, this topology is equivalent to a bus.

17.4 MEDIUM ACCESS CONTROL PROTOCOLS IN LANS

The MAC sublayer differs according to the topology of the LAN. In general, how do different nodes access the medium? There are two possibilities: centralized control, in which a control station grants permission to different nodes, or a decentralized network, in which stations collectively perform medium access control function. In decentralized networks, there are three types of categories: round robin, reservation, contention.

Round robin: Each station is given an opportunity to transmit for a fixed time. Control of the sequence of the nodes may be centralized or distributed. This mechanism is useful when all the stations have data to transmit.

Reservation: Each node is given a time slot (as in TDMA). Reservation of the slots may be centralized or decentralized.

Contention: This is a distributed control mechanism wherein all stations contend for the medium. When two or more stations contend for the medium simultaneously, it may result in collision. CSMA/CD, which was described earlier, is an example of this protocol.

Because multiple nodes have to share the medium, the choice of the MAC protocol is very important in LANs. Though CSMA/CD is the most popular MAC protocol, variations of this protocol have been developed to increase the throughput.

The CSMA/CD protocol is used in bus/tree and star topologies.

The CSMA/CD protocol differs based on whether the LAN is broadband or baseband. In baseband LANs, carrier sensing is done by detecting voltage pulse train. In broadband LANs, carrier sense is done by detecting RF carrier. For collision detection, in baseband LANs, if the signal on the cable at the transmitter tap point exceeds a threshold, collision is detected (because collision produces voltage swings). Because of the attenuation in the cable, a maximum length of the segment is specified in the standard—mainly to ensure collision detection using the threshold. In broadband LANs, head-end can detect the garbled data, or a station can do bit-by-bit comparison between transmitted and received data.

17.5 LAN STANDARDS

There is a wide variety of LANs—different topologies, different transmission media, different data rates, and so forth. The Institute of Electrical and Electronics

The IEEE 802 committee formulated various LAN standards. These standards address only the physical and datalink layers of LANs. Note that the standards also specify the physical medium to be used in different LANs.

Engineers (IEEE) set up a committee known as the 802 committee to develop various LAN standards. These standards together are known as IEEE 802 standards. These standards address only the physical and datalink layers of LANs. They specify the protocols to be used in MAC and LLC sublayers, the physical layer specifications, and the physical medium to be used.

17.5.1 IEEE 802.2 Standard

The IEEE 802.2 standard specifies the LLC sublayer specifications. This sublayer specification is common to all IEEE standards-based LANs.

IEEE 802.2 standard specifies the LLC sublayer, which provides the following services:

Unacknowledged connectionless service (type 1 service): This is a datagram service. There will be no flow control and no error control. Higher layers have to take care of these issues.

Connection-mode service (type 2 service): A logical connection will be set up, and flow control and error control are provided.

Acknowledged connectionless service (type 3 service): This is a datagram service, but with acknowledgements.

This LLC layer is common to all the other IEEE standards–based LANs.

17.5.2 IEEE 802.3 Standard

Based on the popularity of Ethernet, IEEE released a compatible LAN standard that is specified in IEEE 802.3. LANs based on the 802.3 standard have the following characteristics:

- Topology: Bus, tree, or star
- MAC sublayer: CSMA/CD
- Physical layer can be one of the following:
  - Baseband coaxial cable operating at 10Mbps
  - Unshielded twisted pair operating at 10Mbps or 100Mbps
  - Shielded twisted pair operating at 100Mbps
  - Broadband coaxial cable operating at 10Mbps
  - Optical fiber operating at 10Mbps

IEEE 802.3 operating at 10Mbps has six alternatives:

- 10 BASE 5: 10Mbps baseband 500 meter segment length
- 10 BASE 2: 10Mbps baseband, 200 meter segment length
- 10 BASE T: 10Mbps baseband, twisted pair
- 10 BROAD 36: 10Mbps broadband, 3600 meter end-to-end span (1800 meter segment)
- 10 BASE F: 10Mbps baseband, fiber
- 1 BASE T: 1Mbps baseband, twisted pair (now obsolete)

In addition, IEEE 802.3 specifies 100Mbps LAN (fast Ethernet) (known as 100 BASET).

The format of the MAC frame in IEEE 802.3 standard is slightly different from that of the Ethernet frame. The IEEE 802.3 MAC frame format is shown in Figure 17.6.

Bytes :	7	1	2 or 6	2 or 6	2	>=0	>=0	4
	Preamble	SFD	DA	SA	Length	LLC Data	Pad	FCS

FIGURE 17.6 Frame for IEEE 802.3 standard.

The IEEE 802.3 standard is based on the popular Ethernet LAN. The MAC frame formats of Ethernet and IEEE 802.3 are slightly different.

Preamble (7 bytes): The bit pattern 010101... is sent for the receiver to establish synchronization.

SFD (1 byte): Start frame delimiter 10101011 to indicate the actual start of the frame. This enables the receiver to locate the first bit of the rest of the frame.

DA (2 or 6 bytes): Destination address. 48 bits or 16 bits (must be the same for a particular LAN). It can be a node address, group address, or global address.

SA (2 or 6 bytes): Source address—address of the node that sent the frame.

Length (2 bytes): Length of the LLC data field.

LLC data (variable): Data from the LLC layer.

Pad (variable): Bytes added to ensure that frame is long enough for proper operation of the collision detection scheme.

FCS (4 bytes): Frame check sequence is calculated based on all the bits except the preamble, SFD, and FCS (32 bits).

17.5.3 IEEE 802.4 Standard

IEEE 802.4 standard–based LANs have the following characteristics:

- Topology: Bus, tree, or star
- MAC sublayer: Token bus
- The physical layer can be one of the following:
  - Broadband coaxial cable at 1, 5 or 10Mbps
  - Carrier band coaxial cable at 1, 5 or 10Mbps
  - Optical fiber at 5, 10, or 20 Mbps

17.5.4 IEEE 802.5 Standard

IEEE 802.5–based LANs have the following characteristics:

- Topology: Ring
- MAC protocol: Token Ring
- The physical layer can be one of the following:
  - Shielded twisted pair at 4 or 16Mbps, maximum number of repeaters is 250
  - Unshielded twisted pair at 4Mbps, maximum number of repeaters is 72

Because the topology of this LAN is ring, Token Ring protocol is used for MAC.

IEEE 802.5 MAC protocol: A small frame called a token circulates when all the nodes are idle. The node wishing to transmit seizes the token by changing one bit in the token and transforming it into a start-of-frame sequence for a data frame. The node appends the data to construct the data frame. Since there is no token on the ring, all other nodes only listen. The data frame transmitted by the node makes a round trip and returns to the originating node. The node will insert a new token on the ring when it has completed transmission of the frame or the leading edge of the transmitted frame has returned to the node.

Advantages of this MAC protocol are that it provides a flexible control to access the medium and is efficient under heavy load conditions. However, the disadvantages are that maintenance of the token is a problem: if the token is lost, the ring does not operate—so one node acts as monitor. This protocol is inefficient for light load conditions. The following improvements can be made to the Token Ring protocol:

Token Ring priority: Optional priority field and reservation fields in data frame and token (three bits and hence eight levels) are included. A node can transmit if its token priority is higher than the received one (set in the previous data frame). To avoid one or more nodes to having the highest priority all the time, the node that raises its priority in one token must lower its priority subsequently.

An IEEE 802.5 LAN is based on ring topology. Hence, the MAC protocol is the Token Ring protocol. A small frame called a token circulates around the ring when the nodes are idle. A node wishing to transmit will seize the token and transmit its data. Since there is no token on the ring, all other nodes cannot transmit.

Early token release: For efficient ring utilization, a transmitting node can release a token as soon as it completes frame transmission, even if the frame header has not returned to the node.

17.5.5 IEEE 802.12 Standard

IEEE 802.12–based LANs have the following characteristics:

- Topology: Ring
- MAC: Round-robin priority
- Physical layer: Unshielded twisted pair operating at 100Mbps

17.5.6 FDDI LAN

In a fiber distributed data Interface (FDDI) LAN, the MAC protocol is similar to that of IEEE 802.5 except that early token release strategy is followed.

FDDI (fiber distributed data interface)-based LANs have the following characteristics:

- Topology: Dual bus
- MAC Protocol: Token Ring
- Physical layer can be one of the following:
  - Optical fiber operating at 100Mbps, maximum number of repeaters is 100, and the maximum distance between repeaters is 2 km.
  - Unshielded twisted pair at 100Mbps, maximum number of repeaters is 100, and maximum distance between repeaters is 100 meters.

FDDI MAC protocol: It has the same functionality as the IEEE 802.5 MAC protocol except that in 802.5, a bit in the token is reserved to convert it into a data frame. In FDDI, once a token is recognized, it is seized, and the data frame is transmitted. This is done to achieve high data rate support. In FDDI, early token release is followed: the token is released after transmitting the data frame without waiting to receive the leading bit of the data frame.

17.6 LAN BRIDGE

A bridge interconnects two LANs. If the protocols used by the two LANs are different, the bridge will carry out the necessary protocol conversion.

A bridge is used to interconnect two LANs. If both LANs use the same set of protocols, the bridge need not do any protocol conversion. However, if the two LANs run different protocols, the bridge needs to do the necessary protocol conversion. Figure 17.7 shows the protocol conversion required. One LAN is based on IEEE 802.3 standard running CSMA/CD protocol. The other LAN is based on IEEE 802.4 standard running Token Ring protocol. The bridge has to run both the stacks as shown in the figure. It takes a packet from the 802.3 LAN and obtains the LLC frame. The LLC frame is then given to the 802.4 MAC protocol for transfer over the physical medium.

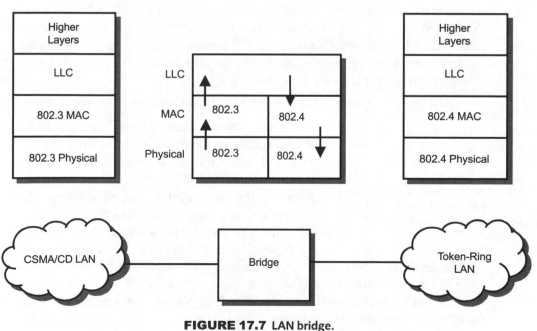

FIGURE 17.7 LAN bridge.

All the LANs mentioned in this section use guided media (twisted pair, coaxial cable, or optical fiber) as the transmission medium. Another set of IEEE standards is available for wireless LANs (WLANs). WLANs have become extremely popular in recent years. These WLAN standards are discussed in the next section.

17.7 WIRELESS LANS

Experiments on WLANs were conducted more than three decades ago. Those WLANs used a protocol known as ALOHA for medium access. In ALOHA, any station can transmit on the air. If collision is detected (every station is always in receive mode), the transmitting station waits for a random time, and retransmits. This is an inefficient access protocol, but 18% is the channel utilization efficiency because there will be a lot of collisions.

Variations of ALOHA such as slotted ALOHA and reservation ALOHA were developed, which are the basis for CSMA/CD LANs.

As wireless technologies matured and the need for mobile computing increased tremendously, WLANs gained popularity. The IEEE 802.11 family of standards specifies the WLAN standards for different applications. These WLANs can be used in offices for networking of different devices such as desktop, laptop, palmtop, printer, fax machine, and so forth. WLANs also can be used at home to network home appliances.

The two configurations in which a WLAN can work are shown in Figure 17.8.

Wireless LANs operate in two configurations. In one configuration, a node will communicate with another node via an access point. In the other configuration, two nodes can communicate directly without any centralized control.

In the configuration shown in Figure 17.8(a), a WLAN node (for example, a laptop with a radio and antenna) can communicate with another node via an access point. A WLAN can contain a number of access points. In the configuration shown in Figure 17.8(b), two nodes can communicate directly with each other without the need for a central relay. Such a configuration is known as an ad hoc network. Two or more devices can form a network when they come nearer to one another without the need for a centralized control. Ad hoc networks are very useful for data synchronization. For instance, when a mobile phone comes near the vicinity of a PDA, the two can form a network, and the address book on the PDA can be transferred to the mobile phone.

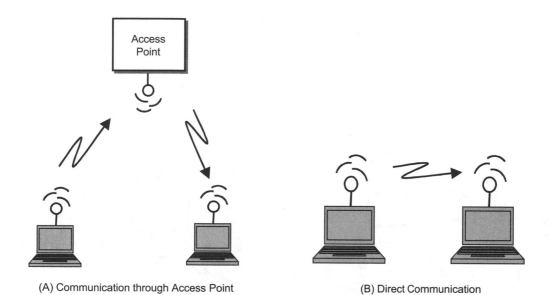

FIGURE 17.8 WLAN configurations.

WLANs based on IEEE 802.11 family standards can be used in offices and homes as well as public places such as airports, hotels, and coffee shops.

IEEE developed the IEEE 802.11 family standards to cater to different industry segment requirements. All the standards in this family use radio as the transmission medium.

17.7.1 IEEE 802.11 Family Standards

The IEEE 802.11 family standards cover the physical and MAC layers of wireless LANs. The LLC layer is the same as discussed earlier. The architecture of the IEEE 802.11 standard for WLANs is shown in Figure 17.9. Each wireless LAN node has a radio and an antenna. All the nodes running the same MAC protocol and competing to access the same medium will form a basic service set (BSS). This BSS can interface to a backbone LAN through an access point (AP). The backbone LAN can be a wired LAN such as Ethernet LAN. Two or more BSSs can be interconnected through the backbone LAN.

In a wireless LAN, there will be a number of access points. These access points are interconnected through a backbone network. The backbone network can be a high-speed network based on Asynchronous Transfer Mode (ATM) protocols.

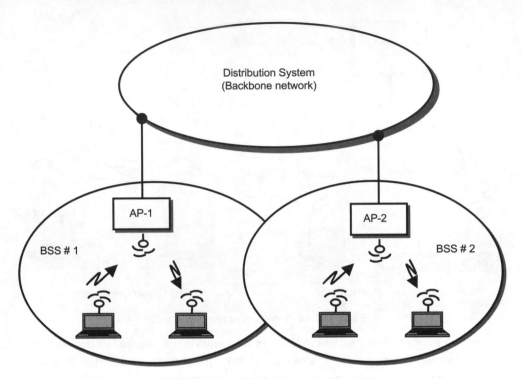

FIGURE 17.9 IEEE 802.11 wireless LAN.

The physical medium specifications for 802.11 WLANs are:

- Diffused infrared with a wavelength between 850 and 950 nm. The data rate supported using this medium is 1Mbps. A 2Mbps data rate is optional.

- Direct sequence spread spectrum operating in 2.4GHz ISM band. Up to seven channels each with a data rate of 1Mbps or 2Mbps can be used.

- Frequency hopping spread spectrum operating at 2.4GHz ISM band with 1Mbps data rate. A 2Mbps data rate is optional.

The three physical medium specifications in IEEE 802.11 wireless LANs are: (a) diffused infrared; (b) direct sequence spread spectrum in 2.4GHz ISM band; and (c) frequency hopping spread spectrum operating in 2.4GHz ISM band.

Extensions to IEEE 802.11 have been developed to support higher data rates. The 802.11b standard has been developed that supports data rates up to 22Mbps at 2.4GHz, with a range of 100 meters. Another extension, 802.11a, operates in the 5GHz frequency band and can support data rates up to 54Mbps with a range of 100 meters.

The Industrial, Scientific and Medical (ISM) band is a free band. No government approvals are required to install radio systems operating in this band.

17.7.2 Medium Access Control

The MAC protocol used in IEEE 802.11 is carrier sense multiple access/collision avoidance (CSMA/CA). A node wishing to transmit senses the channel and, if the channel is free for more than a predefined period, it will transmit its data. If the channel is busy, the node will wait for an additional period called the backoff interval.

The MAC protocol used in 802.11 is called CSMA/CA (carrier sense multiple access with collision avoidance). Before transmitting, a station senses the radio medium and, if the channel is free for a period longer than a predefined value (known as distributed inter frame spacing or DIFS), the station transmits immediately. If the channel is busy, the node keeps sensing the channel. If it is free for a period of DIFS, it waits for another period called the random backoff interval and then transmits its frame. When the destination receives the frame, it has to send an acknowledgment (ACK). To send the ACK, the destination will sense the medium. If it is free for a predefined short time (known as the short inter frame space or SIFS), the ACK is sent. If the ACK does not reach the station, the frame has to be retransmitted using the same procedure. A maximum of seven retransmissions is allowed, after which the frame is discarded. Figure 17.10 depicts the CSMA/CA mechanism.

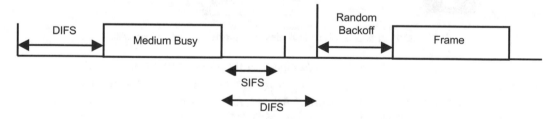

FIGURE 17.10 Medium access in 802.11 LAN.

17.7.3 WiFi

The IEEE 802.11b standard is popularly known as "Wireless Fidelity" (or WiFi) in short. It has become widely popular for wireless LANs in office environments. Proponents of this technology consider it great competition to third generation wireless networks, which also provide high data rate mobile Internet access. WiFi can be used to provide broadband wireless Internet access as shown in Figure 17.11.

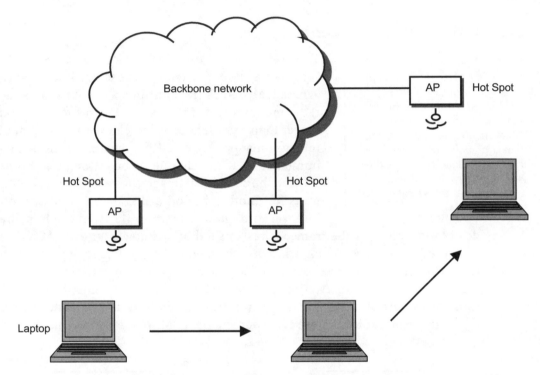

FIGURE 17.11 Broadband wireless access through wireless LAN.

IEEE 802.11b standard is popularly known as Wireless Fideliy or WiFi. The access points are known as hot spots. By installing hot spots at various places in a city, a metropolitan area network can be developed, and this network will be a competitor to 3G wireless networks.

Access Points (APs) can be installed at various locations in the city. The APs are also called "hot spots." All the APs in a city can be interconnected through an ATM-based backbone network. As the wireless device moves from one location to another, the mobile device is connected to the nearest AP.

The proponents of WiFi consider this architecture, for providing broadband Internet access, to be competitive with third generation wireless networks, which support only 2Mbps data rates. Efforts are now being made to make WiFi a highly secure network so that such an architecture can become widespread. Singapore is the first country to provide mobile Internet access using this approach.

WiFi hot spots are being installed in many public places such as hotel lounges, airports, restaurants, and so on to provide wireless access to the Internet.

17.7.4 Mobile IP

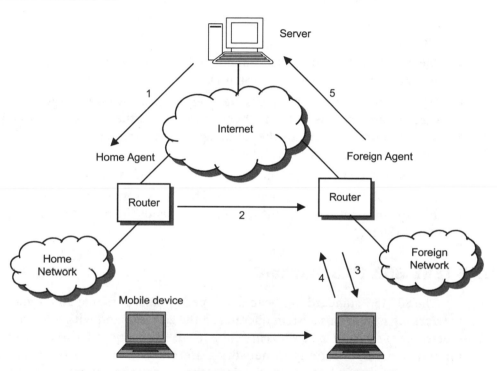

FIGURE 17.12 Mobile IP.

When a mobile device moves from one location to another, the packets have to be sent to the router to which the mobile device is attached. Consider a scenario shown in Figure 17.12. The mobile device initially is connected to its home network. The mobile device initiates a very large file transfer from a server located on the Internet. The server delivers the packets to the router, and the router in turn delivers the packets to the mobile device. Now the mobile device is moving in a car and is reaching another ISP (foreign network). The mobile device now has to attach to the router of the new ISP (known as foreign agent). But the packets keep getting delivered

In mobile IP, the mobile device will have two addresses: home address and care-of address. When the mobile device moves from one network to another network, the packets will be forwarded to the mobile device using those addresses.

to the home agent. The foreign agent assigns a new IP address (called the care-of address) to the mobile device, and this address is made known to the home agent. The home agent forwards all the packets to the foreign agent, which in turn delivers them to the mobile device. If the mobile device has to send packets to the server, it sends them directly via the foreign agent using the care-of address. The requirements for mobile Internet access using mobile IP are:

- The home agent assigns an IP address to the mobile device, called the home address.
- The mobile device initiates a connection to the server and the server sends the packets to the home agent (Step 1).
- As the mobile device approaches the foreign agent, the foreign agent assigns a temporary address to the mobile device, called the care-of address. The care-of address is sent to the home agent. The home agent forwards the packets to the foreign agent (Step 2).
- The foreign agent delivers the packets to the mobile device (Step 3).
- If the mobile agent has to send packets to the server, it sends the packet to the foreign agent (Step 4). The foreign agent sends the packet directly to the server (Step 5).

17.7.5 IEEE 802.15.3 Standard

The 802.15.3 standard has been developed to meet personal and home area networking needs. This system operates in the 2.4GHz band with a range of 10 meters. Data rates up to 55Mbps are supported. This standard also can be used for industrial applications to network different devices in process control systems. For such industrial applications, the operating temperature has to be from –40° C to 60° C.

17.7.6 HiperLAN2

HiperLAN is a wireless LAN standard developed by ETSI. It operates in the 2.4GHz band using TDMA-TDD. Data rates up to 54Mbps are supported to provide broadband multimedia services.

High performance LAN (HiperLAN) is the standard developed by the European Telecommunications Standards Institute (ETSI) for wireless LANs. The HiperLAN1 standard was developed for providing ad hoc connectivity to wireless devices. It uses the CSMA/CA protocol. However, HiperLAN1 was not well suited for applications such as voice/video communication, which require real-time performance. HiperLAN2 is the next standard developed by ETSI. HiperLAN2 supports broadband multimedia services as well.

HiperLAN can be used in offices, homes, and public places such as airports, hotels, and so forth. This standard supports both the configuration shown in Figure 17.8. The salient features of HiperLAN2 are as follows:

Modes of operation: There are two modes of operation—centralized and direct. In centralized mode, a mobile device communicates with another device via the access point (Figure 17.8(a)). This mode is used in office environments. In direct mode, two mobile devices communicate directly with each other. This mode is useful for personal area networking at homes and offices.

Medium access control: The medium access control protocol is TDMA-TDD. The TDMA frame is of 2 milliseconds duration during which AP-to-mobile communication and mobile-to-AP communication take place. The time slots are assigned dynamically, based on a request for a connection.

Data rates: Data rates up to 54Mbps are supported. Broadband multimedia services can be supported such as audio and video communication and video conferencing.

Quality of service: In multimedia applications involving voice and video communication, when real-time communication is required, the quality of service parameters are important. For instance, the user should be able to specify that the delay for a particular communication should be less than 10 milliseconds, or the bit error rate should be very low (one error for 100 million bits). HiperLAN2 allows such quality of service parameters by assuring a certain minimum delay or certain maximum throughput. Compared to 802.11 series standards, this standard has this attractive feature.

Summary

This chapter presented the details of local area network protocols and standards. LANs are used to connect devices within a radius of about 10 km. Ethernet, developed in 1978, has become the most popular LAN. Ethernet operates at 10Mbps using carrier sense multiple access with collision detection (CSMA/CD) protocol for medium access. The IEEE 802.3 standard is based on Ethernet.

The IEEE 802 committee developed a series of standards for LANs using twisted copper pair, coaxial cable, fiber, and radio as the transmission media. The logical link control (LLC) layer, which is common to all the standards, is specified in IEEE 802.2. The IEEE 802.11 family of standards is for wireless LANs.

Wireless LAN based on IEEE 802.11b, popularly known as WiFi (Wireless Fidelity), is the most popular LAN and can be used to provide connectivity in both home and office environments. HiperLAN2 is the wireless LAN standard developed by the European Telecommunications Standards Institute (ETSI). With the availability of wireless LANs that can support up to 54Mbps data rates, wireless LANs are being considered for providing broadband wireless Internet connectivity.

References

R.O. LeMire. "Wireless LANs and Mobile Networking Standards and Future Directions," *IEEE Communications Magazine*. Vol. 34, No. 8, August 1996.

S. Xu and T. Saadawi. "Does the IEEE 802.11 MAC Protocol Work Well in Multihop Wireless and Ad Hoc Networks," *IEEE Communications Magazine*. Vol. 39, No. 6, June 2001.

IEEE Communications Magazine. Vol. 39, No. 11, December 2001. This issue contains a number of articles on wireless personal and home area networks.

J.P. Macker et al. "Mobile and Wireless Internet Services: Putting the Pieces Together". *IEEE Communications Magazine*, Vol. 39, No. 6, June 2001.

P. S. Henry. "WiFi: What's Next?" *IEEE Communications Magazine*, Vol. 40, No. 12, December 2002.

J. Khun-Jush et al. "HiperLAN2: Broadband Wireless Communications at 5 GHz." *IEEE Communications Magazine*, Vol. 40, No. 6, June 2002.

www.ieee.org Web site of IEEE. You can get the IEEE 802 standards documents from this site.

www.hiperlan2.com Web site of the HiperLAN2 Global Forum.

Questions

1. Explain the Ethernet local area network operation, giving the details of medium access protocol and Ethernet frame format.
2. What is the difference between Ethernet MAC frame format and the IEEE 802.3 MAC frame format?
3. What are the different topologies used for LANs?
4. What are the different MAC protocols used for LANs?

Exercise

1. Measure the throughput of the LAN installed in your office/department. You can use the software package available on your server to find out the effective data rate or throughput. Study the effect of traffic on the throughput by increasing the traffic (by invoking many file transfers simultaneously).

2. What are the issues related to security in wireless LANs?

3. Work out a detailed plan for installation of a new LAN on your college campus. You need to study the topology, the expected traffic, and how to interconnect different LAN segments located in different buildings.

4. Survey the commercially available IEEE 802.11 products.

Projects

1. Simulate home agent and a foreign agent on two LAN nodes. You need to study the details of mobile IP to take up this project.

2. Carry out a paper design to develop a wireless LAN for your university campus. You need to do a survey of the various 802.11b products available before doing the design.

3. Interconnect four PCs as a ring using RS232 cables. Develop a Token Ring LAN with these four nodes.

18

Wide Area Networks and X.25 Protocols

In This Chapter

- Wide Area Networking
- X.25 Protocols
- The X.25 Packet Formats
- The Addressing Scheme Used in X.25 Networks

To connect computers or LANs spread over a large geographical area is now the order of the day. These wide area networks (WANs) may be private networks connecting corporate offices spread across the country or the globe, or they may be public networks offering data services to the public. In this chapter, we will study the issues involved in wide area networking, with special emphasis on X.25 protocols used extensively in wide area networks.

18.1 ISSUES IN WIDE AREA NETWORKING

When a computer network is spread over a large geographical area, some special problems are encountered that are not present in the LANs.

- WANs generally do not support very high speeds. Due to lower transmission rates, delays are likely to be higher. If satellite radio is used as the medium, then the delay is much higher and, as a result, special care must be taken in terms of flow control protocols.
- Because of the delay, the response time also will be high. To the extent possible, protocol overheads need to be minimized.

- The communication medium in a WAN environment may not be as reliable as in LANs, and hence the error rate is likely to be higher. This may lead to more retransmissions and more delay.

Wide area networks are characterized by low transmission speeds, high propagation delay, and complex network management.

- Lower transmission rates and higher delays pose problems for real-time voice and video communication. Higher delay causes gaps in voice communication and jerky images in video communication.

The options in transmission media for WANs are dialup/ leased lines, optical fiber, and satellite radio. X.25, Frame Relay, and Asynchronous Transfer Mode (ATM) protocols are used in WANs.

- Network management is more involved and complex as network elements are spread over large geographical areas.

The various options for developing WANs are:

- Dial up lines
- Point-to-point leased lines
- Switched digital networks based on Integrated Services Digital Network (ISDN)
- Switched digital networks based on X.25 standard protocols
- Optical fiber networks based on Frame Relay and Asynchronous Transfer Mode (ATM)

For WANs, X.25 is an important standard. X.25-based WANs have been in place since the 1980s. An overview of the X.25 standard is presented in the following sections.

X.25 protocols are still used extensively in satellite-based wide area networks. However, the protocol overhead is very high in X.25 as compared to Frame Relay.

18.2 OVERVIEW OF X.25

A typical X.25-based packet switched network is shown in Figure 18.1. The network consists of the end systems, called data terminal equipment (DTE), and the X.25 packet switches, called the data circuit terminating equipment (DCE). The packet switches are linked through communication media for transport of data in the form of packets. The packet size is variable, with a maximum limit of 1024 bytes.

An X.25 network consists of packet switches and end systems. The data transmission is done in packet format, the maximum packet size being 1024 bytes.

ITU-T Recommendation X.25 (also referred to as the X.25 standard) is the specification for interface between an end system and packet switch. Most of the public data networks (PDNs) and ISDN packet switching networks use the X.25 standard.

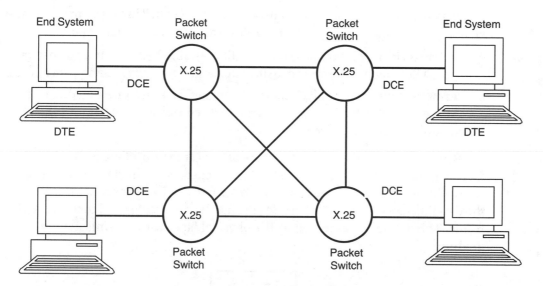

FIGURE 18.1 X.25 packet switched network.

 X.25 protocols address the first three layers of the OSI reference model: physical layer, datalink layer, and network layer.

NOTE

X.25 covers the first three layers in the OSI reference model: physical layer, datalink layer, and network layer.

Physical layer: This layer specifies the physical interface between the end system (DTE) and the link to the switching node (DCE). RS232 can be used as the physical layer.

Datalink layer: This layer ensures reliable data transfer as a sequence of frames. Link access protocol balanced (LAPB), a subset of HDLC, is used at this layer.

The datalink layer in X.25 is called link access protocol balanced (LAPB). LAPB is derived from HDLC.

Network layer: This is also called the packet layer. This layer provides the virtual circuit functionality. X.25 provides two types of virtual circuits:

- **Switched virtual circuit:** A virtual circuit is established dynamically between two DTEs whenever required, using call setup and call clearing procedures.
- **Permanent virtual circuit:** This is a fixed, network-assigned virtual circuit, and hence call setup and call clearing procedures are not required. This is equivalent to a leased line.

Protocol encapsulation in X.25 networks is shown in Figure 18.2. The data from a higher layer is passed on to the X.25 packet layer. The packet layer inserts control information as header and makes a packet. The packet is passed on to the datalink layer (LAPB), which adds the header and trailer and forms an LAPB frame. This frame is passed on to the physical layer for transmission over the medium.

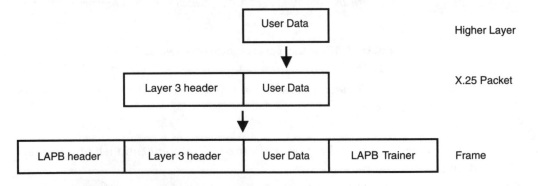

FIGURE 18.2 Protocol encapsulation in X.25.

X.25 supports virtual circuits. A virtual circuit is established between two end systems. The virtual circuit can be set up on a per-call basis or it can be permanent. These are called switched virtual circuit and permanent virtual circuit, respectively.

X.25 supports multiplexing. A DTE can establish 4095 virtual circuits simultaneously with other DTEs over a single DTE-DCE link. Each packet contains a 12-bit virtual circuit number.

The procedure for establishing a call between the source DTE and the destination DTE in an X.25 network is as follows:

1. Source DTE sends a call-request packet to its DCE. The packet includes the source address, destination address, and virtual circuit number.

2. Network routes the packet to the destination's DCE.

3. Destination's DCE sends an incoming-call packet to the destination DTE. This packet has the same frame format as the call-request packet but different virtual circuit number selected by the destination's DCE.

4. Destination DTE sends a call-accepted packet.

5. Source's DCE receives call-accepted packet and sends call-connected packet to the source DTE. Virtual circuit number is same as that in the call-request packet.

6. Source DTE and destination DTE use their respective virtual circuit numbers to exchange data and control packets.

In an X.25 network, communication takes place in three stages: (a) call setup; (b) data transfer; and (c) call disconnection. Hence, X.25 provides a connection-oriented service.

7. Source/destination sends a clear-request packet to terminate the virtual circuit and receives a clear-confirmation packet.

8. Source/destination receives a clear-indication packet and transmits a clear-confirmation packet.

The protocol is to send a set of packets to establish a call, transfer data and control information, and call disconnection. For each type of packet, the packet format is specified. The various packets exchanged are:

Call setup packets: From DTE to DCE, call-request and call-accepted packets

From DCE to DTE, incoming-call and call-connected packets

Call clearing packets: From DTE to DCE, clear-request and clear-confirmation packets

From DCE to DTE, clear-indication and clear-confirmation packets

Data transfer packets: Data packets

Interruption packets: Interrupt and interrupt confirmation packets

Flow control packets: RR (receiver ready) packet

RNR (receiver not ready) packet

REJ (reject) packet

Reset packets: Reset request, reset confirmation packets

Restart packets: From DTE to DCE, restart-request, restart-confirmation packets

From DCE to DTE, restart-indication and restart-confirmation packets

Diagnostics packets: Packets that carry diagnostics information

Registration packets: Request-registration and request-confirmation packets

For each type of packets, the packet formats are specified in the X.25 recommendations. Formats of some of the packets are discussed here.

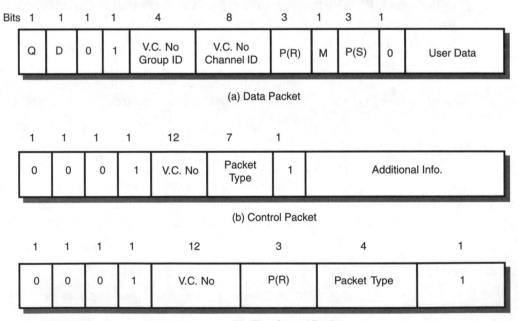

FIGURE 18.3 X.25 Packet Formats.

The large number of packet exchanges indicates that the protocol overhead is very high in X.25 networks. However, this is an excellent protocol to use on transmission media that are characterized by high error rates.

The different types of packets exchanged in an X.25 network are call setup packets, call clearing packets, interruption packets, reset packets, restart packets, flow control packets, registration packets, and data packets.

The various fields in a data packet are Q bit, virtual circuit number, send sequence number, receive sequence number, M bit, and user data.

The format of data packet is shown in Figure 18.3(a). The fields in the data packet are:

Q (1 bit): The Q bit is not defined in the standard—it can be used to define two types of data packets.

Virtual circuit number (12 bits): The 12 bits are divided into 4 bits for group ID and 8 bits for channel ID, which together form the virtual circuit number.

P(R) and P(S): Send sequence number and receive sequence number used for flow control as in HDLC.

D = 0 for acknowledgement between DTE and network (for flow control), and D = 1 for acknowledgement from remote DTE.

M bit: In X.25, two types of packets are defined—A packets and B packets. In A packets, M = 1 and D = 0, indicating that the packet is full; it is of maximum allowable packet length. B packet is a packet that is not an A packet. The packet sequence consists of a number of A packets and one B packet.

The format of a control packet is shown in Figure 18.3(b). This format is used for all packets for call setup and call disconnection. For example, the call-request packet will have additional information containing calling DTE address length (four bits), called DTE address length (four bits), calling DTE address, called DTE address.

The format of flow control packet is shown in Figure 18.3(c). P(R) indicates the number of the next packet expected to be received. Note that flow control is used at both layers 2 and 3 in X.25.

18.3 A SATELLITE-BASED X.25 NETWORK

X.25 is used extensively in satellite-based wide area networks. The typical architecture of the network is shown in Figure 18.4. At the satellite hub (central station), there will be a packet switch (DCE) to which an X.25 host is connected.

A PC add-on card plugged into the server makes the server an X.25 host. At the remotes, there are two possible configurations. Some remotes can have X.25 packet switches, which are connected to X.25 hosts. At other remotes, a special device called a packet assembler/disassembler (PAD) is used. This PAD, as the name suggests, takes the data from a terminal/PC, assembles the packets in the X.25 format, and sends it over the satellite link. The packets received from the satellite are disassembled and given to the terminal/PC. Communication between the PAD and the terminal/PC is through RS232. Two other standards are specified for this configuration. The X.3 standard specifies the parameters of the PAD. The X.28 standard specifies the protocol used by the terminal/PC to communicate with the PAD. The PAD parameters such as the speed of communication, number of data bits, and so on can be set by the terminal/PC.

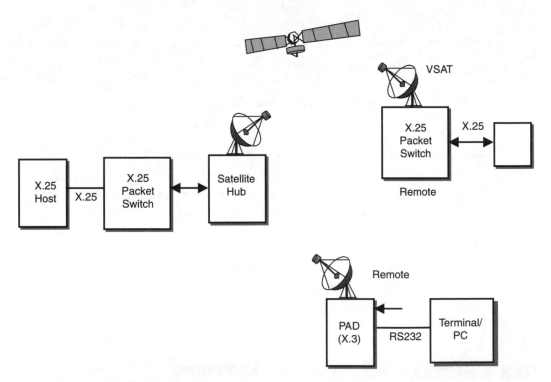

FIGURE 18.4 Satellite-based X.25 wide area network.

When a remote terminal wants to communicate with another remote, a virtual circuit is established between the two terminals. The source sends the packets in X.25 format, and the packet switch at the hub switches the packet to

Data terminals such as PCs can be connected to the X.25 network through a packet assembler/disassembler (PAD). The PAD takes the data from the PC and assembles the packets in X.25 format to send them over the network.

the appropriate destination. In such a case, the X.25 host at the hub does not play any role. Some data has to be broadcast to all the VSAT terminals, then the host is required. The host will use the broadcast address to send the packets over the satellite link, and all the remotes will receive it.

The PAD parameters are specified in ITU-T Recommendation X.3. Recommendation X.28 specifies the protocol used by the PC to communicate with the PAD.

18.4 ADDRESSING IN X.25 NETWORKS

Because X.25 is used for WANs, addressing the different end systems is very important. Addressing in X.25 networks is performed based on the X.121 standard. Figure 18.5 shows the two formats for the X.121 addressing scheme.

X.121 addresses will have 14 digits. In one format, there will be 4 digits of data network identification code (DNIC) and 10 digits that can be specified by the VSAT network operator. In the second format, there will be 3 digits of data country code (DCC) and 11 digits that can be specified by the VSAT network operator. DNIC and DCC will be given by national/international authorities.

(a)
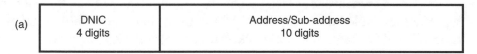

DNIC: Data Network Identification code
DCC: Data Country Code

(b)
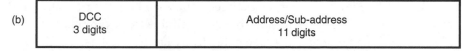

FIGURE 18.5 X.121 Address Formats.

Generally, the address/subaddress portion is divided into logical fields. For instance, in the first format, out of the 11 digits, 2 digits can be given to the state, 3 digits to the district, and the remaining digits to the hosts. A hierarchy can be developed. This complete address directory has to be stored on the host at the central station.

In X.25 networks, the addressing of the various end systems is done per ITU-T Recommendation X.121. X.121 address will have 14 digits.

Though X.25 has been used extensively in public data networks (PDNs), it is a highly complicated protocol because it has been developed to work on transmission media, which are highly susceptible to noise. At each hop (node to node), an exchange of data packet is followed by an acknowledgement packet. In addition, flow control and error control packets are exchanged. Such high overhead is suitable for links with high error rates. With the advent of optical fiber, which is highly immune to noise, such complicated protocols are not necessary. However, in satellite-based data networks, X.25 has been used widely to provide reliable data transfer.

Summary

In this chapter, X.25-based wide area networking is discussed. X.25 provides a connection-oriented service. Hence, when two systems have to communicate with each other, a virtual circuit is established, data transferred, and then the circuit is disconnected. X.25 is a robust protocol and will work even if the transmission medium is not very reliable. X.25 addresses the first three layers of the OSI protocol architecture. The physical layer specifies the interface between the DTE and DCE. The datalink layer is called the LAPB (link access protocol balanced), which is derived from HDLC. The network layer (also called the packet layer) will take care of the virtual circuit establishment. Flow control is done at both layers 2 and 3.

When an asynchronous terminal has to communicate with an X.25 network, a packet assembler disassembler (PAD) is used that takes the data from the terminal and formats it into X.25 packets to send over the X.25 network. Similarly, it disassembles the packets from the X.25 network and presents them to the terminal. In X.25 networks, addressing is done based on the X.121 standard, which specifies a 14-digit address for each system.

Though X.25 is a robust protocol that can be used in networks with unreliable transmission media, as media have become reliable (for example, optical fiber), X.25 has lost its popularity. It is used extensively in satellite-based data networks.

References

W. Stallings. *Computer and Data Communications*. Prentice-Hall, Inc., 1999.

www.farsite.co.uk This site provides useful information on X.25 commercial products.

Questions

1. List the various issues in wide area networking.
2. Describe the architecture of an X.25-based wide area network.
3. Explain the call setup, data transfer, and call disconnection procedures in an X.25 network.
4. Describe the formats of X.121 addressing.
5. Explain the X.25 packet formats.

Exercises

1. Study the X.25 switch, PAD, and X.25 add-on cards supplied by various vendors.
2. Make a comparison of the protocols used in X.25 and Frame Relay. Explain why X.25 is called a heavyweight protocol.
3. Design an addressing scheme based on X.121 for a nationwide network connecting various state governments, district offices, and central government departments.
4. List the important PAD parameters.

Projects

1. Simulate a PAD on a PC. The PAD has to communicate with the PC through an RS232 link. From the PC, you should be able to set the PAD parameters.
2. For an X.25-based wide area network connecting all the universities in the country, you have been asked to design an addressing scheme. Design the addressing scheme based on X.121 formats. You can divide the address/ subaddress field into three portions. The first portion is for the state code, the second portion is for the district code, and the third portion is for the university ID. Think of an alternative that does not use the geographical location. (You can divide the universities into different categories such as technical universities, agricultural universities, and so on.)

19 ▪ Internetworking

In This Chapter

- The Issues Involved in Interconnecting Two or More Networks
- The Importance of the TCP/IP Protocol Stack
- Connection-Oriented and Connectionless Internetworking
- Functions of a Router

To develop a LAN or WAN is straightforward, but how to network different networks is not. Each network has its own protocols, packet sizes and formats, speeds, addressing schemes, and so on. Internetworking or networking of networks is certainly a challenging task, and in this chapter we will discuss the issues involved in internetworking.

19.1 ISSUES IN INTERNETWORKING

Consider the simple case of connecting two networks as shown in Figure 19.1. Network A is a LAN based on Ethernet, and network B is a WAN based on X.25 protocols. It is not possible for a node on the LAN to transmit a packet that can be understood by a node on the WAN because of the following:

- The addressing formats are different.
- The packet sizes are different.
- The medium access protocols are different.
- The speeds of operation are different.
- The protocols used for acknowledgements, flow control, error control, and so on are different.

Interconnection of heterogeneous networks is called internetworking. A router or gateway is used to interconnect two networks.

To achieve connectivity among different networks, we need to solve the problem of interconnecting heterogeneous networks—this is known as *internetwoking*. Two different networks can be connected using a 'router' (or gateway) as shown in Figure 19.1. The connected network as a whole is referred to as an internet (small i) and each network as a subnetwork (or subnet) of the internet. The router operates at layer 3 of the OSI reference model. The router does the necessary protocol translation to make the two subnetworks talk to each other. Note that the router is the name given by equipment vendors; a router is referred to as a gateway in the documents of the Internet standards.

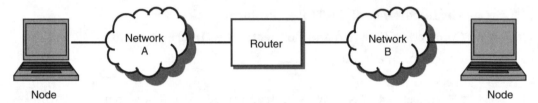

FIGURE 19.1 Internetworking with a router.

The router that interconnects two networks does the necessary protocol conversion. The router operates at layer 3 of the OSI reference model.

NOTE

The functions of the router are as follows:

- Accept the packet from the subnetwork A.
- Translate the packet to a format understood by subnetwork B.
- Transmit the packet to B.

Similarly, the packets from subnetwork B are transferred to subnetwork A after necessary translation or protocol conversion.

Now, consider the internet shown in Figure 19.2, where a number of subnetworks are connected together through routers. A node on subnetwork A has to send packets to a node on subnetwork F. For the packet to reach F, the packet has to traverse through other routers. A router's job is also to transfer the packet to an appropriate router so that the packet reaches the destination. To achieve this, each router has to keep a table, known as the routing table.

This routing table decides to which router (or network) the packet has to be sent next.

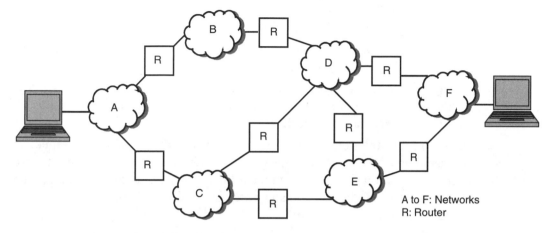

FIGURE 19.2 An internet.

Every router has a routing table. This routing table is used to route the packet to the next router on the network. The routing table is updated periodically to take care of changes in the topology of the network and the traffic on the network.

The requirements of internetworking are:

1. Links between subnetworks.
2. Routing and delivery of packets.
3. Accounting to keep track of use of various networks and routers.
4. To be able to accommodate.
   a. Different addressing schemes
   b. Different packet sizes
   c. Different network access mechanisms
   d. Different timeouts
   e. Different error recovery mechanisms
   f. Different status report mechanisms
   g. Different routing techniques
   h. Connection-oriented and connectionless services

To achieve internetworking, the router is the most important element. To meet all the above requirements, the router has to perform protocol conversion. In addition, the following are required:

- To take care of the differing protocols above the datalink layer, we need another layer of protocol that runs on each router and end system, known as the Internet Protocol (IP).

- Because the addressing formats differ, we need a universal addressing scheme to address each machine uniquely. Each machine on the internet is given a unique address, known as an IP address.

- For two end systems on the different networks to transfer packets irrespective of the underlying networks, we need another layer of communication software that has to run on each end system (or host). This protocol that handles the transportation of packets transparently is known as Transmission Control Protocol (TCP).

- Above the TCP, we need to run different sets of protocols to provide different applications for the end users. We need a separate protocol for each application—for file transfer, electronic mail, remote login, web access; and so on.

> Internetworking is a complex task because different networks have different packet sizes, different address formats, different access protocols, different timeouts, and so on.

These protocols form the TCP/IP protocol suite. The TCP/IP protocol suite is the heart of the world-wide Internet.

The TCP/IP protocol stack runs on each and every end system, whereas the IP protocol runs on every router on the Internet.

Another issue in internetworking is how to connect two or more subnetworks when some subnetworks support connection-oriented service and some networks support connectionless service.

Connection-oriented internetworking: When two networks, both supporting connection-oriented service, have to be connected, it is called connection-oriented internetworking. In this case, the router that is used to connect the two subnetworks appears as a node to the subnetworks to which it is attached. A logical connection is established between the two networks, and this connection is a concatenation of the sequence of logical connections across the subnetworks.

> When two networks, both supporting connection-oriented service, are interconnected, it is called connection-oriented internetworking.

In connectionless internetworking, each packet is handled independently and routed from the source to the destination via routers.

Connectionless internetworking: Consider a situation in which a number of subnetworks are to be interconnected, but some subnetworks support connection-oriented service and some subnetworks support connectionless service. In this situation, the internet can provide datagram service. Each packet is treated independently and routed from the source node to the destination node via the routers and the networks. At each router, an independent routing decision is made. The advantage of this internetworking is that it provides a flexible service because underlying networks can be a mix of connectionless and connection-oriented services. Implementation using this approach is quite easy.

Connectionless internetworking provides lots of flexibility because the networks can be a mix of both connection-oriented and connectionless services. The global Internet provides connectionless internetworking.

The Internet Protocol (IP) and ISO/OSI Connectionless Network Protocol (CLNP) provide connectionless internetworking service (datagram service). However, note that the problems associated with the datagram service have to be taken care of. Packets may be lost, packets may be received with different delays as a result of which they will not be in sequence, and some packets may be received more than once. These problems are taken care of by the TCP layer.

When a LAN is connected to a WAN, the router interconnecting the LAN and WAN has to run the protocol stacks of both LAN and WAN and do the necessary protocol translation.

Figure 19.3 illustrates the protocols that need to run on each node and the router in an internetworking scenario. Two LANs are interconnected through a WAN. Each node on the LAN runs the TCP/IP protocol suite (including the application layer software), and the WAN runs the protocols based on X.25 standards. The router has to run both the LAN and WAN protocols. In addition, it runs the IP software—it takes the packet from the LAN, converts it into packets that can be understood by the WAN protocols, and transmits on the WAN. Similarly, packets received from the other side are converted into a format that can be understood by the LAN. Figure 19.3 depicts the protocol stacks that run on the LAN nodes and the two routers.

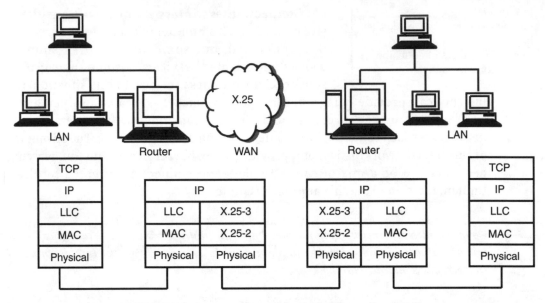

FIGURE 19.3 Internetworking LANs through a WAN.

19.2 THE INTERNET—THE GLOBAL NETWORK OF NETWORKS

The Internet is a global network of computer networks. The TCP/IP protocol suite ensures that the diverse networks (LANs, corporate WANs, public data networks, etc.) can be internetworked for people to exchange information.

The TCP/IP protocol stack provides the capability to interconnect different networks to form a global network of networks, called the Internet. All protocols to be used in the Internet have to be ratified by the Internet Engineering Task Force (IETF).

The entire TCP/IP protocol suite is the result of researchers working in universities and research laboratories. There is no central administrative authority to control the Internet. To ensure that the protocol suite meets the growing requirements, new developments have to take place and new protocols need to be defined. This activity is controlled by the Internet Architecture Board (IAB), which consists of a chairperson and a number of volunteers. IAB has two arms—the Internet Research Task Force (IRTF) and the Internet Engineering Task Force (IETF). IRTF coordinates the research efforts. IETF, consisting of a number of working groups, coordinates the engineering issues. Any new protocol to be implemented has to be approved by IETF. For each protocol, an RFC (Request for Comments) document is released by IETF.

The entire Internet spanning the entire planet is just an extension of the small internet shown in Figure 19.2. The TCP/IP protocol suite enables any machine on the Internet to communicate with any other machine. The Internet is very dynamic—every day, a number of networks are connected, and some are disconnected as well. In spite of this dynamic nature, how do we get connected to any machine and access a Web site? It is the IP and the TCP that do the trick. If internetworking still appears a mystery, we will unravel the mystery in the next chapter, where we will study the TCP/IP protocol suite in detail.

Summary

To connect two or more networks is a challenging task because each network has its own protocols. In this chapter, we studied how a router can be used to network heterogeneous networks. The router's function is to do the necessary protocol conversion. The Internet, a global network interconnecting millions of networks, also uses the same mechanism. The IP and TCP protocols provide the means of achieving global connectivity. The Internet Protocol (IP) has to run on each and every router and each and every end system. The Transmission Control Protocol (TCP) has to run on each and every end system. The router is introduced in this chapter. We will study the details in later chapters.

References

V. Cerf and R. Kahn. "A protocol for Packet Network Interconnection." *IEEE Transactions on Communications*, COM-22, Vol. 5, May 1974. This paper is written by the two persons who laid the foundation for the global Internet.

D.E. Comer and D.L. Stevens. *Internetworking with TCP/IP, Vol. III: Client/Server Programming and Applications*, BSD Socket Version. Prentice Hall Inc., Englewood Cliffs, N.J., 1993. This book gives the complete software for internetworking using TCP/IP.

www.ietf.org The Web site of IETF. You can obtain the RFCs (Request for Comments) that give the complete details of the Internet protocols from this site.

Questions

1. What are the various issues involved in networking of heterogeneous networks?

2. What are the functions of a gateway or router?

3. Draw a diagram that depicts the protocol stacks that need to run on the end systems (hosts) and the routers when an Ethernet LAN is connected to an X.25 WAN.

Exercises

1. In your department/organization, if there are two LANs, study how they are connected and what network elements (bridge/router) interconnect the two LANs.

2. Cisco Corporation is one of the leading suppliers of routers. Study the various internetworking products supplied by them. You can get the information from *www.cisco.com*.

Projects

1. The IP (Internet Protocol) is the heart of internetworking. Study the source code for the IP implementation on a Linux system.

2. Develop a protocol converter that takes RS232 data in serial format and converts it into Ethernet format. The software has to read the data from a serial port on the PC and convert the data into Ethernet packets.

3. Embed the software developed in Project #2 in a processor-based system. You can use an embedded operating system such as Embedded Linux or RTLinux.

20 ▪ TCP/IP Protocol Suite

In This Chapter

- The TCP/IP Protocol Architecture
- The Functions of the TCP and IP Protocol Layers
- Differences Between TCP and UDP
- Flow Control Protocols
- The Interplanetary Internet

The TCP/IP protocol suite was developed during the initial days of research on the Internet and evolved over the years, making it a simple yet efficient architecture for computer networking. The ISO/OSI architecture developed subsequently has not caught on very well because the Internet had spread very fast and the large installation base of TCP/IP-based networks could not be replaced with the ISO/OSI protocol suite.

The TCP/IP protocol suite is now an integral part of most operating systems, making every computer network-ready. Even very small embedded systems are being provided with TCP/IP support to make them network enabled. These systems include Web cameras, Web TVs, and so on. In this chapter, we will study the TCP/IP architecture. A thorough understanding of this architecture is a must for everyone who is interested in the field of computer networking.

20.1 TCP/IP PROTOCOL SUITE

The TCP/IP protocol suite was developed as part of the United States Department of Defense's project ARP Anet (Advanced Research Projects Agency Network), but the standards are publicly available. Due to fast spread of the Internet, the

TCP/IP protocol suite has a very large installation base. The TCP/IP software is now an integral part of most operating systems, including Unix, Linux, Windows. The TCP/IP stack is also being embedded into systems running real-time operating systems such as VxWorks, RTLinux, and OS/9 and handheld operating systems such as Embedded XP, Palm OS, Symbian OS and so on.

The TCP/IP protocol suite is depicted in Figure 20.1. It consists of 5 layers:

■ Physical layer
■ Datalink layer (also referred to as the network layer)
■ Internet Protocol (IP) layer
■ Transport layer (TCP layer and UDP layer)
■ Application layer

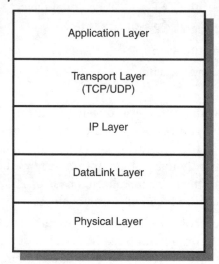

FIGURE 20.1 TCP/IP protocol suite.

Physical layer: This layer defines the characteristics of the transmission such as data rate and signal-encoding scheme.

Datalink layer: This layer defines the logical interface between the end system and the subnetwork.

Internet Protocol (IP) layer: This layer routes the data from source to destination through routers. Addressing is an integral part of the routing mechanism. IP provides an unreliable service: the

The TCP/IP protocol suite consists of five layers: physical layer, datalink layer (also referred to as network layer), Internet Protocol (IP) layer, Transmission Control Protocol (TCP) layer, and application layer.

packets may be lost, arrive out of order, or have variable delay. The IP layer runs on every end system and every router.

Transport layer: This layer is also called the host-to-host layer because it provides end-to-end data transfer service between two hosts (or end systems) connected to the Internet. Because the IP layer does not provide a reliable service, it is the responsibility of the transport layer to incorporate reliability through acknowledgments, retransmissions, and so on. The transport layer software runs on every end system.

For applications that require reliable data transfer (such as most data applications), a connection-oriented transport protocol called Transmission Control Protocol (TCP) is defined. For connectionless service, user datagram protocol (UDP) is defined. Applications such as network management, that do not need very reliable packet transfer use the UDP layer.

Application layer: This layer differs from application to application. Two processes on two end systems communicate with the application layer as the interface.

As in OSI architecture, peer-to-peer communication applies to the TCP/IP architecture as well. The application process (such as for transferring a file) generates an application byte stream, which is divided into TCP segments and sent to the IP layer. The TCP segment is encapsulated in the IP datagram and sent to the datalink layer. The IP datagram is encapsulated in the datalink layer frame. Since datalink layer can be subdivided into LLC layer and MAC layer, IP datagram is encapsulated in the LLC layer and then passed on to the MAC layer. The MAC frame is sent over the physical medium. At the destination, each layer strips off the header, does the necessary processing based on the information in the header, and passes the remaining portion of the data to the higher layer. This mechanism for protocol encapsulation is depicted in Figure 20.2.

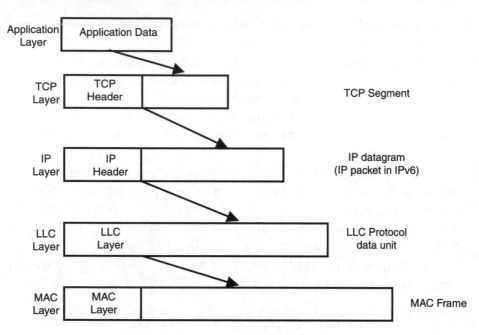

FIGURE 20.2 Protocol encapsulation in TCP/IP.

In the TCP/IP protocol stack, the various application layer protocols are SMTP for e-mail, FTP for file transfer, Telnet for remote login, and HTTP for World Wide Web service.

The complete TCP/IP protocol stack is shown in Figure 20.3 indicating various application layer protocols. The various application layer protocols are

- Simple Mail Transfer Protocol (SMTP), for electronic mail containing ASCII text.

- Multimedia Internet Mail Extension (MIME), for electronic mail with multimedia content.

- File Transfer Protocol (FTP) for file transfer.

- Telnet for remote login.

- Hypertext Transfer Protocol (HTTP) for World Wide Web service.

In addition, the following protocols are also depicted:

- Border gateway protocol (BGP), a routing protocol to exchange routing information between routers. Exterior gateway protocol (EGP) is another routing protocol.

- Internet Control Message Protocol (ICMP), which is at the same level as IP but uses IP service.

- Simple Network Management Protocol (SNMP) for network management. Note that SNMP uses UDP and not TCP.

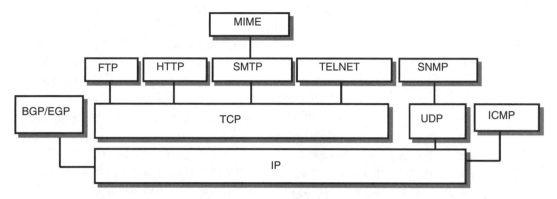

FIGURE 20.3 TCP/IP protocol stack.

 The TCP and IP layer software runs on every end system. The IP layer software runs on every router.

20.2 OPERATION OF TCP AND IP

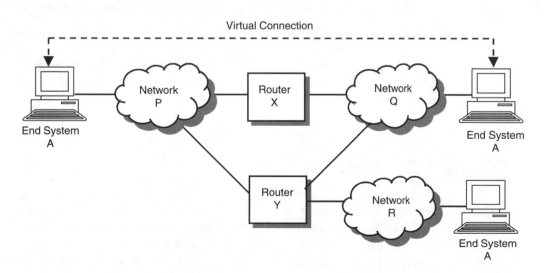

FIGURE 20.4 TCP/IP operation in an internet.

When two end systems have to exchange data using TCP/IP protocol stack, a TCP connection is established, and the data transfer takes place. Though IP layer does not provide a reliable service, it is the TCP layer that will ensure end-to-end reliable transfer of data through error detection and retransmission of packets.

Consider the internet shown in Figure 20.4. Each end system will be running the TCP/IP protocol stack, including the application layer software. Each router will be running the IP layer software. If the networks use different protocols (e.g., one is an Ethernet LAN and another is an X.25 WAN), the router will do the necessary protocol conversion as well. End system A wants to transfer a file to end system B.

Each end system must have a unique address—this address is the IP address. In addition, the process in end system A should establish a connection with the process running in end system B to transfer the file. Another address is assigned to this address, known as the port address. For each application, a specific port address is specified. When port 1 of A wishes to exchange data with port 2 on B, the procedure is:

1. Process on A gives message to its TCP: Send to B port 2.

2. TCP on A gives the message to its IP: Send to host B. (Note: IP layer need not know the port of B.)

3. IP on A gives message to the datalink layer with instructions to send it to router X.

4. Router X examines the IP address and routes it to B.

5. B receives the packet, each layer strips off the header, and finally the message is delivered to the process at port 2.

Though we talk about TCP connection, it needs to be noted that there is no real connection between the end systems. It is a virtual connection. In other words, TCP connection is only an abstraction.

20.3 INTERNET PROTOCOL (IP)

Internet Protocol (IP) is the protocol that enables various networks to talk to each other. IP defines the data formats for transferring data between various networks, and it also specifies the addressing and routing mechanisms. The service delivered by IP is unreliable connectionless packet service. The service is unreliable because there is no guarantee that the packets will be delivered—

The main functions of the IP layer are addressing and routing. Each machine is given an IP address that is unique on the network. The destination address in the IP datagram is used to route the packet from the source to the destination.

packets may be lost if there is congestion, though a best effort is made for the delivery. The packets may not be received in sequence, packets may be duplicated, and packets may arrive at the destination with variable delay. The service is connectionless because each packet is handled independently. IP defines the rules for discarding packets, generating error messages, and how hosts and routers should process the packets.

IP is implemented as software. This software must run on every end system and on every router in any internet using the TCP/IP protocol suite.

In Figure 20.4, the router X may deliver the packet to network Q directly or it may deliver it to router Y, which in turn delivers to network Q. So, the packets may take different routes and arrive at the end system B out of sequence. It is the TCP layer that takes care of presenting the data in proper format to the application layer.

It is important to note that the IP layer does not provide a reliable service. The packets may be lost on the route from the source to the destination if there is congestion in the network. It is the responsibility of the TCP layer to ask for retransmissions and ensure that all the packets are received at the destination.

20.4 TRANSMISSION CONTROL PROTOCOL (TCP)

The TCP layer provides end-to-end reliable transfer of data by taking care of flow control and error control. If packets are received in error, retransmission is requested. If packets are received out of order, they are put in sequence. It appears to the application layer as though everything is fine, but the TCP layer needs to do a lot of work to achieve this.

It is the job of transport layer protocol to ensure that the data is delivered to the application layer without any errors. The functions of the transport layer are:

- To check whether the packets are received in sequence or not. If they are not in sequence, they have to be arranged in sequence.

- To check whether each packet is received without errors using the checksum. If packets are received in error, TCP layer has to ask for retransmissions.

- To check whether all packets are received or whether some packets are lost. It may happen that one of the routers may drop a packet (discard it) because its buffer is full, or the router itself may go faulty. If packets are lost, the TCP layer has to inform the other end system to retransmit the packet. Dropping a packet is generally due to congestion on the network.

Sometimes, one system may send the packets very fast, and the router or end system may not be able to receive the packets at that speed. Flow control is done by the transport layer.

It is the job of the transport layer to provide an end-to-end reliable transfer of data even if the underlying IP layer does not provide reliable service. The Transmission Control Protocol (TCP) does all these functions, through flow control and acknowledgements.

In TCP/IP networks, it is not possible to ensure that all the packets are received at the destination with constant delay. In other words, the delay may vary from packet to packet. Hence, we can say that TCP/IP networks do not guarantee a desired quality of service. This characteristic poses problems to transfer real-time data such as voice or video over TCP/IP networks.

20.4.1 Flow Control and Acknowledgements

To provide a reliable transmission, the acknowledgement policy is used. The two protocols for this mechanism are the stop-and-wait protocol and the sliding window protocol. These protocols take care of lost packets, flow control, and error detection.

Stop-and-Wait Protocol

In stop-and-wait protocol, the source sends a packet and only after the acknowledgement is received from the destination is the next packet sent. This is a simple protocol, but it results in lots of delay, and the bandwidth is not used efficiently.

When the source (end system A) sends the first packet to the destination (end system B), B sends an acknowledgment packet. Then A sends the second packet, and B sends the acknowledgement. This is a very simple protocol. But the problem is that if the acknowledgement for a packet is lost, what has to be done? A sends the first packet and then starts a timer. The destination, after receiving the packet, sends an acknowledgement. If the acknowledgement is received before the timer expires, the source sends the next packet and resets the timer. If the packet sent by the source is lost, or if the acknowledgement sent by the destination is lost, the timer will expire, and the source resends the packet.

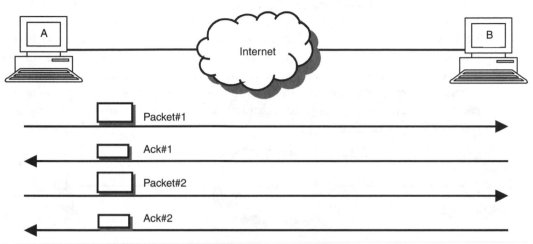

FIGURE 20.5 Stop-and-wait protocol.

This protocol is very simple to implement. However, the drawback is that the throughput will be very poor and the channel bandwidth is not used efficiently. For instance, if this protocol is used in a satellite network, A will send a packet, and after one second it will receive the acknowledgment. During that one second, the satellite channel is free, and the channel is not used effectively. A refinement to this protocol is the sliding window protocol.

Sliding Window Protocol

In this protocol, the source sends a certain number of packets without waiting for the acknowledgements. The destination receives the packets and sends an acknowledgement for each packet. The source will have a timer for each packet and keeps track of the unacknowledged packets. If the timer expires for a particular packet and the acknowledgement is not received, that packet will be resent. This way, the throughput on the network can be increased substantially.

In sliding window protocol, a certain number of packets (say seven) are sent by the source without waiting for an acknowledgement. The destination can send a single acknowledgement for packet 3 indicating that 3 packets are received. Using this approach, the number of acknowledgements can be reduced, and the throughput can be increased.

There are many options as to when to send the acknowledgement. One option is the window size. If the sliding window size is seven, the source can send up to seven packets without waiting for the acknowledgement. The destination can send an acknowledgement after receiving all seven packets. If the destination has not received packet four, it can send an acknowledgement indicating that up to packet three were received. As shown in Figure 20.6, if B sends ACK three, the source knows that up to packetthree were received correctly, and it sends all the packets from four onwards again. Another option in the sliding window protocol is when to send the acknowledgements. A positive acknowledgment can be sent indicating that up to packet #n all packets are received. Alternatively, a negative acknowledgement may be sent indicating that packet #n is not received.

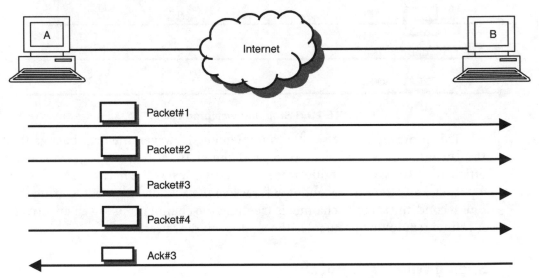

FIGURE 20.6 Sliding window protocol.

The sliding window protocol also addresses flow control. If the destination cannot receive packets, with the speed with which the source sends the packets, the destination can control the packets flow.

Using this simple protocol, TCP layer will take care of flow control error control and tell the source that the packets are being received. We will discuss the details of the IP and TCP layers in the next two chapters.

20.4.2 Congestion Control

On the Internet, many connections get established and closed, so the traffic on the Internet is difficult to predict. If the traffic suddenly goes up, there will be

In TCP/IP networks, congestion control is done through an additive-increase, multiplicative-decrease mechanism. To start with, a congestion window size is fixed. If there is congestion, the window size is reduced to half. If the congestion is reduced, the window size is increased by one.

congestion in the network and, as a result, some packets may be discarded by the routers. Every host has to have some discipline in transmitting its packets—there is no point in pushing packets onto the network if there is congestion.

In TCP, congestion control is done through a mechanism called additive increase/multiplicative decrease. A congestion window size is fixed at the beginning of the transmission, such as 16 packets. If there is suddenly congestion, and a packet loss is detected, the TCP reduces the congestion window size to 8. Even then, if a packet loss is detected, the window size is reduced to 4, and then to Z. The decrease is multiplicative.

If the congestion is reduced on the network, and acknowledgements for the packets are being received by the source, for each acknowledgement received, the TCP increases the window size by 1. If 4 was the earlier window size, it becomes 5, then 6 and so on. The increase is additive.

This simple mechanism for flow control ensures that the transmission channel is used effectively.

20.5 USER DATAGRAM PROTOCOL (UDP)

To provide reliable service, the TCP layer does lots of processing, and hence it is called a heavy-weight protocol. The UDP is another transport layer that provides connectionless service. Processing of UDP packets is much faster.

The TCP provides a reliable service by taking care of error control and flow control. However, the processing required for the TCP layer is very high and is called a heavyweight protocol. In some applications such as real-time voice/video communication and network management, such high processing requirements create problems. Another transport protocol is used for such applications. That is user datagram protocol (UDP). UDP provides a connectionless service. It sends the packets to the destination one after the other, without caring whether they are being received correctly or not. It is the job of the application layer to take care of the problems associated with lack of acknowledgements and error control. Simple Network Management Protocol (SNMP), which is used for network management, runs above the UDP.

In applications such as real-time voice communications, if a packet is lost, there is no point in asking for a retransmission because it causes lots of delay. For such applications, UDP is a better choice than TCP.

20.6 TCP/IP OVER SATELLITE LINKS

The TCP/IP protocol stack can be used in any network—the transmission medium can be cable, optical fiber, terrestrial radio, or satellite radio. However, when TCP/IP is used in satellite networks, the stack poses problems. This is due to the characteristics of the satellite channels. The problems are as follows:

Satellite communication systems are characterized by large propagation delay and high Bit Error Rate. In such systems, using TCP/IP creates problems because of the timeouts for acknowledgements and retransmissions. TCP/IP protocols are suitably modified to work on satellite channels.

- The satellite channel has a large propagation delay. Large delay causes timeouts in the flow control protocol. The source assumes that the packets have not reached the destination and resends the packets. As a result, the destination receives duplicate packets. This causes congestion in the network.
- The satellite channels have a larger Bit Error Rate (BER) than the terrestrial channels. As a result, packet losses will result in more retransmissions. When retransmission of packets is required, TCP automatically reduces the window size, though the network is not congested. As a result, the throughput of the channel goes down.

To overcome these problems, a number of solutions are proposed, which include the following:

- To improve the link performance, error-correcting codes are used. Errors can be corrected at the destination, and retransmissions can be reduced.
- Instead of using the flow control protocols at the transport layer, these protocols can be implemented at the datalink layer, so that the TCP layer does not reduce the window size.
- Instead of using a default window size of 16 bits in the TCP segment, 32 bits can be used to increase the throughput.
- For bulk transfer of information from the source to the destination, multiple TCP connections can be established.

- Another interesting technique used is called spoofing. A small piece of software will run at the source, generating the acknowledgements locally. The local TCP layer is cheated by the spoofing software. The spoofing software in turn receives the actual acknowledgement from the destination and discards it. If a packet is to be retransmitted because it was received in error at the destination, the spoofing software requests the TCP layer to resend the packet.

Many improvements in the TCP/IP protocol layers are required for its use in satellite networks.

20.7 INTERPLANETARY (IPN) INTERNET

The Internet, as we know it, is a network of connected networks spread across the earth. The optical fiber–based backbone (the set of high-capacity, high-availability communication links between network traffic hubs) of the Internet supports very high data rates with negligible delay and negligible error rates, and continuous connectivity is assured. If there is loss of packets, it implies congestion of the network.

Now, imagine Internets on other planets and spacecraft in transit. How do we go about interconnecting these internets with the Earth's Internet? Or think of having an Internet service provider to the entire solar system. The brainchild of Vincent Cerf, the Interplanetary Internet (InterPlaNet or IPN), aims at achieving precisely this. IPN's objective is to define the architecture and protocols to permit interoperation of the Internet on the earth and other remotely located internets situated on other planets and spacecraft in transit, or to build an Internet of Internets.

The deep space communication channels are characterized by high data loss due to errors, transient link outages, asymmetric data rates, unidirectional channels, power constrained end systems. To develop protocols to work in this type of communication environment is a technology challenge; also, such protocols will lead to better solutions even to develop systems on the earth.

The three basic objectives of the IPN are these:

- Deploy low delay internets on other planets and remote spacecraft.
- Connect these distributed (or disconnected) internets through an interplanetary backbone that can handle the high delay.

- Create gateways and relays to interface between high delay and low delay environments.

The TCP/IP protocol suite cannot be used for the IPN for the following reasons:

- Communication capacity is very expensive, every bit counts, and so protocol overhead has to be minimized.
- Interactive protocols do not work and so:
    - Reliable sequential delivery takes too long.
    - Negotiation is not practical.
    - It is difficult to implement flow control and congestion control protocols.
    - Retransmission for error recovery is expensive.
    - Protocols need to be connectionless.

The proposed IPN architecture is shown in Figure 20.7. The thrust areas for implementation of this architecture are:

- Deployment of internets on various planets and spacecraft
- Inter-internet protocols
- Interplanetary gateways (IG)
- Stable backbone
- Security of the user data and the backbone

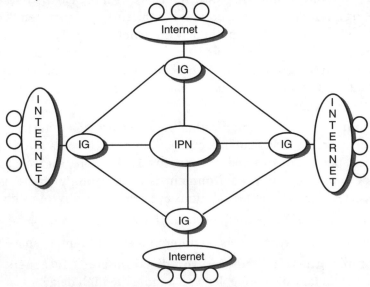

FIGURE 20.7 Architecture of Interplanetary Internet.

The vision of Vincent Cerf, the Interplanetary Internet, will interconnect the various internets located on the Earth and other planets and spacecraft. To develop this network, new protocols need to be designed because TCP/IP protocol stack does not work due to enormous propagation delays, low speeds, and the need for connectionless services.

Presently work is underway by IPNSIG (Interplanetary Internet Special Interest Group: *http://www.ipnsig.org*) to define and test the new set of protocols and the backbone architecture. The work includes defining new layers in the place of TCP and IP, using RF instead of fiber for the backbone network, as well as the addressing issues. (A few years from now, if you have to send mail to someone, you may need to specify *.earth* or *.mars* extension to refer to the internets of Earth and Mars.)

The time frame for the IPN protocol testing is 2003+ for Mars Internet. It is proposed to use schools as test beds for testing the new protocols. Though the IPN may be of practical use perhaps 20 years from now, it is expected that the outcome of this research will solve many of the high-delay problems encountered on the networks on the earth itself.

Summary

This chapter presented an overview of the TCP/IP protocol stack. Above the physical and datalink layers, the Internet Protocol (IP) layer takes care of addressing and routing. IP provides a connectionless service, and there is no guarantee that all the packets will be received. The packets also may be received out of sequence. It is the job of transport layer protocol to take care of these problems. The transport layer provides end-to-end reliable service by taking care of flow control, error control, and acknowledgements. Above the TCP layer, different application layer protocols will be running, such as Simple Mail Transfer Protocol (SMTP) for e-mail, File Transfer Protocol (FTP) for transferring files, and Hypertext Transfer Protocol (HTTP) for the World Wide Web. The user datagram protocol (UDP) also runs above the IP, but it provides a connectionless service. Since the processing involved in UDP is less, it is used for network management and real-time communication applications. The TCP/IP protocol stack presents problems when used in satellite networks because satellite networks have high propagation delay. The TCP layer has to be suitably modified for use in satellite networks.

Another innovative project is the Interplanetary Internet, which plans for interconnection of internets of different planets and spacecrafts. The results of this research will help improve the TCP/IP protocol stack performance in high-delay networks.

References

L.L. Peterson and B.S. Davie. *Computer Networks: A Systems Approach*. Morgan Kaufman Publishers Inc., CA, 2000. This book gives a systems approach, rather than a layered approach to computer networks.

A.S. Tanenbaum. *Computer Networks*. Prentice Hall, Inc., NJ, 1996. This book gives a layered approach to describe the computer networking protocols.

N. Ghani and S. Dixit. "TCP/IP Enhancements for Satellite Networks". *IEEE Communications Magazine*, Vol. 37, No. 1, July 1999.

www.ietf.org The Requests for Comments (RFCs) that give the complete details of the TCP/IP protocol stack can be obtained from this site. Each protocol specification gives the complete details of implementation.

www.ipnsig.org Web site of Interplanetary Internet.

Questions

1. Explain the functions of different layers in the TCP/IP protocol architecture.
2. Explain the operation of TCP and IP.
3. IP does not provide a reliable service, but TCP provides end-to-end reliable service. How?
4. What are the limitations of the TCP/IP protocol stack?
5. Differentiate between TCP and UDP.
6. List the problems associated with running the TCP/IP protocol stack in a satellite network.
7. Explain how congestion is controlled in TCP/IP networks.

Exercises

1. Write a technical report on Interplanetary Internet.
2. Prepare a technical report on running the TCP/IP protocol stack on a satellite network.
3. Two systems, A and B, are connected by a point-to-point link, but the communication is only from A to B. Work out a mechanism to transfer a file from A to B using UDP as the transport protocol.
4. Discuss the benefits of using UDP for data applications if the transmission link is very reliable and if there is no congestion on the network.

5. Compare the performance of stop-and-wait protocol and sliding window protocol in terms of delay and throughput.

Projects

1. Interconnect two PCs using an RS232 link. Simulate a high delay in the network, run the TCP/IP protocol stack, and observe the throughput on the LAN to study the impact of the delay on the TCP/IP stack.

2. Develop software for spoofing—when the acknowledgement receipt is delayed from the other machine, an acknowledgement can be locally generated to cheat the TCP layer.

21 Internet Protocol (IP)

In This Chapter

- The Internet Protocol
- The IP Addressing Formats
- The Limitations of IP Version 4
- The Salient Features of IP Version 6
- The Routing Protocols

The Internet Protocol (IP) is the heart of the Internet. Networks running different protocols are connected together to form the global network because of the IP. In this chapter, we will study IP version 4, which is presently running on most of the routers and end systems. We also will discuss IP Version 6 which is the next version of the IP that is being deployed and that will become predominant in coming years. We also will discuss various routing protocols. Details of Internet Control Message Protocol (ICMP) are also presented in this chapter.

21.1 OVERVIEW OF INTERNET PROTOCOL

The functions of the IP layer are addressing and routing.
IP provides a connectionless service: each packet is handled independently by the router.

Internet Protocol (IP) is the protocol that enables various networks to talk to each other. IP defines the data formats for transferring data between various networks. It also specifies the addressing and routing mechanisms. The service delivered by IP is unreliable connectionless packet service. The service is unreliable because there is no guarantee that the packets will be delivered—packets may be lost if there is congestion, though "best-effort" is made for the delivery. The packets may not be received in sequence, packets may be duplicated, and packets may arrive at the destination with variable delay. The service is connectionless because each packet is handled independently. IP defines the rules for discarding packets, generating error messages, and how hosts and routers should process the packets.

IP is implemented as software. This software must run on every end system and on every router in any internet that uses the TCP/IP protocol suite.

The IP software runs on every router as well as on every end system.

21.2 INTERNET ADDRESSING SCHEME

In IP Version 4, each system on the network is given a unique 32-bit IP address. The address consists of network ID and the host ID.

Each end system on the network has to be uniquely identified. For this, the addressing scheme is very important. Since each end system is a node on a network, the addressing scheme should be such that the address contains both an ID for the network and an ID for the host. This scheme is followed in the IP addressing scheme. Each node on a TCP/IP network is identified by a 32-bit address. The address consists of the network ID and the host ID. IP address can be of five formats, as shown in Figure 21.1.

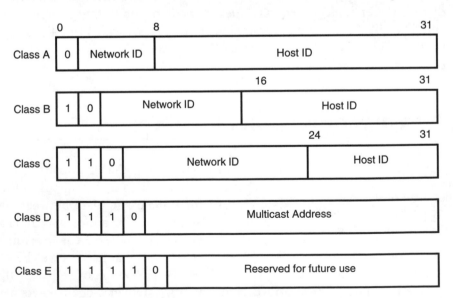

FIGURE 21.1 IP address formats.

Class A addresses: Class A addressing is used when a site contains a small number of networks, and each network has many nodes (more than 65,536). Seven bits are used for network ID and 24 bits for host ID. A class A address has 0 in the first bit.

The IP addresses are divided into five classes: A, B, C, D, and E. The number of bits assigned to the network ID field and the host ID field are different in each class.

The maximum number of class A networks can be 126 (the network addresses 0 and 127 are reserved). Each network can accommodate up to $(2^{24} - 2)$ hosts. Note that two host addresses are reserved (all zeros and all ones).

Class B addresses: Class B addressing is used when a site has a medium number of networks and each network has more than 256 but less than 65,536 hosts. Fourteen bits are allocated for network ID and 16 bits for the host ID. A class B address has 10 for the first two bits.

Class C addresses: Class C addressing is used when a site has a large number of networks with each network having fewer than 256 hosts. Twenty-one bits are allocated to network ID and 8 bits to host ID. A class C address has 110 for the first three bits.

Class D addresses: These addresses are used when multicasting is required, such as when a datagram has to be sent to multiple hosts simultaneously.

Class E addresses: These addresses are reserved for future use.

In the IP address, if the host address bits are all zeros, the IP address represents the network address. If the host address bits are all ones, the IP address is the broadcast address—the packet is addressed to all hosts on the network.

It is possible that a host may be connected to different networks. A router is also connected to different networks. Such computers are known as multihomed hosts. These hosts need multiple IP addresses, each address corresponding to the machine's network connection. Hence, an IP address is given to the network connection.

When the sender wants to communicate over a network but does not know the network ID, network ID is set to all zeros. When that network sends a reply, it contains its network ID, which is recorded by the sender for future use.

Some of the drawbacks of this addressing scheme are:

- In a class C network, if the number of hosts increases to more than 256, the whole addressing scheme has to be changed because network ID has to change. This calls for lots of work for the system administrator.

■ If a host is disconnected from one network and connected to another network, the IP address has to change.

The addressing format of IP Version 4 is not an efficient addressing scheme because changing from one class of address to another class is very difficult.

An important point to be noted while transmitting the IP address is that integers are sent most significant byte first ("Big-Endian style") so that all the machines can interpret the correct address.

If a datagram has to be sent to multiple hosts, a multicast addressing scheme specified by class D addresses is used. IP multicast addresses can be assigned by a central authority (called well-known addresses) or temporarily created (called transient multicast groups). Multicast addressing is useful for applications such as audio and video conferencing.

The multicast address is used to send a datagram to multiple hosts. This addressing mechanism is required for applications such as audio/video conferencing.

In a class D addressing scheme, the first four bits are 1110. The remaining 28 bits identify the multicast address. Obviously, this address can be used only in the destination IP address and not in the source IP address.

An IP multicast address is mapped to the Ethernet multicast address by placing the lower 23 bits of the IP address into the low-order 23 bits of the Ethernet multicast address. However, note that this mapping is not unique, because five bits in the IP address are ignored. Hence, it is possible that some hosts on the Ethernet that are not part of the multicast group may receive a datagram erroneously, and it is the host's responsibility to discard the datagram.

Special routers called multicast routers are used to route datagrams with multicast addresses. Internet's multicast backbone (MBONE) has multicast routers that route the multicast traffic over the Internet. If a router does not support multicast routing, the mulitcast datagram is encapsulated in the normal unicast IP datagram, and the receiver has to interpret the multicast address.

21.2.1 Dotted Decimal Notation

Because it is difficult for us to read the IP address if it is written in 32-bit format, dotted decimal notation is used. If the IP address is

11111110 01111111 10000001 10101010

it can be represented as *254.127.129.170* for easy readability.

The 32-bit IP address is represented as a.b.c.d where the values of a, b, c and d can be between 0 and 255. This notation is called dotted decimal notation.

Internet Network Information Center (InterNIC) is the central authority to assign IP addresses. InterNIC gives the network ID and the organization is free to assign host addresses. Many organizations assign local IP addresses without obtaining the network ID from InterNIC. This is OK if the network remains isolated, but if it is connected to the Internet later on, there may be an address clash.

21.3 ADDRESS RESOLUTION PROTOCOL

Address resolution protocol (ARP) is used when the router does not know the IP address of the node to which the packet has to be sent. The router broadcasts the packet, and the node with the corresponding IP address responds with its physical address.

Consider a case when a router connected to a LAN receives a packet for one of the nodes on the LAN. The packet is routed up to the router based on the network ID. How does the router send the packet to that specific node? Remember, the packet has to be sent to the node using the physical address, and the received packet contains only the IP address.

The ARP solves this problem. The router connected to the LAN sends a broadcast message with the IP address of the node as the destination address. All the nodes receive this packet, and the node with that IP address responds with its physical address. Subsequently, the packets will be transmitted to that node with that physical address. Since broadcasting is a costly proposition, the router can keep a cache in which the physical addresses corresponding to the IP addresses are stored for subsequent use. Since the physical address may change, the router keeps the physical address in the cache only, because it is likely to be valid only for a certain period of time.

In the Ethernet frame, the ARP message is embedded in the data field. The type field of the Ethernet frame contains the code *0806* (hex) to indicate that it is an ARP message.

21.4 REVERSE ADDRESS RESOLUTION PROTOCOL

Machines that do not have secondary storage (diskless machines) do not know their IP addresses. At the time of bootup, the machine has to get its own IP

In reverse address resolution protocol, a server stores the IP addresses and the corresponding network addresses of the nodes. This protocol is used when the nodes are diskless machines and do not know their IP addresses.

address. However, the physical address of the machine is known, because it is a part of the hardware, the network card. As you can see, this problem is the opposite of the earlier problem, and hence reverse address resolution protocol (RARP) was developed.

In RARP, the machine that wants to find out its IP address broadcasts a packet with its own network address. A RARP server, which stores the IP addresses corresponding to all the network addresses (in secondary storage), receives the packet and then sends a reply to the machine with the information on its IP address. A requirement for RARP to succeed is that a designated RARP server should be present that contains the lookup table. Sometimes, for providing a reliable service, primary and secondary RARP servers will be installed in LANs. If the primary RARP server fails, the secondary RARP server will take over and provide the IP addresses to the nodes on the LAN.

21.5 IP DATAGRAM FORMAT

The IP that runs on most of the hosts and routers on the Internet is IP Version 4 (IPv4). Gradually, the IP software will be upgraded to IP Version 6 (IPv6). We will first study the IPv4 and then study IPv6.

The basic unit of data transfer is called datagram in IP Version 4 and a packet in IP Version 6. We will use the words datagram and packet interchangeably in the following discussion. The datagram of IPv4 has a header and the data fields. The detailed format of the datagram is shown in Figure 21.2. Each of the fields is explained.

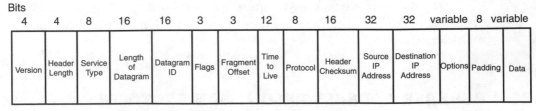

FIGURE 21.2 IPv4 datagram format.

Version number (4 bits): Version number of the IP. The version presently running in most of the systems is Version 4. Version 6 is now slowly being deployed. This field ensures that the correct version of the software is used to process the datagram.

Header length (4 bits): Length of the IP header in 32-bit words. Note that some fields in the header are variable, and so the IP datagram header length may vary. The minimum length of the header is 20 bytes, and so the minimum value of this field is 5.

Service type (8 bits): These bits specify how the datagram has to be handled by systems. The first three bits specify the precedence—0 indicating normal precedence and 7 network control. Most routers ignore these bits, but if all the routers implement it, these bits can be used for providing precedence to send control information such as congestion control. The 4th, 5th, and 6th bits are called D, T, and R bits. Setting D bit is to request low delay, setting T bit is to request high throughput, and setting R bit is to request high reliability. However, it is only a request; there is no guarantee that the request will be honored. Note that these bits are to set the quality of service (QoS) parameters. there is no guarantee that the required QoS will be provided. That is the reason we keep hearing that sentence "IP does not guarantee a desired QoS."

Length (16 bits): Total length of the datagram in bytes including header and data. The length of the data field is calculated by subtracting the header length from the value of this field. Since 16 bits is allotted to this field, the maximum size of an IP datagram is limited to 65,535 bytes.

Note that the IP datagram size is much larger than can be accommodated by a LAN that can handle only 1526 bytes in one frame for instance. In such a case, the datagram has to be fragmented and sent over the network. The minimum datagram size that every host and router must handle is 576 bytes. Each fragment contains most of the original datagram header. The fragmentation is done at routers, but the reassembly is done at the destination.

The minimum datagram size every host and router must handle is 576 bytes. If larger datagrams are received, they may need to be fragmented and reassembled.

Identification (16 bits): Unique ID to identify the datagram. This field lets the destination know which fragment belongs to which datagaram. The source address, destination address, and identification together uniquely identify the datagram on the Internet.

The time-to-live field is used to ensure that a datagram does not go round and round in the network without reaching its destination. Every router decrements this field by 1, and if a router receives a datagram with this field value of 0, the datagram is discarded.

Flags (3 bits): If the first bit, called the do-not-fragment bit, is set to 1, this is to indicate that the datagram should not be fragmented. If a router cannot handle the datagram without fragmenting it, the datagram is discarded, and an error message is sent to the source. The second bit is called the more-fragments bit and is set to 0 to indicate that this is the last fragment. The third bit remains unused.

Fragment offset (3 bits): Specifies the offset of the fragment in the datagram, measured in 8 octets (one octet is 8 bits), starting at offset 0. The destination receives all the fragments and reassembles the fragments using the offset value, starting with 0 to the highest value.

Time-to-live (12 bits): It may happen that a packet on the Internet may go round and round without reaching the destination, particularly when using the dynamic routing algorithms. To avoid such unnecessary traffic, this field is very useful. This field contains the number of hops a packet can travel. At every router, this field is decremented by 1, and either the packet reaches the destination before the field becomes 0 or, if it reaches 0 earlier, it is discarded. The default hop count is 64—the packet can traverse at most through 64 routers.

It is possible for some packets to be discarded by a router and IP is said to provide an unreliable service. The TCP has to take care of the lost datagrams by asking for retransmission.

Protocol (8 bits): The protocol field specifies which higher layer protocol is encapsulated in the data area of the IP packet. The higher layer protocol can be TCP or UDP.

Header checksum (16 bits): The bit pattern is considered as a 16-bit integer, and these bit patterns are added using one's complement arithmetic. The one's complement of the result is taken as the header checksum. Note that checksum is calculated only for the header, and this value has to be calculated at each router because some of the fields in the header change at each router. For calculation of the checksum, this field is considered to be having all zeros.

Source IP address (32 bits): This field contains the IP address of the source that is sending the datagram.

Destination IP address (32 bits): This field contains the IP address of the final destination.

Options: This is a variable-length field. This field contains data for network testing and debugging. The format for the option field is option code of one octet followed by data for the option. These options are for operations such as recording the route of a datagram, timestamping a datagram along the route, and source routing that specifies the route to be taken by the datagram.

When the record route option is set, each router adds its IP address in the options field and then forwards it. When the source route option is set and the IP addresses of all the hops are mentioned in the options field, the datagram takes only that route. This is not of any significance from an end user point of view but network administrators use it for finding the throughput of the specific paths. However, one needs to know the topology of the network to specify the source routing.

When the timestamp option is set, each router inserts its IP address (32 bits, which again is optional) and timestamp (32 bits) in the options field, which can be analyzed at the destination.

Padding (8 bits): To make the IP header an exact multiple of 32 bits, the padding bits are added if required.

Data (variable): This is a variable field whose length is specified in the datagram header. It should be an integer multiple of 8 bits.

21.6 SUBNET ADDRESSING

The IP addressing scheme discussed earlier has a main problem: it is likely that the IP addresses will get exhausted fast. To overcome this problem, subnet addressing and supernet addressing are used.

In subnet addressing, a site that has a number of networks will have a single IP address. Consider a site with three physical networks. Instead of assigning three IP addresses, only one IP address can be assigned.

To make better use of the IP addresses, subnet addressing is used. In subnet addressing, a site that has a number of networks uses a single IP address. A packet received with the subnet addressing is routed to the correct destination using subnet mask.

In subnetting, the IP address is divided into two portions—network portion and local portion. The network portion is for the external network consumption, and the local portion is for the local administrator. The local portion is divided into the physical network portion and the host portion. In essence, we are creating a hierarchical network with the Internet as part of the address and the local part containing the physical address portion and the host portion. For instance, in decimal notation, the first two fields can be used for the Internet portion, the third field for local network ID, and the fourth for the host, so that of the 32 bits, 16 bits are for network ID, 8 bits are for the local network ID, and 8 bits are for the host ID.

For the other routers to know that subnet addressing scheme is used at a site, there should be a way to let others know this subnet addressing scheme. This is done through a 32-bit subnet mask. Bits in the subnet mask are set to 1 if the subnet treats the corresponding IP address bit as network address bit and 0 if the subnet does not treat the corresponding IP address bit as network address bit. As an example, a site may have the first 16 bits for the network ID, 8 bits for the local network ID and the remaining 8 bits for host ID. In such a case, the subnet mask is

> 11111111 11111111 00000000 00000000

The subnet mask is generally represented in dotted decimal notation as *255.255.0.0* .

21.7 SUPERNET ADDRESSING

In supernet addressing, a site is allocated a block of class C addresses instead of one class B address. However, this results in a large routing table. To solve this problem, classless inter-domain routing technique is used.

In the Internet addressing scheme, it is likely that the class B addresses get exhausted faster than class A or class C addresses. To overcome this problem, the supernet addressing scheme is introduced. Instead of allocating a class B address, a site can be allocated a chunk of class C addresses. At the site, the physical networks can be allocated these class C addresses.

In addition to making better use of address space, this supernet addressing scheme allows a hierarchy to be developed at a site. For instance, an Internet service provider (ISP) can be given a block of class C addresses that he can use to create a supernet with the networks of the ISP's subscribers.

Instead of one class B address, if a large number of class C addresses are given to a site, the routing table becomes very large. To solve this problem, the classless inter-domain routing (CIDR) technique is used, in which the routing table contains the entries in the format

> (network address, count)

where network address is the starting of the IP address block and count is the number of IP addresses in that block.

If each ISP is given a block of class C addresses with one entry in the routing table, the routing to that ISP becomes very easy.

21.8 LIMITATIONS OF IP VERSION 4

The drawbacks of IP Version 4 are limited addressing space, inability to support real-time communication services such as voice and video, lack of enough security features, and the need for high-processing power at the routers to analyze the IP datagram header.

The IP as we know it is running since the late 1970s. During those early days of the Internet, there were no PCs, and computers were either super computers, mainframes, or minis. With the advent of PCs, there has been a tremendous growth in the use of computers and the need to network them, and above all to be on the Internet to access worldwide resources. In the 1990s, the need was felt to revise the IP protocol to deal with the exponential growth of the Internet, to provide new services that require better security, and to provide real-time services for audio and video conferencing. IP Version 4 has the following limitations:

- The main drawback of IP Version 4 is its limited address space due to the address length of 32 bits. Nearly 4 billion addresses are possible with this address length, which appears very high (with a population of 6 billion and a large percentage of the population in the developing world never having seen a computer). But now we want every TV to be connected to the Internet and we want Internet-enabled appliances such as refrigerators, cameras, and so on. This makes the present address length of 32 bits insufficient, and it needs to be expanded.

- The present IP format does not provide the necessary mechanisms to transmit audio and video packets that require priority processing at the routers so that they can be received at the destination with constant delay, not variable delay. The Internet is being used extensively for voice and video communications, and the need for change in the format of the IP datagram is urgent.

- Applications such as e-commerce require high security—both in terms of maintaining secrecy while transmitting and authentication of the sender. IP Version 4 has very limited security features.

- The IP datagram has a fixed header with variable options, because of which each router has to do lots of processing, which calls for high processing power of the routers and also lots of delay in processing.

Due to these limitations of IP Version 4, IP Version 6 has been developed. IP Version 5 has been used on an experimental basis for some time, but it was not widely deployed.

21.9 FEATURES OF IP VERSION 6

The important features of IP Version 6 are:

- Increased address space: instead of 32 bits, IPv6 uses an address length of 128 bits.

- Increased security features.

- Modified header format to reduce processing at the routers.

- Capability for resource allocation to support real-time audio and video applications.

- Support for unicast, multicast, and anycast addressing formats.

In IP Version 6, the address length is 128 bits. In the future, every desktop, laptop, mobile phone, TV set, etc. can be given a unique IP address.

- New options to provide additional facilities.

- Compatibility with IPv4. However, translator software is required for conversion of IPv4 datagrams into IPv6 packets.

 IP Version 6 provides backward compatibility with IP Version 4 because a very large number of routers have IP Version 4 software, and it will be many more years before all routers run IP Version 6 software.

NOTE

21.9.1 IPv6 Packet Format

In IPv4, the header contains many fields because of which the router has to do lots of processing. In IPv6, some header fields have been dropped or made optional for faster processing of the packets by the router. Optional headers (called extension headers) have been included that provide greater flexibility. Flow labeling capability has been added for real-time transmission applications such as audio and video.

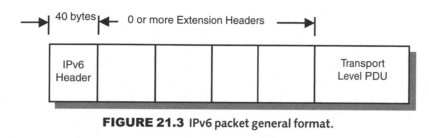

FIGURE 21.3 IPv6 packet general format.

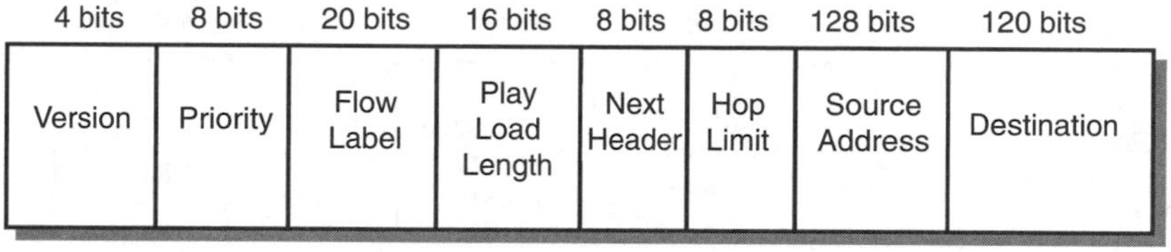

FIGURE 21.4 IPv6 packet header.

In IP Version 6, there are many optional headers. Optional headers provide the necessary flexibility. If these are absent, the processing of the header information can be done very fast at the routers.

The IPv6 packet consists of the IPv6 header, a number of optional headers, and the data. The IPv6 packet general format is shown in Figure 21.3, and the packet header format is shown in Figure 21.4.

Version (4 bits): Specifies the version number 6.

Priority or traffic class (8 bits): This field specifies the type of data being sent in the data field. Priority is indicated through this field.

Flow label (20 bits): Audio and video data needs to be handled as special packets for real-time applications. This field specifies the special handling of such type of data.

Payload length (16-bit unsigned integer): Length in octets, and extension headers are also considered part of the length.

Next header (8 bits): This field identifies the type of header following the IPv6 header.

Hop limit (8-bit unsigned integer): This field is decremented by 1 at each node that forwards the packet. When the value in this field becomes 0, the packet is discarded.

Source address (128 bits): This field specifies the source address of the packet.

Destination address (128 bits): This field specifies the destination address of the packet.

IPv6 extension headers: Depending on the requirement, the IPv6 header can be followed by a number of optional headers, known as *extension headers*. Extension headers are not normally processed by any node except the hop-by-hop option header. If this header is present, it must immediately follow the IPv6 header only. Extension headers must be processed in the order given in Figure 21.5. Figure 21.6 gives the format of different extension headers from which the functionality of each header is obvious, along with the number of bits required for each field.

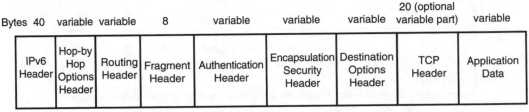

FIGURE 21.5 IPv6 packet with all extension headers.

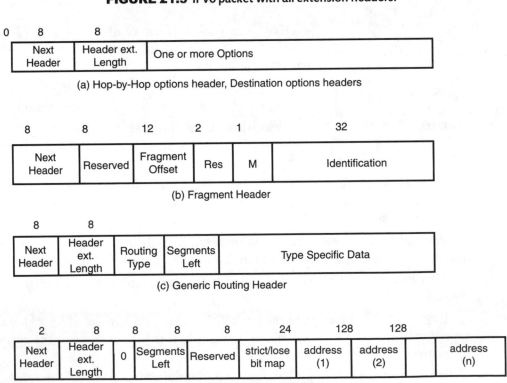

FIGURE 21.6 IPv6 routing header formats.

The optional headers are called extension headers. If no extension headers are present, the IP packet has the IP header with the *next header* field as TCP.

If no extension headers are present, the IP packet has an IPv6 header with the `next header` field containing TCP, followed by the TCP header and data.

If the routing header is to be included as an extension header, the IPv6 header contains `Routing` in the next header field, followed by the routing header. The `next header` field of the routing header contains TCP. The routing header is then followed by the TCP header and the data.

The IPv6 format appears complicated, with very long address fields and many extension headers. However, this reduces a lot of the processing burden on the routers, and so the routers can do the packet switching very fast. IPv6 will provide the necessary functionality for real-time audio and video transmission as well as improved security.

However, very few routers and end systems are presently running IPv6 software, and it will be many more years before IPv6 becomes universal. Microsoft Windows XP supports IPv6, and soon most of the operating systems and routers will be upgraded to support IPv6.

21.10 ROUTING PROTOCOLS

When a router receives a packet, the header information has to be analyzed to find the destination address, and the packet has to be forwarded toward the destination. The router does this job with the help of the routing tables.

When a packet arrives at a router, the router has to make a decision as to which node/router the packet has to be sent to. This is done using the routing protocols.

Each end system and router maintains a routing table. The routing table can be static or dynamic. In static routing, each system maintains a routing table that does not change often. In dynamic routing, the routing table is upgraded periodically. Recording of the route taken by a packet can be supported—each router appends its IP address to the datagram. This is used mainly for testing and debugging of the network. While routing the packet, the decision has to be made with the following issues in mind.

- Datagram lifetime
- Segmentation and reassembly

- Error control
- Flow control

If there is congestion in the network, the router may forward the packet through an alternate route. However, if source routing is specified in the packet, then the packet has to be routed according to these instructions.

Datagram lifetime: If dynamic routing is used, the possibility for a datagram to loop indefinitely through the Internet exists. To avoid this, each datagram can be marked with a lifetime—a hop count. When a datagram is passed through a router, the count is decreased. Alternatively, a true measure of time can be used, which requires a global clocking mechanism. When a datagram is received by a router with a hop count of zero, then the datagram is discarded.

Segmentation and reassembly: When a router receives a datagram and if it has to be routed to a network that supports a smaller packet size, the datagram has to be broken into small fragments, which is known as *segmentation*. Segmentation is done by routers. Reassembly is done by end systems and not routers because a large buffer is required at the router and all datagrams must pass through the same router, in which case dynamic routing is not possible.

Error control: Datagrams may be discarded by a router due to lifetime expiration, congestion, or error in the datagram. Hence, IP does not guarantee successful delivery of datagrams.

Flow control: If a router receives datagrams at a very fast rate and cannot process the datagrams at that rate, the router may send flow control packets requesting reduced data flow.

Consider an internet in which a number of networks are interconnected through routers. When a datagram has to be sent from one node on one network to a node on another network, the path to be taken by the datagram is decided by the routing algorithms. Each router receives the datagram and makes a decision as to the next hop. This is a complicated mechanism because:

- If the IP datagram specifies the option for source routing, the datagram has to be routed per the instructions.
- The datagram length may vary and, based on the length, the datagram may need to be forwarded to different routers because not all routers may be able to handle large datagrams. (Though they can fragment it, it may result in too much delay, or the do-not-fragment bit may have been set.)
- The network may get congested, and some routes may be experiencing traffic jams. In such cases, to ensure no packet loss, alternate routes may need to be chosen.

- The path to be chosen has to be the shortest, to reduce load on the network and the delay.

In this section, we will study the routing protocols. Note that, though routing is the function of a router, hosts also may be involved in routing under special circumstances. For instance, a network may have two routers that provide connections to other networks. A host on this network can forward the datagram to one of the routers. (Which router is the issue.) Similarly, multihomed hosts are involved in routing decisions. In the following subsections, we will discuss routing as applied to the routers.

21.10.1 Direct Delivery

If the source and the destination have the same network ID, the datagram is directly delivered to the destination. This is known as direct delivery.

Consider a LAN running TCP/IP protocol suite. Delivering an IP datagram to a node is straightforward; it does not involve any routers. Because the IP address has the network ID as a field, and the source and the destination have the same network ID, the datagram is delivered directly to the destination. This is known as *direct delivery*.

Consider a case in which a LAN is connected to the Internet through a router. After the datagram travels across the Internet and reaches the final destination's router, the direct delivery mechanism is used for handing over the datagram to the destination.

21.10.2 Indirect Delivery

In indirect delivery, between the source and the destination, there may be a number of routers. In such a case, each router has to forward the datagram to another router. This is known as indirect delivery.

On the Internet, between the source and the destination, there are many routers. Each router has to get the datagram and forward it to the next router toward the destination. This is known as *indirect delivery*. The routing has to be such that finally there should be one router that will deliver the datagram to the destination using the direct delivery mechanism. There are a number of mechanisms for indirect routing.

Routing Tables and Next-Hop Routing

Each router and host can have a table, known as the *routing table*, that contains the destination network address (not the host address) and the router to which the datagram has to be forwarded. When a datagram arrives, the network ID portion is examined, and the routing table is consulted to find the next router to which the datagram is to be forwarded. This mechanism reduces the size of the routing table because we need not store the destination host addresses; only the network addresses are required. Each entry in a routing table is in the format.

Destination network IP address: IP address of the next router to which datagram is to be forwarded

Each entry in a routing table will be in the format, destination network IP address: IP address of the next router to which the datagram is to be forwarded.

The next router is called the next-hop router, and routing based on this type of routing tables is called next-hop routing. If the routing table does not contain an entry for a destination network address, the datagram is sent to a default router.

Adaptive Routing

In this routing mechanism, the routing of packets is based on information about failure of links and congestion. The advantages of this mechanism are improved performance of the network and less congestion. However, the disadvantages are processing burden on routers and traffic burden on network to send routing information and network status. Information about the failure of links and congestion is shared between routers passing the necessary messages.

In adaptive routing, the routing tables are periodically updated based on the traffic information. The routers need to exchange this traffic information periodically.

First generation Internet (ARPAnet) implemented an adaptive algorithm using estimated delay as the criterion. Each node sends to all the neighbors the estimated delay (or the queue length). The line speed is not considered, and so it is not an accurate algorithm. The routing algorithm was subsequently modified in the second generation Internet, and an adaptive algorithm based on actual delay was implemented. Instead of queue length, delay is measured. When a packet arrives, it is timestamped, and when it is retransmitted the time is noted. Hence, delay is calculated exactly. The average delay is calculated every 10 seconds and sent to all neighboring nodes (known as *flooding*). The disadvantage with this routing algorithm is that every node tries to get the best route. Subsequently, the routing algorithm has been further refined. The algorithms uses a line utilization parameter for the adaptive algorithm. Instead of the best route, a threshold value is used so that every node will not try to get the best route.

For the routing algorithms, the concept of autonomous systems is important. An internet connected by homogeneous routers, with all routers under the administrative control of a single entity, is called an autonomous system. An autonomous system can use a routing protocol of its choice, but when autonomous systems are interconnected (as in the Internet), a well-defined protocol is a must.

21.11 AUTONOMOUS SYSTEM

An autonomous system is a network of computer networks within one administrative control. An autonomous system can use its own routing protocols.

A site may have a large number of networks within one administrative control. Such a system is called an autonomous system. This system can be connected to the external Internet through one or more routers. To provide efficient routing, the entire system appears as a single entity on the external Internet and not as multiple networks. Within the system, how the routing is to be done can be decided by the administrative authority. To provide the necessary routing information to the external Internet, one or more routers will be acting as the interface.

Each autonomous system is given an autonomous system number by a central authority.

21.12 ROUTING WITHIN AUTONOMOUS SYSTEMS

Routing protocols used within the autonomous system are called interior gateway protocols. Open shortest path first (OSPF) is one such protocol.

If the autonomous system consists of a small number of networks, the administrator can manually change the routing tables. If the system is very large and changes in topology (addition and deletion of networks) are frequent, an automatic procedure is required. Routing protocols used within an autonomous system are called interior gateway protocols (IGPs). A number of IGPs are used because the autonomous systems vary widely—it may be a small system with few LANs or a WAN with a large number of networks. Open shortest path first (OSPF) is the most efficient protocol used for routing within an autonomous system. In OSPF, routing is done based on the shortest path—the shortest path can be calculated between two nodes using criteria such as distance, dollar cost, and so on.

21.12.1 Open Shortest Path First (OSPF)

In OSPF protocol, the shortest path is calculated from the source to the destination, and routing is done using this path. The criteria for calculation of the shortest path can be distance or cost.

Interior router protocol (IRP) passes routing information between routers within an autonomous system. Open shortest path first (OSPF) protocol is used as the interior router protocol in TCP/IP networks. OSPF routing is a dynamic routing technique. As the name implies, the OSPF protocol computes a route through the Internet that incurs the least cost. It is analogous to a salesman finding the shortest path from one location to another location. The shortest path can be calculated based on the distance to be traveled or based on the cost of travel. Similarly, the cost of routing the packets can be based on a measure that can be configured by the administrator—it can be a function of delay, data rate, or simply dollar cost. Each router maintains the database of costs. Any change in costs is shared with other routers.

21.12.2 Flooding

In flooding, each router sends the packets to all the other routers connected to it. Routing information is shared among the routers using this protocol.

Flooding is another protocol that can be used for routers to share the routing information. In flooding, each router sends its routing information to all the other routers to which it is directly connected. This information is shared again by these routers with their neighbors. This is a very simple and easy-to-implement protocol. However, the traffic on the network will be very high, particularly if the flooding is done very frequently.

21.13 ROUTING BETWEEN AUTONOMOUS SYSTEMS

The Internet is a network of a large number of autonomous systems. A host connected to one autonomous system can communicate with another host connected to another autonomous system only when the routing information is

On the Internet, routing between two autonomous systems is done through exterior gateway protocol (EGP) or border gateway protocol (BGP).

shared between the two autonomous systems. The protocols used for sharing the routing information are exterior gateway protocol (EGP) and border gateway protocol (BGP).

21.13.1 Exterior Gateway Protocol

As shown in Figure 21.7, two routers each belonging to a different autonomous system share routing information using the exterior gateway protocol (EGP). These routers are called *exterior routers*. Conceptually, EGP works as seen in Figure 21.7.

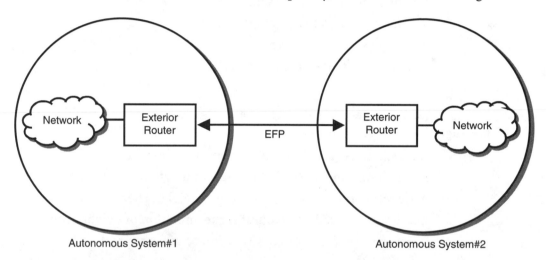

FIGURE 21.7 Exterior gateway protocol.

- A router acquires a neighbor by sending a hello message. The neighbor responds to the hello.
- The two routers periodically exchange messages to check whether the neighbor is alive. The Internet being very dynamic, there is no guarantee that today's neighbor is there tomorrow. Whenever there is any change in the routing information within the autonomous system, the information is conveyed to the other router through routing update messages.

  EGP has the following restrictions:

- An exterior router conveys only reachability information. It does not ensure that the path is the shortest.

- An exterior router can announce only the routing information to reach the networks within the autonomous system, not some other autonomous system.
- EGP can advertise only one path to a given network, even if alternate paths are available.

The EGP message consists of a header and parameters fields. The header has the following fields:

Version (8 bits): Specifies the version number of EGP.

Type (8 bits): Specifies the type of message. The message types are:

- Neighbor acquisition messages (type 3)
- Neighbor reachability messages (type 5)
- Poll request messages (type 2)
- Routing update messages (type 1)
- Error

Code (16 bits): Specifies the subtype of the message. For example, type 3 messages have the following codes:

- Acquisition request (code 0) to request for neighborhood
- Acquisition confirm (code 1) to accept neighborhood
- Acquisition refusal (code 2) to refuse neighborhood
- Cease request (code 3), request for termination of neighborhood
- Cease confirm (code 4), confirmation to cease request

Neighbor reachability messages (type 5) messages have the following codes:

- Hello request message (code 0)
- I heard you message (code 1)

Periodically, the neighbors exchange the routing tables using the type 1 message format, with code as 0. The parameters field contains a list of all routers on the network and the distance of the destination for each.

Using this simple protocol, the Internet attained its great flexibility. An autonomous system can enter into an arrangement with a neighbor to share the routing information without the need for a central authority to control the connectivity between various internets. However, because of the limitations mentioned, a new protocol, border gateway protocol, has been developed.

In EGP, two routers establish a neighborhood relationship and keep exchanging the routing information. Hence, there is no need for a centralized system to send routing information to different routers.

21.13.2 Border Gateway Protocol

Routers connected to different autonomous systems can exchange routing information using BGP, the latest version being BGP Version 4. As compared to EGP, BGP is a complex protocol. We will illustrate the functioning of BGP through an example.

In BGP, each autonomous system will have a BGP speaker that is authorized to advertise its routing information. This BGP speaker sends information about all the autonomous systems that can be reached via its routers.

Figure 21.8 shows how BGP is used between autonomous systems. Assume that AS#1 and AS#2 are two autonomous systems that need to share routing information. AS#2 is connected to two other autonomous systems: AS#3 and AS#4. AS#1 will have one router designated as a BGP speaker, which has the authority to advertise its routing information with other routers. Similarly, AS#2 will have a BGP speaker.

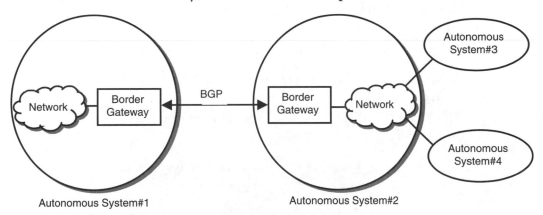

FIGURE 21.8 Border gateway protocol.

The BGP speaker of AS#2 will send the reachability information to the BGP speaker of AS #1. This reachability information includes all the other autonomous systems that can be reached via its routers—in our example, the autonomous systems AS#3 and AS#4. It is this feature that gives complexity to the BGP, but the advantage is that the complete reachability information is obtained by AS#1.

It may happen that because of the failure of a link between AS#2 and AS#3, for example, AS#3 becomes unreachable temporarily. The BGP will also give negative advertisement by sending information about the routes that have been withdrawn.

It is the BGP that is making the Internet a dynamic network. When new networks are added to the Internet, a user can access the new network with the BGP without a centralized administrative authority.

21.14 INTERNET CONTROL MESSAGE PROTOCOL

The Internet is a dynamic network. Networks get connected and removed without the knowledge of the senders, who can be anywhere on the earth. Also, each router functions autonomously, routing the datagrams based on the destination IP address and the routing table. It is likely that a datagram cannot be forwarded or delivered to its destination, or due to congestion the datagram may have to be dropped. There must be a procedure to report errors to the source whenever there is a problem. The Internet Control Message Protocol (ICMP) is used for this purpose. Every IP implementation must support ICMP also. ICMP messages are a part of the data field of the IP datagram.

An ICMP message, which is encapsulated in the data field of the IP datagram, consists of three fields:

- Message type to identify the type of message (8 bits)
- Code field to provide further information on the message type (8 bits)
- Checksum field (16 bits), which uses the same algorithm as the IP checksum

In addition, an ICMP message contains the first 64 data bits of the datagram that caused the problem.

The ICMP type fields and corresponding message types are given in the following table:

Type Field	ICMP Message Type
0	Echo reply
3	Destination unreachable
4	Source quench
5	Change the route (redirect)
8	Echo request
11	Time to live expired for datagram
12	Parameter problem on datagram
13	Timestamp request
14	Timestamp reply
17	Address mask request
18	Address mask reply

Echo request and echo reply messages are used to test the reachability of a system. Source sends an echo request to the receiver, and the receiver sends an echo reply to the sender. The request can contain an optional data field that is returned along with the echo reply. When an echo reply is received by the sender, this is an indication that the entire route to the destination is OK, and the destination is reachable. The system command `ping` is used to send the ICMP echo requests and display the reply messages.

Reports of an unreachable destination are sent by a router to the source. The possible reason is also sent in the ICMP message. The reason can be the following:

- Network unreachable
- Host unreachable
- Protocol unreachable
- Port unreachable
- Fragmentation needed but do-not-fragment bit is set
- Source routing failed
- Destination network unknown
- Source host isolated
- Communication with the destination network prohibited for administrative reasons
- Network unreachable for type of service, and host unreachable for the type of service

The source quench message is sent by the router to the source when congestion is experienced. This message is to inform the source to reduce its rate of datagrams because the router cannot handle such a high speed. When a router cannot handle the incoming datagrams, it discards the datagram and sends the source quench message to the source.

ICMP is used to report errors to the source. Information such as network or host unreachable, time-to-live field expired, fragmentation needed but do-not-fragment bit is set, and so on, are sent to the source using this protocol.

A change the route (redirect) message is sent by a router to the host. Generally, it is the router's function and responsibility to update its routes, and hosts keep minimal routing information. However, when a router detects that a host is using a nonoptimal route, it sends an ICMP message to the host to change the route. This message contains the IP address of the router the host has to use for the routing.

A time-to-live-expired message is sent by a router when the hop count becomes zero. As mentioned earlier, to ensure that datagrams do not keep on circulating between routers endlessly, this hop count is introduced. When hop count becomes zero, the datagram is discarded and the message is sent to the source. This message is also sent when fragment reassembly time exceeds the threshold value.

For any other problem because of which the datagram has to be discarded, the router sends a parameter problem message to the source.

Timestamp request and reply messages are used between systems to obtain the time information. Because each system acts independently, there is no mechanism for synchronizing the clocks. A system can send a timestamp request to another system and obtain the timestamp reply. This information can be used to compute the delays on the network and also to synchronize the clocks of the two systems.

Subnet address mask request and reply messages are exchanged between machines. In some IP addresses, a portion of the host address corresponds to the subnet address. The information required to interpret this address is represented by 32 bits called the subnet mask. For example, if a host wants to know the subnet mask used by a LAN, the host sends a subnet address mask request to the router on the LAN or broadcasts the message on the LAN. The subnet address mask reply will contain the subnet address mask.

In the next chapter, we will study the details of the TCP layer that runs above the IP layer.

Every IP implementation must support ICMP. ICMP messages are included in the data field of the IP datagram.

Summary

The Internet Protocol, the heart of the global Internet, is presented in this chapter. The present version of IP, running on end systems and routers, is IP Version 4. The two main functions of IP are addressing and routing. IP Version 4 has an address length of 32 bits. Each system is given a 32-bit IP address that uniquely identifies the system globally. This addressing scheme can cover at most 4 billion addresses. This turns out to be a small number for the future, particularly when we would like to connect even consumer items such as TVs, mobile phones, and such to the Internet. The latest version of IP, IP Version 6, has an address length of 128 bits.

In addition to the address length, IP Version 4 has limitations for handling secure applications and real-time applications. The detailed formats of IP Version 4 and Version 6 are presented, which bring out the salient features of both versions.

Another important function of IP is routing. We studied the routing protocols within an autonomous system and between autonomous systems. An autonomous system is a network within the administrative control of an organization. Routers within an autonomous system can share routing information using protocols such as open shortest path first (OSPF) and flooding. Routers connected to different autonomous systems can share routing information using exterior gateway protocol or border gateway protocol.

The Internet Control Message Protocol (ICMP) is used to report errors and to send management information between routers. ICMP messages are sent as part of the IP datagram. The details of ICMP are also presented in this chapter.

References

W. R. Stevens. *TCP/IP Illustrated Vol. I: The Protocols*, Addison Wesley, Reading, MA, 1994. This book gives a complete description of TCP/IP and its implementation in Berkeley Unix.

D.E. Comer and D.L. Stevens. *Internetworking with TCP/IP Vol. III: Client/Server Programming and Applications*. BSD Socket Version. Prentice Hall, Englewood Cliffs, N.J., 1993.

Questions

1. Describe the format of an IP Version 4 datagram.
2. What are the limitations of IP Version 4 and explain how IP Version 6 addresses these limitations.
3. Describe the format of an IP Version 6 packet.
4. Describe the protocols used for routing within autonomous systems.
5. Describe the protocols used for routing between autonomous systems.
6. Explain the ICMP protocol and its functionality.

Exercises

1. Find out the IP address of your computer.
2. Calculate the total number of addresses supported by class B IP address format. Note that 14 bits are used for network ID and 16 bits for host ID.
3. Calculate the total number of addresses supported by class C IP address format. Note that 24 bits are used for network ID and 8 bits for host ID.
4. Calculate the maximum number of addresses supported by IP Version 6.
5. Write a technical paper on IP Version 6.

Projects

1. Write a program that captures all the packets transmitted over the LAN and displays the source IP address and the destination IP address of each packet. You can use the packet driver software that comes with the Windows Device Driver Kit to develop this program.
2. Extend the software that is written in Project #1 so that all the packets corresponding to a particular destination IP address are stored in a file. Use this software to find out the passwords of different users of the LAN. Because this software is nothing but hacking software, obtain the permission of your system administrator before testing this software. You can refine this software to develop a firewall that filters out all the packets with a specific IP address.

22

Transport Layer Protocols: TCP and UDP

In This Chapter

- The Services Provided by the TCP Layer
- The TCP Segment Format
- The UDP Datagram Format
- Security-Related Issues and Protocols

As we saw in the previous chapter, IP does not provide a reliable service—it does not guarantee successful delivery of packets. However, from an end user's point of view, reliable service is a must—when you transfer a file, you want the file to be delivered intact at the destination. The TCP layer provides the mechanism for ensuring reliable service between two end systems. In this chapter, we will study the details of TCP. We will also discuss UDP, another transport layer protocol that provides less overhead than TCP.

22.1 SERVICES OF THE TRANSPORT LAYER

When we use a network for a service (such as a file transfer), we expect reliable service so that the file reaches the destination intact. However, IP does not provide reliable service. The packets may be lost, they may arrive out of order, they might arrive with variable delay, and so on. The transport layer protocols provide end-to-end transmission by taking care of all these problems; it provides a reliable data transfer mechanism to the higher layers.

Suppose you want to transfer a file from one system to another system. You invoke a file transfer application on your system. The networking software first has to establish a connection with the other end system, transfer the file, and then remove the connection. The transport layer provides the necessary functionality to establish the connection, transfer the packets to the other system reliably even if there are packet losses because of the IP's limitations, and then remove the connection. The transport layer protocol does this job by providing the following services:

- Type of service
- Quality of service
- Data transfer
- Connection management
- Expedited delivery
- Status reporting

Type of service: The service can be connection-oriented or connectionless. In connection-oriented service, flow control and error control are incorporated so that the service is more reliable and sequence of packets is maintained. However, there is an overhead to connection establishment. In connectionless service, there are no overheads for connection establishment and so it is efficient, but reliable data transfer is not guaranteed. Connectionless service is used for applications such as telemetry, real-time applications such as audio/video communication, and so on. On the Internet, TCP provides connection-oriented service, and UDP provides connectionless service.

Quality of service: Depending on the application, the quality of service parameters have to be defined, such as acceptable error levels, desired average and maximum delay, desired average and minimum throughput, and priority levels. Note that IP also provides the services of priority, delay, and such. However, in the TCP/IP networks, the required quality of service is not guaranteed. The quality of service parameters are very important when transmitting voice or video data because delay, packet loss, and so on will have an impact on the quality of the speech or video.

The quality of service parameters can be set and will be honored in systems using Asynchronous Transfer Mode (ATM) and Frame Relay.

Data transfer: This service defines whether the data transfer is full duplex, half duplex, or simplex. Transfer of data and control information is the function of this service.

The communication can be full duplex, half duplex, or simplex. In some communication systems, the communication can be only one way (simplex). In such a case, it is not possible to transmit acknowledgements at all. UDP can be used in such systems. An example of such a system is a satellite communication system in which the VSATs can receive data but cannot transmit.

The services provided by the transport layer are connection management and data transfer. In addition, this layer handles issues related to quality of service, status reporting, and delivery of urgent data.

Connection management: In connection-oriented service, the transport layer is responsible for establishment and termination of the connection. In case of abrupt termination of a connection, data in transit may be lost. In case of graceful termination, termination is prevented until data has been completely delivered.

Expedited delivery: When some urgent data is to be sent, the interrupt mechanism is used to transfer the urgent data.

Status reports: Status reporting of performance characteristics is also done by this layer. The performance characteristics are delay, throughput, addresses (network address or port address), current timer values, and degradation in quality of service.

In TCP/IP networks, though quality of service parameters can be set, the quality of service cannot be guaranteed.

NOTE

22.2 TRANSMISSION CONTROL PROTOCOL

Transmission Control Protocol (TCP) is a connection-oriented protocol that provides reliable communication across an internet. TCP is equivalent to ISO transport layer protocol.

The units of data exchanged between two end systems are called TCP segments. Ordinarily, TCP waits for sufficient data to be accumulated to create a TCP segment. This TCP segment is given to the IP layer. The TCP user (the application layer) can request the TCP to transmit data with a push flag. At the receiving end, TCP will deliver the data in the same manner. This mechanism is known as data stream push.

When urgent data is to be transmitted, urgent data signaling is used, which is a means of informing the destination TCP user that urgent data is being transmitted. It is up to the destination user to determine appropriate action. In the TCP segment, there is a field called urgent pointer to indicate that the data is urgent.

When a TCP connection is to be established between two end systems, commands are used from the TCP user (the higher layer protocol), which are known as TCP service request primitives. Responses, which are known as TCP service response primitives, are sent from the TCP to the TCP user.

The TCP service request primitives are:

- **"Fully specified passive open":** Listen for connection attempt at specified security and precedence from specified destination. The following parameters are the arguments: source port, destination port, destination address, "time out", "time out action", "precedence", "security range". Parameters within the parentheses are optional.

- **"Active open":** Request connection at a particular level of security and precedence to a specified destination.

- **"Active open with data":** Request connection as in "Active Open" and transmit data with the request.

- **"Send":** Transfer data across the named connection.

- **"Close":** Close connection gracefully.

- **"Abort":** Close connection abruptly.

- **"Status":** Query connection status.

The TCP service response primitives (issued by TCP to local TCP user) are:

- **"Open ID":** Inform TCP user of the connection name assigned to pending connection request.

- **"Open failure":** Reports failure of an active open request.

- **"Deliver":** Reports arrival of data.

- **"Closing":** Reports that remote TCP user has issued a close and that all data sent by remote TCP user is delivered.

- **"Terminate":** Reports that connection has been terminated. Reason for termination is provided.

- **"Response":** Reports current status of connection.

- **"Error":** Reports internal error.

The TCP layer provides a connection-oriented service. A TCP connection is established between two end systems, and a series of service request primitives and service response primitives is exchanged for connection management, data transfer, and error reporting.

Whenever a connection has to be established, the service request primitives are sent to the TCP, and the TCP sends the service response primitives. If there are no errors, a connection is established with the other TCP peer running on another machine, and data transfer takes place. After successful transfer of data, the application layer is informed by the TCP layer that the data has been successfully transferred.

22.2.1 Virtual Circuit Connection

Before two end systems transfer data, a TCP connection is established. This is a virtual circuit connection. A connection is established between the TCP protocol ports, which identify the ultimate destination within the machines. This connection is full duplex, so the data flows in both directions. Note that the TCP connection is only an abstraction; it is not a real connection because the IP datagrams take different routes to reach the destination.

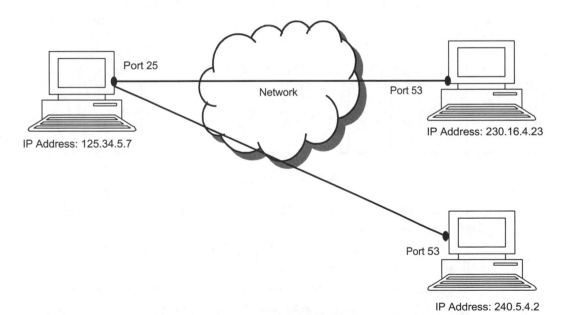

FIGURE 22.1 Multiple TCP connections.

A TCP connection is identified by a pair of end points. Each end point is defined by the IP address (host number) and the port number. Port number is a predefined small integer.

Each connection is identified by a pair of end points. Each end point is defined by the host number and the port number. The host number is the IP address, and the port number is a predefined small integer. The port number is a process ID running on end systems—there will be no physical port. For instance, a connection can be represented by the two end points

(125.34.5.7, 25) and (230.16.4.23, 53)

It is possible to establish multiple connections between two end points as shown in Figure 22.1. For instance, one of the above end points can have another connection such as

(125.34.5.7, 25) and (240.5.4.2, 53)

Hence, a given TCP port number can be shared by multiple connections.

22.2.2 TCP Segment Format

The TCP segment consists of the TCP header and protocol data unit (PDU) data. The format of the TCP header is shown in Figure 22.2. The minimum header length is 20 bytes.

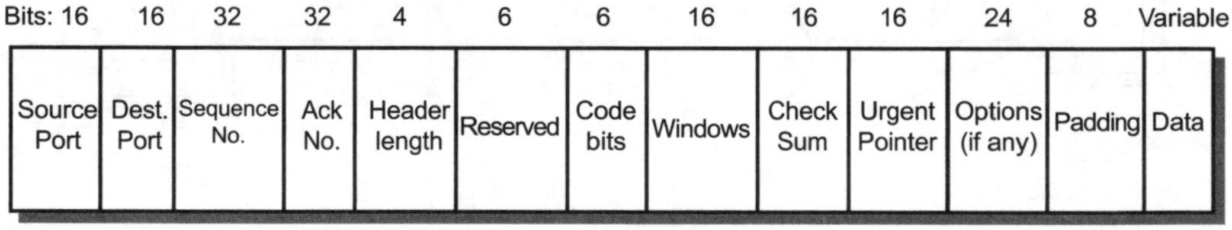

Bits: 16	16	32	32	4	6	6	16	16	16	24	8	Variable
Source Port	Dest. Port	Sequence No.	Ack No.	Header length	Reserved	Code bits	Windows	Check Sum	Urgent Pointer	Options (if any)	Padding	Data

FIGURE 22.2 TCP segment format.

Source port address (16 bits): The port number that identifies the application program of the source.

Destination port address (16 bits): The port number that identifies the application program of the destination.

Sequence number (32 bits): The sequence number of the TCP segment.

Acknowledgement number (32 bits): The number of the octet that the source expects to receive next.

Header length (4 bits): The length of the header in 32-bit words. This is required because the header length varies due to the presence of the options field.

Reserved (6 bits): Reserved for future use.

Code bits (6 bits): Specify the purpose and content of the segment. The six bits are URG, ACK, PSH, RST, SYN and FIN. If these bits are set, the following information is conveyed:

URG	Urgent pointer field is valid.
ACK	Acknowledgment field is valid.
PSH	The segment requests a push.
RST	Reset the connection.
SYN	Synchronize the sequence numbers.
FIN	Sender has reached end of its byte stream.

The TCP segment header consists of the following fields: source port address, destination port address, sequence number, acknowledgement number, header length, reserved bits, code bits, window size, checksum, urgent pointer and options. The minimum header length is 20 bytes.

Window (16 bits): Specifies the buffer size, which indicates how much data it is ready to accept, beginning with the byte indicated in ACK field. This is flow control information.

Checksum (16 bits): Checksum is calculated from the header fields and the data fields. For checksum computation, the pseudoheader is used. The pseudoheader consists of 96 bits—32 bits of source IP address, 32 bits of destination IP address, 8 bits of zeros, 8 bits of protocol, and 16 bits of TCP length. This data is obtained from the IP layer software, and checksum is calculated using the same one's complement of 16-bit words and taking the one's complement of the result. The protocol field is to specify which underlying protocol is used; for IP, the value is 6.

Urgent pointer (16 bits): This field is used to indicate that the data segment is urgent. The destination has to process this segment even if there are other segments to be processed. This is required in such applications as remote login, when the user has to abort a program without waiting any further.

Options, if any (variable): This field is used to negotiate the maximum TCP segment size. The TCP software can indicate the maximum segment size (MSS) in this field. Otherwise, a segment size of 536 bytes is used. This value is 576 bytes of default IP datagram minus the TCP and IP header lengths.

Padding (8 bits): Padding for making the TCP segment complete.

Data (variable): User data, which is of variable length.

22.2.3 TCP Mechanism

For data transfer, there will be three phases: connection establishment, data transfer, and connection termination.

Connection Establishment

The connection is determined by the source and destination ports. Some of the important currently assigned TCP port numbers are given in the following table:

Port number	Application	Description
21	FTP	File Transfer Protocol
23	Telnet	remote login
25	SMTP	Simple Mail Transfer Protocol
37	time	time
42	nameserver	hostname server
53	domain	Domain Name Server
79	Finger	Finger
80	HTTP	World Wide Web
103	x400	X.400 messaging service
113	auth	authentication service

Only one TCP connection is established between a pair of ports. However, one port can support multiple connections, each with a different partner port as shown in Figure 22.1. Handshaking is used for establishing the connection using the following procedure:

- Sender sends a request for connection with sequence number.
- Receiver responds with request for connection with ACK flag set.
- Sender responds with ACK flag set.

Data Transfer

The data is transferred in segments but viewed as a byte stream. Every byte is numbered using modulo 2^{32}. Each segment contains the sequence number of the first byte in the data field. Flow control is specified in the number of bytes. Data is buffered by the sender and the receiver, and when to construct a segment is at the discretion of the TCP (except when push flag is used). For priority data, an urgent flag is used. If a segment arrives at a host and the segment is not meant for it, an rst flag is set.

Connection Termination

Each TCP user issues a close primitive. TCP sets the fin flag on the last segment. If the user issues an abort primitive, abrupt termination is done. All data in buffers is discarded and an rst segment is sent.

To implement the TCP, there are a few implementation options, which are briefly discussed.

Send policy: When does the TCP layer start sending the data to the layer below? One option is to buffer the data and construct the TCP segment. The other option is not to buffer the data and construct the TCP segment when some data is available to be sent to the layer below.

A TCP connection is established between two TCP ports on two end systems. Only one connection can be established between a pair of ports. However, one port can support multiple connections, each with a different partner port.

Delivery policy: After the data is received from the layer below, when to transfer the data to the upper layer (the TCP user) is another issue. One option is to buffer the TCP segment and send to the TCP user. The other option is to send without buffering. When to deliver is a performance consideration.

Accept policy: If segments are received out of sequence, there are two options:

- In-order: accept only segments received in order, discard segments that are received out of order. This is a simple implementation but a burden on the network.
- In-window: accept all segments that are within the receive window. This reduces the number of transmissions, but a buffering scheme is required at the hosts.

Retransmit policy: Three retransmission strategies are possible.

- First only: The sender maintains a timer. If ACK is received, the corresponding segment is removed from the buffer and the timer is reset. If the timer expires, the segment at the front of the queue is retransmitted and the timer is reset.

- Batch: The sender maintains one timer for retransmission for the entire queue. If ACK is received, segments are removed from the buffer and the timer is reset. If the timer expires, all segments in the queue are retransmitted and the timer is reset.

- Individual: The sender maintains one timer for each segment in the queue. If ACK is received, the segments are removed from the queue and the timers are reset. If any timer expires, the corresponding segment is retransmitted individually and the timer is reset.

Acknowledgement policy: There are two options here.

- Immediate: When data is accepted, immediately transmit empty segment with ACK number. This is simple but involves extra transmissions.

- Cumulative: When data is accepted, wait for outbound segment with data in which ACK is also sent (called piggybacking the ACK). To avoid long delay, a window timer is set. if the timer expires before ACK is sent, an empty segment containing ACK is transmitted. This involves more processing but is used extensively due to a smaller number of transmissions.

  Implementation of TCP in software gives the software developer these choices. Based on the delay considerations and bandwidth considerations, one choice may be better than the other.

The port number is a predefined integer for each application that runs above the TCP layer. Some important port numbers are: 21 for FTP, 25 for SMTP, and 80 for HTTP.

22.3 USER DATAGRAM PROTOCOL

User datagram protocol (UDP) is another transport layer protocol that can be run above the IP layer. In the place of TCP, UDP can be used to transport a message from one machine to another machine, using the IP as the underlying protocol. UDP provides an unreliable connectionless delivery service. It is unreliable because packets may be lost, arrive out of sequenceor or be duplicated.

It is the higher layer software that has to take care of these problems. The advantage of UDP is that it has low protocol overhead compared to TCP. A UDP message is called a *user datagram*.

The format of the user datagram is shown in Figure 22.3.

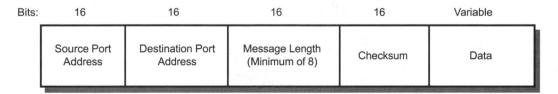

Bits:	16	16	16	16	Variable
	Source Port Address	Destination Port Address	Message Length (Minimum of 8)	Checksum	Data

FIGURE 22.3 UDP datagram format.

Source port address (16 bits): Specifies the protocol port address from which data is originating. This is an optional field. If present, this is to indicate the address to which a response has to be sent by the destination port. If absent, the field should contain all zeros.

Destination port address (16 bits): Specifies the protocol port address to which data is intended.

Message length (16 bits): Specifies the length of the user datagram, including the header and data in octets.

Checksum (16 bits): The data is divided into 16-bit words, the one's complement sum is computed, and its one's complement is taken. However, checksum calculation is optional, and can be avoided by keeping zeros in this field to reduce computational overhead.

Data (variable length): The actual data of the user datagram.

This user datagram is encapsulated in the IP datagram's data field and then encapsulated in the frame of the datalink layer and sent over the network.

Though UDP is unreliable, it reduces the protocol overhead as compared to TCP. Hence, UDP is used for network management. Simple Network Management Protocol (SNMP) runs above the UDP. UDP is also used for voice/fax/video communication on IP networks to enable real-time transmission. This aspect is addressed in the chapter on multimedia communication over IP networks.

The UDP datagram has the following fields: source port address, destination port address, message length, checksum, and data. The checksum field is optional.

Another use of UDP is in networks where it is not possible to implement acknowledgements. For example, consider a satellite network (shown in

Figure 22.4) working in star configuration with a hub (central location) and a number of receive-only VSATs. The server located at the hub (central station) has to transmit a file to the VSATs. It is not possible to use the TCP protocol in such a network because the VSAT cannot send an acknowledgement. We can use UDP to transmit the packets one after another from the computer at the hub; the computer at the VSAT location assembles all the UDP packets to get the file. The application layer running above UDP has to take care of assembling the file by joining the UDP datagrams in sequence.

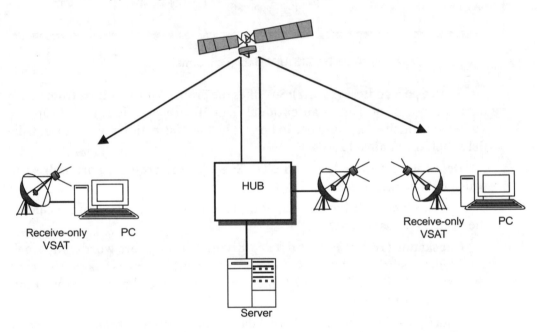

FIGURE 22.4 File transfer from hub to receive-only VSATs using UDP.

The header of UDP is much smaller than the header of TCP. As a result, processing of UDP user datagrams is much faster.

22.4 TRANSPORT LAYER SECURITY

The TCP/IP stack does not provide the necessary security features—and security has become the most important issue, particularly when the Internet has to be used for e-commerce transactions. Users type in their credit card numbers, bank account numbers, and other confidential personal information. This data can be stolen and misused. When confidential documents are transmitted over the Internet, it has to be ensured that unauthorized persons do not receive them. To achieve the desired security, a special layer software runs on the TCP to provide privacy and confidentiality of the data over the Internet. This layer is known as transport layer security (or Secure Socket Layer).

TLS consists of two sublayers—TLS record protocol, which runs above the TCP, and TLS handshake protocol, which runs above the TLS record protocol. TLS record protocol provides connection security using cryptographic techniques such as DES (Data Encryption Standard), proposed by the U.S. Department of Defense. For each session, a unique key is generated, and all the data is encrypted using the chosen algorithm. The key is valid only for that session. For a new session, a new key has to be generated. The key is negotiated between the two systems (for exmaple, client and server) using TLS handshake protocol.

The process for providing secure communication is as follows: When a client and server have to exchange information using a secure communication link, the TLS handshake protocol enables the client and the server to authenticate each other and negotiate the encryption algorithm and encryption keys before the application process transmits or receives the first byte of data. Once the client and the server agree on the algorithm and the keys, the TLS protocol encrypts the data with the TLS record protocol and passes the encrypted data to the transport layer.

To provide security while transferring the data, a layer called transport layer security (TLS) is introduced between the transport layer and the application layer. TLS consists of two sublayers: TLS record protocol and TLS handshake protocol.

Above the transport layer protocol (TCP or UDP), the application layer protocol will be running. We will discuss various application layer protocols in the next chapter.

To provide secure communication, the two important features to be incorporated are authentication and encryption of data. Authentication ensures the genuineness of the users. Encryption transforms the bit stream using an encryption key, and the data can be decoded only if the encryption key is known to the receiver.

Summary

This chapter presented the details of the transport layer protocols used on the Internet. The transport layer is an end-to-end protocol—it is the responsibility of the transport layer to ensure that all the packets are put in sequence, to retransmit the packets if there are errors, and to report the status information. The Transmission Control Protocol (TCP) is a transport layer protocol that provides connection-oriented service. Between two end systems, a virtual connection is established by specifying the IP address and the port address. After the connection is established, data transfer takes place, and then the connection is removed. To provide the necessary security features, transport layer security (TLS) is another layer of software that can be run to provide the necessary encryption of data. The user datagram protocol (UDP) provides a connectionless service. The UDP datagram can contain only the source address, destination address, message length, and message. As a result, the UDP is a very light-weight protocol, and high processing power is not required to analyze the UDP header. Hence, UDP is used for applications such as real-time voice/video communication. The formats of TCP segment and UDP datagram are also presented in this chapter.

References

W.R. Stevens. *TCP/IP Illustrated, Vol. I; The Protocols*. Addison Wesley, Reading, MA, 1994. This book gives a complete description of TCP and its implementation in Berkeley Unix.

D. E. Comer and D.L. Stevens. *Internetworking with TCP/IP Vol. III: Client/Server Programming and Applications*. BSD Socket Version, Prentice Hall Inc., Englewood Cliffs, 1993.

www.ietf.org IETF home page. You can obtain the Requests for Comments (RFCs) that give the complete details of the protocols from this site.

Questions

1. Explain the services of the transport layer protocol.
2. Describe the TCP segment format.
3. Describe the UDP datagram format.
4. Explain the differences between TCP and UDP.
5. Explain the transport layer security mechanism.
6. Explain why TCP is not well suited for real-time communication.

Exercises

1. Obtain the source code for TCP/IP stack implementation from the Internet and study the code.

2. Write a Java program to implement the UDP datagram.

3. What is the silly window syndrome?

4. In a satellite communication system, a file has to be transferred from the central station to a number of VSATs, but the VSATs are receive-only, and there is no communication from VSAT to the hub. Work out a procedure for the file transfer.

5. Study the intricacies of the sliding-window protocol used in the TCP layer.

Projects

1. Using open source TCP/IP software, develop a LAN analyzer. The LAN analyzer has to capture each packet that is broadcast by the nodes and calculate the number of packets transmitted per second. It also has to display the traffic matrix, which indicates the number of packets transmitted from one node to another node. You need to analyze each packet for its source address and destination address.

2. Write software that captures the packets being transmitted on the LAN and checks whether the user data portion of the packet contains a keyword. The GUI should facilitate giving a keyword. For instance, if the keyword is specified as "Professor," the software has to check whether the word "Professor" is present in the user data portion of the packet.

23 ■ Distributed Applications

In This Chapter

- Application Layer Protocols Used on the Internet
- The SMTP, MIME, and HTTP Protocols
- How Directory Service is Implemented Using LDAP
- The SNMP Network Management Protocol

All the applications available on the Internet—electronic mail, file transfer, remote login, World Wide Web, and so on—are based on simple application layer protocols that run on the TCP/IP protocols. In this chapter, we will study the most widely used application layer protocols: SMTP and MIME for electronic mail, HTTP for the World Wide Web, and LDAP for directory services. We also will discuss SNMP for network management.

23.1 SIMPLE MAIL TRANSFER PROTOCOL (SMTP)

Simple Mail Transfer Protocol (SMTP) is used for electronic mail. E-mail consists of two parts: header and message body. The header is of the format

```
To: kvkk.Prasad@ieee.org
Cc: wdt@bol.net.in
Sub: hello
Bcc: icsplted@hd1.vsnl.net.in
```

The body consists of the actual message and attachments, if any. In general, the attachments can be text or binary files, graphics, and audio or video clips. However, SMTP supports only text messages.

SMTP uses information written on the envelope (message header) but does not look at the contents. However, SMTP handles e-mail with the restriction that the message character set must be 7-bit ASCII. MIME gives SMTP the capability to add multimedia features, as discussed in the next section. SMTP adds log information to the start of the delivered message that indicates the path the message took for delivery at the destination.

Every e-mail message contains a header that gives the details of the paths traveled by the message to reach the destination.

NOTE

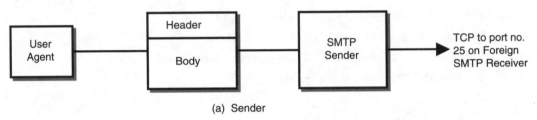

(a) Sender

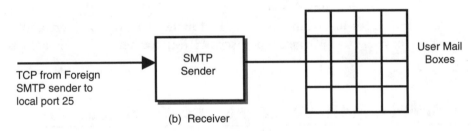

(b) Receiver

FIGURE 23.1 SMTP.

SMTP is the application layer protocol for e-mail. SMTP handles mail with the restriction that the message should contain only ASCII text. Multimedia mail messages are handled by MIME, an extension to SMTP.

The SMTP mechanism is shown in Figure 23.1. There will be an SMTP sender and an SMTP receiver. A TCP connection is first established between the sender and the receiver.

SMTP sender: A user composes the mail message consisting of the header and the body and gives the message to the SMTP sender. The TCP connection between the sender and the receiver uses port 25 on the target host. When a mail is for multiple hosts, the SMTP sender sends mail to the destinations and deletes it from the queue, transfers all copies using a single connection, and then closes the connection.

If the destination is unreachable (a TCP connection cannot be established, the host is unreachable, or the message has a wrong address), the message is requeued for later delivery, and the SMTP sender gives up after a certain number of retries and returns an error message if attempts to retransmit fail.

Note that in the protocol, the SMTP sender's responsibility ends when message is transferred to the SMTP receiver (not when the user sees the message).

The format for the text message used in SMTP consists of the message ID, a header containing the date, from, to, cc, and bcc addresses, and the message body in ASCII text. This format is specified in RFC 822.

SMTP receiver: SMTP receiver accepts each message from the SMTP sender and puts the mail in the corresponding mailbox. If the SMTP receiver is a forwarding machine, it keeps the message in the outgoing queue. SMTP receiver should be capable of handling errors such as disk full or wrong user address.

SMTP protocol: The SMTP protocol uses the TCP connection between the sender and the receiver to transfer the messages. After the messages are transferred, the sender and receiver can switch roles: the sender can become the receiver to get the messages. SMTP does not include an acknowledgement to the message originator.

The format to be used for text messages in SMTP is specified in RFC 822. The format consists of:

Message ID: a unique identifier

Header: date, from, subject, to, cc, bcc

Message body in ASCII text

Web-based mail is used extensively mainly because it is provided as a free service by a number of service providers such as Hotmail and Yahoo!. In Web-based mailing systems, the HTTP protocol is used to connect to the server.

SMTP works as follows: A TCP connection is established between the sender and receiver using port 25. SMTP sender and receiver exchange a series of commands/responses for connection setup, mail transfer, and disconnection. Sender and receiver can switch roles for exchange of the mail.

Application software to generate mail messages is not part of SMTP. It is an application program that runs above the SMTP.

SMTP commands are sent by the SMTP sender and replies are sent by the SMTP receiver.

Some SMTP commands are:

HELO <space><domain><CR><LF>   to send identification

RCPT <space>To:<path><CR><LF>   identifies recipient of the mail

DATA <CR><LF>   to transfer message text

QUIT <CR><LFalso>   to close TCP connection

TURN <CR><LF>   reverse role of sender and receiver

MAIL <space>FROM<><CR><LF>   to identify originator

(CR stands for carriage return and LF for line feed.)

Some SMTP replies corresponding to these commands are given below.

Positive completion reply messages

      220   <domain> Service ready

      250   Requested mail action okay, completed the task

      251   User not local, will forward mail

Positive intermediate reply

      354   Start mail input, end with <CR><LF>

Transient negative replies

      450   Requested mail action not taken, mailbox not available

      451   Requested action not taken, insufficient disk space

Permanent negative replies

      500   Syntax error, command unrecognized

      501   Command not implemented

Using the above commands and replies, a typical SMTP session would be as follows. The session consists of three phases: connection setup, mail transfer, and disconnection.

Connection setup:

1. Sender opens TCP connection with target host.
2. Once connection is established, receiver identifies itself 220 service ready.

3. Sender identifies itself with HELO command.

4. Receiver accepts sender's identification: 250 OK.

   Mail transfer:

1. MAIL command identifies the originator of the message.

2. RCPT command identifies the recipients of the message.

3. DAT command transfers the message text.

   If receiver accepts mail, 250 is sent.

   If receiver has to forward the mail, 251 is sent.

   If mailbox does not exist, 550 is sent.

   Disconnection:

1. Sender sends QUIT command and waits for reply.

2. Sender initiates TCP close operation.

Because SMTP messages have to be in the format specified in RFC 822, the SMTP mails will have the following limitations:

- Executable files cannot be transmitted as attachments.

- Text other than 7-bit ASCII cannot be transmitted.

- SMTP servers may reject a mail message if it exceeds a certain size.

- If two machines use different codes (such as ASCII and EBCDIC), translation problems are encountered.

- SMTP cannot operate with X.400 MHS (message handling system), the standards specified for electronic mail in the ISO/OSI protocol suite.

- There are nonstandard implementations of SMTP/RFC 822:

    - Deletion, addition, reordering of <CR< and <LF>

    - Truncation or wrapping lines longer than 76 characters

    - Padding of lines to the same length

    - Conversion of a Tab to multiple spaces

To overcome these limitations, the MIME protocol has been defined and runs above the SMTP.

X.400 message handling system is the standard application layer software for e-mail in the ISO/OSI protocol suite. It is a very sophisticated protocol that can handle registered mail, acknowledgments, and so on: SMTP, as the name implies, is simple and hence has limited features.

23.2 MULTIPURPOSE INTERNET MAIL EXTENSION (MIME)

MIME is defined in RFC 1521 and 1522. The specifications include:

- Five new message header fields that provide information about the body of the message.
- A number of content formats to support multimedia mail.

The five header fields are:

- MIME Version 1.0 (conforms to RFC 1521 and 1522).
- Content type that describes data contained in the body.
- Content transfer encoding that indicates the type of transformation that has been used to represent the body of the message.
- Content ID to uniquely identify MIME content types.
- Content description: plain text description of object, such as audio, text, or video clip.

Some of the MIME content types are listed here.

Type	Subtype	Description
Text	Plain	
Multipart	Mixed	To be presented to the user in the order they appear.
	Parallel	No order is required.
	Alternative	Alternative version, best to be displayed.
	Digest	Similar to mixed, but RFC 822.
Message	RFC822	
Partial	Fragment	Large mails.
	External body	Pointer to object
Image	JPEG	
	GIF	
Video	MPEG	
Audio	Basic	8-bit at 8kHz
Application	PostScript	
	Octet stream	Binary data with 8-bit types.

An example MIME header looks like this:

```
From: kvkk.prasad@ieee.org
To: kvravi@ece.nus.sg
Sub: demo
MIME Version 1.0
Content type: multipart/mixed
Content type: video/MPEG
```

MIME is an extension to SMTP to provide multimedia message content to e-mails. In addition to plain text, images, executable files, audio, and video clips can be attached to the e-mail. MIME is defined in RFCs 1521 and 1522.

Hence, MIME is an extension to the SMTP to provide multimedia support in the message of the mail using the various MIME types.

23.3 HYPERTEXT TRANSFER PROTOCOL (HTTP)

The Uniform Resource Locator (URL) has six fields: protocol, domain name, port address, directory path, object name, and the specific spot in the object. The server on which the object is located is called the origin server.

To access the World Wide Web service, HTTP protocol is used, the latest version being HTTP 1.1. HTTP uses URL (Uniform Resource Locator), which represents the location and access method for a resource available on the Internet. For example:

http://www.iseeyes.com

ftp://rtfm.mit.edu/pub/Index.README

The format of a URL is

http://www.elearn.cdacindia.com:80:/hypertext/www/Addressing/Addressing.html#spot

| 1 | 2 | 3 | 4 | 5 | 6 |

(1) Protocol (such as HTTP, FTP)

(2) Domain name

(3) Port address

(4) Directory path

(5) Object name

(6) To reach a specific spot (or link)

Some of the most commonly used protocols along with port addresses are

FTP	21 (port address)
HTTP	80
Telnet	23

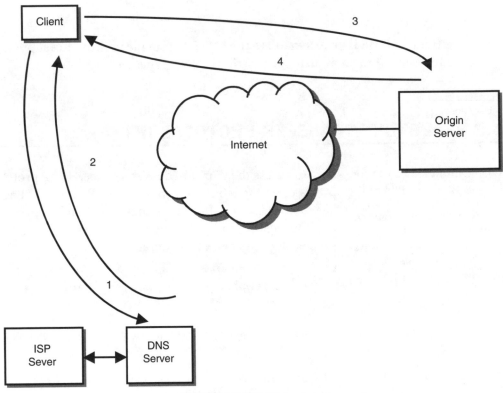

FIGURE 23.2 Web access.

When a user wants to access a resource by giving a URL, the IP address of the origin server is obtained using a Domain Name System. Then a TCP connection is established between the client and the origin server. The origin server sends the HTML document corresponding to that URL to the client.

Figure 23.2 shows the procedure for accessing the resource corresponding to a URL. The client invokes the browser (such as Internet Explorer or Netscape Communicator) and specifies the URL. The URL is passed to a Domain Name System (DNS) server (Step 1), which gives the IP address of the server (Step 2) that has the resource corresponding to that URL. The server that has the resource is known as the origin server. Using that IP address, the HTTP request is sent to the Web server (Step 3) and the Web server gives the response in the form of an HTML document to the client via the ISP server (Step 4). Sometimes Web servers do not allow access to every user due to security reasons. In such cases, a proxy server is used between the client and the origin server. The proxy server acts as a client to the origin server and as a server to the actual client. When there is no restriction on accessing an origin server, the servers between will act as tunnels. A tunnel is an intermediate program that acts as a blind relay between two connections.

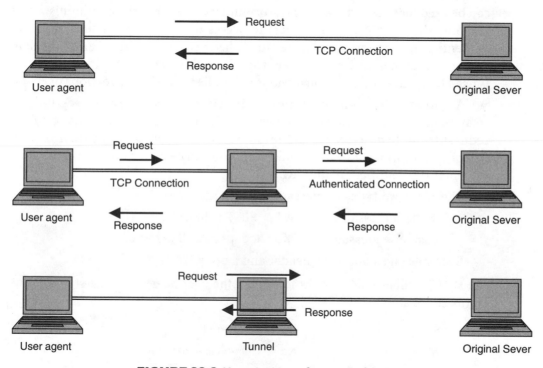

FIGURE 23.3 Hypertext transfer protocol (HTTP).

HTTP is a transaction-oriented client/server protocol. The client is the Web browser, and the server is the web server. The mechanism for HTTP protocol is shown in Figure 23.3. The origin server is the server on which the resource is located. There are three possibilities for interaction between the client and the origin server:

■ Direct connection

■ Proxy

■ Relay

A proxy is used when the Internet server does not permit direct access to the users. The proxy acts as a client to the server and as a server to the client.

In direct connection, a TCP connection is established between the client (user agent) and the origin server. The user agent sends an HTTP request, and the origin server sends the response.

A proxy acts as a server in interacting with clients. The proxy server acts as a client to the origin server. The user agent sends the request to the proxy, and the proxy in turns sends the request to the origin server and gets the response, which is forwarded to the client. The proxy server can be used for different reasons. The origin server, for security or administrative reasons, allows only certain servers to access its resources. In such cases, an authenticated connection has to be established between the proxy and the origin server. Sometimes, different servers run different versions of HTTP. The proxy is used to do the necessary conversions to handle the different versions of HTTP.

A tunnel performs no operation on HTTP requests and responses. It is the relay point between two TCP connections. HTTP messages are of two types, request from client to server and response from server to the client. A request will be in the format GET <URL>, and a response will be a block of data containing information identified by the URL.

There are two types of messages:

Request messages: GET, POST, DELETE

Response messages: OK, Accepted, Use Proxy

Each message contains a header and a body.

HTTP defines status codes to inform the client/server of the status. Some status codes are:

OK	Request successful.
No content	No information to send back.
Moved permanently	Resource URL has changed permanently.
Moved temporarily	Resource URL has changed temporarily.
Use Proxy	Resource must be accessed through proxy.
Unauthorized	Access control is denied due to security reasons.

When a user wants to access a Web site, the URL given in the address field of the browser is sent to the ISP server, which gets the IP address corresponding to the URL. Then a TCP connection is established and a GET command is sent. The origin server sends back the resource (the HTML file). The HTML code containing

HTTP is a simple transaction-oriented client/server protocol. The client sends a GET request along with a URL, and the server responds with a POST response containing the HTML document corresponding to that URL. In case there is a problem such as nonavailability of the content, nonavailability of the server, and so on, an error message is sent.

the tags will be interpreted by the browser and displayed to the user. In case any error is encountered, the error message is displayed based on the status code received by the browser.

The World Wide Web has become the most attractive service on the Internet because of HTTP. The information corresponding to a URL is obtained from the origin server in the form of Hypertext Markup Language (HTML), which is a simple text file with tags (called markup). These tags specify how the content has to be formatted and displayed (bold letters, underline, in table format, and so on). The HTML file also contains special tags called *anchor tags*. These anchor tags provide links to other HTML pages. When you click on the link, the HTML page corresponding to that link will be displayed. This new HTML page can reside on the same server or another server located in another part of the world. As a result, you can access information without the need to know where the information is physically located. That is the power of the Web.

23.4 LIGHTWEIGHT DIRECTORY ACCESS PROTOCOL (LDAP)

To access network services, directories play an important role. We keep information about contact addresses in a directory (name, address, e-mail, phone numbers and so on). Information about registered users is kept in directories at the servers, DNS servers keep the information about the IP addresses in a directory, information obtained from the finger command (the users who are presently logged on to the network) is kept in a directory and so on.

Presently, we store directory information in different formats on the desktop, on the palmtop, on the mobile phone, and so on. That is the reason we cannot transfer directory information from one device to another device. If there is a standard mechanism for storing and retrieving directory information, it would be of great help, LDAP is meant exactly for this.

If directory information is kept using standard mechanisms, it can be accessed through standard protocols, and the directories are accessible universally. To facilitate this, ISO/OSI protocol suite defines X.500 directory service standards. However, implementation of these directory services and protocols is difficult, particularly on computers with less processing power such as a PC. Hence, LDAP has been formulated as a lightweight protocol that can run on PCs having less processing power.

 X.500 is the standard for directory service. This standard specifies how directory information is organized and how to search for information. It uses Directory Access Protocol (DAP) to access the directory. However, it is not well suited to run on small computers such as PCs, and hence a lighter version was developed called LDAP.

LDAP uses the X.500 directory service framework and the Directory Access Protocol (DAP) defined in X.500. X.500 defines the following:

- Information model gives the format in which the information has to be stored in the directory.
- Namespace specifies how the information has to be referenced and organized.
- Functional model defines operations that can be performed on the information.
- Authentication model defines how to secure the information stored in the directory.
- Distributed operational model defines how data is distributed and how the operations are carried out.

LDAP defines the format for organizing directory information using a hierarchical naming scheme. It also specifies the operations to be performed on the directory to update the directory and search/find records in the directory.

Since DAP is a heavyweight protocol, LDAP has been developed. LDAP gives the same functionality without much overhead.

LDAP is a client/server protocol. The LDAP client sends a request, and the LDAP server sends the response. The request can be to search and find an entry in the directory, to modify an entry in the directory, or to delete an entry in the directory. The LDAP server provides the information in the form of an LDAP server response. If the server has to contact some other server to obtain the information (called the references), it is the server's responsibility to do that.

LDAP also defines how the information has to be organized in the directory. Each entry in the directory is given a unique name; the collection of names is called the *namespace*. Each entry will have a number of attributes. The entries are organized hierarchically. The various operations that can be performed on the directory are search and find, modify, and add records to the directory.

Many application software packages such as e-mail packages, browsers, and such have LDAP functionality.

When you use multiple devices (palmtop, mobile phone, and so forth), an important requirement is that the address book be the same in both the devices—the data in both devices should be synchronized. Synchronization is easy if a standard, such as LDAP, is followed for storing the directory information. To synchronize the data, a markup language called SyncML has been standardized.

23.5 SIMPLE NETWORK MANAGEMENT PROTOCOL (SNMP)

For any network, network management software is extremely important to provide reliable service, to tune the performance of the network, to upgrade the network based on traffic requirements, and to take care of topological changes. Network administrators have to obtain the information about the various network elements such as servers, routers, bridges, and so on. The network administrator should also be able to configure these network elements through simple commands.

Two types of network management are possible: centralized and decentralized, depending on the requirements of the network. In centralized network management, a central station carries out the entire management of the network, whereas in decentralized network management, the management functions are distributed.

Figure 23.4 shows a typical network with various elements for network management. Each subnetwork to be managed will have a management agent. These agents are controlled by a set of managers. These managers in turn will act as agents to a centralized management server. The agents can be different network elements such as a bridge, a router, or an end system. These agents run the agent software and the managers run the manager software.

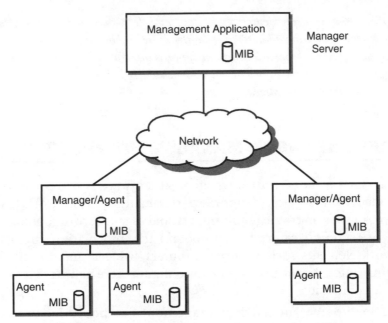

FIGURE 23.4 Network management using SNMP.

The key elements of network management are:

Management station: The management station provides the human machine interface (HMI) to monitor and control the network. It contains the set of applications for data analysis and fault recovery. It provides a graphical user interface to manage the network through simple commands or mouse-clicks. It has the complete database of network management information.

Agents: This is the software residing in the bridges, routers, hosts, and such for remote management. This software will send the necessary data based on the commands from the management station.

Management information base (MIB): Data variables (objects) represent one aspect of a managed agent. The collection of objects is the MIB. For example, for a packet switch, the buffer size, the number of packets dropped, the delay, and so on can be the variables.

Network management protocol: This is SNMP (Simple Network Management Protocol) in TCP/IP networks and CMIP (common management information protocol) in ISO/OSI networks. SNMP runs on the UDP and not TCP. Because the network management information is not of very high priority, UDP is used instead of TCP.

The present version of SNMP running on the Internet is SNMP Version 2. SNMP is capable of three operations: get, set, and notify. Using the get operation, the management station retrieves values of objects at the agent. Using the set operation, the management station sets values of objects at the agent. Using the notify operation, the agent notifies management station of any significant events.

In a network, there will be network management workstations, which collect the data and present it to the users, and SNMP agent packages that run on the elements to be managed. SNMP does not by itself provide network management, it provides the infrastructure for network management. That is, network management applications can be built using the components. Applications such as fault management, performance monitoring, accounting, and traffic analysis are outside the scope of SNMP.

SNMP v2 is used to exchange management information using a set of data types, structures, and so forth. As shown in Figure 23.4, there will be an MIB at each manager/agent that is used to do the network management through application software. An MIB is organized as a hierarchy of objects. The SNMP protocol runs as follows:

- The manager issues a get request PDU (protocol data unit) with a list of object names for which values are requested.

- The agent will send a response PDU with values of objects. Partial responses are allowed. Error codes are sent if an object name is incorrect.

- A get next request PDU is sent if more object values are required. The next value in the hierarchy is sent back by the agent.

- A get bulk request PDU (supported by SNMP v2) is used to avoid a large number of data exchanges. A PDU is sent with a request to send information about a set of variables, and the information is returned in one PDU. This strategy reduces the traffic on the network for network management.

SNMP provides the basic framework required to carry out network management by defining how the management information has to be organized and how the information has to be retrieved without dealing with the specific information to be handled. If the network administrator has to obtain the information on the number of packets dropped in a router, the object corresponding to this information has to be created in the router, through a get command from the manager, the information is obtained. Of course, network equipment vendors provide the necessary objects for network management, so the network administrator's job is only to write the graphical user interface to process the information and present it in a user-friendly manner.

The Internet is such a great platform for exciting applications thanks to the TCP/IP protocols and the application layer protocols described in this chapter.

SNMP uses UDP and not TCP to reduce the communication overhead in the networks. Note that SNMP provides only the necessary framework for network management. To generate information such as traffic analysis, fault reports, and such is not in the scope of SNMP—these have to be done by application software.

Summary

This chapter presented the most important application layer protocols: Simple Mail Transfer Protocol (SMTP), Multipurpose Internet Mail Extension (MIME), Hypertext Transfer Protocol (HTTP), Lightweight Directory Access Protocol (LDAP), and Simple Network Management Protocol (SNMP). SMTP facilitates mail transfer but with the restriction that only ASCII text messages can be sent. The MIME protocol enhances the capability of SMTP by supporting multimedia. HTTP is a client/server protocol that allows sending a URL as a request and getting back the resource corresponding to that URL from the origin server. LDAP defines a standard mechanism for storing, retrieving, and modifying directory information. SNMP provides the necessary framework to manage computer networks by defining a simple protocol to obtain management information from network elements such as hosts, servers, routers, and bridges.

References

L.L. Peterson and B.S. Davie. *Computer Networks: A Systems Approach*, Morgan Kaufmann Publishers, 2000. Chapter 9 of this book deals with application layer protocols.

www.apache.org You can obtain the details of Apache HTTP server from this site.

www.ietf.org You can download the Requests for Comments (RFCs) referred to in this chapter from this site.

www.w3.org Web site of the World Wide Web Consortium.

Questions

1. Explain the operation of SMTP.

2. What is the protocol used for supporting multimedia e-mail content? Explain the protocol.

3. Explain how Web content is accessed through HTTP.

4. What is LDAP? In what way does it help end users in directory service?

5. Explain SNMP.

Exercises

1. Study the FTP protocol used for file transfer.

2. Interconnect two PCs using an RS232 link and implement SMTP to exchange mail between the two PCs.

3. What are the problems associated with having proprietary directories? How does LDAP solve these problems?

4. List some of the standard protocols for file transfer.

5. List the network management information to be collected to monitor performance as well as to plan expansion.

Projects

1. Develop an intranet messaging system. On the LAN, the server should have different folders for the users of the LAN. Any user should be able to log on to the server and check the mailbox for received messages and also send messages to others. Effectively, you need to implement a mailing system similar to Hotmail or Yahoo!.

2. Download Apache Web server software and port it onto your system.

3. Using an RDBMS package such as MS Access or MS SQL or Oracle, implement LDAP functionality.

24 | The Wired Internet

In This Chapter

- Internet Architecture and Applications
- Domain Name System
- Information Security
- Internet Access Products
- The Architecture of Virtual Private Networks

The Internet is the network of computer networks spread across the globe. An outcome of the research funded by Advanced Research Projects Agency (ARPA) of the U.S. Department of Defense, the Internet now interconnects networks in almost every country with millions of hosts providing information. Internet access is now a daily activity for most of us. In this chapter, we will study in detail the services offered by the Internet and the architecture of the Internet.

Many organizations with multiple offices are no longer building their own dedicated corporate networks. Instead, they are using the Internet infrastructure to create networks known as virtual private networks (VPNs). We will discuss the architecture of VPNs and the security issues involved in VPNs in particular and the Internet in general.

24.1 INTERNET APPLICATIONS

From an end user point of view, the Internet provides the following applications:

Electronic mail (e-mail): Users can send messages to one another—the message can be text along with attachments of different types—graphics, audio or video clips, and executable programs (and unfortunately, viruses as well). E-mail is now used extensively for business correspondence in addition to sending personal messages.

File transfer: Users can download or upload files (documents or executable programs) from one computer to another. Nowdays, many software packages are distributed using this mechanism instead of distributing the software on floppies, CD-ROMs, or tapes.

Remote login: Users can log in to another computer and use its resources; of course, the necessary permissions are required to log into the other machine. This provides the scope for collaborative working—a researcher in India can log in to the computer of a U.S. university to carry out research in highly specialized areas.

Interactive applications: Different users can log in to a server for interactive applications such as chat and games. Nowdays, the Internet is also being used for interaction among people using voice and desktop video. Software packages such as Microsoft NetMeeting provide the facility to chat using text, voice, and video and hold a business meeting.

World Wide Web: The World Wide Web (in short, the Web) has revolutionized Internet use. The Web provides the means of accessing information without knowing where the resource is physically present. With the Web site address *www.iseeyes.com*, we get information about the company with that address. Where that information is located (on which server) is not even known to us. The information that is presented on your desktop may contain links to other resources and, when you click on a link, you will automatically get that information. The Web provides the information not just in text, but with multimedia content. In many parts of the world, though, this poses problems because downloading voice and video takes a lot of time. In those parts of the world, WWW is called *world wide wait*.

Initially, Web pages were noninteractive, but now interactivity is provided whereby the user can input information to be processed by the server, and output is given to the user. This led to many interesting applications, particularly in e-commerce. Business to business commerce and business to consumer commerce are now the most promising applications. Supply chain management, wherein a business can do the paperwork for obtaining raw material and for supplying finished goods through the Internet, will be the next killer application. Workflow automation, wherein the entire movement of information within an organization can be automated, can lead to true paperless office. All these exciting applications are now paving the way for an Internet economy—the whole economy is now centered around the Internet. "Web services" is the term used to provide such applications using standard Web protocols and development tools.

Presently, we access Web services through our desktop PCs. Voice-enabled Web, whereby Web services can be accessed through telephones, is now on the drawing board, using speech recognition and text-to-speech conversion technologies.

The Java programming language and eXtensible Markup Language (XML) are the two important tools for creating content on the Internet. Most of the existing content on the Internet is written using Hypertext Markup Language (HTML), though XML is catching up fast. For accessing the content from mobile devices, Wireless Markup Language (WML) is used.

Distributed applications: The Internet is a repository of knowledge. One can obtain information available on millions of servers without knowing the exact location of the information. One can use a search engine (such as Google, Yahoo!, or AltaVista) to get this information. If the search engine is provided with keywords, the engine will search various servers and present the user with the various resources corresponding to that keyword. Of course, this is now leading to an information explosion. If you search for the keyword Artificial Intelligence and are provided with a million resources, which one do you select? Intelligent search engines will soon provide more focused information for your requirement.

Searching for images, audio clips, and video clips is still a major problem and is presently an active research area. Intelligent search engines are being developed that give relevant information so that we can avoid the information explosion.

Broadcasting: Internet can be used to broadcast voice and video programs as well, though bandwidth is a limitation presently. One can tune one's desktop to an Internet radio station (by giving the Web address) and listen to music or watch a video clip.

For audio and video communication over the Internet, low-bit rate coding techniques are used. The coding is done based on the H.323 series of standards.

In addition to the basic services such as e-mail, file transfer, remote login, and the World Wide Web, a number of interactive and distributed applications are supported by the Internet, including audio/video broadcasting, two-way and multiparty audio/video conferencing, distributed database access, e-commerce, and Web-based learning.

Web-based learning: The Internet is also revolutionizing education. We no longer need to attend traditional classroom lectures. We can sit in the comfort of our living rooms and attend lectures delivered through the Internet and also interact with the lecturer. We can take online examinations and obtain certification. Distance education using the Internet infrastructure is now catching up and is a threat to all the second-rate universities. Many western countries offer formal degree programs using this approach.

24.2 INTERNET ARCHITECTURE

The Internet is a complex network, and has no fixed topology. Figure 24.1 shows the simplified architecture. Internet service providers (ISPs) are connected to the backbone network. In fact, there are multiple backbone networks. Individual/corporate users access the Internet through dialup or leased lines. Content providers set up servers and connect to ISPs.

The Internet is a connection of various internets, connected through routers (also called gateways). Also, the Internet is a very dynamic network—networks are added and removed almost on a daily basis. Web sites crop up and go down also on a daily basis. When a user invokes his browser and enters a URL, how does he get the information? The first requirement is that the server where the resource corresponding to the URL is located has to be known, and the request has to be forwarded to that server. Secondly, when the packets are transmitted from one node to another, the packets have to be routed properly.

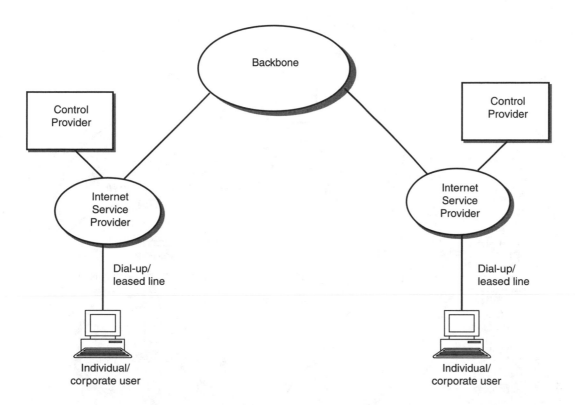

FIGURE 24.1 Internet architecture.

The Internet has multiple backbone networks. Internet service providers (ISPs) connect their networks to one of these backbone networks. The Internet is a network of networks, but it has no fixed topology.

To get connected to the Internet, we need to obtain an account from a local Internet service provider (ISP). As shown in Figure 24.2, the ISP has various servers such as mail server, domain name server, Web server, and authentication server. We can get access to the ISP server either through a dial-up line or a leased line. We log in to the server at the ISP, and the ISP provides access to the rest of the Internet through a router/gateway.

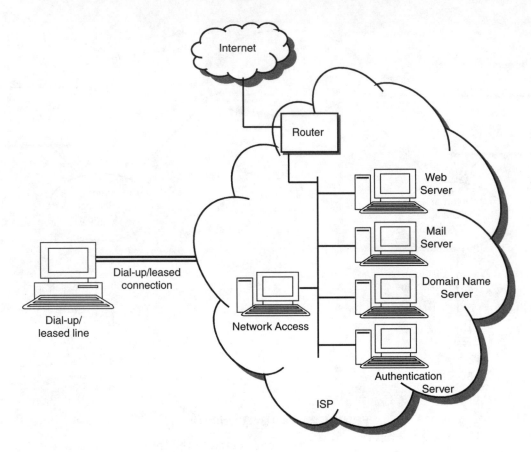

FIGURE 24.2 Accessing the Internet through ISP.

24.3 DOMAIN NAME SYSTEM

To identify the computers on the Internet, a 32-bit IP address is assigned to each computer. To identify computers by an address of this format is not user friendly—an easy to remember, easy to pronounce name would be a better choice. Domain Name System (DNS) provides this facility. However, this naming system should be universal—a computer on the Internet should be uniquely identified by the name, and there should be unique mapping between the name and the IP address.

The Domain Name System facilitates translation of a user-friendly address (such as *www.iseeyes.com*) to the IP address. Domain name servers provide this facility. A hierarchical naming scheme is used to assign domain names. The top-level domain name can reflect the name of the country or the type of organization.

To have one central authority to assign names becomes a complex task. Therefore, a hierarchical naming scheme has been devised.

Assigning a naming scheme is analogous to the telephone numbering scheme. At the highest level, each country is assigned a country code (which is done centrally). Within the country, each region is assigned a code, and this assignment is done by a central authority in the respective countries. Within each region, the numbering scheme is decentralized.

On the Internet, a typical domain name can be:

cs.jntu.edu

where *edu* is the top-level domain name, *cs.jntu.edu* is the secondary domain, and *cs.jntu.edu* is the tertiary domain. The top-level domain names have to follow a standard specified by the International Organization for Standardization (ISO). Some of the accepted top-level domain names are:

com	commercial organizations
biz	commercial organizations
edu	educational institutions
net	major network support centers /commercial organizations
org	non-profit organizations
gov	government organizations
mil	military organizations
name	individuals

In addition, the top-level domain names can denote the name of the country:

us	United States of America
in	India
uk	United Kingdom
tw	Taiwan
sg	Singapore
de	Germany

Some top-level domain names have been introduced recently. These include .biz, .name, and .tv. If you want to have your own URL, you can check its availability at the site www.register.com.

24.3.1 Domain Name Servers

The server that does the translation between the domain name and the corresponding IP address is called the domain name server. The client runs name resolution software to obtain an IP address. There are two ways for name resolution:

- Recursive resolution
- Nonrecursive resolution or iterative resolution

In recursive resolution, if a name server cannot resolve the IP address of a name, it will contact another name server and obtain the information, which will be sent to the client. In nonrecursive or iterative resolution, if the name server cannot resolve the address, it will send a message to the client, giving information as to which name server to contact for obtaining the IP address.

When you get connected to the Internet, you are assigned a temporary IP address called the Dynamic IP address. This IP address is given by the ISP server. When you access a URL, the IP address of the Origin Server is obtained by the Domain Name Server using recursive resolution or iterative resolution.

Figure 24.3 illustrates how the URL is translated to an IP address. When the user logs in to the ISP's network access server through a dial-up connection, a temporary IP address is given to his computer (known as the dynamic IP address). This address is given by the ISP server from a pool of dynamic IP addresses. The user invokes a browser and gives a URL. This URL is sent to the domain name server. The domain name server obtains the IP address and sends it to the user's application. The application will establish a TCP connection (through the IP layer) with the origin server in which the resource corresponding to the URL is located. Then the origin server sends the HTML page corresponding to the URL.

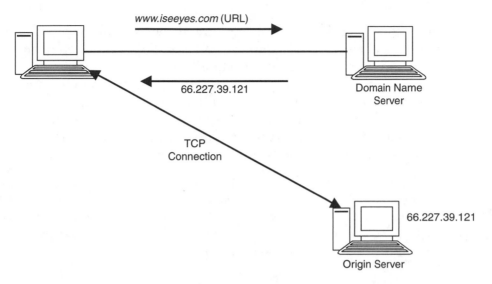

FIGURE 24.3 URL translation to IP address.

Once a TCP connection is established between the client and the origin server, the packets are exchanged between the client and the server, using the TCP/IP protocol. The routing algorithms discussed in Chapter 21, "Internet Protocol (IP)" are used to route the packets.

 The resource corresponding to a URL is a Web page written in HTML. The HTML file is a text file containing tags. This HTML file is downloaded on to the client's computer. The browser, such as Internet Explorer, interprets this HTML file and presents the content to the user.

24.4 SECURITY ISSUES ON THE INTERNET

To provide security, authentication and encryption are the two most important measures. Authentication is a process to verify the genuineness of the user of the application through a username and a password. Encryption of data involves modifying the bit stream using an algorithm and an encryption key. For decoding the data, the receiver should know the encryption key.

Information is power. Nowadays, all organizations keep their information on computers, and these computers are networked. To ensure that unauthorized persons do not have access to this information and that the information is not manipulated or erased is of paramount importance. In addition, the virus menace is creating havoc—an innocent-looking program or e-mail can damage the information of a corporation. To provide information security is crucial for the survival of an organization. In this section, we will discuss the various issues involved in providing security.

To provide security, the two measures taken are authentication and encryption. Authentication is a mechanism used to verify the genuineness of the user of the application. Authentication is done through a username and password. Encryption is a mechanism in which the data is modified using a predefined bit stream known as an encryption key; the data can be decoded at the destination only if the encryption key is known at the receiving end. For security, each ISP installs a server for authentication purposes. The RADIUS server is the standard way of providing security by ISPs.

Security continues to be a major issue. Hackers and antisocial elements continue to devise innovative methods to find the security loopholes on servers. Both Windows and Linux operating systems have security loopholes; operating system security is fundamental to providing highly secure applications over the Internet.

24.4.1 RADIUS Server

Remote Authentication Dial In User Service (RADIUS) is now widely used by all ISPs to provide:

- Authentication: to determine who the user is.
- Authorization: to determine what services the user is permitted to access
- Accounting: to track the type of services and duration of the services accessed by the user for billing purposes.

The RADIUS server has a database engine that stores all the information related to users and usage statistics. When a user tries to connect to the network via the network access server (NAS), an access request is sent to the RADIUS server by the NAS. The RADIUS server queries the authentication database and checks whether the user is permitted to access the network and, if so, what types of services are permitted. This information is given to the NAS. NAS also sends an accounting message to the RADIUS server at the beginning of the session and at the end of the session. The RADIUS server keeps a log file for accounting purposes.

> Remote Authentication Dial In User Service (RADIUS) is widely used by Internet service providers to provide authentication, authorization, and accounting services.

24.4.2 Firewalls

A firewall can be a separate server or it can be built into the NAS as software. A firewall restricts access to a specific service. For instance, a simple firewall can be a URL filter. URLs related to a specific topic (say, pornography) can be filtered out so that if a user gives a URL that is on the forbidden list, access to that URL is denied by the firewall. However, note that it is difficult to implement such filters because new URLs keep coming up daily on the Internet. Similarly, an ISP can disallow Internet telephony by filtering the URLs that provide voice communication service over the Internet.

> A firewall is used to restrict the use of some services. Firewalls can be implemented on dedicated servers, or they can be implemented on an existing server. A firewall can filter e-mails or restrict access to specific URLs such as adult sites.

One can develop a packet filter that analyzes the packets from the users and checks the destination IP address. Based on the destination IP address, the packet can be filtered out (not transmitted further). Similarly, packets coming from servers also can be filtered out.

24.5 INTERNET ACCESS PRODUCTS

> Dial-up lines and ISDN lines are used extensively for Internet access. The family of digital subscriber lines such as HDSL and VDSL provide dedicated high-speed connection between the subscriber and the Internet service provider. Present access speeds using the DSL technologies are up to 50Mbps.

To access the Internet, the user has to be connected to the server of the local ISP. This access can be through a dial-up line, a digital subscriber line, or wireless links.

Dial-up lines: The dial-up connection can be a regular PSTN line, in which case the data rate is generally limited to about 56kbps. Alternatively, ISDN lines can be used that support higher data rates up to 128kbps.

Digital subscriber lines (DSL): Nowdays, DSLs are being provided between the subscriber premises and the ISP premises. DSLs with various speeds are available, which are collectively referred to as xDSL. Using HDSL (high data rate digital subscriber line), data rates up to 2Mbps are now possible. DSL can provide a dedicated connection between the ISP and the subscriber, and so the subscriber is always connected to the Internet. The latest in the DSL family is VDSL (very high speed DSL). Using VDSL, speeds up to 48Mbps can be achieved if the distance is no more than 300 meters and up to 25Mbps if the distance is no more than 1500 meters, with 0.5mm copper cable. Soon, fiber to home will be more widespread, which will support much higher data rates.

Wireless access to the Internet: With the popularity of mobile phones and PDAs, accessing the mobile network through wireless devices is catching up very fast. Though access speed with the present wireless networks is about 14.4kbps, in the near future access speeds up to 2Mbps are feasible with the third generation (3G) wireless networks.

Wireless Internet access is now catching up fast. Using third generation wireless networks, the access speed can be up to 2Mbps. Using IEEE 802.11, known popularly as WiFi, access speeds up to 50Mbps will be possible in the near future.

The Internet Protocol (IP) used for accessing the Internet through desktops is not suitable in a wireless environment if the access device is mobile. A new protocol, Mobile IP, has been developed for accessing the Internet through mobile devices such as laptops, PDAs, and mobile phones.

24.6 VIRTUAL PRIVATE NETWORKS

A virtual private network (VPN) is a network of corporate sites using the existing infrastructure of the Internet. VPN provides a cost-effective solution as compared to having a dedicated corporate network.

If an organization has many offices, these offices are connected through a private network wherein the organization builds its own infrastructure for wide area networking. For example, if an organization has offices spread throughout the country, it can develop its own corporate network, perhaps using VSATs with a hub at the corporate headquarters. Such a network provides the necessary security and is fully within the administrative control of the organization. Since it is a dedicated network, the network can be optimized for the requirements of the organization. However, this approach has many disadvantages, which are as follows:

- High installation and maintenance costs: The cost of training the employees and having a dedicated workforce, particularly when the core business of the organization is not communications, is enormous. For instance, a pharmaceutical company does not like to keep a large workforce to maintain its communication network—their business is pharmaceuticals, not computer networks!

- Lack of flexibility and scalability: When new sites have to be added (when a new branch is opened) or when a new customer is to be provided access to the organization's information, additional network equipment has to be installed. Also, the operational costs go up.
- The recurring communication costs will be high as the organization has to pay for the communication facilities on a monthly or yearly basis.

With the Internet having spread itself globally, an alternative approach to interconnecting corporate sites is through virtual private networks (VPNs). A VPN can be defined as a network of corporate sites using the open, distributed infrastructure of the Internet. The VPN has many advantages:

- It is economical because the Internet's existing infrastructure will be used. The organization need not invest in costly communication equipment. In addition, the recurring expenditure also will be less—the expenditure will be based on the usage.
- Minimal training costs: there is no need for special training for users, because the Internet tools will be extensively used, and users are already familiar with the technology.
- Increased flexibility and scalability: expansion of the network or reconfiguration of the sites will be easier.
- Minimal maintenance costs: there is no need for a dedicated team of professionals for maintenance of the network because the Internet service provider will take care of maintenance issues.
- Value added service provisioning: employees, suppliers, and customers can be provided with access to the intranet/extranet resources very easily. Mobility can be supported very easily—an employee on the move can always be in touch with the corporate office.
- On-demand bandwidth: as the communication requirements vary from time to time, the ISP can provide on-demand bandwidth, so there is no need to decide on the capacities of the communication system at the beginning.

The advantages of a virtual private network are low infrastructure costs, flexibility, expandability at low incremental cost, low maintenance cost, and availability of on-demand bandwidth.

However, the TCP/IP architecture does not take care of two important issues: security and performance. In VPNs, these two issues need to be addressed to provide a communication facility that is comparable to that of a dedicated network, but with additional functionality and flexibility. Lots of work is presently being carried out on performance issues. For providing security, a number of protocols have been defined, which are discussed in the next section.

Security remains a major issue for virtual private networks. Banks, security agencies, and transportation agencies continue to have their own dedicated networks precisely due to this reason. Things are likely to change, though, in the future.

24.6.1 VPN Security Requirements

To provide secure communications in a VPN, the following are the requirements:

- Authentication: to ensure that the originator of data is genuine.
- Access control: to restrict unauthorized entry to access the network.
- Confidentiality: to ensure that data is received by the intended recipient only.
- Data integrity: to ensure that the data is not tampered/modified while it is traveling on the public Internet.

Authentication, access control, confidentiality, and data integrity are the main issues related to security in virtual private networks. Security products need to implement these features.

For authentication and access control, a number of mechanisms are available, such as password-based systems, Challenge Handshake Authentication Protocol (CHAP), Remote Authentication Dial-in User Service (RADIUS), and digital certificates. Encryption is used to achieve confidentiality.

For authentication and access control, biometric techniques such as fingerprint recognition are being used. In the past, fingerprint recognition was used to verify the authenticity of the illiterate!

24.6.2 VPN Architecture and Protocols

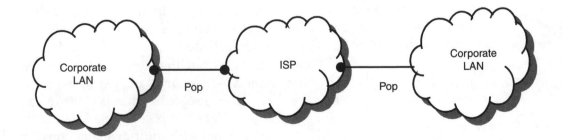

FIGURE 24.4 Virtual private network.

A simplified VPN architecture is shown in Figure 24.4. Each corporate LAN will be connected to the ISP's point of presence (PoP). In VPN, permanent links are not maintained between two end points—the links are created dynamically based on the requirement. When leased lines are used to interconnect sites, these connections are dedicated to the traffic from a single corporate customer; this is called *tunneling*. To extend the concept of tunneling to the Internet-based VPN, protocols are defined to create tunnels. Tunneling allows senders to encapsulate their data in the IP packets that hide the underlying routing and switching infrastructure of the Internet. These packets can be encrypted to achieve end-to-end confidentiality. The concept of tunneling is illustrated in Figure 24.5. Even though network A and network B are interconnected through a large internet, a virtual link is created between the two routers.

If a host on network A wants to send data to a host on network B, the host will send each packet (containing the destination address of the host on network B) to router P. Router P will create an IP datagram that will have the source address P and destination address Q, and the data portion will be the packet sent by the host on network A. This IP datagram will reach router Q in the normal fashion. Router Q will take out the encapsulated packet, obtain the destination address of the host on network B, and then forward it to the host. This encapsulation of the IP datagram is equivalent to creating a tunnel between router P and router Q.

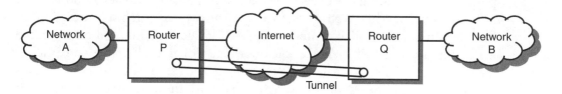

FIGURE 24.5 Tunneling.

IPSec is a layer 3 protocol used in VPNs to provide security. IPSec works in two modes: transport mode and tunnel mode. In transport mode, only the transport layer segment is encrypted. In tunnel mode, the entire IP packet including header is encrypted.

To provide security in a VPN, IP Security (IPSec) protocol is used extensively.

IPSecurity (IPSec): IPSec is a layer 3 protocol. IPSec is the outcome of work on IPv6, but it can be used on IPv4 as well. It allows the sender (a client or a router) to authenticate and/or encrypt each IP packet. IPSec can be used in two modes.

Transport mode: In this mode, only the transport layer segment of the IP packet is encrypted.

Tunnel mode: In this mode, the entire IP packet including the header is encrypted. Hence, this mode gives better protection because the source and destination addresses can be decoded only when the encryption key is available.

IPSec is considered the best protocol for VPN implementation because it provides a complete solution for providing the required security. IPSec is based on standard encryption mechanisms such as:

- Data Encryption Standard (DES) for encryption.
- Digital certificates for validating public keys.
- Key exchange using both manual and the Internet Key Exchange (IKE) automatic method.

However, at this time security remains a major issue for VPNs; commercial security products are very costly. Standards are evolving for various aspects of security and, in the future VPNs will provide the most cost-effective means of corporate networking.

Summary

The wired Internet, within a span of about 30 years, has become an excellent platform for a wide variety of services such as e-mail, file transfer, remote login, and the World Wide Web. Interesting applications such as audio/video broadcasting, audio/video conferencing, Web-based learning, and e-commerce are being supported, which makes the Internet a global marketplace and a global learning center. Though the present mechanism of accessing the Internet through dial-up or leased lines limits the access speed, with the availability of very high speed digital subscriber lines (VDSL), speeds up to nearly 50Mbps can be achieved in due course.

Major corporations can now use the Internet infrastructure to build virtual private networks (VPNs), which result in major cost savings. However, security is a major issue to be tackled for VPNs to become widespread.

References

M.D. Nava and C. Del-Toso. "A Short Overview of the VDSL System Requirements." *IEEE Communications Magazine*, Vol. 40, No. 12, December 2002.

www.techonline.com Web site for online learning on hardware-oriented topics.

www.webCT.com Portal for e-learning.

www.w3.org Web site of the World Wide Web Consortium.

Questions

1. Explain the architecture of the public Internet.
2. Explain the Domain Name System.
3. What are the advantages of a virtual private network? How is the security aspect addressed in VPNs?
4. Explain the concept of tunneling.
5. List the various services and applications supported by the Internet.

Exercises

1. If you want to start your own organization, you need to have a Web presence. To have Web presence, you need to have a domain name registered. Go to a domain name registration site (for instance, *www.register.com*) and determine if the domain name you want is available.
2. For an already registered Web site (e.g., *www.iseeyes.com*), find out who is the owner.
3. Create a Web site with multimedia content (text, graphics, voice clippings, and video clippings) using XHTML.
4. Make a comparison of the various search engines. What are the criteria based on which search engines can be compared?
5. Make a list of the Web-based learning portals.

Projects

1. Develop an intranet for your department/organization. Any authorized user of the LAN should be able to access the information provided on the intranet. The user has to log in to the intranet and access a home page. The home page should contain the links for a syllabus, examination schedules, information about the professors, information about the project, works done by earlier batch students, and so on.

2. Develop an e-learning portal. The portal should contain the lecture material, study material for a few topics (such as compiler design and database management system) and also an online examination using objective (multiple choice) tests.

3. Prepare a technical report on the various digital subscriber line technologies.

25

Network Computing

In This Chapter

- The Latest Trends in Software Development
- The Architecture of a Network Computer
- The Importance of Dynamic Distributed Systems
- The Architecture of Jini
- Emerging Internet-Based Applications

The idea of using computers in standalone mode is now obsolete. Every computer is connected to a network for resource sharing. Sun Microsystems' slogan "The network is the computer," whether you like it or not, is a reality. This is a very important paradigm shift in the field of computer science—the computer of tomorrow is going to be different from the computer as we know it today. We will see revolutionary changes in computer architecture, operating systems, application software usage, and software development tools.

In this chapter, we will review how computer networking is leading to new computer architecture through network computers. We will peep into some of the emerging applications such as application service provisioning (ASP), workflow automation, supply chain management, dynamic distributed systems, and Web-based learning—all of which use the network-computing paradigm. We will discuss the concept of dynamic distributed systems and how Jini helps in creating dynamic distributed systems.

25.1 SOFTWARE DEVELOPMENT IN THE NETWORKING ERA

For nearly two decades, we used the PC extensively. The reference model of the PC is shown in Figure 25.1. The PC runs on hardware dominated by one company:

Intel; above the hardware runs the OS, dominated by another company: Microsoft. Above the OS run the applications that can be developed using any of the programming languages—C, or C++, or Java. As shown in Figure 25.2, we can achieve language-independent software development. However, note that this framework works for one platform—the Intel-Microsoft platform.

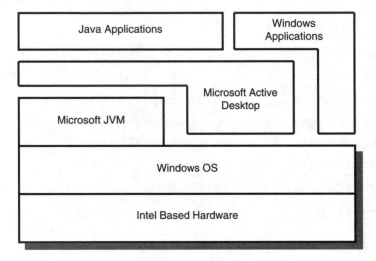

FIGURE 25.1 PC reference model.

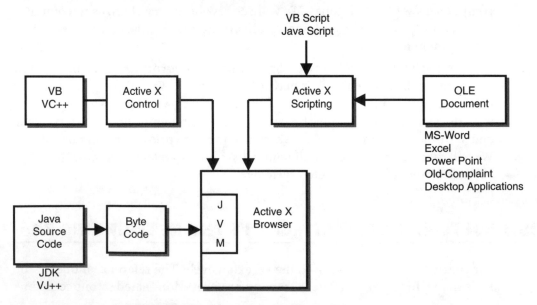

FIGURE 25.2 Language-independent software development.

Prior to the Java era, software development involved lots of porting. If an application has to run on different platforms, the source code has to be modified on the new platform and then executed. Java, through its platform independence, has given a new paradigm in software development: "write once, run anywhere." As shown in Figure 25.3, the Java compiler converts the source code into bytecode, and the bytecode is platform independent. This bytecode can be interpreted by a Java Virtual Machine (JVM) that runs on Windows, Unix, Mac, and so on. We no longer need to port the source code; all that is required is to have a JVM for different platforms. However, note that this framework works for one language—the Java language.

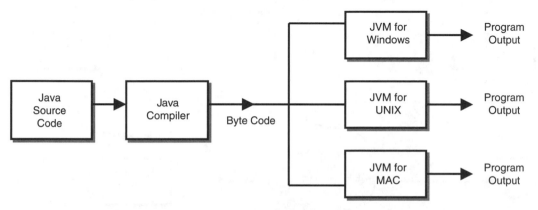

FIGURE 25.3 Platform-independent software development.

The Java programming language has revolutionized software development. Java is the driving force behind Sun Microsystems' slogan "The network is the computer." Java facilitates "write once, run anywhere."

NOTE

Software developed on one platform can run on another platform if the Java programming language is used. The bytecode generated by a Java compiler is platform independent. This bytecode is interpreted by the Java Virtual Machine that runs on different platforms.

Java's main attraction is that the bytecode can be downloaded from the network and executed on the local machine. This is an important paradigm shift—the code moves on the network. However, when the code moves from the network to the machine, we need to ensure that there is no security violation—the code should not corrupt the data in the local machine. As shown in Figure 25.4, the bytecode verifier will ensure that there is no security violation, and the bytecode interpreter (which can run on the browser) will interpret the bytecode. Instead of a bytecode interpreter, which is slow, a Just In Time (JIT) compiler can be used. This framework, coupled with object-oriented features of Java, has brought on revolutionary changes in software development.

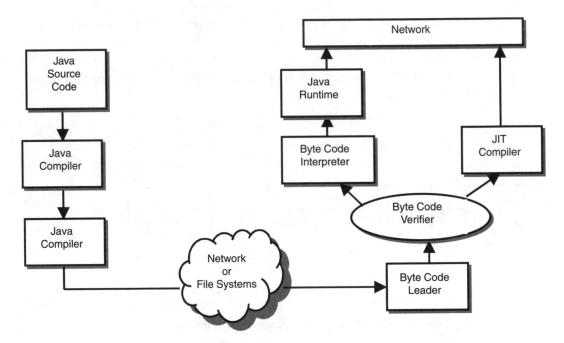

FIGURE 25.4 Executing a Java applet.

The software component–based software development is now catching up. In hardware, if we have to create a design, we look for readily available components and interconnect them to obtain the desired hardware functionality—the days of discrete design are gone. Such an approach has not been there in software development—for every application to be developed, we do the development from scratch. With the advent of software components, we can select the readily available components and interconnect them to obtain the necessary software functionality. A software component is shown in Figure 25.5. The component interfaces with the other components

Software development can be done very fast using the concept of software components. Developing a software package is just the integration of different components—conceptually it is similar to hardware development, wherein we use readily available components to develop a system. Each software component is defined by its external interfaces.

through properties, methods, and events. The internal structure of the component is not visible to the external entities. The software component can be completely defined by its external interface. This component framework has an important advantage—the components can be anywhere on the network, and it is not necessary to have them only on a single machine or on a server. Distributed component platforms such as DCOM and CORBA use this framework for software development.

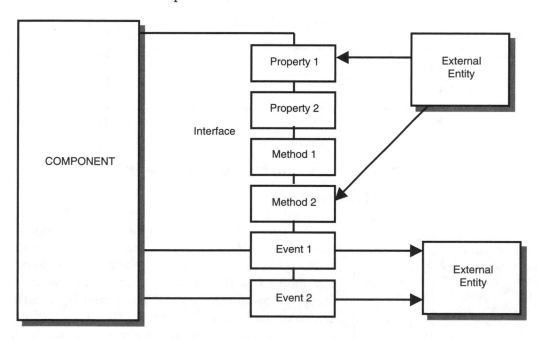

FIGURE 25.5 Software component.

25.2 NETWORK COMPUTER

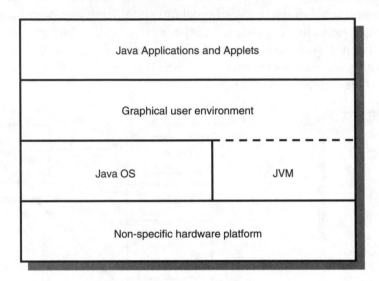

FIGURE 25.6 NC reference model.

The network computer architecture consists of hardware and software that are independent of the manufacturer. The NC need not have any secondary storage. All applications will reside on the server only. The NC gets connected to the server, downloads the bootup code, and then runs the application.

With the advent of the Java programming language, we are now witnessing a revolutionary change in the architecture of computers through the network computer (NC). The NC reference model is shown in Figure 25.6. The beauty of this model is that the hardware platform is independent of the manufacturer. The Java operating system can run on any hardware platform. NC will be connected to the network and access all the resources (including the bootup code) from the network. The attractions of NC are:

- Since there is no need for secondary storage, the cost of a NC will be much lower than the PC, and the maintenance costs are much lower.

- There is no need to store even application software locally; any resource required is obtained from the network. This results in enormous savings in software costs.

- Software renting is possible. This again is a new concept in software usage. Generally, when we buy application software, we use hardly 10% of the features, and also we use the software not so frequently. In such a case, there is no need to keep the software package on the hard disk (and spend a considerable amount of money on it). Instead, the software package can be available on a server on the network and, whenever required, it can be downloaded and only the usage charges are paid to the application service provider (ASP). Software renting—for games, business applications, and so on—is already catching up in many developed countries.

The NC is creating new avenues of business both for services and for software development. Software development using component technology is the order of the day and, in the future, software development can be much faster and also will result in more reliable systems.

The Network Computer is paving the way for a brand new set of applications such as software renting, application service provisioning, and so on.

25.3 DYNAMIC DISTRIBUTED SYSTEMS

It is very difficult to maintain today's computer networks—lots of administrative work is involved to change the topology, add or remove nodes, add new services, and so on. Dynamic distributed systems are networks in which plug-and-play features are provided to the devices connected to the networks. They aim to develop administration-free networks.

The computer networks we use today have many drawbacks: the network is prone to failures and the topology of the network changes when new nodes are added or removed, and so lots of administrative work is required to keep the network up and running. These systems are called *dynamic distributed systems* because as time goes, their topology, number of nodes connected, services available on the network, and so on keep changing. Ideally, these systems should be administration free, and the network must be reliable and robust. Also, it must be very easy to install software, and the network must run without much maintenance. Today, even for a corporate LAN, a dedicated administrator has to do a lot of work when the topology changes, when new software or hardware is installed, or when the network is upgraded. Even connecting a new printer and making the print service available to the users involves a lot of manual procedures. When a service becomes unavailable (say, when the printer is out of order), the users will not even be aware of it. Using the Jini connection technology of Sun Microsystems; all these problems can be solved. Jini provides the technology to develop dynamic distributed systems by providing plug-and-play features to the various devices connected through a network, and there is no need for an administrator to do manual work to add new services to the network or to inform the users when a service becomes unavailable. In Jini terminology, a set of devices can form a "spontaneous community" when they are networked, and one device can make its services available to other devices automatically.

25.3.1 Jini

Jini is a technology developed by Sun Microsystems to create dynamic distributed systems. Jini is based on Java and remote method invocation (RMI). A device can be Jini enabled by running a small piece of software on the device.

In January 1999, Sun Microsystems introduced Jini. Using Jini, dynamic distributed systems can be built. Jini is built on Java and uses remote method invocation (RMI) to create spontaneous communities of devices. Every device to be connected to the network has to be Jini-enabled by running a piece of software on it. When a Jini-enabled device comes in the vicinity of a network of other Jini-enabled devices, the device can find out, through a defined set of protocols, what services are available on the network and use these services. It also can tell the other devices what services it can offer for others to avail of them. Not only the nodes of a LAN, but the peripherals such as printer, scanner, PDA, digital camera, and mobile phones can be Jini enabled. The requirement for a device is that it should be able to be connected to a TCP/IP network, it must have an IP address, it must be able to send and receive multicast addresses, and Jini software must run on the device. Jini software can be run on a PC with a JVM. If a device such as an embedded system with limited resources cannot run the Jini software, another system such as a PC can act as a proxy on its behalf.

Two or more Jini-enabled devices form a spontaneous community. A Jini-enabled device can find out the details of services available with other Jini-enabled services and make use of these services. The communication is done through a standard set of protocols.

Consider a LAN in which the devices are Jini enabled. When a desktop wants a print service (when you want to take a printout), the desktop will look for the availability of the print service, and the server will give a list of printers available, along with the capabilities. Then the desktop will reserve the print service for a fixed duration, use the service, and then free it. The service can be a hardware service such as a print service or it can be a software service such as a program to convert a BMP file to a JPEG file. Using Jini on a general-purpose computer is shown in Figure 25.7. The service provider can be a PC to which a printer is attached. On behalf of the printer, the PC will register itself with the lookup service making the print service available over the network. The service consumer, another PC, will look up the print service, connect to the other PC through the TCP/IP network, and use the print service. If for some reason the printer is not available, the lookup service is told this so that the service consumer can look for another available printer on the network.

To make a device Jini enabled, the device should have an IP address, it should be able to connect to a TCP/IP network, and Jini software should be running on the device.

Dynamic distributed systems can be created using Jini through five key concepts: lookup, discovery, leasing, remote events, and transactions.

The architecture of Jini is shown in Figure 25.8. Jini is based on five key concepts: lookup, discovery, leasing, remote events, and transactions.

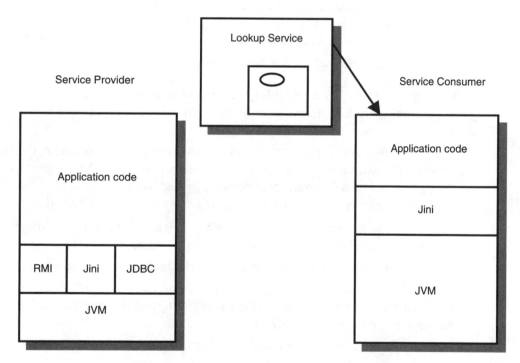

FIGURE 25.7 Using Jini on a general purpose computer.

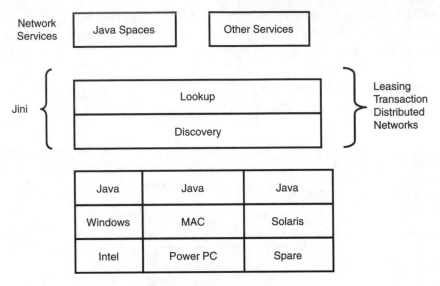

FIGURE 25.8 Jini architecture.

Lookup: Lookup service enables a device to search and find the desired service within the Jini community. In other words, the lookup service provides the directory services within a Jini community.

Discovery: Discovery is the process of finding communities on the network and joining a community. The discovery process gives Jini the property of building a spontaneous community of devices. To discover the services available/ provided on the network, the following discovery protocols are defined.

- Multicast request protocol: used when a service becomes active and the service needs to find nearby lookup services that may be active.
- Multicast announcement protocol: used by the lookup service to announce its presence.
- Unicast discovery protocol: used when a service already knows the particular lookup service it wishes to talk to.
- Unicast lookup protocol: used to create a static connection between services.

Leasing: When a device wants to use a service, it takes the service on "lease" for a fixed duration—by default five minutes. Service lease can be renewed or, canceled early or the lease period can be negotiated. Third-party leasing is also allowed. This leasing is similar to leasing of a house by a tenant—the tenant negotiates with the owner on the initial lease period and subsequently the lease can be renewed or canceled.

Remote events: In a network, it is possible that some service may be withdrawn (such as a scanner being removed from the network). The remote events allow services to notify each other of any changes in their state.

Transactions: Transactions are mechanisms to ensure data integrity and consistency. These mechanisms are very important because they ensure that the transactions are either fully complete or none is completed. Consider the example of a money transaction, where money has to be withdrawn from a bank account and deposited to a credit card account. If the first transaction is completed and then the network fails, it will be a loss of money to the user who initiated the transaction. Hence, it is necessary to ensure that either both transactions are complete or none is completed. to maintain data consistency and integrity. Using two phase commit (2PC), it is ensured that both operations take place or the first one is canceled if the second operation is not successful for some reason.

Two phase commit (2PC) is a very important concept in all database applications. It ensures data integrity and data consistency. It is based on simple logic: if a transaction involves a number of operations, all the operations should be performed or none of the operations should be performed.

25.4 EMERGING APPLICATIONS

Network computing brought in revolutionary changes in the history of computing. Many new exciting applications are now emerging. Some of these are:

- Application service provisioning
- Workflow automation
- Supply chain management
- Web-based learning

Using the concept of application service provisioning, an organization can outsource its complete IT-related work to a service provider. The service provider will host the application software on his server, and the organization only needs to use the application software to update the information and generate reports.

Application service provisioning: Organizations, whose core competence is not IT are now outsourcing their IT activity. For example, a hotel may not like to keep a separate MIS department. Instead, a third party can keep the hotel's MIS software in a server, and the hotel staff use the server to update the information and also obtain the reports. The hotel can have just network computers (NCs) to access the data, and the total maintenance of the server/application software is the responsibility of the application service provider.

Business process outsourcing (BPO) is the catch phrase for the Indian software industry nowdays. BPO uses the concept of application service provisioning.

Workflow automation: In every organization, movement of paper files is common. With the advent of network computing, a paperless office is a distinct possibility as the complete workflow is automated.

Supply chain management: Organizations can link their suppliers and customers through their extranets to automate the entire supply chain. The procurement of raw material and also the delivery of finished goods can be done through a portal.

Web-based learning: In recent years, Web-based learning has caught up to supplement traditional "chalk-and-talk" classroom teaching. One can log on to an education portal and learn at one's own pace. Education portals are available for all age groups, and some universities are awarding formal degrees. Web-based learning is very useful for lifelong learning and work-based learning. The advantages of Web-based learning are:

- One can learn at one's own pace.

- Educational programs can be designed by recognized international experts, and this expertise can be made available across the globe.

- In a traditional classroom, people with different levels and some without the necessary background will be present, and it is difficult for the teacher to carry the entire class along. This problem is not present in Web-based learning because the necessary background can be specified, and only those students who have the background will be taking the class.

The various components that form a Web-based learning system are lecture material along with lecture notes with or without voice, e-mail, chat, bulletin board, and online examination modules.

- Web-based learning modules can be made interactive by providing voice, video, and chat facilities for online interaction between the instructor and the students.

- Shy students can open up and interact with the instructor because students are faceless in such interactions.

Because of all these advantages, Web-based learning is now being used extensively for continuing education programs. Corporations are using their intranets effectively for providing corporate training in customer support, sales, and so on.

The various modules that comprise the Web-based learning software are: e-mail, chat, intranet radio, and intranet telephony. In addition, a "learning without lectures" module will have the complete presentation material for the module along with the voice annotation. An objective-testing module is used to test the comprehension of the students periodically. The students can carry out the administrative formalities such as registration for a course through the administration module. The databases contain complete information about the students, the faculty, the courses being offered, and the registration details. The complete course material will also be available as a database.

Normally, Web-based learning can be used for offline learning so that, whenever a student is interested, he can log in and go through the course material, send mail to the instructor, and so on. Periodically, online sessions can be organized at predetermined hours at which the instructor will be available for interaction through chat, voice over IP, or even video.

Web-based learning is now becoming more sophisticated with modules for building learning communities wherein a group of people with common interests can form a community and carry out discussions.

For Web-based learning of engineering subjects, laboratory work is very important. Using virtual laboratories, students can carry out different experiments without having the necessary infrastructure in their own premises.

Summary

Computer networks are paving the way for new developments in computer architecture software development and for creating new applications. In this chapter, we studied the various aspects of network computing. The PC reference model, which is now predominant, is slowly being replaced by the network computer (NC) reference model. The main attraction of NC is that it is independent of the hardware platform. Even if secondary storage is not available on the NC, the NC can use the services available on the network. Applications such as software renting and application service provisioning are possible with NC. Another innovation that is the result of network computing is the software component framework. Even if the software components are available at different locations on the network, these components can be glued together or combined to form a software package. Software components help in faster software development.

Dynamic distributed systems are another innovation that facilitate administration-free computer networks. The Jini architecture developed by Sun Microsystems facilitates development of dynamic distributed systems.

Network computing is leading to revolutionary changes—in computer architecture, in software development, in application use and in software distribution, and it is also creating avenues to develop new applications.

References

Dreamtech Software Team. *Programming for Embedded Systems*, Wiley Dreamtech India Pvt. Ltd., 2002. This book contains a case study for implementation of a dynamic distributed system using Jini.

www.hp.com/go/embedded HP Chai Platform enables development of Internet-enabled appliances and embedded systems. You can download the HP Chai Development kit from this link.

www.javasoft.com/products/jini You can get the details of Jini and the Jini Development Kit from this link.

Questions

1. Compare the PC reference model with the NC reference model.
2. What is a dynamic distributed system? Explain how a dynamic distributed system can be implemented using the Jini framework.
3. Explain the frameworks for language-independent software development and platform-independent software development.
4. What is a software component? Explain how the software component framework helps in software development. (Hint: Reusable components).

Exercises

1. Study an e-learning portal available on the Internet (for example, *www.techonline.com*). Write down the software requirements of the e-learning portal. Prepare a technical paper on e-learning that gives the pros and cons of e-learning.

2. List the various modules of an e-learning portal.

3. Virtual laboratories are now available on the Internet. Access the virtual laboratory available on the Internet at *www.techonline.com* and prepare a technical paper on virtual laboratories.

4. "Web services" is now a hot topic. Microsoft, Sun Microsystems, IBM, Oracle, and others offer many interesting products and development tools for Web services. Search the Internet and prepare a document on Web services technologies and standards.

5. What are J2EE, J2SE, and J2ME?

6. What is XML? List the various markup languages that are derived from XML.

7. What is a networked information appliance? Explain how an embedded system can be network enabled.

8. E-governance is now being implemented by a number of governments. List the various services that can be offered to citizens through e-governance.

9. Internet voting is now being planned wherein the citizens can elect their representatives by voting through the Internet instead of going to polling booths. Discuss the pros and cons of Internet voting.

Projects

1. Develop workflow automation software for your department over the LAN. The software should facilitate the automation of the complete leave application process. A student should electronically submit a leave letter, the professor should approve it electronically, and the student should be informed. A database of leaves used and leaves remaining should be kept.

2. Develop a Jini-enabled PC that will provide a software service such as conversion of a BMP image to a JPEG image. The software package for the conversion available on a PC should be Jini enabled. Another PC should be able to use this service.

3. Develop an Internet-enabled appliance using the HP Chai Development Kit. You can simulate an appliance (for giving weather information) on a PC. This appliance should be accessed through another PC connected to the LAN.

4. Develop an e-learning portal based on the explanation given in this chapter.

5. Develop a prototype for Internet voting. The voting system should facilitate election of the student representatives for your college students union, using the intranet of your college.

26 ∎ Signaling System No. 7

In This Chapter

- The Drawbacks of PSTN Signaling
- The Architecture of Signaling System No. 7
- Using SS7
- The Protocol Architecture of SS7
- Value-Added Services Using SS7

A communication system has to handle two types of data—user data and signaling data. The signaling data is used for setting up a call, disconnecting the call, and other things such as billing. The signaling used in the PSTN is not very efficient, so Signaling System No. 7 has been developed. This signaling system uses a separate data network for carrying signaling information. In this chapter, we will discuss the architecture of the Signaling System No. 7.

26.1 DRAWBACKS OF PSTN SIGNALING

In the PSTN, the following signaling information is transmitted from the subscriber to the switch and between switches:

- When a subscriber wants to make a telephone call, the off-hook information is sent to the switch.
- When the subscriber dials a number, the dialed digits are sent to the switch.
- The switch analyzes the dialed digits to route the call to the destination.
- The switch uses the signaling information to allocate trunks.

- The switch monitors the on-hook condition and generates billing information.
- Tones such as a ringing tone and a busy signal are sent to subscribers.

The signaling information can be divided into two types: (a) signaling between the subscriber terminal and the switch; and (b) signaling between switches.

In the PSTN, signaling information is exchanged to set up and disconnect a call, to exchange billing information, and to exchange network management information.

As we discussed in the chapter on PSTN, in-band signaling and channel associated signaling are used for transferring the signaling information between the subscriber and the switch and between the switches. This type of signaling has the following drawbacks:

- The same communication link is used for both user data and signaling, so the communication resources are not used efficiently.
- Call setup and disconnection are very slow.
- The tones used for signaling will interfere with human speech because the tones are in the voice frequency band.

The signaling used in the PSTN is inefficient because the same communication channel is used for both signaling and data transfer.

Due to these drawbacks, a new standard by ITU-T has been developed, and it is called Signaling System No. 7. This system is used extensively in cellular mobile communications, Integrated Services Digital Network (ISDN), and PSTN in many developed countries.

26.2 OVERVIEW OF SIGNALING SYSTEM NO. 7

Signaling System No. 7 (SS7) defines the architecture for a data communication network to exchange signaling information between switches. SS7 facilitates fast and reliable exchange of signaling information.

Signaling System No. 7 (SS7) defines the architecture of the data network to perform out-of-band signaling for call establishment, billing, and routing information of the switches in PSTN/ISDN/GSM. Out-of-band signaling is much more efficient than in-band signaling and also allows value-added services to be provided to the subscribers, and it is now being deployed extensively all over the world.

The signaling information between the network users and the network elements contains the dialed digits, providing various tones such as the dial tone and the call waiting tone. To establish connection between two subscribers, network resources have to be assigned, and this is also a part of signaling. SS7 is the means by which network elements exchange information for call management. This information is exchanged in the form of messages. SS7 can be viewed as a data network that provides the signaling service to the PSTN/ISDN/GSM. The important features of SS7 are high-speed packet data exchange for signaling information and out-of-band signaling. Separate paths, called signaling links, are used to exchange the signaling messages. The advantages of SS7 are:

- Signaling information can be exchanged even when a call is in progress.

- Messages are transmitted at very high speeds.

- It enables signaling between network elements that do not have trunk connections between themselves.

In SS7, a separate data network is used for signaling. This network consists of three components, as shown in Figure 26.1.

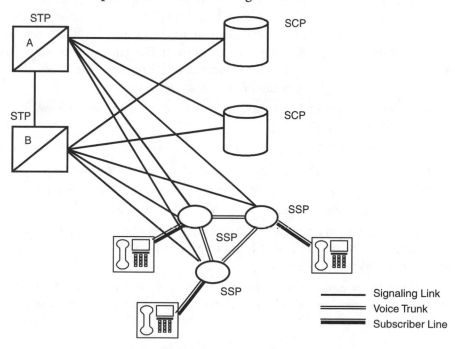

FIGURE 26.1 A smple SS7 network.

An SS7 network consists of three network elements: (a) signaling switching points (SSPs); (b) signaling transfer points (STPs); and (c) signaling control points (SCPs).

Signaling switching points (SSPs): These are telephone switches with SS7 software and terminating signaling links.

Signaling transfer points (STPs): These are packet switches of the SS7 network. They receive signaling messages and route them to the appropriate destination.

Signaling control points (SCPs): These are databases that provide necessary information for advanced call processing capabilities such as toll-free numbers (800 numbers).

Since continuous operation of SS7 networks is a must even for setting up calls, SS7 networks must have 100% availability. Redundancy is required in the network elements. STPs and SCPs are deployed in pairs for redundancy. Also, the communication links between the network elements are provided with redundancy. A typical SS7 network with SSPs, STPs, and SCPs is shown in Figure 26.1, along with voice trunks and signaling links. Note that in this figure, the STPs A and B perform the same function for redundancy. These STPs are called mated pairs. Each SSP has two links, one for each STP of a mated pair. STPs of a mated pair are joined with a link or a set of links. Two mated pairs of STPs are connected by four links, which are called quads. SCPs also are deployed in pairs, but they are not joined directly by a pair of links. Each link is generally a 56kbps or 64kbps bidirectional link. These links are classified into the following categories:

STPs and SCPs are deployed in pairs to provide redundancy so that the signaling network will have 100% availability.

A links (access links): Interconnect STP and either STP or SCP and interconnect SSP and STP.

B links (bridge links): Quad of links interconnecting peer pairs of STPs.

C links (cross links): Interconnect mated STPs for redundancy.

D links (diagonal links): Quad of links interconnecting mated pairs of STPs at different hierarchical levels.

E links (extended links): SSP is connected to its home STP by a set of A links. To ensure reliability, additional links to a second STP pair called E links are provided. E links are optional.

F links (fully associated links): These links directly connect two signaling end points. F links allow only associated signaling. F links generally are not deployed.

26.3 CALL SETUP USING SS7

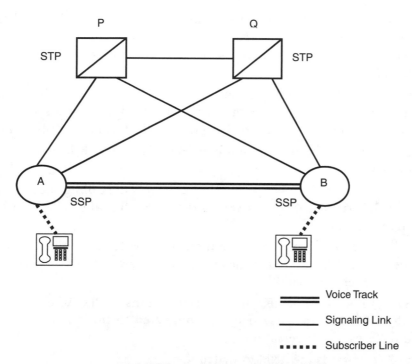

FIGURE 26.2 Call setup using SS7.

We will illustrate call setup using a simple SS7 network shown in Figure 26.2. Assume that a subscriber connected to switch A would like to make a call to a subscriber connected to switch B. The steps for call setup are as follows:

1. Switch A analyzes the dialed digits and determines that the call is meant for a subscriber of switch B.

2. Switch A selects an idle trunk between A and B and then formulates an initial address message (IAM). This message identifies the initiator switch (A), the destination switch (B), the trunk selected, the called number, and the calling number.

3. Switch A selects an A link and sends the message on the link for routing to switch B.

4. STP P receives the message, checks the routing label and, based on the label, sends it on a link (PB) to B.

5. Switch B receives the message and checks whether the called number is idle.

6. Switch B formulates the address complete message (ACM) to indicate that IAM has reached the correct destination. The ACM message contains recipient switch (A), sending switch (B) addresses, and the selected trunk number.

7. Switch B picks an A link (BQ) and sends the ACM to switch A. Simultaneously, it completes the call path in the direction toward A and sends a ringing tone over the trunk toward A and also a ring on the line of the called subscriber.

8. STP Q receives the message and, based on the routing label, sends the message to switch A.

9. Switch A receives the ACM and then connects the calling subscriber to the selected trunk in the backward direction so that the caller can hear the ringing sent by B.

10. When the called party picks up the phone, switch B formulates an answer message (ANM) that identifies the recipient switch (A), originator switch (B), and selected trunk.

11. Switch B uses the same A link it used to transmit the ACM and sends the ANM. By this time, the trunk is connected to the called party in both directions for conversation.

12. STP Q forwards the ANM to switch A.

13. Switch A ensures that the calling subscriber is connected to the trunk in both directions for the conversation.

14. If the calling subscriber goes on hook, switch A will generate a release (REL) message to switch B.

15. STP P receives the message and forwards it to switch B.

For call setup using SS7, a series of packets is exchanged between the switches through the STPs using high-speed datalinks. Because the voice trunks do not carry any signaling information, they will be optimally utilized.

16. Switch B receives the REL message, disconnects trunks, returns the trunk to idle state, and generates a release complete (RLC) message addressed to A. RLC identifies the trunk used for the call.

17. On receiving the RLC, switch A makes the identified trunk idle.

Because of the high-speed datalinks, call setup and disconnection are very fast.

26.4 DATABASE QUERY USING SS7

One of the most attractive features of SS7 is its ability to provide value-added services to subscribers such as toll-free numbers (800 numbers) and calling card facilities. For instance, a subscriber can call the toll-free number of an organization and obtain product information without paying for the call. In such services, information has to be obtained from databases. For instance, the toll-free number is generally not assigned to any particular subscriber line. It is in fact a virtual telephone number. When a subscriber dials an 800 number such as 800-123-4567, the switch gets the actual telephone number from a database. Alternatively, the database may indicate to which network the call has to be routed. Depending on the calling number, time of the day, day of the week, and so on, the actual telephone number can be different. The procedure to query a database is described with reference to Figure 26.3.

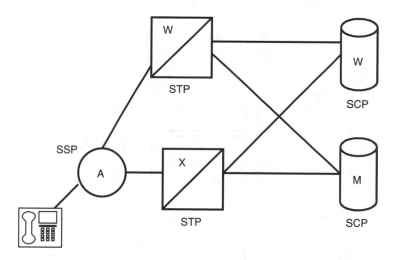

FIGURE 26.3 Querying a database using SS7.

Value-added services such as toll-free numbers and prepaid calling cards can be provided using the SS7 network.

1. Subscriber of switch A calls an 800 number.
2. Switch A formulates an 800 message that contains the calling number and the called number and forwards it to the STP over link AX.
3. STP selects a suitable database for responding to the query (M) and forwards the 800 message to the database over the link MX.

4. SCP M processes the query and selects the real telephone number or network or both and forwards the result to the STP over the link MX.

5. STP receives the message and routes it to switch A.

6. Switch A then identifies the number to be called and generates the IAM message as in the previous example for call setup, and then the call setup is carried out.

When a subscriber dials a toll-free number, the actual telephone number corresponding to that toll-free number is obtained from a database. Hence, the toll-free number is only a virtual telephone number.

26.5 SS7 PROTOCOL LAYERS

The protocols in SS7 follow layered architecture. The protocol layers are shown in Figure 26.4. The functions of each layer follow.

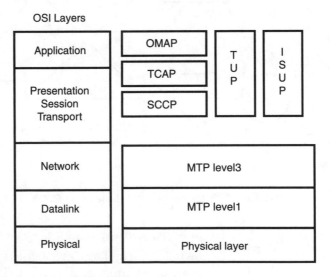

FIGURE 26.4 SS7 Protocol layers.

Physical layer: The physical layer defines the physical and electrical characteristics of the link. The links carry raw signaling data at 56 or 64kbps.

The bottom three layers in the SS7 protocol architecture are: (a) physical layer that supports 56 or 64kbps data rate; (b) message transfer part level 2, which provides error checking, flow control, and sequence checking; and (c) message transfer part level 3, which provides addressing, routing, and congestion control.

Message transfer part level 2 (MTP level 2): This layer provides link layer functionality to ensure that the messages are reliably exchanged. The functions are error checking, flow control, and sequence checking.

Message transfer part level 3 (MTP level 3): This layer provides network layer functionality to ensure delivery even if signaling points are not directly connected. The functions include node addressing, routing, alternate routing, and congestion control.

MTP level 2 and level 3 together are known as MTP.

Signaling connection control part (SCCP): SCCP provides capability to address applications within a signaling point. MTP can only receive and deliver messages from a node, but it does not deal with software applications within a node. MTP messages handle call setup and network management messages. Other messages are used by separate applications (referred to as subsystems) within the node. Examples are 800 call processing, calling card processing, advanced intelligent network (AIN), repeat dialing, and call return services. SCCP allows these subsystems to be addressed explicitly.

SCCP's second function is to perform incremental routing using global title translation (GTT) capability. Because it is practically impossible for an originating signaling point to know all the possible destinations to route a message, GTT is introduced. A switch can originate a query and address it to its STP along with a request for GTT. For example, consider the case of a calling card query. It has to be routed to a particular SCP that contains the calling card data. It is not possible for each STP to know this SCP, so the originating STP will route the message to a specific STP, which in turn routes the message to the destination STP/SCP.

Telephone user part (TUP) and ISDN user part (ISUP): TUP and ISUP define the messages and protocols to establish and disconnect voice and data calls over public switched networks such as PSTN/ISDN and to manage the trunks.

SS7 is now being used in the PSTN as well. To establish and disconnect voice and data calls over the PSTN and ISDN, telephone user part and ISDN user part protocols are defined.

Transactions capabilities application part (TCAP): TCAP defines messages and protocols used to communicate between applications such as 800 services, calling card services in the nodes, and the switches. Since TCAP messages must be delivered to individual applications with the node they address, these messages use SCCP for transport.

Value-added services such as toll-free number service and calling card service are handled by the transaction capabilities application part (TCAP) protocols.

Operations, maintenance, and administration part (OMAP): OMAP defines messages and protocols to administer SS7 networks. The functions include validation of route tables and diagnosis of links. OMAP is an application that uses the TCAP and SCCP services.

26.6 ADDRESSING AND SIGNALING UNITS

Each signaling point in the SS7 network has an address in the format network number, cluster number, and member number. Each of these numbers is between 0 and 255.

Addressing: Each signaling point is assigned an address in a three-level hierarchy:

Network number, cluster number, member number

Each signaling point is assigned to a cluster of signaling points. Each cluster is part of a network. Each of these numbers is an 8-bit digit (0 to 255). The three-level address is known as the point code of signaling point. Network numbers are assigned by a national authority for each operator.

Signaling units: Signaling messages exchanged over the signaling links are called signaling units (SUs). In SS7 protocols, there are three types of SUs.

- Message signal units (MSUs)
- Link Status signal units (LSSUs)
- Fill-in signal units (FISUs)

The three types of signaling units are message signaling units, link status signaling units and fill-in signaling units.

These SUs are transmitted continuously in both directions. FISUs are fill-in signal units—if a signaling point does not have MSUs or LSSUs to send, it will send FISUs continuously to facilitate link monitoring.

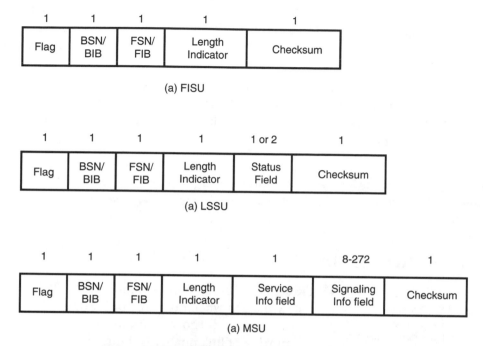

FIGURE 26.5 Signaling unit formats.

SU formats: The formats of the SUs are shown in Figure 26.5. The following fields used by MTP level 2 are common:

- Flag: Marks the end of one SU and the beginning of the next SU. Flag will be of the form 01111110.

- Checksum: Each signaling point calculates the checksum and asks for retransmission if the checksum does not match.

- Length indicator: This is the number of bytes (octets) between itself and checksum. This is 0 for FISUs, 1 or 2 for LSSUs, and greater than 2 for MSUs. Hence, the type of SU is known from this field. If the length is greater than 63, the value 63 is placed in this field.

- BSN/BIB and FSN/FIB: The backward sequence number (BSN), backward indication bit (BIN), forward sequence number (FSN), and forward indication bit (FIB) are used to confirm receipt of SUs and to ensure that they are received in order as well as for flow control.

MSUs and LSSUs are assigned sequence numbers, which are put in the FSN of an outgoing SU. The SU is stored until acknowledgement is received by the signaling point. Since only seven bits are assigned, a signaling point can send only 128 unacknowledged SUs.

Signaling points acknowledge the SUs by placing the sequence number of the last correctly received SU in the BSN. The forward and backward indication bits are used to indicate sequencing or data corruption errors and to request retransmission.

The fill-in signaling units are exchanged between signaling points continuously if there is no other data to transmit. These signaling units facilitate continuous monitoring of the status of the links.

NOTE

26.6.1 Functions of SUs

Line status signaling units are exchanged between the nodes to exchange status of the link. The signaling unit need not contain the address because the links are point to point.

FISUs: FISUs have no information payload. These SUs are exchanged when no LSSUs or MSUs are to be sent. These SUs undergo error checking, so constant monitoring of link quality is done.

LSSUs: LSSUs are used to communicate information about signaling links between the nodes on either end of the link. The information is sent in the status field. This is to indicate initiation of link alignment, quality of received signal traffic, and the status of processors at either end. No addressing information is required because the SUs are exchanged between the two signaling points.

MSUs: In addition to the common fields indicated above, the MSU consists of the service information field, which contains three types of information:

- Four bits to indicate the type of information in the signaling information field, called the service indicator. A value of 0 indicates signaling network management, 1 indicates signaling network testing and maintenance, 3 indicates SCCP, and 5 indicates ISUP.

- Two bits to indicate whether the message is for national or international network (generally 2 for national network).

- Two bits for priority. The priority value can be 0 to 3, with 3 indicating the highest priority. The priority field is used only when there is network congestion. The field indicates whether the message can be discarded or merits transmission in case of congestion.

Message signaling units are used to exchange signaling information to set up and disconnect the calls.

The format of the MSU is based on the service indicator, with the first portion of the signaling information containing the routing label of seven octets. Three octets are for destination point code (address of the destination node), three octets for the message originator point code (address of the originating node), and one octet for signaling link selection to distribute load among the redundant nodes.

The SS7 networks are now being deployed throughout the world by basic telecom operators as well as cellular operators to provide efficient signaling and value-added services.

SS7 is used in mobile communication systems and ISDN. SS7 is also now being deployed in the PSTN.

NOTE

Summary

Signaling System No. 7 is a data network that provides an efficient signaling mechanism in telephone networks. The network consists of signaling switching points (SSP), signaling transfer points (STPs) and signaling control points (SCPs), which are interconnected through high-speed links (56kbps or 64kbps). Using a separate dedicated data network for signaling results in fast signaling. To provide 100% availability of the signaling network, the STPs and SCPs and the links are provided with the necessary redundancy.

In addition to providing efficient signaling, SS7 allows value-added services such as toll-free numbers and calling card facilities to be provided. Due to these advantages, SS7 is used extensively in the ISDN and cellular mobile networks. SS7 signaling is also now being used in the PSTN.

References

Guy Redmill. *An Introduction to SS7*. Brooktrout Technology, July 2001.

www.openss7.org You can get information about open source software from Linux on this site.

www.cisco.com You can get excellent technical literature on SS7 and product information from this site.

Questions

1. Explain the drawbacks of the present PSTN signaling.
2. What are the advantages of signaling based on SS7?
3. Explain the SS7 protocol architecture.
4. Explain the procedure for call establishment using SS7 signaling.
5. Explain how a toll-free number can be accessed using SS7 signaling.

Exercises

1. You can buy a prepaid telephone calling card to make telephone calls. The procedure is to dial a specific number given on the calling card, give your calling card number, and then dial the telephone number. SS7 signaling is used for verification of the calling card number and how much money is still left. Explain how the processing is done using a database query.
2. SS7 signaling is used on the GSM network. Study the details of the signaling in mobile communication networks such as GSM and prepare a technical paper.
3. SS7 is now being used on ATM networks. Write a technical paper on SS7 over ATM.
4. Search the Internet for freely available source code for the SS7 protocol stack.

Projects

1. Obtain freely available source code for SS7 protocol stack, by searching the Internet. Run the stack on your system and study the source code to understand implementation aspects of it.
2. Implement an SS7 network over a LAN. You can implement a database query operation by implementing the STP and SCP over different nodes. Two nodes on the LAN can be considered subscriber telephones, one node as STP, and one node as SCP. When one subscriber makes a toll-free call, the number has to be passed to the STP. The STP has to send the information to the SCP, which has a database of the toll-free numbers and the corresponding actual telephone numbers. The SCP has to tell the STP the actual telephone number. The STP in turn has to tell the telephone subscriber (the designated node) the actual number, and then the message from the subscriber has to be sent to the other subscriber.

27 Integrated Services Digital Network

In This Chapter

- The Objectives of ISDN
- The ISDN Architecture
- Differences Between Narrowband ISDN and Broadband ISDN
- The ISDN Protocol Architecture

When you buy a house, the house has standard electric power sockets into which you can plug any electric device and start using it. How about a similar setup for telecommunication as well? Imagine a standard communication socket into which we can plug a device for sending data, voice, fax, or video? Technology fiction? No. Integrated Services Digital Network (ISDN) is aimed at achieving precisely this objective—to develop standard interfaces and standard terminals to integrate data, voice, and video services, all in digital format. The standardization work on ISDN was initiated in the 1970s. Today, ISDN has a good installation base in the developed world, and the developing world is catching up fast. This chapter covers the principles and standards of ISDN.

27.1 OBJECTIVES OF ISDN

The main objective of ISDN is to replace the Public Switched Telephone Network (PSTN), which is based on analog technology. As the name implies, ISDN is a fully digital network to provide integrated services—voice, data, and video—through standard interfaces. To provide services to meet the requirements of different users, narrowband ISDN and broadband ISDN standards have been

The objective of Integrated Services Digital Network (ISDN) is to provide data, voice, and video services with standard interfaces using completely digital technology.

developed. Narrowband ISDN supports services up to 2.048Mbps, and broadband ISDN supports services up to 622.08Mbps. International Telecommunications Union Telecommunications Sector (ITU-T) I-series recommendations specify the ISDN standards. The objectives of ISDN are:

- To support voice and nonvoice applications using a set of standard interfaces.
- To support switched and nonswitched applications.
- To have a layered architecture so that the advantages of layered architecture can be exploited.
- To support a variety of configurations to meet varying user requirements.
- To provide intelligence in the network.
- To provide integration with the existing networks such as Public Switched Telephone Network (PSTN) and Public Data Network (PDN) and to rely on 64kbps connections.
- To standardize all the interfaces for access equipment so that there will be competition among the equipment suppliers, which will benefit the subscribers.
- To support transparency so that the services and applications are independent of the underlying network.
- To develop a tariff strategy that is based on the volume of the traffic and not the type of data. Whether the user is transmitting voice or video should not matter; the tariff should be based only on the amount of data sent.

ISDN can cater to different user requirements. Users at homes or small offices can subscribe to low-speed ISDN services, and large organizations can subscribe to high-speed services.

27.2 PRINCIPLES OF ISDN

ISDN caters to different user requirements. At homes or small offices, users require low-speed voice and data services. In small and medium-size organizations, users require voice and data services at high speeds to cater to the requirements of a number of employees. For large organizations as well as

Narrowband ISDN provides 144kbps and 2.048Mbps data rates. Broadband ISDN provides 155.52Mbps and 622.08Mbps data rates.

for backbone networks, support for very high data rates is required. Accordingly, ISDN has many versions.

Narrowband ISDN (N-ISDN) caters to homes and small offices by providing 144kbps and 2.048Mbps data rates. Broadband ISDN supports very high data rates: 155.52Mbps and 622.08Mbps. Research in broadband ISDN led to the development of Frame Relay and Asynchronous Transfer Mode (ATM) technologies. We will discuss Frame Relay and ATM in subsequent chapters and focus on narrowband ISDN in this chapter.

In narrowband ISDN, basic rate interface (BRI) provides 144kbps data rate, and primary rate interface (PRI) provides 2.048Mbps data rate.

Figure 27.1 shows the narrowband ISDN configuration. Subscribers who require low data rate services will have basic rate interface (BRI), which provides 144kbps of data rate. Subscribers who require higher data rates will have primary rate interface (PRI), which provides 2.048Mbps data rate. As in the case of a PSTN local loop, a dedicated copper cable will provide the physical connection between the ISDN switch and the subscriber equipment.

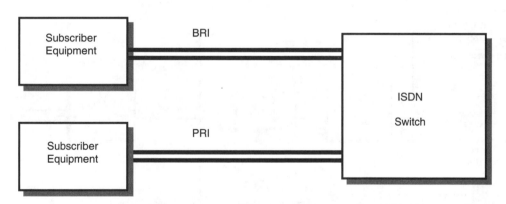

FIGURE 27.1 Narrowband ISDN configurations.

To obtain ISDN services, the subscribers can buy ISDN terminals such as an ISDN phone and a ISDN-compatible PC. Alternatively, the existing telephone/ desktop PC can be used. However, special equipment called a terminal adapter is required to make existing equipment ISDN compatible. The ISDN architecture is designed in such a way that even if the subscriber does not have ISDN-compatible equipment, the present equipment can be used.

To obtain ISDN service, the subscriber can install ISDN-compatible equipment such as an ISDN telephone or computer. Alternatively, existing subscriber equipment can be connected to the ISDN switch through a terminal adapter.

27.3 ISDN ARCHITECTURE

In the ISDN architecture, four reference points are defined: R, S, T, and U interfaces. These are conceptual points to describe the interfaces between various equipment.

ISDN system architecture is shown in Figure 27.2. If the subscriber has an ISDN telephone, an ISDN terminal, and an ISDN PBX, they are connected to the network termination-1 (NT1), and NT1 is in turn connected to the ISDN switch. Non-ISDN equipment such as a PSTN telephone, or a normal computer can be connected to the ISDN interfaces through a terminal adapter (TA).

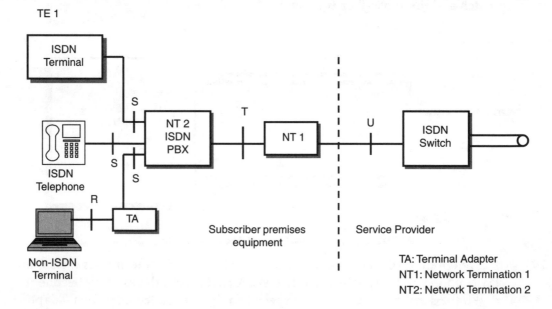

FIGURE 27.2 ISDN architecture.

To ensure that the ISDN-compliant equipment as well as the legacy equipment can be connected through standard interfaces, various interfaces are defined as shown in Figure 27.2. R, S, T, and U interfaces are called the reference points. These reference points are conceptual points to describe the interfaces between various equipment. The advantages of this approach are:

- Interface standards can be developed at each reference point.

- Improvements/modifications on one piece of equipment do not have an effect on the other equipment.

- The subscriber is free to procure equipment from different suppliers.

Three types of network terminations are defined in the ISDN architecture: (a) NT1 forms the boundary to the network and is owned by the service provider; (b) NT2 is an intelligent device that can be connected to the ISDN switch and is owned by the subscriber; and (c) NT12 combines the functions of NT1 and NT2 and is owned by the service provider.

Terminal equipment: Terminal equipment is of two types—TE1 and TE2. TE1 devices support standard ISDN interfaces such as a digital telephone, integrated voice/data terminals, and digital fax machines. TE2 devices are the present non-ISDN equipment such as PC with RS232 interface computer with an X.25 interface. A terminal adapter (TA) is required to interface the TE2 devices with the ISDN network.

Network terminations: Three types of network terminations are defined—NT1, NT2, and NT12. NT1 includes functions associated with physical and electrical terminations of ISDN at user premises. This corresponds to OSI layer 1. NT1 may be controlled by an ISDN service provider and forms a boundary to the network. NT1 also performs line maintenance functions such as loop-back testing and performance monitoring. At NT1, the bit streams from different terminals are multiplexed using synchronous TDM. NT1 can support multiple devices in a multidrop arrangement.

NT2 is an intelligent device that can perform switching functions. It includes functionality up to layer 3 of the OSI model. Equipment such as a digital PBX and a local area network (LAN) are examples of NT2 devices.

NT12 is a single piece of equipment that combines functions of NT1 and NT2. The ISDN service provider owns NT12.

27.3.1 Reference Points

Reference point T (terminal): This reference point corresponds to a minimal ISDN network termination at the subscriber premises. It separates the ISDN service provider's equipment from the user's equipment.

Reference point S (system): This reference point corresponds to the interface of individual ISDN terminals and separates the user terminal equipment from network-related communication functions.

Reference point R (rate): This reference point provides a non-ISDN interface between user equipment that is not ISDN compatible and adapter equipment (such as an RS232 interface to connect an existing PC to ISDN through a terminal adapter).

Reference point U: This reference point provides the interface between the ISDN switch and the network termination-1.

The interfaces at the reference points are well defined in the standards, so it is possible to integrate equipments supplied by different vendors.

27.3.2 Narrowband ISDN Channels

In narrowband ISDN, two types of channels are defined—B channel and D channel. B channel carries user data, and D channel carries signaling data.

In N-ISDN, two types of channels are defined: B channel and D channel. B channel carries user data such as voice/low bit rate video/fax/data. D channel carries signaling data and is also used for low data rate applications such as alarms for the house.

An ISDN channel can be viewed as a pipe carrying the B channels and D channel. Figure 27.3 shows the ISDN pipes for basic rate interface and primary rate interface.

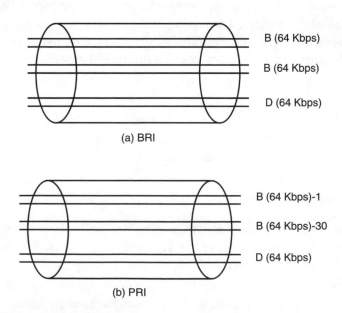

FIGURE 27.3 ISDN pipe: (a) BRI (b) PRI.

For basic rate interface (BRI), there will be two B channels and one D channel. Each B channel supports 64kbps data rate. The D channel supports 16kbps data rate. Basic ISDN service using two B channels and one D channel at 16kbps is referred to as 2B+D ISDN. The total data rate for 2B + D is $2 \times 64 + 16 = 144$kbps. With additional overhead of 48kbps, the aggregate data rate will be 192kbps.

In BRI, there will be two B channels and one D channel. The total data rate is 144kbps. With additional overhead of 48kbps, the aggregate data rate is 192kbps. This interface is referred to as 2B+D ISDN.

For primary rate interface (PRI), there will be 30 B channels of 64kbps each and one D channel of 64kbps. ISDN service using 30 B channels and one D channel at 64kbps is referred to as 30B+D ISDN. This is the European standard. In North America, the corresponding standard is 23B+D. The aggregate data rate for 30B+D ISDN is 2.048Mbps.

Note that the B and D channels are only logical channels, from the subscriber premises, a single cable will carry the entire data load.

In primary rate interface (PRI), there will be 30 B channels of 64kbps each and one D channel of 64kbps. The aggregate data rate is 2.048Mbps. This interface is referred to as 30B+D ISDN.

B Channel

The B channel carries user data such as voice and slow-scan TV. Both circuit-switched calls and packet-switched calls can be setup on the B channel.

This is the basic user channel to carry PCM voice, data, or fax messages or slow-scan TV data. The different types of connections that can be set up over a B channel are:

- **Circuit-switched calls:** These calls are similar to calls of PSTN, but signaling is sent over the D channel.
- **Semipermanent connections:** These calls are similar to leased lines of PSTN, and no call establishment procedure is required.
- **Packet-switched calls:** The user is connected to a packet-switched network, and data is exchanged using X.25 protocols.

D Channel

D channel carries signaling information. In addition, low speed packet-switched data services can be supported, such as fire alarm data, data from energy meters, and such.

This channel carries signaling information to control circuit-switched calls. In addition, this channel supports low-speed (up to 100bps) packet-switched data services. Data from energy meters, fire alarms, and such can be sent over the D channel.

During the initial days of standardization, ISDN was expected to replace analog PSTN through digital technology using BRI and PRI. However, the present installation base of ISDN is very low.

27.4 ISDN STANDARDS

ISDN standards are specified in I series recommendations.

- I.100 series specifies the general concepts. I.110 gives general introduction and structure, I.120 gives overall description of ISDN and evolution, and I.130 gives terminology and concepts to specify services.

- I.200 series specifies service capabilities and services to be provided to the users.

- I.300 series gives the network aspects, including protocol reference model, common channel signaling, numbering, and addressing.

- I.400 series specifies user-network interfaces, including functional grouping and reference points, transmission rates to be offered to users, and protocol specifications at OSI layers 1, 2, and 3.

- I.500 series recommendations specify internetwork interfaces for interworking between ISDN and Public Switched Telephone Network (PSTN), Public Data Network (PDN), Packet Switched Public Data Network (PSPDN), and so on.

The I series of recommendations of International Telecommunications Union Telecommunications Sector (ITU-T) specifies the ISDN standards.

- I.600 series recommendations specify maintenance principles to test for failure localization, verification, and so on.

27.5 ISDN PROTOCOL ARCHITECTURE

ISDN supports both circuit-switched and packet-switched operations, and the protocol architecture takes care of both types of operations. In addition, signaling-related protocols are also specified. In ISDN, the signaling between the ISDN switches is done through SS7 signaling, discussed in the previous chapter.

The ISDN protocol architecture is shown in Figure 27.4.

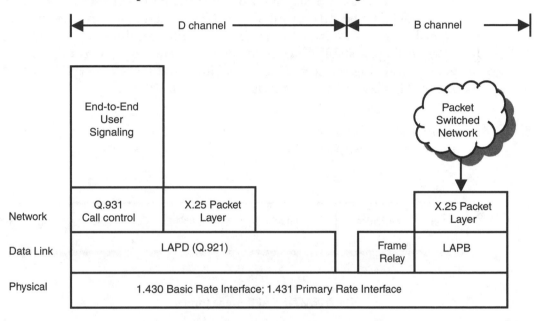

FIGURE 27.4 ISDN protocol architecture.

Physical Layer

The physical interface is based on X.21 for interfacing to a public circuit switching network through an 8-pin connector. Specifications provide for capabilities to transfer power across the interface (generally from network to terminal). Just as the PSTN telephone is supplied the power (-48 V) from the switch, ISDN switch is also capable of this function. ITU-T Recommendation I.430 specifies the physical layer functionality for BRI, and Recommendation I.431 specifies the physical layer functionality for PRI.

Data Encoding

Basic rate is 192kbps using the pseudoternary encoding scheme, in which binary 1 is represented by an absence of voltage and 0 is represented by a positive or negative pulse of 750 mV with 10% tolerance. Primary rate is 2.048Mbps, with 2.048Mbps with AMI with HDB3 encoding scheme.

Datalink Layer

The protocol structure is different for B and D channels. For D channel, LAPD, based on HDLC, is used. All transmission on D channel is based on LAPD frames, exchanged between subscriber equipment and switch for three applications: (a) control signaling to establish, maintain, and terminate connection on B channel; (b) packet switching services to the subscriber; and (c) telemetry. The LAPD frame format, shown in Figure 27.5, is derived from HDLC.

The datalink layer protocol used for D channel is called LAPD (link access protocol D channel) which is derived from HDLC.

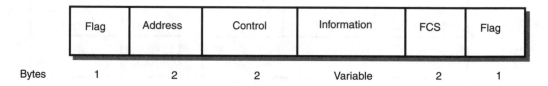

Flag	Address	Control	Information	FCS	Flag

Bytes   1   2   2   Variable   2   1

FIGURE 27.5 LAPD frame format.

The LAPD frame consists of one byte of flag, two bytes of address, two bytes of control information, a variable number of bytes of user information, two bytes of FCS, and one byte of flag.

The differences between HDLC and LAPD are:

- LAPD always uses 7-bit sequence numbers (3-bit sequences are not allowed).
- FCS (frame check sequence) is always a 16-bit field for CRC.
- The address field for LAPD is a 16-bit field that contains two subaddresses: one to identify one of possibly multiple devices on the user side of the interface and the other to identify one of possibly multiple logical users of LAPD on the user side of the interface.

ISDN protocol architecture addresses the first three layers of the ISO/OSI model. Layers 4 to 7 are not the concern of this protocol architecture. Using these protocols, circuit-switched and packet-switched calls can be established.

LAPD is specified in Recommendation Q.931.

For B channel, circuit switching and semi-permanent circuits are supported. From the ISDN viewpoint, for circuit switching, there is no need for layers 2 to 7 and hence they are not specified.

For packet switching, once a circuit is set up through the D channel, X.25 layers 2 and 3 can be used for virtual circuit data transfer. At the datalink layer in the protocol architecture, Frame Relay is also shown. This is used in the broadband ISDN applications.

Network Layer

The network layer for packet switching applications is based on X.25. For call control (signaling) the Q.931 recommendation is followed.

In ISDN protocol architecture, layers 4 to 7 of the ISO/OSI model are not specified, because ISDN is not concerned about these details. ISDN provides facility only to establish circuit-switched and packet-switched calls at the specified data rates. Any type of user application can be developed using this infrastructure.

27.6 ISDN CONNECTIONS

ISDN supports the following types of calls: circuit-switched and packet-switched calls over the B channel and packet-switched calls over the D channel.

The four types of connections supported in ISDN are:

- Circuit-switched calls over B channel
- Semipermanent connections over B channel
- Packet-switched calls over B channel
- Packet-switched calls over D channel

Circuit-switched calls: Circuit-switched calls involve both B and D channels. The procedure is as follows:

1. User requests a circuit over D channel.
2. User is informed of the circuit establishment over D channel.
3. Data is transferred over B channel.
4. Call is terminated through signaling on D channel.

Semipermanent connection: For semipermanent connection, only layer 1 functionality is provided by the network interface. No call control protocol is required because a connection already exists.

Packet-switched calls over B channel: The procedure for establishing packet-switched calls is as follows:

- The user makes a request over the D channel for a connection on B channel using the call control procedure.

- The connection is set up, and the user is notified on the D channel using the call control procedure.

- The user sets up a virtual circuit to another user using X.25 call establishment procedure on B channel.

- The user terminates the virtual circuit using X.25 on B channel.

- After one or more virtual calls on the B channel, the user signals through the D channel to terminate connection to the packet-switching node.

In ISDN, signaling information between switches is exchanged through a standard called Signaling System No. 7 (SS7).

Packet-switched calls over the D channel: Low-speed data is sent over the D channel, for which X.25 level 3 protocol using LAPD frames is employed.

Signaling in ISDN

Layer 3 protocol for signaling is defined in I.451 recommendations. Messages are exchanged between the user and the switch over the D channel for establishing, controlling, and terminating a call. Signaling between the switches is through an SS7 signaling network, which was discussed in the previous chapter.

27.7 BROADBAND ISDN

The fast packet switching technologies—Frame Relay and Asynchronous Transfer Mode (ATM)—have been developed as a result of the research in broadband ISDN.

Narrowband ISDN did not become successful for a long time. With the hope that if users are offered very high data rates to support services such as high definition TV (HDTV), multimedia communication, video conferencing, and so on, broadband ISDN development work was initiated. Broadband ISDN supports 155.52Mbps and 622.08Mbps data rates. However, to support such high data rates, optical fiber has to be used as the transmission medium, and the switching has to be done very fast. The fast packet switching technologies developed as a result of research on broadband ISDN are Frame Relay and ATM, which will be discussed in subsequent chapters.

References

R. Horak. *Communications Systems and Networks*. Wiley-Dreamtech India Pvt. Ltd., 2002.

www.frforum.com The Web site of the Frame Relay Forum. Frame Relay is an outcome of research in broadband ISDN.

www.atmforum.com The Web site of the ATM Forum. The ATM technology is an outcome of research in broadband ISDN.

www.itu.int The Web site of International Telecommunications Union.

Questions

1. Explain the ISDN architecture and the various reference points used in the architecture.
2. Explain basic rate interface (BRI) of ISDN.
3. What is primary rate interface (PRI) of ISDN?
4. Explain the ISDN protocol architecture.
5. Explain the LAPD format.

Exercises

1. List the various services that can be supported by basic rate interface (BRI).
2. List the various services that can be supported by primary rate interface (PRI).
3. "Telemedicine" or medicine from a distance is a service offered for remote rural areas. Study the data rate requirements for providing telemedicine service.
4. Study the architecture of commercially available terminal adapters.

Projects

1. Telemedicine is now becoming an important service being provided to rural areas, for which ISDN is extensively used in Europe. Prepare a technical report on telemedicine. Implement a telemedicine system on a LAN.

2. Prepare a technical report on broadband ISDN.

28 ∷ Frame Relay

Frame Relay provides fast switching of packets by reducing the protocol overhead. Due to fast packet switching, multimedia services at very high data rates can be supported. Frame Relay networks are now becoming popular for wide area networking. In this chapter, we will study the details of Frame Relay.

28.1 ADVANTAGES OF FRAME RELAY

In Frame Relay networks, the protocol overhead is reduced by eliminating the sequence number, acknowledgements, flow control, and error control. Hence, switching of packets is done very fast to support multimedia services.

Using X.25 protocols for wide area networks (WANs) results in delay mainly because X.25 protocols have lots of overheads. The packet sequence number, acknowledgements, supervisory frames, and so on create processing overhead at each packet switch. In Frame Relay, this protocol overhead is reduced to achieve fast switching.

In Frame Relay networks, each packet is called a frame. The frame contains user information, a 2-Byte Frame Relay header, a frame check sequence, and a flag. The protocol overhead is eliminated using the following strategy:

- There is no sequence number for the frames.

- The Frame Relay switch does not send an acknowledgement after receiving a frame.
- Multiplexing and switching of logical channels is done at layer 2, hence there is no need for layer 3 protocol. One complete layer of processing is eliminated.
- Call control signaling is done on a separate logical connection, so intermediate switches need not process messages related to call processing.
- There is no flow control or error control. End-to-end flow control and error control need to be done by higher layers. If a switch receives a frame and there are errors, the frame is discarded.

Frame Relay protocols are well suited for transmission media that are less noisy, such as optical fiber.

28.2 FRAME RELAY NETWORK

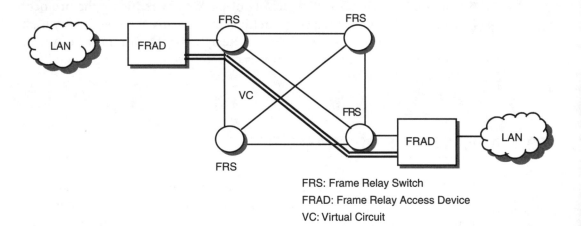

FRS: Frame Relay Switch
FRAD: Frame Relay Access Device
VC: Virtual Circuit

FIGURE 28.1 Frame Relay network.

A Frame Relay network consists of a number of Frame Relay switches. The subscriber equipment is connected to the switches through a Frame Relay access device (FRAD).

The architecture of a Frame Relay network is shown in Figure 28.1. The network consists of a number of Frame Relay switches. Subscriber equipment is connected to the Frame Relay switch through a Frame Relay access device (FRAD). PCs, servers, and such can be connected to the FRAD. Bridges and routers can be provided with Frame Relay interfaces so that they can be connected to the Frame Relay switch.

Frame Relay provides a connection-oriented service, just like X.25. When two end systems want to exchange data, a virtual connection (VC) is established between the two end points, and data transfer takes place.

A virtual connection is of two types:

- Permanent virtual connection (PVC)
- Switched virtual connection (SVC)

Frame Relay provides a connection-oriented service. A virtual connection is established between two end systems. The virtual connection can be switched or permanent, called SVC and PVC, respectively.

Permanent virtual connection (PVC): The Frame Relay operator provides the PVC between two end points. This is done by programming the Frame Relay switches in the network to assign a series of links between the switches to form the PVC between the two end points. The PVC is permanently available to the two parties at both ends.

Switched virtual connection (SVC): When the connection is established on a call-by-call basis, it is called a switched virtual connection. SVC requires some overhead for call establishment and disconnection. When Frame Relay was first developed, only PVC was supported, but SVC was added later to meet user requirements.

In Figure 28.1, if an organization has two offices in two cities separated by a large distance, the organization can take a PVC from the Frame Relay service provider if the traffic is high. If the traffic is low, the organization may opt for a SVC and pay only on the basis of use.

A Frame Relay virtual connection is identified by a number called a data link connection identifier (DLCI). Each link can support multiple connections.

The data transfer involves:

A virtual connection is identified by a number called data link connection identifier (DLCI). The procedure for communication between two systems is to establish a logical connection, exchange data frames, and release the connection.

- Establishing a logical connection between end points and assigning a DLCI.
- Exchanging data frames with each frame containing the assigned DLCI.
- Releasing the logical connection.

To establish a logical connection, messages are exchanged over a dedicated logical connection for which DLCI = 0. Four messages, SETUP, CONNECT, RELEASE, and RELEASE COMPLETE, are exchanged to set up and disconnect a logical connection. In the case of PVC, this procedure is not required.

28.3 FRAME FORMAT

The link layer protocol used in Frame Relay is called link access protocol for Frame Relay, abbreviated LAPF. LAPF is derived from HDLC.

The Frame Relay frame format is derived from link access protocol balanced (LAPB), which in turn is derived from HDLC. The link layer protocol in Frame Relay is called link access protocol for Frame Relay (LAPF). LAPF frame format is shown in Figure 28.2.

The difference between LAPB (or HDLC) and the LAPF is that there is no control field in LAPF. Hence, there is only one type of frame that carries user data, so control information cannot be carried. It is not possible to carry out in-band signaling. Also, flow control and error control cannot be done because there is no sequence number. The Frame Relay header contains the DLCI. Remaining fields have the same significance and functionality as in HDLC.

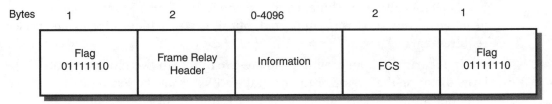

FIGURE 28.2 LAPF frame format.

Flag: This 8-bit field has a fixed bit pattern 01111110 to identify the beginning of a frame. The same bit pattern appears at the end of the frame.

Frame Relay header: This is a 16-bit field. The details are explained in the next section.

Information: This field contains the user data obtained from the higher layer. The maximum size is 4096 bytes. Though the standard specifies 4096 bytes, 1600 bytes is the ANSI (American National Standards Intstitute) recommendation. Most of the Frame Relay networks are connected to LANs, so 1600 Bytes is recommended because it is compatible with the Ethernet frame size.

FCS: Frame check sequence. Sixteen bits of CRC is used for header error control.

The LAPF frame consists of the following fields: flag, frame relay header, information, and frame check sequence. The maximum size of the information field is 4096 bytes, as per the standards. However, 1600 bytes is the recommended size to make the frame compatible with the Ethernet frame size.

28.3.1 Frame Relay Header

The details of the Frame Relay header are seen in Figure 28.3.

Bits	6	1	1	4	1	1	1	1
	DLCI	C/R	EA	DLCI	FECN	BECN	DE	EA

FIGURE 28.3 Frame Relay header.

When a frame is received by a Frame Relay switch, the switch checks whether the data in the frame is correctly received using the frame check sequence field. If there is an error, the frame is discarded.

DLCI: Data link connection identifier is a 10-bit field split into two portions, one portion containing 6 bits and another portion containing 4 bits.

C/R: Command/response field bit. This bit is used by the FRAD to send a command and receive a response. This command/response is required by some protocols such as SNA (system network architecture) of IBM for signaling and control purposes.

FECN: Forward explicit congestion notification. This bit is used by the switch to inform other switches about congestion.

BECN: Backward explicit congestion notification. This bit is also used by the switch to inform the other switches about congestion.

DE: Discard eligibility indicator. If this bit is set, the frame can be discarded by a switch.

EA: Extension of address. The present Frame Relay header field is restricted to 2 Bytes, with 10 bits for DLCI. If Frame Relay popularity grows and the address capability needs to be extended, the EA bits are required. Using these EA bits, an indication can be sent whether the header is of 3 Bytes or 4 Bytes. This is for future use.

As you can see, the frame format is very simple, and fast switching is achieved. However, because flow control and error control are not supported at the datalink layer, if the frames are received in error at the switch, what needs to be done?

When a frame is received by the switch, it checks the FCS field and verifies whether or not there is an error in the frame. If there is no error, it uses the routing table to forward the frame toward the destination based on the DLCI. If the frame is in error, the frame is discarded. If a frame is received and the DLCI field is wrong (not defined for the link), the frame is discarded. The logic here is simple—if anything is wrong, discard the frame.

If frames are discarded like this, how do we ensure end-to-end reliable data delivery? Frame Relay makes switching faster by transferring flow control and error control to the higher layers. For example, in TCP/IP networks, the TCP layer takes care of flow control and error control, so there is no need to duplicate the work at the link layer. In Frame Relay networks it is the responsibility of higher layers to take care of loss of frames and controlling the flow through acknowledgements by incorporating the necessary protocols.

There is no flow control and error control in the link layer of Frame Relay, and it is the responsibility of the higher layer to take care of these issues. For instance, if TCP/IP protocol stack is used, the TCP layer will do so.

28.4 CONGESTION CONTROL

If there is a sudden increase in traffic, it is possible that a switch may experience congestion; its buffer may be full or it may not have enough memory to process any more frames. In such a case, the two bits FECN and BECN are used to tell the other switches about the congestion.

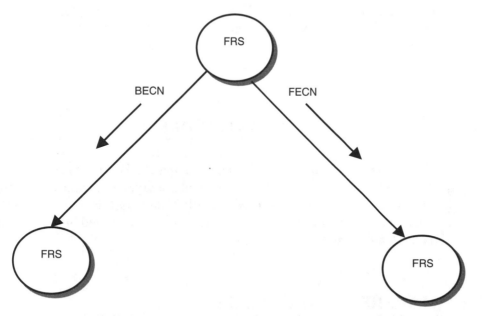

FRS: Frame Relay Switch

FIGURE 28.4 Congestion control through FECN and BECN.

If there is congestion in a Frame Relay switch, the two fields FECN and BECN are used to notify the other switches so that those switches will reduce their speed.

When a switch experiences congestion, it changes the FECN bit from 0 to 1 and sends the frame to the switch toward the destination. It also changes the BECN bit from 0 to 1 and sends the frame toward the source to notify the source that congestion is being experienced. Note that it is the responsibility of the higher layer to reduce the number of frames if the congestion notification is obtained.

28.4.1 Committed Information Rate

Committed information rate (CIR) is the minimum data rate that has to be supported by a virtual connection. At the time of buying the Frame Relay service, this parameter is specified by the subscriber.

When a virtual connection is established, the average data capacity of the VC can be specified in terms of the bit rate that is assured by the service provider. At the time of buying the Frame Relay service, the subscriber generally specifies the committed information rate (CIR) to the service provider. The VC must maintain the minimum CIR. If the subscriber is pumping data at a rate faster than CIR, the switch can discard some frames. This is done based on the DE (discard eligibility) bit. The DE bit is set to 1 by the switch if the frame is above the CIR. If DE =1, the frame can be discarded by any of the other switches on route to the destination if that switch experiences congestion.

28.5 LOCAL MANAGEMENT INTERFACE

To reduce overheads little management information is exchanged between the switch and the FRAD. However, using special management frames with unique DLCI address, connection status information such as the status of a VC and whether there is any problem with the interface is exchanged between a switch and a FRAD such as a router.

28.6 FRAME RELAY STANDARDS

Frame Relay standards emerged out of the standardization activities of Integrated Services Digital Network (ISDN). ANSI, ITU-T, and the Frame Relay Forum are the major contributors to these standards.

American National Standards Institute (ANSI) and International Telecommunications Union Telecommunications Sector (ITU-T) are the two major standardization bodies that developed the Frame Relay standards. The Frame Relay Forum, a nonprofit organization established in 1991 with many equipment vendors as members, also contributed significantly to the standardization activities. The important ITU-T standards for frame relay are:

- Architecture and service description: I.233
- Datalink layer core aspects: I.922 Annex A
- Switched virtual circuit management: Q.933
- Permanent virtual circuit management: Q.933 Annex A
- Congestion management: I.370

28.7 FRAME RELAY VERSUS X.25

X.25 was used extensively for wide area networks for many years. With the introduction of Frame Relay for wide area networks, the significance of X.25

reduced drastically. A comparison of the X.25 and Frame Relay protocol architectures reveals the reasons.

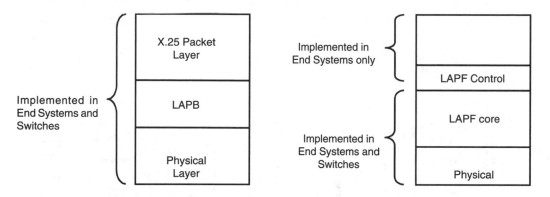

FIGURE 28.5 Comparison of X.25 and Frame Relay protocol standards.

Compared to X.25 protocols, Frame Relay protocols have less overhead, so switching is very fast in Frame Relay networks. However, Frame Relay is more suited for less noisy transmission media such as optical fiber.

In X.25, above the physical layer, LABP and X.25 packet layer protocols will be running that will take care of flow control and error control. A number of acknowledgements are exchanged between the end systems and the switches. Lots of supervisory information also is sent over the network. In the case of Frame Relay, the datalink layer contains only the LAPF protocol, which is a very lightweight protocol with little overhead. Flow control and error control need to be taken care of by the higher layers, such as the TCP layer in TCP/IP networks. The end systems have the necessary processing power so this will not be a problem. Secondly, in Frame Relay networks the physical medium is generally optical fiber, so there are fewer errors.

Frame Relay protocols are much more simplified than X.25 and, if used over transmission media with high bandwidths and low errors, they will be very efficient. Frame Relay–based networks are becoming very popular with the advent of optical fiber communications.

28.8 REPRESENTATIVE FRAME RELAY NETWORKS

Frame Relay networks can be used not only for data services, but for voice and video as well. Because committed information rate (CIR) can be specified by the subscriber, a desired quality of service (QoS) can be maintained.

Figure 28.6 shows a Frame Relay network that supports both voice and data services. The FRAD can interface with the PBX through an E1 trunk (2.048Mbps), and the LAN can be interfaced through a router. The FRAD will do the necessary framing for both data and voice. However, for voice applications, instead of coding the voice signals at 64kbps, low bit rate codecs are used that are an integral part of the FRAD. Many commercial systems use G.729-based voice coding technique to reduce the voice data rate to 8kbps.

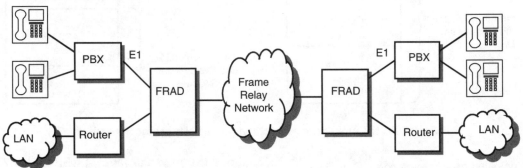

FIGURE 28.6 Voice/data services over Frame Relay.

Figure 28.7 shows how Frame Relay and ISDN networks can be used by an organization. FRAD with ISDN interfaces are available to facilitate connecting the ISDN BRI or PRI equipment to the FRAD. Organizations can use the ISDN network for their regular voice and data traffic; whenever bulk data transfers are involved or when video conferencing has to be set up, they can use the Frame Relay network with switched virtual connections.

A Frame Relay network supports a variety of access speeds from 64kbps to 34.368Mbps (E3). Frame Relay access devices are available with 64kbps interfaces,

Frame Relay networks have to maintain a committed information rate and are well suited for multimedia communication. Many corporate networks are based on Frame Relay.

E1 interface, ISDN basic rate interface, ISDN primary rate interface, and others.

Frame Relay networks are gaining popularity in recent years. However, there is stiff competition from Asynchronous Transfer Mode (ATM) networks. We will study ATM in the next chapter.

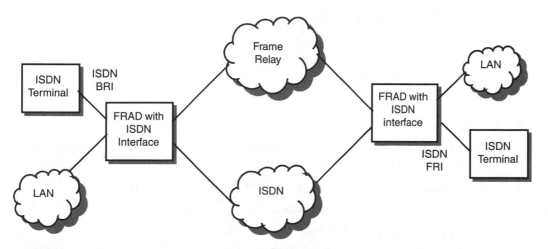

FIGURE 28.7 Integration of Frame Relay and ISDN.

Summary

This chapter presented the details of Frame Relay networks. Frame Relay offers very fast packet switching by eliminating the protocol overhead, as compared to X.25. Frame Relay provides virtual connection service. Two types of virtual connections can be set up. Permanent virtual connection (PVC) is set up between end points to provide permanent connectivity, such as leased line. Switched virtual connection (SVC) is established on a call-by-call basis. In Frame Relay, the packet is called a frame and can contain up to 4096 Bytes of user information. The datalink layer protocol is called link access protocol for Frame Relay (LAPF), derived from LAPB. Two bytes of the frame Relay header contain information about the virtual connection identification called data link connection identifier (DLCI). There are two fields to notify about congestion in the switch and one bit to indicate whether the packet can be discarded. If the subscriber exceeds the committed information rate, then the frame can be discarded. Frame Relay networks are gaining popularity due to fast switching and support for multimedia services.

References

R. Horak. *Communications Systems and Networks*. Wiley-Dreamtech India Pvt. Ltd., 2002. Chapter 10 covers broadband networks including Frame Relay.

www.frforum.com Web site of the Frame Relay Forum.

www.ansi.org Web site of the American National Standards Institute.

www.cisco.com Cisco is a leading supplier of Frame Relay equipment. You can obtain product information from this site.

Questions

1. What are the advantages of Frame Relay as compared to X.25? Compare the protocol stacks of X.25 and Frame Relay networks.
2. Explain how fast switching is achieved in Frame Relay.
3. Describe the LAPF frame format.
4. Explain the procedure for congestion control in Frame Relay.
5. What is committed information rate? What does the Frame Relay switch do if this rate is exceeded?

Exercises

1. Study the features of Frame Relay commercial products and prepare a technical report.
2. Study the various interfaces provided by a commercial FRAD (for instance, that of Cisco). Study the voice compression techniques used for voice over Frame Relay.
3. Compare the number of packets/frames exchanged in an X.25 network and a Frame Relay network.
4. As compared to X.25, Frame Relay protocols are a better choice when optical fiber is the transmission medium. Justify this statement.
5. Is it necessary to use low bit rate coding of voice/video in Frame Relay networks? If so, why?

Projects

1. Develop a software package that simulates the LAPF frame format.
2. Carry out a paper design to develop a Frame Relay network for an organization that has five offices spread over the country. Each office has a LAN, a PBX with E1 interface for trunking, and basic rate interface (BRI) ISDN equipment. Identify the commercial products that have these interfaces.

29 ■ Asynchronous Transfer Mode

In This Chapter

- The Architecture of ATM Networks
- The ATM Protocol Architecture
- Call Establishment Procedure in ATM Networks
- The Need for ATM Adaptation Layer to Support Multimedia Services

The concept of packet switching is fundamental to data communication. An important design issue in packet switching networks is the size of the packet. If the packet is very large, large buffers are required at the packet switches, the switching is very slow, and delay is high. On the other hand, if the packet is very small, switching is fast but the overheads are high because each packet has to carry control information. Variable packet size generally is used in packet switching networks, but variable size leads to computation overhead and variable buffer size. Asynchronous Transfer Mode (ATM) is an attractive packet switching technology that achieves fast packet switching with small packets of fixed size. ATM technology is an outcome of the developments in broadband ISDN. ATM now holds great promise for multimedia communication, and this technology is now being deployed extensively in both LANs and WANs. In this chapter, we will study the concepts and protocol architecture of ATM.

29.1 CONCEPTS OF ATM

In ATM networks, a special node called an ATM switch is used to which computers are connected. Because of the high speeds supported by ATM, a pair of multimode

An ATM network consists of a number of ATM switches. The end systems are connected to the ATM switch through an interface card, which contains the interfaces for optical fiber. optical fibers is used to interconnect the computers with the ATM switch. An ATM interface card is plugged into the computer; it contains an LED or laser that converts the data into light signals that are sent over the fiber. The ATM interface card also contains a sensor that converts the light signals back to data to be given to the computer. An ATM network is formed by interconnecting two or more ATM switches as shown in Figure 29.1.

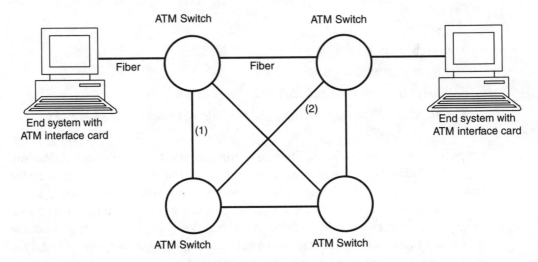

FIGURE 29.1 An ATM network.

In ATM, each packet is called a cell. The cell size is 53 bytes—5 bytes of header and 48 bytes of data. The small cell size facilitates fast switching of the cells. In ATM, each packet is called a cell. The cell size is 53 bytes—a 5-byte header and 48 bytes of data. The small and fixed size of the cell ensures faster processing of the cells and hence faster switching and simple hardware implementation.

In an ATM network, for two computers to exchange data, a connection has to be established, data transferred, and connection released. Hence, ATM provides a connection-oriented service.

There are two types of connections in ATM:

- Switched virtual circuit
- Permanent virtual circuit

When a host wants to communicate with another host, the sender makes a request to its switch, and the switch in turn establishes a connection with the other host. A 24-bit identifier known as virtual circuit identifier is used to identify the connection. Then all the packets are transmitted over this connection. This is known as *switched virtual circuit*. At the time of connection, each ATM switch allocates the necessary resources, such as buffers, incoming port, and outgoing port, for routing the cells.

Virtual circuits that are dedicated connections between two end systems are called *permanent virtual circuits*.

In an ATM network, two types of connections can be formed: Switched Virtual Circuit (SVC) and permanent virtual circuit (PVC). SVC is formed on a per-call basis, whereas PVC is a permanent connection.

The virtual circuit is identified by a 24-bit field, out of which 8 bits are to identify the virtual path and 16 bits are to identify the virtual circuit.

The 24-bit identifier for a virtual circuit is divided into two portions—virtual path identifier (VPI) of 8 bits and virtual circuit identifier (VCI) of 16 bits.

The most attractive feature of an ATM network is that it provides the required quality of service (QoS), unlike the TCP/IP networks that do not guarantee the QoS. For instance, if video conferencing has to be established between two end systems, a certain bandwidth is required for the video and audio transmission. The delay also should be minimum. Such QoS parameters can be set at the time of establishing the virtual circuit.

29.2 ATM LOGICAL CONNECTIONS

The logical connection between two end nodes is called virtual channel connection (VCC). A bundle of VCCs that have the same end points is called the virtual path connection (VPC).

ATM provides fast packet switching and so can be used for multimedia communication. In voice and video applications, the packets have to reach the destination in sequence and without variable delay. In other words, the quality of service is of paramount importance for supporting multimedia services. To achieve this objective, in ATM networks the concepts of virtual channel connection and virtual path connection are introduced.

The logical connection between two end nodes is called a virtual channel connection (VCC). This is analogous to a virtual circuit in X.25. VCC is set up between two end users, a user and a network (for control signaling), or network to network (for network management and routing).

A bundle of VCCs that have the same end point is a called virtual path connection (VPC). All the cells flowing over all of the VCCs in a single VPC are switched together. The concept of virtual channel and virtual path is illustrated in Figure 29.2.

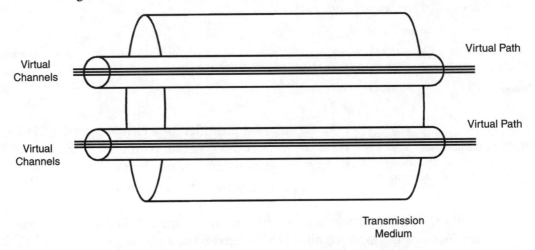

FIGURE 29.2 ATM virtual channels and virtual path.

Using the virtual path has the following advantages:

The concept of virtual path facilitates faster switching of the cells, simplified network architecture and reduced processing for connection setup. Moreover, the desired quality of service can be ensured.

1. Once a virtual path is established, establishing new virtual channel connections is easy. capacity can be reserved in anticipation of later call arrivals. No call processing is required at intermediate switches, so adding virtual channels involves minimal processing. Hence, the processing is reduced and connection setup time is much less.

2. The switching is very fast and, as a result, network performance and reliability will increase.

3. The network architecture will be simplified.

In all packet switching networks, to ensure that the network provides the desired quality of service is of paramount importance. The QoS can be defined in terms of maximum delay allowed or loss of packets. The concept of VPC and VCC allows ensuring a desired QoS to the maximum possible extent.

29.3 CALL ESTABLISHMENT IN ATM NETWORKS

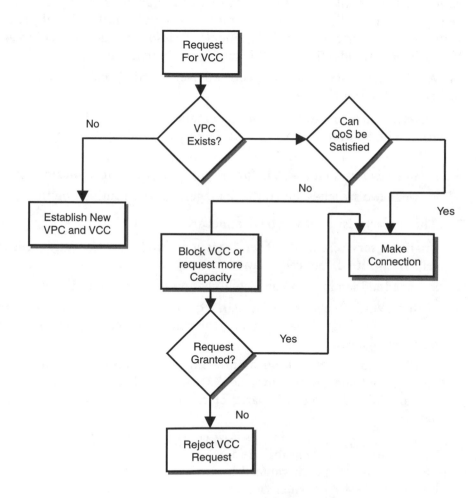

FIGURE 29.3 ATM call establishment procedure.

A virtual channel connection can be established between end users, between the end user and the switch, or between two switches.

The flow chart for establishment of a call in an ATM is shown in Figure 29.3. To set up a virtual channel, there must be a virtual path connection to the destination node with sufficient available capacity to support the virtual channel, with the required quality of service. Virtual path control mechanisms include calculating routes, allocating capacity, and storing connection state information. Virtual path connection (VPC) is a concatenation of VC links between various switches (for example, the concatenation of link 1 and link 2 in Figure 29.1). Cell integrity is preserved for cells belonging to the same VCC. A unique ID is given to the virtual path and virtual channel. Virtual channel identifier (VCI) is for a particular link in a given VPC. Virtual path identifier (VPI) is the ID for a particular link.

A VCC can be established between end users, between end user and switch, or between two switches.

- **Between end users:** to carry user data/control signaling. VPC provides the users with an overall capacity. The set of VCCs should not exceed the VPC capacity.
- **Between end user and switch:** for user to network control information.
- **Between two switches:** for traffic management and routing functions.

The characteristics of virtual channel are:

- Quality of service: Quality of service is specified by cell loss ratio (cells lost/ cells transmitted) and cell delay variation.
- Switched and semipermanent virtual channel connections can be established.
- Within a VCC, all cells are received in sequence. If the VCC carries voice, then at the destination, it is easy to replay the voice cells because there is no need for rearrangement.
- Traffic parameters and usage monitoring are negotiated for each VCC. The network continuously monitors the VCC to ensure that negotiated parameters are not violated. The traffic parameters can be average data rate and peak data rate.

Within a VCC, all the cells are received in sequence. The quality of service parameters such as cell loss ratio, and cell delay variation will be maintained within a VCC.

In the flow chart given in Figure 29.3, a VCC is established when the QoS parameters can be guaranteed. It may happen that new VCCs are requested when QoS cannot be guaranteed. So, to manage the traffic parameters, strategies need to be worked out.

The strategies can be:

- Deny request for new VCCs (the best choice)
- Discard cells
- Terminate existing connections (worst)

The characteristics of virtual path connection are
- To match the desired quality of service.
- To set up switched and semipermanent virtual path connections.
- To negotiate traffic parameters and to monitor the usage.
- To maintain cell sequence.

Note that the characteristics are duplicated in VPC and VCC. The advantages of duplication of characteristics are that the VCC has to function within certain boundary limits and after VPC is set up, new VCCs may be set up. Both these aspects introduce discipline into the network.

When a desired quality of service cannot be maintained, one of the following strategies is used: (a) deny requests for new VCC, which is the best choice, (b) discard some cells, and (c) terminate the existing connections which is the worst choice.

NOTE

29.4 ATM PROTOCOL ARCHITECTURE

The ATM protocol architecture is shown in Figure 29.4.

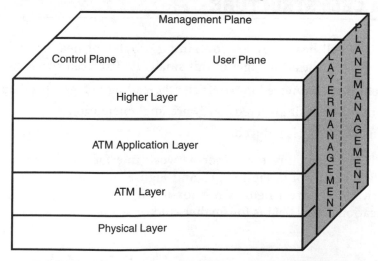

FIGURE 29.4 ATM protocol architecture (ITU-T standard).

The ATM protocol layers are physical layer, ATM layer and ATM adaptation layer. In addition, user plane, control plane, and management plane are defined to take care of different control and management functions.

Physical layer: This layer specifies the transmission data rates and encoding scheme. Data rates of 155.52 and 622.08Mbps are supported. Frame generation and reassembly are done in this layer.

ATM layer: Transmission of data in fixed cells using logical connections. The functions of this layer include cell header generation/extraction, flow control, and cell multiplexing and demultiplexing.

ATM adaptation layer: This layer is used to support information transfer protocols that are not based on ATM (such as PCM over ATM). In such cases, segmentation/reassembly of the data is required, which is the function of this layer.

User plane: This plane provides for user information transfer, along with associated controls (such as flow control and error control).

Control plane: This plane performs call control and connection control functions.

Management plane: Functions of this plane include plane management and layer management. Plane management functions are related to the system as a whole and between planes. The layer management module performs management functions related to the resources in the network.

29.5 ATM CELL STRUCTURE

The ATM cell has 53 bytes, containing 5 bytes of header and 48 bytes of information. The advantages of this small and fixed cell size are:

- Cells can be switched more efficiently, leading to high-speed data transfer.
- Since the cell size is fixed, hardware implementation is easier.
- Queuing delay is reduced.

The cell structure for a user-network interface and a network-network interface are given in Figure 29.5 and Figure 29.6, respectively. Note that the generic flow control field is not for network-network interface. The exact application of this field is for further study.

The size of the cell is small and fixed, so the ATM switches the cells very fast. Moreover, queuing delay is reduced, and hardware implementation is very easy. For these reasons, ATM is an excellent choice for multimedia communications.

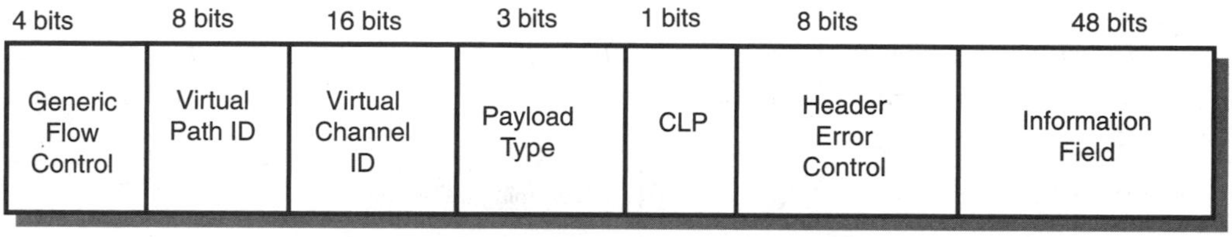

4 bits	8 bits	16 bits	3 bits	1 bits	8 bits	48 bits
Generic Flow Control	Virtual Path ID	Virtual Channel ID	Payload Type	CLP	Header Error Control	Information Field

FIGURE 29.5 ATM cell: user–network interface.

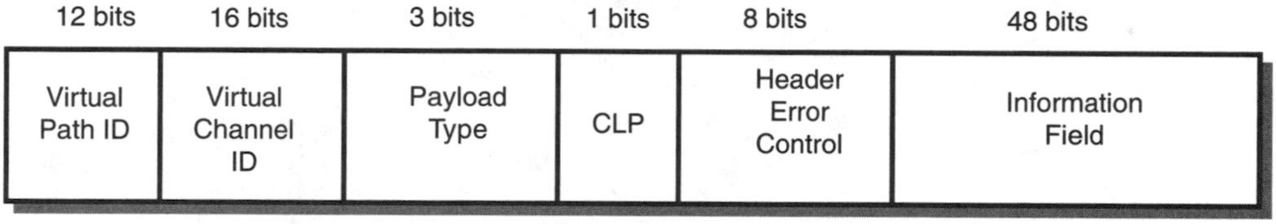

12 bits	16 bits	3 bits	1 bits	8 bits	48 bits
Virtual Path ID	Virtual Channel ID	Payload Type	CLP	Header Error Control	Information Field

FIGURE 29.6 ATM cell: network–network interface.

The ATM cell contains the following fields: virtual path identifier, virtual circuit identifier, payload type, cell loss priority, header error control, and user information.

VPI (8 bits and 12 bits): Virtual path identifier. It specifies the routing field for the network.

VCI (16 bits and 16 bits): Virtual circuit identifier. It specifies routing to and from the end user.

Payload type: Payload type consists of three bits: P, Q, and R.

The P bit is 0 for user information, 1 for network management/maintenance information.

The Q bit is 0 if there is no congestion, 1 if there is.

The R bit can be used to convey information between users.

CLP: cell loss priority. CLP is set to 0 to indicate high priority and to 1 to indicate less priority. CLP = 0 implies that the cell should be discarded only if there is no other option, and CLP = 1 indicates that the cell can be discarded.

Header error control (8 bits): Header error control (HEC) bits are calculated based on the remaining 32 bits of header using the polynomial $x^8 + x^2 + x + 1$. Since there are 32 data bits and 8 error control bits, error correction is possible.

Using this simple cell structure, ATM achieves fast switching. This fast switching allows support for multimedia services.

The flow chart for detecting errors/rejecting cells is shown in Figure 29.7. If errors are detected in the HEC bits, correction is attempted. If correction is successful, the cell is taken as valid; otherwise cells are discarded. Single-bit errors are corrected, but cells with burst errors are discarded.

Using the header error control (HEC) bits, single-bit errors can be corrected. If more than one bit is in error, the cell is discarded by the switch.

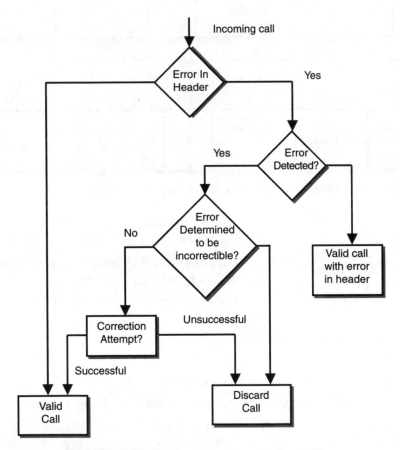

FIGURE 29.7 ATM header error control asgorithm.

29.6 TRANSMISSION OF ATM CELLS

In an ATM network, the cells are transmitted as a stream. Hence, each switch has to identify the beginning of each cell. This is done through the cell delineation algorithm.

In ATM, no framing is imposed and a continuous stream of cells is transmitted. Hence synchronization is required—the receiver has to know where the cell starts. Header error control (HEC) field provides synchronization. A cell delineation algorithm is required to detect the boundaries of the cells. The algorithm is as follows: match the received HEC with the calculated HEC with an assumed cell structure. The cell delineation algorithm is successful if HEC matches x times consecutively, and it is lost if HEC mismatches y times continuously. The values of x and y are design parameters between 5 and 10. Note that greater x implies longer delay in synchronization but robustness against false delineation; greater y implies longer delay in recognizing a misalignment but robustness against false misalignment.

29.7 ATM FOR MULTIMEDIA

The functions of the ATM adaptation layer are segmentation and reassembly, flow control, timing control, handling transmission errors, and handling lost cells.

When ATM is used for applications such as audio/video transmission in real time, it is necessary that at the receiving end, the cells are received without variable delay; the cells should be received at a constant speed. For example, PCM-coded speech produces a continuous stream of bits at the rate of 64kbps. To transmit PCM over ATM, the PCM data stream has to be assembled into cells and transmitted over the network; at the receiver, the cells should be received with constant delay. To achieve this, special protocols are required. The ATM adaptation layer (AAL) has been defined to provide the necessary services to handle such applications. The AAL provides the following services:

- Segmentation and reassembly
- Flow control and timing control
- Handling of transmission errors
- Handling of lost cells

ITU-T defined four classes of services and five types of AAL protocols:

Class A services require synchronization the source and the destination, have a constant bit rate, and are connection-oriented services. An example of a class A service is voice transmission. Type 1 AAL protocol is used for this class of services.

Class B services require synchronization between the source and the destination, but the data rate is variable, and they are connection-oriented services. An example of class B services is variable rate video transmission for video conferencing. Type 2 AAL protocol is used for this class of services.

Class C services require no synchronization between the source and the destination, the data rate is variable, and they are connection-oriented services. Data applications fall under this category. Type 3, 4 and Type 5 AAL protocols are used for this class of service.

Class D services require no synchronization between the source and the destination, the data rate is variable, and they are connectionless services. Connectionless data applications fall under this category. Type 3and 4 AAL protocol as are used for this class of service.

Networks based on ATM are now becoming popular because ATM provides simple yet efficient protocols to handle multimedia services.

ATM adaptation layer provides support for multimedia services by defining five types of AAL protocols that support fixed and variable data rate services.

Summary

This chapter presented the ATM technology. Asynchronous Transfer Mode (ATM) is based on small packets to facilitate the fast switching required for multimedia services. ATM provides connection-oriented service. ATM packets are called cells. Each cell will be of 53 bytes, with 5 bytes of header and 48 bytes of user data. The main attraction of ATM is that it guarantees quality of service. The quality of service parameters such as delay and packet loss can be set. To achieve the desired QoS, virtual channel and virtual path are used. A virtual channel is established between two end systems. A bundle of virtual channels having the same end points is called virtual path. All cells belonging to a virtual path are switched together.

To support multimedia services in the ATM, the ATM adaptation layer is defined. The ATM adaptation layer has five classes of services to support voice, video, and data applications. Support for quality of service parameters and fast switching are the main attractions of ATM. Hence, ATM gained popularity in both LANs and WANs. In the future, ATM is likely to be used extensively in backbone networks.

References

R. Horak. *Communications Systems and Networks*, Wiley Dreamtech India Pvt. Ltd., 2003. Chapter 19 covers the details of broadband services; ATM is also covered.

www.atmforum.com Web site of ATM Forum. An excellent resource for ATM technology.

www.iec.org/online/tutorials You can get online tutorials on ATM from this link.

Questions

1. What are the advantages of a small packet size?
2. Explain the two types of virtual connections in an ATM network.
3. Explain the ATM cell structure.
4. Explain the functionality of different layers in the ATM protocol architecture.
5. Explain the different classes of services in the ATM adaptation layer.

Exercises

1. Study the details of the ATM adaptation layer for supporting real-time voice communication and video conferencing.
2. Study the ATM products of commercial vendors such as Cisco and Nortel.
3. To support voice services with 64kbps PCM over ATM, which AAL protocol is used?
4. In a network, both variable rate coding and fixed rate coding need to be used for video transmission. Which AAL protocols need to be used to support both these services?
5. Calculate the overhead in the ATM cells.

Projects

1. Simulate the ATM cell. Write a program in C language that creates an ATM cell with the header for user-network interface.

2. Prepare a technical report on using ATM as the backbone for IP-based networks.

III ░ Mobile Computing and Convergence Technologies

In spite of the many developments that have taken place in the field of telecommunications, we are yet to witness any revolutionary change in the way we communicate and the way we interact. Telecommunications is still very costly, it is still very difficult (keeping in view how much training we need to use a new telecom device), we need a variety of gadgets to access different services, and we need to subscribe to a variety of networks.

In this part of the book, we discuss the new technologies and systems that have been conceived and commercialized in the last decade of the 20th century. These technologies are paving the way for the 'anywhere anytime' communication. The digital mobile communication systems have been widely deployed in the 1990's and now using these systems for Internet access is catching up through technologies such as WAP, GPRS, 3G etc. The Internet is now emerging as the platform for voice, fax and video transmission as well. Computers and telephones were worlds apart earlier, but not any longer—Computer Telephony Integration is now paving the way for unified messaging—a technology that makes our lives easier by making computers speak and understand our speech. Wireless Personal Area Networks will soon become popular wherein devices in office and home can communicate with one another without the need for wires— thanks to the technologies such as Bluetooth and IrDA. We will study these exciting technologies in this part of the book.

This part of the book contains 10 chapters covering the details of mobile computing and convergence technologies. The last chapter gives a glimpse of the services and applications that will be available to us in the next decade, which will make the utopia of "Global Village" a reality.

30 Radio Paging

In This Chapter

- A Radio Paging System
- Types of Pagers
- Paging Protocols
- Recent Advances in Paging Systems

For people on the move, to be in constant touch with their family members, business associates, and customers is of paramount importance. Radio paging technology provides a very low-cost mechanism for sending information to a person on the move. The information can be numeric digits (for example, a telephone number), alphanumeric text messages, or even voice messages. By keeping the bandwidth of the network low and restricting the functionality of the user terminals, a very low-cost paging service can be provided. In this chapter, we discuss the architecture of the radio paging system and the protocols used for paging. New developments that are taking place in paging technology such as voice paging, automated paging, and two-way paging, are also discussed.

30.1 PAGING SYSTEM ARCHITECTURE AND OPERATION

In a simulcast paging system, a paging controller controls a number of base stations. The message is broadcast by all the base stations.

The architecture of a paging system is shown in Figure 30.1. The paging system consists of a base station that broadcasts the paging messages to a service area. The base station is interfaced to a paging controller. The message is composed by an operator who receives the message from a user through a telephone. The paging controller has the necessary hardware and software for handling the paging protocols. The message is broadcast through the base station using radio as the transmission medium.

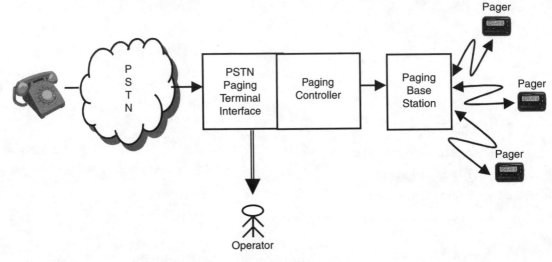

FIGURE 30.1 Paging system.

If the service area is very large, one base station may not be enough. In such a case, the paging controller will control a number of base stations as shown in Figure 30.2. Such systems are called *simulcast paging systems*.

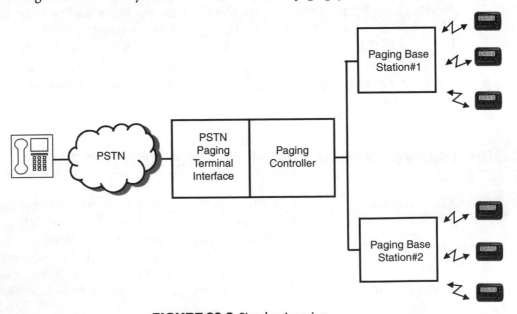

FIGURE 30.2 Simulcast paging.

In a paging system, the message is broadcast from the paging control system. All the paging terminals (or pagers) receive the message, but only the pager whose address matches with the received address displays the message.

Operation: A PSTN subscriber calls an operator at the paging control system and gives the pager number and the message. The operator broadcasts the message, with the pager number included in the format. The message is broadcast in the service area, and all the pagers in the area receive the data. Each pager finds out whether the message is addressed to it by matching its own address with the address received. If the received address matches the pager's own address, then the message will be displayed on the pager; otherwise it is discarded by the pager.

Capacity of the paging system: The following parameters decide the capacity of the paging system. When designing a paging system, these parameters need to be considered.

Number of trunks from the paging controller to the PSTN (which decides the number of persons who can call the paging controller simultaneously)

Data rate (50bps, 100bps, 300bps, etc.)

Code capacity (number of pagers)

Support for different types of pagers

NOTE

A large city can be served by a single paging system. However, the capacity of the paging system is decided by the data rates supported and the various types of pagers.

30.2 TYPES OF PAGERS

Pagers can be divided into beep pagers, numeric pagers, alphanumeric pagers, and voice pagers.

Pagers can be divided into the following categories:

Beep pager: This type of pager gives only a small beep when someone tries to contact the subscriber. This is a very low-cost pager, and the bandwidth occupied will be much less. Hence, for a given frequency, a large number of pagers can be supported.

Numeric pager: Numeric pager displays only numeric data (such as a telephone number). The bandwidth requirement will be low in this case also, but more than the bandwidth required for beep pagers.

Alphanumeric pager: An alphanumeric pager displays alphanumeric text and so is more useful because normal text messages can be sent to the pager. However, the bandwidth requirement will be higher.

Voice pager: In paging systems that support voice paging, voice messages can be transmitted. The voice will be replayed at the pager. The bandwidth requirement will be the highest for supporting this service.

30.3 FREQUENCIES OF OPERATION

Different frequency bands are allocated for beep, numeric, and alphanumeric pagers and voice pagers. The bandwidth requirement is higher for voice pagers.

A number of frequency bands are allocated for paging services. The following bands are generally used for paging.

For beep, numeric, and alphanumeric paging systems, the following frequency bands are used:

32.02–35.68 MHz

43.20–43.68 MHz

152.51–152.84 MHz

157.77–158.07 MHz

158.49–158.64 MHz

459.025–459.625 MHz

929.0125–945.9875 MHz

For voice paging systems, the following frequency bands are used:

152.01–152.21 MHz

453.025–453.125 MHz

454.025–454.65 MHz

462.75–462.925 MHz

Paging systems operate in the VHF and UHF bands so that large service areas can be covered.

30.4 PAGING PROTOCOLS

Post Office Code Standardization Advisory Group (POCSAG) formulated a paging protocol, known as the POCSAG protocol, which is used extensively in most paging systems.

Though a number of paging protocols are available, the most widely used protocol is the one standardized by POCSAG (Post Office Code Standardization Advisory group). This POCSAG protocol was standardized by the British post office and used worldwide in both commercial paging systems and amateur paging systems.

POCSAG supports three data rates: 512bps, 1200bps, and 2400bps. FSK modulation is used with ±4.5kHz frequency shift. Data is NRZ (non-return to zero) coded with higher frequency representing a 1 and lower frequency representing a 0. While transmitting the data, the most significant bit is sent first.

The basic unit of data is called a *codeword*. Each codeword is 32 bits long. The first bit of the codeword (bit 31) indicates whether the codeword is pager address or message—if it is 0, the codeword is the address; if it is 1, the codeword is the message.

Message codewords use bits 30-11 (20 bits) as message data, bits 10 to 1 are used for error correction, and bit 0 is the even parity bit. Alphanumeric text is coded in ASCII format in the message.

Address codewords use bits 30-13 (18 bits) as address, bits 12, 11 indicate the type and format of the pager, and bits 10 to 1 are used for error correction. Bit 0 is the even parity bit.

For both types of codewords, 1- or 2-bit errors can be corrected. Codewords are sent in batches of 16, each batch preceded by a special seventeenth codeword that contains the fixed frame synchronization pattern. Batches of codewords are preceded by a preamble (101010...), which must be at least 576 bits long.

When the paging controller sends the codewords continuously in the above pattern, each pager has to synchronize with the received data, using the preamble and synchronization pattern, and then decode the address codeword. If it matches its own address, the corresponding message is taken.

30.5 ADVANCES IN RADIO PAGING

One-way radio paging provides a very low-cost messaging system. To make the paging system more efficient and useful, a number of improvements have been made for the basic paging system described previously.

Voice paging: Instead of sending an alphanumeric message, a voice message can be sent to the pager. This avoids taking a message from the user by the operator and then typing the message and sending it to the base station. The voice message will be sent to the voice pager, and the paging subscriber can listen to the message. However, voice paging requires higher bandwidth. To reduce the bandwidth, low bit rate voice coding techniques are used.

For example, voice can be coded at 2.4kbps using a low bit rate coding technique such as linear prediction coding (LPC). If the voice message is for 60 seconds, the total data rate is 2.4kbps × 60 = 144kbps. If 10 such messages have to be stored in the voice pager, the total memory requirement will be 1440kbits (1.44Mbits or 0.18MBytes).

To reduce the transmission data rate as well as the memory required in the pagers, voice is coded at low bit rates in voice paging systems.

Automated paging systems: To avoid the operator at the paging controller, automated paging systems are used wherein users can log in to a computer at the paging controller and send a text message that will be broadcast over the paging network. With unified messaging, it is possible to send a message to a pager by sending an e-mail to the pager. The e-mail will be forwarded by the Internet service provider to the paging service provider.

Voice paging, two-way paging and multi-lingual paging are the major advances in paging systems. Operator-less paging systems are also on the anvil.

Two-way paging systems: One-way paging systems have the disadvantage that the subscriber cannot send back a message or reply to a received message. Two-way paging allows a paging subscriber to send a message. The pager contains a keyboard and a transmitter. Hence, the cost of the pager goes up.

Multi-lingual paging systems: In many Asian countries, the paging systems need to support local languages in addition to English. Paging systems that support multiple languages are known as multi-lingual paging systems. The pagers used in such systems need to support graphic characters. Presently, multilingual pagers are being developed using Unicode. Unicode facilitates representation of any character of any of the world language using 16-bits.

With the advent of cellular mobile communication systems that provide two-way voice communication as well as Internet access through the mobile phone, paging systems have lost their popularity in recent years. In the next chapter, we will study cellular mobile communication systems.

Summary

In this chapter, the details of radio paging systems are presented. Radio paging is a low-cost service that can be provided to people on the move. From a radio paging base station, a short message will be broadcast to all users in the service area. This message also contains the address of the pager. All the pagers in the service area will receive the message, but only the pager whose address matches the address received will display the message. Pagers can be divided into different categories: beep pagers, numeric pagers, alphanumeric pagers, and voice pagers. Presently, two-way pagers are available that allow transmission of short messages from the pager. With the advances in mobile communications that support two-way voice communication, the importance of paging systems has decreased considerably. However, even in mobile voice communications, paging is used to locate the mobile user, and therefore an understanding of the concepts of paging is very important.

References

R. Horak. *Communication Systems and Networks*. Third Edition, Wiley-Dreamtech India Pvt. Ltd., 2002.

www.bearnet.demon.co.uk You can get more information on the POCSAG standard from this site.

www.motorola.com Motorola is one of the leading manufacturers of paging equipment. You can obtain information about commercial paging systems from this site.

Questions

1. Explain the operation of a paging system.
2. What is simulcast paging? Explain its operation.

3. What are the different categories of pagers? Give the different frequency bands in which paging systems operate.

Exercises

1. Is it possible to eliminate the need for an operator at the paging terminal interface (refer to Figure 30.1)? If so, work out the framework for achieving it. (Note: Any PSTN subscriber should be able to dial in to a computer located at the paging terminal and leave a message that will be broadcast to all the pagers. This calls for automatic speech recognition software that is not reliable. An alternative is to use a computer to send the message through a modem.)

2. Calculate the storage requirement for a voice mail pager if the pager has to store 10 messages. Assume that each message is 30 seconds long and voice coding is done at 5.3kbps.

3. For a voice paging system that uses 2.4kbps voice coding, what is the memory requirement on a voice pager that has to store 20 messages, each 30 seconds long?

4. Search the Internet for POCSAG and obtain the decoder software that runs on the PC.

Projects

1. An Ethernet LAN can be considered a paging system because it also operates in broadcast mode. Simulate a paging system over the LAN. Use the POCSAG format for generating the paging message. The paging controller can be simulated on the LAN server.

2. Simulate a simulcast paging system on the LAN. The nodes on the LAN have to be separated into two groups. If your LAN has 20 nodes, assign 10 nodes to one group and 10 nodes to another group. Broadcast the paging message to the two groups separately.

31 Cellular Mobile Communication Systems

In This Chapter

- The Principles of Cellular Mobile Communications
- Standards for Mobile Communications
- The Architecture of Global System for Mobile Communication (GSM)
- Mobile Satellite Systems

To be able to communicate while on the move is no longer a luxury, it has become a necessity in this fast world. Two-way voice communication service for people on the move has become very popular, and the technology has matured considerably in the past four decades. Though the initial systems were all analog, in the 1990s digital cellular systems made mobile communications reliable, secure and, because of standardization efforts, less costly.

Though the initial mobile communication systems were mostly proprietary, standardization activities in Europe resulted in the Global System for Mobile Communications (GSM) standards, and many countries in Asia and Africa also adopted GSM. In this chapter, we study the various aspects of mobile communications technology, with special emphasis on GSM standards. The killer application on mobile systems continues to be voice, but enhancements to the GSM standards have been proposed to provide value-added services including data and fax services on GSM networks, and these enhancements also are discussed briefly.

Many projects have been initiated to provide mobile communications using satellites. These systems are known as mobile satellite systems. The details of these systems are also presented in this chapter.

31.1 PRINCIPLES OF CELLULAR MOBILE COMMUNICATIONS

The growth of mobile communications during the last decade of the twentieth century has been exponential. Throughout the world, analog mobile networks slowly are being replaced by digital mobile networks. Though voice communication continues to be the main application in mobile systems, data services, particularly accessing the Internet, are catching up fast and, technology forecasters say that wireless Internet access devices will outnumber wired Internet access devices in the near future.

In cellular mobile communication systems, there are many standards, mainly because the systems evolved from analog systems in different countries. Cellular systems in the U.S., Japan, and Europe are based on different standards. In this section, we will study the general principles of cellular mobile communication systems and study the Global System for Mobile Communications (GSM), which has been the widely accepted standard in Europe, Asia, and Africa.

> Though the killer application on mobile communication systems continues to be voice communication, data applications are now catching up very fast.

31.1.1 Single-Cell Systems

> In single-cell systems, the entire service area is covered by one base station. The drawbacks of this approach are high-power transmitters are required, power consumption of the mobile phone will be high, and expansion of the system is difficult.

The first mobile communication systems were similar to TV broadcasting systems. A base station was located at the highest point in a service area. The base station had a very powerful transmitter. The base station catered to an area of about 50 km radius. The mobile terminals, consisting of antenna, radio transmitter, receiver, and associated control circuitry, were car mounted. For communication by one mobile terminal, one channel is used. A channel consists of two frequencies: one frequency for communication from the base station to the mobile terminal (called downlink) and one frequency for communication from the mobile terminal to the base station (called uplink). Each base station is assigned a number of channels, based on the subscriber density in the region.

These traditional mobile systems can be called single-cell systems because the entire coverage area is only one cell. The disadvantages of this type of system are:

- Very powerful transmitters are required at the base station and the mobile terminals—high-power transmitters are costly.

- The capacity of the system will be very low because only a fixed number of channels are available for a service area (because of the limited radio spectrum).

- The number of subscribers who can make calls simultaneously also will be limited.

- The size of the mobile terminals will be large because of the high-power transmitters.

- Due to high-power radio devices on the mobile devices, the power consumption will be very high, calling for frequent recharging of the batteries of the mobile devices.

- Expansion of the system to cater to a higher number of subscribers will be very difficult.

    To overcome these limitations, multicell systems have been developed.

31.1.2 Multicell Systems

In multicell systems, the service area is divided into small regions called cells. Each cell will have a base station with a low-power transmitter. Adjacent cells cannot use the same frequencies.

Bell Laboratories introduced the concept of multicell systems in the early 1970s. The Nordic countries were the first to introduce commercial multicell mobile systems in 1981. In a multicell system, the service area is divided into cells as shown in Figure 31.1. Cell is the basic geographic unit in a mobile system. Each cell is represented by a hexagon. Each cell has a base station with a low-power transmitter. The size of the cell may vary, depending on the terrain: natural terrain such as mountains and lakes or manmade terrain such as buildings. Each cell is allocated some channels, and all the mobiles, when they are in that cell, use these channels for communication. The main attraction of this approach is the use of very low-power transmitters at the base stations as well as in the mobile phones.

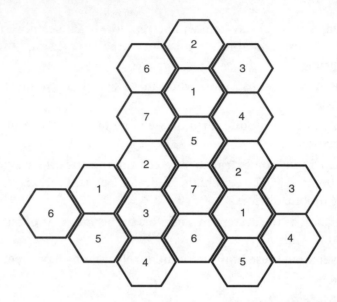

FIGURE 31.1 Multicell system with service area divided into cells (seven-cell cluster).

In a multicell system, two adjacent cells cannot use the same channel because there will be interference. When a mobile subscriber moves from one cell to another cell while the call is in progress, there are two options: either the call has to be dropped or the mobile terminal has to switch to the channel used by the new cell. Since dropping a call is not acceptable, the other option is followed. When the mobile terminal is at the edge of one cell, the signal strength goes down, and the mobile terminal monitors the signal strengths of the channels in the adjacent cells and switches to the channel for which signal strength is high. The call will not be dropped, and conversation can continue using the new channel. This process is called *handoff* or *handover*. Certainly, handover introduces complexity in cellular mobile communication, but it has many advantages:

- Because of the low power required at each base station and the mobile terminals, low-cost systems can be developed. The size of the mobile terminals also will be smaller.
- Depending on the distance between the mobile terminal and the base station, variable power levels can be used for communication, reducing the power requirements and hence the battery requirements of the mobile terminals.
- Based on the number of channels allocated for each cell, there will be a limit on the number of simultaneous calls. If the subscriber number or the traffic

increases in a cell over a period of time, a cell can be split and new base stations can be installed.

■ The cell size is not fixed, so cells can be of different sizes. In urban areas with high subscriber density, cell size can be small, and in rural areas cell size can be large.

In a multicell system, when a mobile device moves from one cell to another cell, the frequency of operation will change. This process is called handover.

The advantages of multicell systems are that only low-power transmitters are required at base stations and mobile terminals, frequencies can be reused, and expansion of the network is easy.

Frequency reuse: Every cellular service provider will be allocated a fixed number of channels for use in a service area. The service provider has to make best use of the channels to provide the maximum number of simultaneous calls. Though adjacent cells cannot use the same channels, the same channels can be reused in other cells, provided there is a minimum separation distance between the cells using the same channels. The concept of clusters is of importance here. A cluster is a group of cells, and no channels are reused within a cluster.

In frequency reuse, each cell is assigned a group of radio channels. The same channels can be reused in another cluster of cells. In Figure 31.1, a 7-cell cluster is shown. Cells denoted by 1 in all three clusters can use the same set of channels.

Cell splitting: For economic reasons, at the beginning, the cellular service provider will not design the cellular system with cells of small size. The service provider may have large cells to start with and, as the subscriber load increases, the cells will be split and more base stations will be installed, and the frequency reuse pattern is reworked.

In multicell systems, each cell is assigned a group of radio channels. The same radio channels can be reused in another cell, provided a minimum distance is maintained between the two cells using the same radio channels.

31.2 CELLULAR SYSTEM DESIGN ISSUES

While designing a cellular communication system, there are many issues to be addressed because designing these systems is a complicated task. Some important issues are discussed in the following:

Radio engineering: Based on the terrain of the service area (keeping in view hills, lakes, high-rise buildings, and so on) and the likely subscriber density in different areas, the service area has to be divided into different cells. The hexagonal cell is only a theoretical representation, in practice, there may be overlaps of the cells, and some small regions may not be covered by the radio. A radio survey is to be carried out to find out the best location for base station installations and to ensure maximum possible coverage.

Frequency allocation: For each base station located in a cell, a number of channels (pairs of frequencies) have to be assigned. The number of channels is based on the subscriber capacity in that locality and the maximum number of simultaneous calls that should be supported. The allocation of the frequencies for different cells should keep in mind the frequency reuse distance. A mobile operator has to obtain the frequency allocation from a national authority.

PSTN connectivity: The mobile network is connected to the PSTN to enable normal telephone subscribers to be contacted by mobile subscribers and vice versa. The trunk capacity between the mobile switching system and the PSTN switch has to be decided based on the traffic considerations.

Cell splitting: As the subscriber density increases in a cell, that cell has to be split, an additional base station has to be installed, and new frequencies have to be allocated. This has to be done keeping in view the frequency reuse patterns.

Shadow regions: Regions that do not receive the radio signals are called *shadow regions*. Because of the terrain, some areas may not have radio coverage, as a result of which mobile calls will be dropped when the subscriber is in the shadow region. The cellular operator has to ensure that there are no shadow regions or at least that their area is minimized. Generally, in hilly areas there will be many shadow regions. The basements of multistoried apartments are also shadow regions. To provide coverage for such areas, repeaters are used.

GSM phones do not receive strong signals in indoor environments and in basements. In such places, GSM repeaters are installed that will enhance the signal strength.

In designing multicell systems, the important considerations are a radio survey of the service area, frequency allocation for different cells, ensuring that shadow regions are minimized, and a traffic analysis to identify the number of cells.

Traffic analysis: Once the mobile communication system is operational, the operator has to monitor the traffic continuously to check whether there is any network congestion. If there is congestion, users will get network busy signals on their mobile phones. In case of network congestion, capacities have to be enhanced.

31.3 MOBILE COMMUNICATIONS STANDARDS

The mobile communications systems deployed in various countries follow different standards, so universal mobility is still a problem. Nowdays, dual-band mobile handsets are available so that a user can use different mobile systems, particularly when moving from one country to another. The various systems deployed for mobile communications are:

- Nordic Mobile Telephony (NMT 450) system, operating in 450MHz band since 1981.
- Advanced Mobile Phone System (AMPS), operating in 800/900MHz band since 1983.
- Total Access Communication System (TACS), operational since 1985.
- Nordic Mobile Telephony (NMT 900), operating in 900MHz band since 1986.
- American Digital Cellular (ADC) system, operational since 1991.
- Global System for Mobile Communications (GSM), operational since 1991.

All these systems are operational in many parts of the world, though the analog systems—NMT, AMPS, and TACS—are now outdated. The drawbacks of these analog cellular systems are low calling capacity (about 55 calls/cell), limited spectrum, poor data communication support, and lack of privacy. Subsequently, a number of digital cellular systems were deployed, out of which GSM became the most widely deployed system in Europe, Africa, and Asia. In the whole world, at the end of the year 2002, GSM subscribers numbered about 787 million.

Over the last 30 years, a number of standards have been developed for cellular mobile communications. Presently predominant standards are GSM and ADC.

31.4 GSM ARCHITECTURE

The Global System for Mobile Communications (GSM) standard was developed by the European Standards Telecommunications Institute (ETSI). GSM systems are now installed in Asia, Africa, and the Middle East, in addition to Europe.

To facilitate roaming from one country to another within Europe using the same mobile terminal, in 1983 the European Telecommunications Standards Institute (ETSI) formed Groupe Speciale Mobile (GSM) to develop standards for mobile communication system. Because the standard has been adapted widely by many countries in Asia, Africa, and the Middle East in addition to Europe, GSM is now known as Global System for Mobile Communications.

The Memorandum of Understanding for GSM was signed by 17 European operators and manufacturers in 1987. The first trial version of GSM was developed in 1991, and the commercial systems were launched in 1992. During the initial days of GSM, many people were skeptical about its commercial viability because of the complex protocol architecture. GSM used to be called "God, Send Mobiles" because of the complexity involved in developing handsets. Thanks to the advances in microelectronics, GSM handsets now go into pockets.

31.4.1 Salient Features of GSM

The salient features of GSM are:

- It is based on digital technology, so security can be built into the system easily and has all the advantages of the digital communication systems.
- Since the interfaces are standardized, network elements manufactured by different equipment vendors can work with one another, paving the way for competition, and hence the network operator and the subscriber will be benefited.
- A higher calling capacity per cell (about 125 calls per cell) as compared to analog systems.
- Support for international roaming.
- In addition to voice services, data services are also supported.

The broad specifications of the GSM system are:

- **Frequency band:** 900MHz band (890–915 MHz for uplink and 935–960 MHz for downlink). As the 900MHz band become congested, 1800MHz band has been allocated with 1710–1785 MHz for uplink and 1805–1880 MHz for

GSM is a digital cellular system operating in the 900MHz band. It uses TDMA access technology. A 13kbps voice codec is used to transmit digitized voice.

downlink. The systems operating in 1800MHz band are referred to as DCS 1800.

- Duplex distance (distance between uplink and downlink frequencies): 45MHz.
- Channel spacing (between adjacent carrier frequencies): 200kHz.

- Modulation: Gaussian minimum shift keying (GMSK). GMSK is a special form of FSK, ones and zeros are represented by shifting the RF carrier plus or minus 67.708kHz. FSK modulation in which the bit rate is exactly four times the frequency shift is called minimum shift keying (MSK). As the modulation spectrum is reduced by applying a Gaussian premodulation filter to avoid spreading energy into adjacent channels, the modulation is called Gaussian MSK (GMSK).

- Transmit data rate (over the air bit rate): 270.833kbps, which is exactly four times the RF frequency shift.

- Access method: TDMA with eight time slots.

- Speech codec: 13kbps using residually excited linear prediction (RELP), a derivative of the LPC coding technique.

- Signaling: Signaling System No. 7 (SS7) is used to carry out signaling. Hence, the radio channels are used efficiently for speech transmission.

31.4.2 GSM Services

In addition to the voice service, GSM supports the following services: short messaging service (SMS), with which messages with a maximum of 160 characters can be sent, Group 3 fax, voice mail, and fax mail.

GSM services are divided into telephony services (referred to as teleservices) and data services (referred to as bearer services). In addition to the normal telephony services, the following services are also supported:

- Group 3 facsimile transmission through a special interface.

- Short messaging service (SMS) to transmit a maximum of 160 alphanumeric characters. If the handset is turned off or is out of the coverage area, the message will be stored in a message center and sent to the handset when it is turned on or when it is within the coverage area.

- Cell broadcast to transmit maximum of 93 characters to all the handsets in a particular cell. This service can be used to transmit information regarding traffic congestion, accident information, and so on.

- Voice mail.
- Fax mail.

 The GSM system also supports the following supplementary services:

- Call forwarding, to forward a call to another mobile handset or a land line.
- Barring outgoing calls.
- Barring incoming calls. All incoming calls can be barred or only incoming calls when roaming outside a PLMN can be barred.
- Advice of charge, which gives an estimate of the call charges based on the call duration.
- Call hold, to interrupt a call and then reestablish it again.
- Call waiting, to announce an incoming call when a conversation is in progress.
- Multiparty service to provide conferencing facility.

The supplementary services supported by GSM are call forwarding, call barring, call holding, call waiting, calling line identification, advice of charge, and closed user groups.

- Calling Line Identification Presentation (CLIP) to display the telephone number of the calling party.
- Closed user groups (CUGs), which emulate the function of a PBX. A predefined group of mobile terminals will form the equivalent of a PBX.

In Closed user group service, a number of mobile subscribers can communicate among themselves at reduced charges.

NOTE

31.4.3 GSM System Architecture

The GSM architecture is shown in Figure 31.2.

 A mobile communications service provider operates in a given geographical region. The mobile network of the entire region is known as a public land mobile network (PLMN). The PLMN will be in the administrative control of one operator. The PLMN consists of mobile stations (MS), base station subsystems (BSS), and network switching subsystem (NSS).

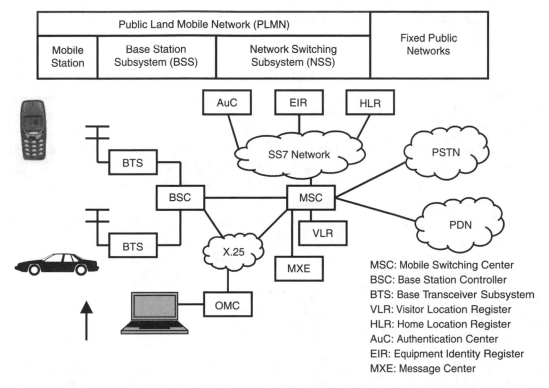

FIGURE 31.2 GSM architecture.

The GSM system consists of mobile stations, base transceiver subsystem (BTS), base station controller (BSC), and mobile switching center (MSC). Public networks such as PSTN and ISDN are connected to the MSC.

The MS can be handheld or car mounted. The BSS consists of base station controller (BSC) and base transceiver subsystem (BTS). The NSS consists of mobile switching center (MSC), home location register (HLR), equipment identity register (EIR), authentication center (AuC) and visitor location register (VLR). In addition to these elements, there will be an operation and management center (OMC) that provides the man-machine interface to carry out administrative functionality such as subscriber management, network management, billing, and so on.

The PLMN is connected to the Public Switched Telephone Network (PSTN) or Public Data Network (PDN) or Integrated Services Digital Network (ISDN) at the MSC. The functions of each element of the GSM system are described in the following:

The home location register (HLR), visitor location register (VLR), equipment identity register (EIR), and authentication center (AuC) are databases that will be accessed by the MSC. These databases are implemented using an RDBMS engine such as Oracle or Informix.

BSC contains the transcoders that convert the PCM-coded speech into 13kbps data for sending it to the BTS. In the reverse direction, the 13kbps-coded speech is converted into 64kbps PCM data to send it to the MSC.

Mobile station (MS): Also known as mobile handset or hand phone, this is the subscriber terminal. Nowdays, mobile terminals are coming with many features—voice dialing, whereby one can use voice to dial out a number, powerful batteries to provide at least 6 hours of talk time and 4 to 5 days of standby time, and such. The power transmitted by the MS is in the range 0.8–20 watts.

The MS is identified by a number known as MS-ISDN (the mobile phone number). Each MS is also uniquely identified by IMSI (international mobile subscriber identity). The MS contains a SIM (subscriber identity module). SIM is a smart card inserted in the handset. It is protected by a personal identity number (PIN). PIN is checked locally and not transmitted over the radio link. SIM contains IMSI. To identify a handset hardware uniquely, IMEI (international mobile equipment identity) is used, which is a number given by the manufacturer to the handset.

BSS: The BSS acts as a relay between the NSS and the mobile stations. The BSS consists of BSC and BTSs. The service area is arranged into cells, and each cell will have a BTS. Each cell can vary from 350 meters to 35 kilometers, depending on the terrain and the subscriber density. Multiple BSSs can be controlled by one BSC.

BSC: BSC handles radio management functions. BSC arranges new radio link connections to mobile stations when handover is required. It is connected to the MSC through landlines, normally 2Mbps links using PCM for voice transmission. To reduce radio bandwidth, GSM uses low bit rate coding of speech at 13kbps between the MS and the BSC. BSC does the transcoding—conversion of the 13kbps speech to PCM and vice versa. Each BSC controls a number of BTSs, typically up to 40.

BTS: BTS is the radio interface between the MS and the BSC. Communication between the MS and the BTS is through one channel consisting of a pair of frequencies—one for uplink and one for downlink. The frequency allocation for GSM in 900MHz band is depicted in Figure 31.3. Carriers are separated by 200kHz with 16kbps as the aggregate data rate per channel. To overcome signal fading and reflections due to highrise buildings and such, frequency hopping is used. Depending on the shape of the cell, omnidirectional or sectoral antennas are used at each BTS. The maximum power transmitted by the BTS is in the range 0.25–320 watts. To conserve radio spectrum, 13kbps low bit rate voice coding is used. BTS uses TDMA for multiple access, with 8 slots per channel. The TDMA frame format is shown in Figure 31.4. This is a very simplified format— TDMA slots are also used to carry out all the signaling between the BSC and the MS.

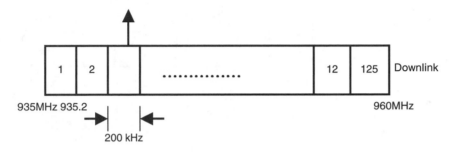

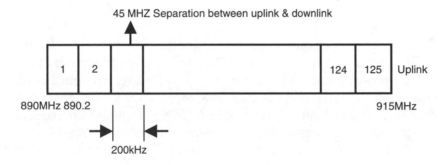

FIGURE 31.3 Frequency allocation for GSM.

Some slots in the TDMA frame are exclusively for transmitting the signaling information. Short messages are also sent in the signaling slots.

TDMA format: Each data bit is of 3.692 microseconds duration. Each time slot has a time period equal to 156.25 data bits. There are eight time slots per frame, and hence frame period is 4.615 milliseconds. Twenty-six or 51 frames are grouped together to make a multiframe. A superframe consists of 51 or 26 multiframes. This complex frame and multiframe structure is used to transmit control information, to carry out synchronization, and of course, to carry speech data.

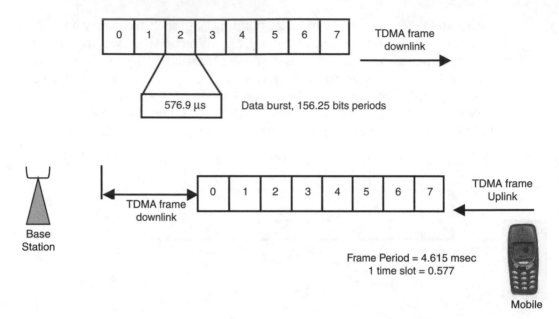

FIGURE 31.4 TDMA frame format in GSM.

TCH (traffic channel) carries the bidirectional speech data between the mobile station and the base station. Each base station produces a BCH (broadcast channel), which acts as a beacon signal to find service and decode network information. BCH occupies time slot zero. Each cell is given a number of frequency pairs (channels) denoted by ARFCN (absolute radio frequency channel numbers). If a cell has one ARFCN, there will be one BCH and seven time slots for TCH. If there are two ARFCN's in one cell, there will be one BCH and 15 time slots for TCH.

Home location register (HLR): HLR is a centralized database to manage the subscriber data. It is a standalone system connected to GSM network subsystems with Signaling System No. 7. This database contains:

- Subscriber information
- Subscriber rights and privileges (what types of calls are permitted)
- Location information
- Activity status

HLR permanently knows the location of the subscriber. When an MS receives a call, the HLR is consulted, and the database translates the MS-ISDN number to

IMSI number. HLR reroutes incoming calls to the MSC target or ISDN number when call forwarding is requested.

Authentication center (AuC): AuC contains security functions such as IMSI, the encryption key, and the algorithm to be used for encryption. AuC provides the data to verify the identity of each user and to provide confidentiality of the conversation/data.

HLR/AuC are administered by man-machine interface (MMI) commands from the OMC.

Visitor location register (VLR): VLR contains information about all the mobile subscribers currently located in the MSC service area. VLR is generally integrated into MSC. When a mobile station roams into a new MSC service area, the VLR connected to that MSC gets the data about the mobile station from the HLR and stores it. Temporary mobile subscriber identity (TMSI) is assigned by VLR and is used for call establishment. VLR is responsible for the current location of the user.

Equipment identity register (EIR): EIR contains information about the mobile equipment. Each MS is uniquely identified by IMEI (international mobile equipment identity). When a mobile handset is lost, the subscriber informs the customer support center, and this information is stored in the EIR. When the lost mobile is used for making a call, the EIR will not permit the call. Because EIR also provides the security, EIR and AuC can be combined into one computer.

The mobile phone hardware is uniquely identified by the international mobile equipment identity (IMEI). Hence, it is possible to trace a stolen mobile phone if the thief is using it.

Mobile switching center (MSC): MSC provides complete switching functionality for the entire network, so all call control functions are built in the MSC. It also provides the interface to the rest of the world—PSTN, PDN, ISDN, and so on. MSC is also connected to the OMC through which the configuration of the network, entry and modification of the subscriber data, traffic analysis, billing, and other network management functions can be carried out.

A service area can have more than one MSC. In such a case, one of the MSCs is designated as gateway MSC. Public networks such as PSTN are connected to the gateway MSC.

The message center is a computer that is used to store and forward SMS, voice mail, fax mail, and so on. Many value-added services such as examination results announcements and astrological services can be provided through the message center.

Operation and maintenance center (OMC): The OMC is used to carry out network management activities such as fault diagnosis of various network elements, traffic analysis, billing, performance management, configuration management (such as adding a new BSC and cell splitting), as well as managing subscriber information.

The communication between the MSC and the databases (HLR, EIR, AuC) is through an SS7 network, because only signaling information is exchanged between these entities. The communication between the OMC and the MSC/BSC is through a packet switching network based on X.25 standards.

In addition, the GSM system can contain the following network elements:

Message center: Message center is a node that provides voice, data, and fax messaging. It handles the SMS, cell broadcast, voice mail, fax mail, and e-mail messaging. Separate servers are required to handle these messaging systems, which are connected to the MSC.

Gateway MSC: When a PLMN contains more than one MSC, one of the MSCs is designated as a gateway MSC to interconnect with other networks such as PSTN and ISDN. If the PLMN contains only one MSC, that MSC itself can act as a gateway MSC.

31.4.4 GSM Network Areas

The area covered by one network operator is called the PLMN. The area covered by one MSC is called the MSC/VLR area. The area paged to locate a subscriber is called location area. The area covered by one BTS is called a cell.

In a GSM network, the following areas are defined: cell, location area, service area, and PLMN.

Cell: Cell is the basic service area; one BTS covers one cell. Each cell is given a cell global identity (CGI), a number that uniquely identifies the cell.

Location area: A group of cells form a location area. This is the area that is paged when a subscriber gets an incoming call. Each location area is assigned a location area identity (LAI). Each location area is served by one or more BSCs.

MSC/VLR service area: The area covered by one MSC is called the MSC/VLR service area.

PLMN: The area covered by one network operator is called PLMN. A PLMN can contain one or more MSCs.

Figure 31.5 shows a GSM PLMN serving two cities. Each city will have a number of BTSs and one BSC. The two BSCs are connected to the MSC. MSC also acts as the gateway to the PSTN. The number of BTSs in a city depends on the subscriber density and the terrain. Presently, each Indian city is covered by about 100 to 200 BTSs. SingTel, the cellular operator in Singapore, installed 1800 BTSs—this is because of the very high subscriber density in Singapore.

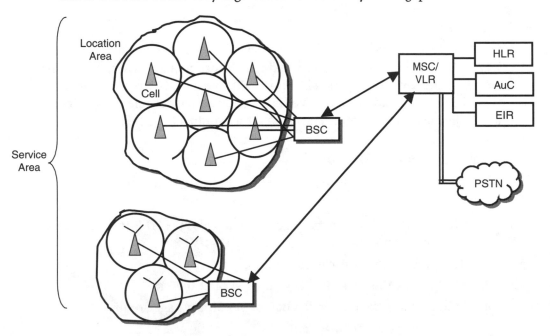

FIGURE 31.5 GSM PLMN serving two cities.

31.4.5 GSM Operation

The operation of the GSM system can be understood by studying the sequence of events that takes place when a call is initiated.

Call from Mobile Station

When a mobile phone initiates a call, the MSC verifies the authenticity of the mobile device and initiates call setup with the PSTN. A frequency and a voice slot are allocated to the mobile device.

When a mobile subscriber makes a call to a PSTN telephone subscriber, the following sequence of events takes place:

1. The MSC/VLR receives the message of a call request.

2. The MSC/VLR checks if the mobile station is authorized to access the network. If so, the mobile station is activated.

3. MSC/VLR analyzes the number and initiates a call setup with the PSTN.

4. MSC/VLR asks the corresponding BSC to allocate a traffic channel (a radio channel and a time slot).

5. BSC allocates the traffic channel and passes the information to the mobile station.

6. Called party answers the call, and the conversation takes place.

7. The mobile station keeps on taking measurements of the radio channels in the present cell and neighboring cells and passes the information to the BSC. BSC decides if handover is required and, if so, a new traffic channel is allocated to the mobile station and the handover is performed.

Call to a Mobile Station

When a PSTN subscriber calls a mobile station, the sequence of events is as follows:

1. The gateway MSC receives the call and queries the HLR for the information needed to route the call to the serving MSC/VLR.

2. The GMSC routes the call to the MSC/VLR.

3. MSC checks the VLR for the location area of the MS.

4. MSC contacts the MS via the BSC by sending a pager request.

5. MS responds to the page request.

6. BSC allocates a traffic channel and sends a message to the MS to tune to the channel. The MS generates a ringing signal and, when the subscriber answers, the speech connection is established.

7. Handover, if required, takes place, as discussed in the earlier case.

Note that the MS codes the speech at 13kbps for transmission over the radio channel in the given time slot. The BSC transcodes the speech to 64kbps and sends it over a land link or radio link to the MSC. MSC then forwards the speech data to the PSTN. In the reverse direction, the speech is received at 64kbps rate at the BSC, and the BSC does the transcoding to 13kbps for radio transmission.

31.5 ENHANCEMENTS OF GSM STANDARDS

In its original form for data applications, GSM supports 9.6kbps data rate. The data is transmitted in the TDMA time slot. Over the last few years, many enhancements were made to the GSM standards to increase capacity and provide value added services, which are briefly discussed here.

31.5.1 GSM Phase 2

Because the spectrum allocated in the 900MHz band was not sufficient, the 1800MHz band has been allocated for GSM operation. Most of the present mobile phones are dual-band phones, so they can operate in both bands.

Phase 2 GSM services include advice of charge, calling line identification, and closed user groups. In addition, 1800MHz band has been allocated to GSM systems in areas where the 900MHz band is saturated.

GSM Phase 2+ addresses nearly 90 enhancements to the existing standards to introduce new services, particularly data services. The major enhancements are in the areas of speech quality improvements and enhanced data services.

In a service area, if there are four operators, two operators are allotted 900MHz band and the other two operators are allotted the 1800MHz band.

NOTE

31.5.2 Speech Quality Improvements

Presently in the GSM system, speech is encoded at 13kbps. Half-rate codecs will encode the speech at 6.5kbps. As a result, the capacity of the GSM system can be doubled.

GSM Phase I supports 13kbps speech codec—the speech quality being far from good. Speech codec for GSM was finalized in 1987, and significant advances have been made in speech coding technology in recent years. So, GSM Phase 2+ now will introduce a new codec (developed by Nokia and University of Sherbrooke, Canada) whose quality will be comparable to PSTN speech quality.

Half-rate codec, a part of GSM Phase 1 standard but not introduced so far, can be introduced and will have the same voice quality as the existing GSM's voice quality.

Cellular operators will have two choices—improve the speech quality by introducing the new codecs in the full-rate speech channels or, using half-rate codecs, increase the capacity with the same quality of speech as that of the present full-rate codecs.

31.5.3 Enhanced Data Services

GSM Phase 2 enhancements include support for high data rate services. Using these enhancements, users can access the Internet through mobile phones.

The 64kbps access speed is very common for Internet access using the PSTN. GSM users expect the same data rate support over their mobile data terminals. Particularly, such high data rate is a must to access Internet services such as multimedia Web services and large file transfers. In GSM Phase 2+, high speed circuit switched data (HSCSD) service and general packet radio service (GPRS) have been proposed, which provide high-speed data services over GSM. These are add-on features to the GSM network, with minor software-only upgrades. HSCSD is based on multislotting—dynamically combine two to eight time slots to give maximum data rate up to 64kbps for a single user. This is a connection-oriented service. If all the eight time slots of one TDMA frame are allotted to one user, it gives 76.8kbps (8×9.6 kbps) of theoretical data rate. In the case of GPRS, packet switching is done, and hence it is 76.8kbps of instantaneous data rate. We will study data services in detail in the chapter on wireless Internet.

Using data compression techniques (based on V.42 standard), the effective data rate can be increased up to 300kbps for textual data. This data rate perhaps is sufficient for accessing the existing widely used data services over the Internet such as the Web, large file transfers, audio, and so on. But then, the data users' demands are rising very fast, and already there is enough demand for data rates such as 384kbps and even 2Mbps.

To summarize, mobile communications services have become very popular in the last decade, and though at present voice is the main application, in the future wireless data services using the mobile communication systems will catch up. Anywhere, anytime communication is now a reality—thanks to cellular mobile communication systems.

31.6 MOBILE SATELLITE SYSTEMS

To provide voice communication, the PSTN and the PLMN have been widely deployed all over the world. However, their reach is mainly limited to cities and towns in developed countries. In developing countries, particularly in Asia and Africa, remote and rural areas remain isolated with no communication facilities.

A number of mobile communication systems have been developed in which low Earth orbiting (LEO) and medium Earth orbiting (MEO) satellites are used. However, a few such projects did not become a commercial success.

To lay cable to villages, where the subscriber density will be low and also the traffic is low will, not be a good business proposition because the return on investment will be very low. Satellite communications, its distance insensitivity, provide a great scope to provide communications anywhere on the earth. A number of projects have been initiated to provide voice communications using a satellite as the medium. Geostationary satellites—low Earth orbiting (LEO) satellites and medium Earth orbiting (MEO) satellites—can be used for providing satellite-based telephony services, also known as mobile satellite services (MSS). An overview of the various projects taken up for MSS is given in this section.

31.6.1 Iridium

The Iridium system was conceived and developed by Motorola. It used 66 satellites to provide global mobile communication. Though it was a great technological marvel, it was a big commercial failure.

A very ambitious program launched by Motorola, the leading manufacturer of mobile equipment, is Iridium. Iridium's plan was to establish a global mobile personal communication system (GMPCS) through a constellation of 66 low Earth orbiting (LEO) satellites to provide telephony services anywhere on the Earth. The operations of Iridium began in January 1999, but the project ran into troubles—technical snags and lack of enough customers because of high prices. Initially the Iridium's phone was priced at about US$3000, with connection charges of US$7 per minute. Subsequently, prices were slashed, and the subscriber base increased, but Iridium could never get over the financial debts and by mid-2000, Iridium became bankrupt. "Geography is now History" was the slogan of Iridium, but Iridium became history. All financial institutions now look at MSS with skepticism. However, MSS still provides hope, but we need to learn many lessons from Iridium.

The salient features of Iridium are:

- Altitude: 780 km
- Visibility of the satellite: 11.1 minutes
- Access method: FDMA/TDMA
- Frequency of operation: 1.6138–1.6265 GHz
- Intersatellite link frequency of operation: 23.18–23.38 GHz

- Number of channels: 4,000
- Services supported: voice, data, fax, and location services

31.6.2 Globalstar

Globalstar is a mobile satellite system that uses CDMA as the access technique. It provides global connectivity. However, it is yet to become a commercial success.

Globalstar was officially launched in October 1999. Globalstar consists of a total of 48 satellites orbiting at an altitude of 1389 km. With an orbit inclination of 52° and eight orbital planes, there are six satellites per orbital plane. The frequency bands of operation are 1610–1626.5 MHz and 2483.5–2500 MHz. The multiple access technique used is CDMA. The handsets available on the market are dual-band handsets that can be used with both the land mobile cellular system as well as the satellite-based system.

Though the present subscriber base is very low and the first year operations resulted in losses for Globalstar, there is still hope for Globalstar to succeed. The salient features of Globalstar are:

- Number of satellites: 48
- Altitude: 1,389 km
- Access method: CDMA
- Number of channels: 2,700
- Frequency of operation: 1.61–1.6265 GHz

    2.4835–2.5 GHz

31.6.3 ICO/Teledesic

ICO/Teledesic is called "Internet in the sky" because it aims at providing Internet connectivity with speeds ranging from 19.2kbps to 2.048Mbps.

ICO Global Communications Holding Ltd. was established in 1995 to develop satellite-based mobile telephony services. ICO stands for intermediate circular orbit to reflect the location of the satellites—the orbit is also known as medium Earth orbit (MEO).

ICO went almost bankrupt in 1999 but reemerged in June 2000, with support from financial institutions and Craig McCaw, a pioneer in mobile communications. The name has been changed to New ICO to reflect a new business plan and new architecture so that not just voice but wireless Internet services can be provided through the system.

There is now a proposal to merge New ICO and Teledesic Corporation to form ICO-Teledesic Global. If the merger is approved, New ICO and Teledesic will be subsidiaries of ICO-Teledesic Global. New ICO plans to offer its services, a satellite equivalent of 3G services, worldwide in 2004.

The salient features of Teledesic are:

- Number of satellites: 288
- Altitude: 700 km
- Access method: FDMA/TDMA
- Frequency of operation: 28.6–29.1 GHz
                            18.8–19.3 GHz
- Bit rates: 64Mbps downlink; 19.2kbps–2.048Mbps uplink

The mobile satellite systems' objective is to provide telephone connectivity to remote and rural areas. Most of the projects did not become a commercial success because in many rural areas, two square meals are more important than communication.

Technically, mobile satellite services are the best bet for providing rural telecommunications. However, economics plays the most important role—in a village where the average daily income of a family is less than $2, if the communication cost is $2 per minute, it is not an attractive proposition. The need to fill the stomach takes priority, not communicating!

Summary

Cellular mobile communications systems provide voice communication services while on the move. In this chapter, the details of the digital cellular system based on GSM standards are presented. The analog mobile communication systems were replaced by the digital systems in the 1990s. Among the digital systems, the system based on GSM standards attained popularity not only in Europe, but in Asia and Africa as well. The various network elements of the GSM system are the mobile station carried by the subscribers, the base transceiver subsystem (BTS) that consists of the radio modules, the base station controller (BSC), which controls multiple base transceivers, and the mobile switching center (MSC), which carries out the switching operations. The MSC contains a number of databases. The HLR is a database that stores the subscriber information. VLR is a database that contains the current users in the service area. The EIR database contains the mobile equipment identity. AuC is the database that contains information to authenticate the user. The MSC is connected to the PSTN. In addition to voice communication, GSM supports short messaging service, with which short messages limited to 160 bytes can be sent using a store-and-forward mechanism. GSM systems are now being upgraded to provide high data rate services to access the Internet.

The salient features of mobile satellite communication systems are also presented in this chapter. Iridium, ICO Globalstar, and Teledesic are the major projects for providing global mobile communications. However, their success is yet to be ascertained.

References

Dreamtech Software Team. *WAP, Bluetooth and 3G Programming*. Wiley Dreamtech India Pvt. Ltd. This book contains an overview of the GSM system and also third generation wireless networks.

www.3gpp.org The official site of the 3G Partnership Programme. You can obtain the GSM standards documents from this site.

www.cdg.org Web site of CDMA development group. The competition to GSM is CDMA-based systems, which are extensively deployed in the U.S. You can get the details from this site.

www.qualcomm.com Qualcomm Corporation is a pioneer in CDMA technology. Qualcomm is the leading supplier of CDMA-based cellular mobile communication systems. You can get the product details from this site.

Questions

1. Explain the advantages of multicell systems for mobile communications as compared to single-cell systems.

2. Describe the architecture of a GSM-based mobile communication system. Explain the functionality of each subsystem.

3. What are the additional enhancements for the original GSM standards? What additional services can be provided through these enhancements?

4. Explain the operation of mobile satellite systems. List the salient features of Iridium, Globalstar, and Teledesic.

Exercises

1. In the GSM system, speech is coded at 13Kbps data rate. Obtain freely available source code for coding the speech at this data rate and experiment with it.
2. Study the details of the TDMA time slots of the GSM system.
3. Write a technical paper on the mobile communication systems based on CDMA used in North America and China.
4. Explain how a stolen mobile phone can be traced.

Projects

1. Develop a database of about 100 mobile subscribers using any RDBMS package (such as Oracle or MS SQL). The database should contain the name and address of the subscriber, the mobile number, and the details of the calls made by them. The details of the calls include the called number, time and date of initiating the call, and the duration. Using this database, develop a billing system. The billing system should have provision for normal rates, half the normal rates during night hours, and one-fourth the normal rates during Sundays and public holidays. Note that the database generated by you manually is automatically generated by the MSC. It is called the call details record (CDR). The CDR is processed offline to generate the bills. Your job is to develop the billing system. Make appropriate assumptions while developing the software.
2. Carry out the system design for providing a mobile service for your city. Assume that 10% of the population will use mobile phones. Based on the mobile subscribers density in different areas of the city, you can divide the city into different cells. Use the city map to decide the cells and location of BTS. If there are hills or lakes in the city, you need to consider the terrain-related issues while planning the number of cells.

32 Global Positioning System

In the past sailors used to look at the stars in the sky to navigate their ships. Subsequently, a number of electromagnetic and electronic navigational aids were developed to determine the location of moving as well as stationary objects. Perhaps the greatest innovation in navigational aids is the Global Positioning System (GPS), in which electronic gadgets look at artificial satellites to determine the location and speed of an object. The most attractive feature of the GPS is that it is a reliable navigational aid anywhere on the earth, operating in all weather conditions 24 hours a day, and can be used by maritime and land users. In this chapter, we study the GPS architecture and the applications of GPS technology. The features of commercial GPS receivers are discussed briefly along with methodology for developing applications using the GPS technology.

32.1 GPS ARCHITECTURE

The GPS technology was developed by the U.S. Department of Defense in 1983 at a cost of US$12 billion. The GPS consists of three segments, as shown in Figure. 32.1.

Space segment

Control segment

User segment

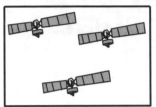

Space Segment

Control Segment

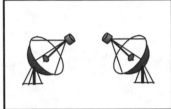

Airplane

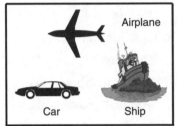

User Segment

Car   Ship

FIGURE 32.1 Global Positioning System.

The space segment of GPS consists of 24 NAVSTAR satellites, each with an orbital period of 12 hours. These satellites orbit at 20,200 km above the surface of the Earth.

Space segment: The space segment of GPS consists of 24 NAVSTAR satellites (including three spare satellites) operational since August 1995. These satellites orbit at 20,200 km above the surface of the Earth in six circular orbital planes with a 12-hour orbital period each. These planes are such that from any location on the earth, at least five satellites can be viewed. These satellites operate in the L1 frequency band (1.5742GHz) and continuously broadcast navigational signals. This navigational data can be received by anyone anywhere on the earth free of cost for decoding and finding navigational parameters: location (longitude, latitude and altitude), velocity and time.

Control segment: The control segment of GPS consists of a master control station (MCS) located at Colorado Springs in the U.S. and a number of smaller Earth stations located at different places around the world. The monitoring stations track GPS satellites and pass the measured data to the MCS. The MCS computes the satellite parameters (called *ephemeris*) and sends them back to the satellite. The satellites broadcast these parameters to all the GPS receivers.

The user segment of the GPS consists of receivers, which calculate the location information using the signals received from the satellites.

User segment: The user segment consists of all objects with GPS receivers. The U.S. Department of Defense has not put any restrictions on the use of the

GPS receivers; anyone, anywhere on the earth can receive the navigational data through a GPS receiver.

Another global positioning system, called Galileo, is being planned by the European countries. It will be operational in 2007.

32.2 GPS RECEIVER

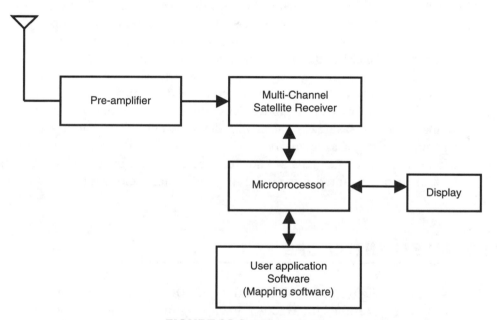

FIGURE 32.2 GPS receiver.

As shown in Figure 32.2, the GPS receiver is a microprocessor-based (or DSP-based) system along with a multichannel satellite receiver, which receives the signals from NAVSTAR satellites visible at that location. The data sent by each satellite, known as *almanac*, consists of coded time and location of the satellite. The GPS receiver calculates its own location parameters from a minimum of four such signals. These parameters are calculated once per second and sent to the display unit. Optionally, the location data can be integrated with an application software package such as mapping software.

The GPS receiver has to carry out very complex calculations to obtain the position parameters using the signals received from the satellites. A digital signal processor (DSP) is generally used in the GPS receiver.

An accuracy of about 20 meters can be obtained using the GPS receiver. The accuracy can be improved using differential GPS.

GPS accuracy: In GPS, errors in measurement can occur due to propagation delays or clock variations. In coarse acquisition, the GPS receiver gives an accuracy variance of less than 100 meters 95% of the time and 300 meters 99.99% of the time. It may happen that your car could be shown to be moving on a lake when you are traveling on a road adjacent to a lake! The U.S. government used to degrade the GPS signals available to the public, intentionally and this was known as selective availability (SA). On 1st May 2000, the U.S. government discontinued SA, and hence the GPS signals are no longer intentionally degraded. This resulted in an improved accuracy of about 20 meters. This measure was taken after demonstrating the capability to deny GPS signals selectively on a regional basis. This was intended to show that the government can ensure that GPS signals are not received in a specific area. If more accurate location data is required, differential GPS (DGPS) technology is used.

32.3 DIFFERENTIAL GPS

As shown in Figure 32.3, in differential GPS (DGPS), there is a reference station (a surveyed location with exactly known longitude, latitude, and altitude). This reference station has a GPS receiver as well as an FM transmitter. This station receives the GPS signals and computes its location parameters. These parameters are compared with its own known location parameters to calculate the error of the GPS receiver. The error correction to be applied to the results of the GPS signals is thus calculated, and this information is broadcast to all the GPS receivers around the reference station using the FM transmitter.

The differential GPS receiver consists of a GPS receiver that gets the signals from the GPS satellites and an FM receiver that gets the signals from the FM transmitter located at the DGPS base station. Based on these two signals, the DGPS receiver calculates its position parameters accurately.

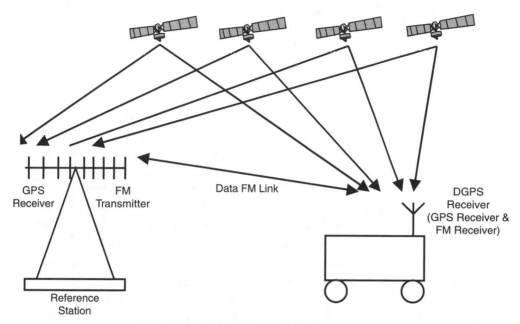

FIGURE 32.3 Differential GPS system.

The DGPS receivers of all users are fitted with FM receivers in addition to the GPS receivers. The DGPS receiver obtains the location parameters from the GPS data and then applies the correction based on the information received on the FM channel. This results in obtaining more accurate location parameters.

Using differential GPS, one meter accuracy can be obtained. Such high accuracy is required for applications such as surveying and missile guidance.

Using differential GPS, location accuracy up to one meter can be achieved. For applications that require high accuracy, such as surveying or defense applications, differential GPS needs to be used as instead of normal GPS.

32.4 APPLICATIONS OF GPS

The GPS receiver comes in handy whenever the location of any object is to be determined accurately. It is the most important navigational aid for nearly 17 million ships and boats. The GPS receiver also is becoming a common gadget in many cars.

If the GPS receiver displays the raw data of longitude and latitude, it is not of much use. The GPS receiver generally is integrated with a digitized map of the location in which the vehicle is moving. The digitized map contains the minute details of the roads, the important landmarks of the locality, and so on. The positional information obtained from the GPS receiver is correlated with the digitized map so that the user can get the information about where he is.

GPS receivers integrated with laptops and palmtops are paving the way for many fascinating applications. For example, the laptop can contain the digitized map of the area in which a car is moving. As the car moves, the route of the travel is displayed dynamically on the monitor. The user can feed in information regarding a destination, and the system automatically navigates the driver by displaying the route to be taken (one product has interfaced a speech synthesizer that will give verbal instructions such as "take a right turn at the next crossing").

Because of its small size and low cost, the GPS receiver is now becoming an integral part of cellular phones, pagers, and even digital wristwatches. GPS technology also finds applications in surveying, remote sensing, accident relief operations by traffic police, and ambulance service for hospitals. Needless to say, applications in the defense sector are many, such as determining tanker movements and target identification.

The most attractive application of GPS is navigation systems used in cars, aircrafts, ships, and missiles.

GPS technology is very useful only if accurate digitized maps are available. To digitize the maps of different cities itself is a gigantic task.

NOTE

32.5 COMMERCIAL GPS RECEIVERS

Commercial GPS receiver hardware is available with an RS232 interface or a USB interface. GPS receiver chips are also being integrated into consumer items such as watches and mobile phones.

Initially, when the U.S. Department of Defense introduced the GPS, the cost of the receiver was nearly US$30,000. Over the past decade, the cost has come down drastically due to the availability of chipsets from major semiconductor vendors. Now the cost of a chipset is very low, and this has paved the way for GPS receivers to become a consumer item. This low cost is drawing many solution providers into developing interesting applications. GPS receivers are now being integrated with laptops, cellular phones, pagers, and even wristwatches to display positional information.

To develop application software using the GPS, GPS receivers come in three different forms:

- A PC add-on card that can be plugged into the motherboard.

- GPS receiver hardware with an RS232 interface. This receiver can be connected to the RS232 port of the PC, and the port can be programmed to decode the GPS data for subsequent processing.

- GPS receiver hardware with a USB interface. This receiver can be connected to the USB port of the PC, and the port can be programmed to decode the GPS data for subsequent processing.

Using the driver software and the associated application software, GPS receiver data can be used for developing software to meet user requirements.

32.6 DEVELOPING APPLICATIONS USING GPS RECEIVERS

The raw data obtained from the GPS receiver—longitude, latitude, altitude, velocity and time—are not of much use unless the positional information is integrated with other software to provide useful information to the user. For instance, the digitized map of a city can be stored in the computer and, based on the GPS data obtained, the information as to where you are can be displayed on the screen. Another application can be to retrieve from a database the hotels/restaurants/gas stations in the location. For instance, when you enter a new city in your car, the GPS receiver fitted to your mobile phone can send the location data to a local server, and the server will retrieve information about the nearby hotels from the database and send that information to your mobile phone.

To develop a navigation system using GPS technology, the data obtained from the GPS receiver has to be mapped on to a digitized map.

Developing such applications requires integration of the GPS data with digitized maps, databases, and other application software. Commercial software packages such as MapInfo, with interface to popular databases such as Oracle, are available, with which developing applications will be extremely easy. Systems capable of storing, manipulating, and displaying geographic information are called geographic information systems (GIS). GIS in conjunction with GPS is used extensively for automatic vehicle tracking and to provide location-based services.

For providing location-based services, a GPS receiver can be integrated into a mobile phone. The mobile phone can send the location data to a server, and the server will send the information related to that location to the mobile phone.

Location-based services are catching up very fast, and GPS technology will be an important component for such services.

Summary

In this chapter, the details of the Global Positioning System are presented. The GPS consists of space segment, control segment, and user segment. The space segment has 24 NAVSTAR satellites launched by the U.S. Department of Defense. The space segment is controlled by a master control station and a number of earth stations located all over the world. The user segment consists of GPS receivers located on cars, mobile phones, and even wristwatches. The GPS receiver obtains the positional data from the satellites and, based on this data, calculates its own positional parameters: longitude, latitude, and altitude. In addition, the velocity of the moving vehicle and the time can be displayed on the GPS receiver. The location can be displayed up to an accuracy of about 20 meters. In case more accurate location information is required, differential GPS (DGPS) is used. In DGPS, at a surveyed location (whose longitude and latitude are known exactly), there will be an FM transmitter in addition to the GPS receiver. At this surveyed location, from the GPS receiver data, the error in the positional parameters is calculated and the correction factor is obtained. This correction factor is transmitted over the FM link. The DGPS receiver located on a car, for instance will have a GPS receiver and an FM receiver. The DGPS receiver calculates its positional parameters using the GPS receiver data and then applies the correction factor based on the data received through the FM receiver. Accuracies up to one meter can be achieved using DGPS.

GPS receivers will be very useful as navigational aids. The navigation systems are now being fitted in cars, trucks, ambulances, and so forth. Geographical information systems (GIS) are now being used extensively for a variety of application such as surveying, oil exploration, and defense. In GIS, the positional parameters obtained from the GPS receiver are displayed on a digitized map of a region.

A GPS receiver has now become a consumer item with the availability of low-cost hardware. Location-based services are now considered the next killer applications in mobile communication systems.

References

Dreamtech Software Team. *Programming for Embedded Systems*. Wiley-Dreamtech India Pvt. Ltd., 2002. This book has one chapter that contains the source code for development of an embedded navigation system using a GPS receiver.

Dreamtech Software Team. *WAP, Bluetooth and 3G Programming*. Wiley-Dreamtech India Pvt. Ltd., 2002. This book contains source code for developing location-based services for mobile phones.

Mobile phone manufacturers such as Nokia, Motorola, Samsung, and Ericsson have mobile phones with GPS receivers. You can get the information from their Web sites. The following Web sites provide a wealth of information on GPS:

www.gpsworld.com: Gives information about GPS hardware, software, mapping software, and such.

www.conexant.com: Gives information about the Conexant GPS receiver.

www.locationforum.org: Web site of the location interoperability forum. This forum has been formed to ensure interoperability of various location-based applications developed by different vendors.

Questions

1. What are the three segments of the GPS system? Explain each segment.
2. What is differential GPS? Explain how the accuracy of location information can be improved using differential GPS.
3. List the various uses of GPS in defense applications and consumer applications.
4. Explain the architecture of a GPS receiver.
5. List the important features of commercial GPS receivers.

Exercises

1. Study the format in which location data is received from an RS232 interface of a commercial GPS receiver.
2. Derive the formula to calculate the distance between two points when the longitude and latitude of the two points are known.
3. Study the features of commercially available databases that facilitate development of Geological information systems.

4. The concept used in differential GPS can be used to track a person continuously. Work out a framework for this type of tracking. Discuss the pros and cons of such a technology. Think of applications of such a technology for public safety and also the applications that are against the privacy of individuals.

Projects

1. If you have access to a GPS receiver, develop software that reads the data from the GPS receiver (through RS232 or USB port). Interface this software to mapping software. If the software is loaded on a laptop and you travel in a car, the location of the car has to be displayed on the map.

2. Generate a digital map of the city in which you live. You can take a laptop fitted with a GPS receiver and go around in the city and note the longitude and latitude values of the important places and bus stops. This information can be transferred onto the map. This map can be used for developing location-based services, a navigation system, and so forth. And, you can sell such a map for a good price too!

3. Develop a Web site that has location-based information. When a user visits this Web site, a form can be presented to the user asking him to input the longitude and latitude of the place in which he is located. Based on this information, the location-based information (nearby hotels, restaurants, theaters, places of historical importance) needs to be provided to the user.

33 ∷ Wireless Internet

In This Chapter

- Challenges in Accessing the Internet Through Wireless Devices
- 2.5G and 3G Wireless Networks
- Wireless Internet Applications
- The Wireless Application Protocol architecture

Accessing the Internet services from our desktops at the office or at home is now taken for granted. Through a dial-up line or a leased line, we are connected to the server of the Internet service provider and access the various services. The drawback of this access method is that we are constrained to one place—there is no mobility. If we can access the Internet services while on the move—while traveling in a car, bus, train or airplane, it would be an immense value addition. The wireless device can be a mobile phone, a laptop, or a palmtop. However, to achieve this, we face lots of challenges—the mobile phone has a small display and small memory, the mobile networks support very low data rates, and delays are high in the wireless networks. But then, the need for mobile access to the Internet is growing rapidly, and it is expected that the number of wireless devices accessing the Internet will exceed the wired devices. This market demand is paving the way for exciting developments in wireless Internet services and technologies, which are discussed in this chapter. With the present- day mobile networks, the data rates supported are very low but, in the next few years, the mobile networks are expected to support high data rates, paving the way for multimedia content to be accessed by mobile devices. This chapter discusses the roadmap for achieving anywhere, anytime access to Internet services.

33.1 WIRELESS ACCESS TO INTERNET

At present, most of us access the Internet through our desktop computers or our corporate LANs through wired connections (dial-up or leased lines). Our desktops have high processing capability, large primary storage (RAM), and secondary storage. So, we can run a browser such as Internet Explorer or Netscape Navigator, which requires huge system resources (hard disk and RAM). The monitors are capable of displaying high-resolution color graphics. And, the wired connection can support high data rates, anywhere between 64kbps and 2Mbps. Hence, accessing the Internet is lots of fun, with lots of multimedia content, fast downloading of files, and fast and easy navigation.

The drawback of wired access to the Internet is the lack of mobility. Even when we move with our laptops, we have to look around for a telephone jack—in a hotel room or at an airport. If wireless access to the Internet is provided, users can be mobile, and accessing the Internet anywhere, anytime would be possible. The mobile device can be just a mobile phone, a personal digital assistant (PDA), a laptop, or a palmtop. But then, providing the Internet services over these mobile devices is a pretty challenging task. The following section will explain this.

> Desktops have high processing power, large memory, and a large screen, so accessing the Internet through wired connections that support high-speeds is a lot of fun. The only drawback is the lack of mobility.

33.1.1 Challenges in Wireless Access to the Internet

Accessing the Internet (particularly the Web service) has many problems at present. These problems are:

- There are various protocols for the wireless networks, such as TDMA, CDMA, GSM, and PDC etc. the protocols for wireless access to the Internet need to be independent of the underlying cellular network protocols.

- Present wireless networks support very low data rates, ranging from 300bps to 14.4kbps. Accessing a Web page with multimedia will take ages! The round-trip delay is also very high in wireless networks.

- The wireless devices (mobile phone, palmtop and such) have limited capabilities:

  - Small screen, generally 4 lines with 8 to 12 characters per line.

  - Screen with low resolution and no support for color.

- Low power because the device has to operate from a battery.

- Keypad with a very limited functionality. Hence, inputting text is difficult.

- The wireless devices vary widely—different processors (8-bit, 16-bit, or 32-bit), different operating systems, different memory capacities, and so on.

Accessing Internet content through wireless devices poses many challenges due to the variety of wireless network protocols, low-speed connections, high delay, and limited capabilities of the mobile devices.

Due to these problems, developing protocols and applications for wireless access of Internet services is a real technical challenge. The software industry created a lot of hype in the late 1990s for wireless access to the Internet, but it has been realized that lots of work needs to be done in enhancing the speeds of the wireless networks, developing efficient protocols, and creating useful and appealing applications.

 Wireless devices are characterized by low processing power, small displays, and keypads with very limited functionality. As a result, new protocols and new markup languages are required for accessing Internet content through mobile devices.

33.2 WIRELESS INTERNET ROADMAP: GENERATIONS OF WIRELESS NETWORKS

The first Generation wireless networks were analog systems that supported only voice communication. These systems have been replaced by digital systems in most countries.

First generation wireless networks: The first generation wireless networks were analog systems such as AMPS (Advanced Mobile Phone System), TACS (Total Access Control System), and NMT (Nordic Mobile Telephony). which had a large installation base in North America and Europe. These systems supported mainly voice communication and had very limited data capability. Because of the limited, radio spectrum, the capacities of these systems were limited, and expansion was a problem. Slowly, these systems were replaced by digital systems.

Second generation wireless networks: The 2G (second generation) networks, which are presently in operation in all countries, are digital systems. These systems include:

- Global System for Mobile Communications (GSM)–based networks which were initially deployed in Europe. Subsequently, GSM standards have been adopted by many other countries in Asia and Africa. GSM systems also are being introduced in North America.

- TDMA systems based on IS 136 and IS 41 (IS stands for interim standard)
- CDMA systems based on IS 95 and IS 41 deployed extensively in the U.S., China, and Korea.

The second generation wireless networks are digital systems that support data services with data rates up to 28.8kbps in addition to voice services. GSM, TDMA, and CDMA networks presently in operation in most countries are examples of 2G wireless networks.

These systems typically support data rates less than 28.8 kbps. To access the Internet at these data rates is not a good experience. However, to retrieve focused information (for example, stock quotes, bank information, sports/entertainment news, and travel information these data rates will suffice. The TCP/IP protocol suite cannot run and hence on the mobile devices because of its high protocol overhead, so new protocols such as wireless application protocol (WAP) and i-mode have been developed. The WAP stack with a microbrowser runs on the mobile device. A special markup language known as Wireless Markup Language (WML) has been developed to create content that can be accessed through the WAP protocols from mobile devices.

2.5G wireless networks: The two-and-a-half generation (2.5G) wireless networks will evolve from the 2G networks to support data rates in the range 64–144 kbps, using technologies such as general packet radio service (GPRS) and cdma2000-1x. With the higher data rates, reasonably high-quality graphics and reasonably good quality video can be supported. With 2.5G systems, the mobile devices can access multimedia content.

Third generation wireless networks: The 3G wireless networks will be capable of supporting very high data rates in the range 384kbps–2.048Mbps to provide

2.5G wireless networks support data services with data rates in the range 64–144 kbps. General packet radio service (GPRS) and cdma2000-1x are examples of 2.5G networks.

3G wireless networks will be capable of supporting data rates in the range 384–2.048 kbps to provide multimedia services to mobile users. 3G networks will evolve from the present GSM and CDMA networks.

multimedia services to mobile users. With these data rates, applications such as two-way video conferencing, high-quality audio (music) downloading, and high-resolution graphics can be supported. The 3G systems will evolve from the GSM or CDMA system.

The roadmap for evolution to 3G is shown in Figure 33.1. It needs to be noted that standardization process has not reached a stage wherein a single technology will be chosen for 3G. This is because the 3G systems should have backward compatibility with the existing cellular infrastructure, and there is wide range of technologies today. Hence, it is likely that there will be variety of technologies under the 3G umbrella. Regional standardization bodies such as ETSI in Europe, CWTS in China, T1 in the U.S., and TTA in Korea formed the 3G Partnership Project (3GPP) group to bring harmony to the various conflicting proposals submitted for standardization.

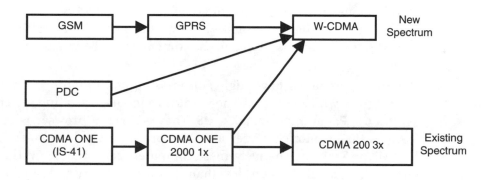

FIGURE 33.1 Evolution to 3G.

For deploying 3G systems, there are many issues to be resolved, such as spectrum allocations, the standards for protocol stacks, and the standards for developing content to be accessed by mobile devices. During the next few years, there will be a lot of activity in this direction. In the following sections, we will review the present trends on these aspects.

There is no single international standard for 3G networks mainly because the 3G networks evolve from the present mobile networks, which follow different standards in different countries.

NOTE

33.2.1 Wireless Access Devices

The wireless Internet access device can be any of the following: mobile phone (cell phone or hand phone, as it is called in some countries), pagers, two-way radios, personal communicators, smart phones, desktops, laptops, palmtops, Personal Digital Assistants, WebTV, and Internet kiosks installed in rural or remote areas. The important point to note here is that each of these devices has different processing capabilities and different platform technologies. All these devices vary in terms of processing power, primary and secondary storage capacity, display size and resolution, battery capacity, input device capability and so on. Not all devices have a full-fledged operating system residing on them—some only have a mobile operating system (such as Win CE, Palm OS, OS-9, Symbian or Java OS). Some use 32-bit processors, and some have only 8-or 16-bit microcontrollers.

Because of the variety of wireless access devices, the mechanism for accessing the content from the Internet also varies:

- Mobile phones, two-way pagers, and such have limited processing capability, limited memory, and very small display. They need to access the content using protocols such as WAP, and the content has to be in a format such as WML.

- Handheld computers have higher processing power, more memory, and color display with larger size. They can run a mobile operating system such as Win CE, PalmOS, or Symbian OS. They can run a browser with better features than a microbrowser and interpret markup languages such as XHTML or XML. Alternatively, these computers can run a KVM (kilobytes virtual machine) that occupy less than 256KB of memory and can download Java code and interpret it.

- Laptop computers that can run a full-fledged desktop operating system and access Internet content the way we access the content from desktops.

Handheld computers are becoming more and more powerful, with a full-fledged operating system, color display, handwritten character recognition software, and compact secondary storage devices capable of storing large amounts of data. In a few years they will have processing capabilities comparable to today's desktops, and wireless access of Internet content will be a much better experience. However, cost will continue to be the important factor for users in selecting a mobile device,

Wireless Internet access device can be any of the following: mobile phone, pager, personal digital assistant, laptop, palmtop, smart phone, or Internet kiosk installed in a remote/rural area.

and a large number of users will continue to have less powerful mobile devices for accessing the Internet.

The capabilities (processing power, memory, display size, keypad and such) vary widely for mobile devices. Hence, developing the content to take care of these varying capabilities is the biggest challenge in wireless Internet.

33.3 WIRELESS INTERNET APPLICATIONS

Because of the limited capabilities of mobile devices, applications such as browsing can be limited. Hence, developing applications/content for mobile devices is a challenging task. The various applications include e-mail, mobile commerce, entertainment, business services, news and weather information, and so on.

As the wireless networks support higher data rates and as the capabilities of the mobile devices increase, the applications can support high-resolution graphics, and audio and video streaming applications as well. As a result, users can carry out audio and video conferencing while on the move. Because audio and video communication requires real-time streaming, the mobile devices should be running a powerful mobile/real-time operating system with capability to do signal processing for encoding and decoding of audio and video signals.

The applications that are supported for wireless Internet access must be highly focused. E-mail, mobile commerce, entertainment services, and tourism information are the most widely used applications.

Applications such as e-learning, telemedicine, and location-based services will provide users with an enhanced quality of life largely because of the anywhere, anytime communication capability.

33.3.1 Pull Model and Push Model

Normally, when we access the Internet, we use the pull model. From the desktop, we make a request in the form of a URL, and we obtain the response from the server in the form of content in text format, which is interpreted by the browser running on the desktop. The pull model can be used by mobile devices to access information from the Internet servers.

In the pull model, we give a request to a server by specifying a URL, and the server downloads the content onto the desktop. This is called the pull model because we pull the information by specifically giving a URL.

Another model, the push model, also can be used effectively in a mobile environment. In this model, the server will send the content to the mobile device without the user explicitly making a request by giving a URL. For example, if a user is interested in obtaining the details of a baseball game, he can register with a Web site by giving his mobile number and how often he would like to receive the information (such as hourly). Then the server will automatically send the scores every hour. To ensure that the pushed content does not intrude into the work of the user, the following procedure is followed:

(a) The user registers with a Web server, giving the details of the mobile number and the frequency with which he would like to receive the information.

(b) The server sends a small message called a service indication. This message is an alert to inform the user that the scores of the game are being sent and asking the user if he would like to view the information now. The service indication also contains the URL from which the content is to be obtained.

(c) The user can select yes or no. If the response is yes, the URL in the service indication is invoked, and the content is presented to the user. If the user response is no, then the service indication can be stored in the cache memory of the mobile device, and the user can retrieve it later to see the content.

This procedure is followed to avoid intrusion. It would be annoying if a message is received when the user is composing an e-mail for instance.

In push model, the user need not make a request repeatedly to obtain the information. The server can automatically send the information periodically to the user.

The push model is useful for such applications as obtaining stock quotes periodically, for obtaining sports information periodically, and for pushing advertisements of the user's choice.

In the push model, to avoid intrusion, the server first sends a small message called service indication, which is an alert to the user that the requested information is to follow. The user can either view the information immediately or postpone it to a later time.

33.3.2 Location-Based Services

Location-based services are now catching up very fast. These services provide location-dependent information to the users when they are on the move. For

In location-based services, the location of the mobile device is sent to the Web server. The Web server sends the information specific to the area in which the user is located, such as nearby hotels, tourist spots, and such.

instance, when a person enters a new city, he can obtain local information such as the availability of nearby hospitals, hotels, and restaurants. This is achieved by sending the location information (longitude and latitude) of the mobile device to the server. A GPS receiver can be integrated into the mobile device, and the mobile device automatically sends the location information to the server along with the request for, say, a hotel in the vicinity. The server will use the location information to find the hotels in that locality and send the hotel information to the mobile device.

To exploit the potential of location-based services and at the same time to ensure interoperability between mobile positioning systems developed by different organizations, the Location Interoperability Forum (LIF) was founded by Ericsson, Motorola, and Nokia in September 2000.

Location-based services will be of immense use in public safety and emergency services. New, exciting personalized services also can be developed using this technology: a user can locate friends when visiting a place, or he can be alerted on his mobile phone when a friend (or an enemy!) is in the user's vicinity. These services will be made available on both 2G and 3G networks.

To provide location-based services, the mobile device has to be fitted with a GPS receiver. The GPS receiver obtains the longitude and latitude of the mobile device, and the information is sent over the wireless network to the Web server.

33.4 WIRELESS APPLICATION PROTOCOL

Wireless application protocol (WAP) specifies the protocol conversion between an IP network and a cellular network. WAP works efficiently on second generation wireless networks, which are characterized by low data rates.

To provide wireless access to the Internet via 2G wireless networks, the WAP Forum was launched in December 1997 by Ericsson, Motorola, Nokia, and Phone.com. THe WAP Forum defined the wireless application protocol (WAP) specifications. WAP specifies the protocol conversion between IP and cellular networks. The WAP protocols are lightweight; that is, the protocol overhead is reduced for wireless networks with low bandwidths and high delays. WAP 1.1 was released in June 1999, Version 1.2 in November 1999, and Version 2.0 in August 2001.

33.4.1 WAP Overview

WAP bridges the wireless network and the Internet by providing the necessary protocol conversions between the mobile network and the IP network. The need for a new protocol arose because of the limitations of the mobile networks—low data rates and high delay of course, the TCP/IP protocol is a heavyweight protocol that cannot run on small mobile devices. As shown in Figure 33.2, there will be a WAP gateway that bridges the wireless network and the Internet. The Internet runs the TCP/IP protocol stack (and HTTP for the Web). The mobile device runs the WAP protocol stack and communicates to the WAP gateway. The WAP gateway does the necessary protocol conversion. From the handset, a Web page is requested by giving a URL. The WAP gateway gets the URL and talks to an origin server where the resource corresponding to the URL is located. The content of the URL is transferred to the WAP gateway, which in turn transfers it to the mobile device.

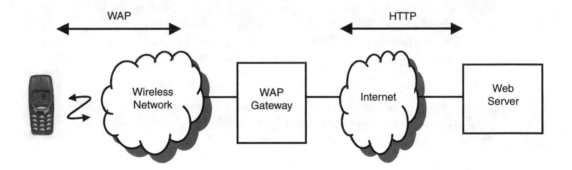

FIGURE 33.2 WAP architecture.

To access Internet content through a mobile device using WAP, the mobile device should be WAP enabled; that is, the mobile device should have the WAP protocol stack and a microbrowser.

The microbrowser that runs on the WAP-enabled mobile device is similar to the browsers that run on the desktop. However, the microbrowser interprets only Wireless Markup Language (WML) content.

NOTE

The WAP programming model is shown in Figure 33.3. A WAP-enabled mobile device (a mobile phone or a PDA for instance) sends a URL request. This request

The WAP gateway acts as the interface between the wireless network and the Internet. The WAP gateway runs the WAP protocol stack as well the as TCP/IP protocol stack and does the protocol conversion.

is binary encoded and sent over the mobile network to a WAP gateway. The WAP gateway decodes the request and converts it to an HTTP request and sends it to the Web server. The Web server sends the response (either in WML content or HTML content) to the WAP gateway. The WAP gateway encodes the content and sends it over the mobile network to the user agent. If the content is in WML format, it is encoded and sent to the mobile device. If the content is in HTML format, it is converted into WML format and then encoded and sent. The mobile device runs a microbrowser, which interprets the content and displays it to the user. In this model, there is a WAP gateway that has mainly two functions: (a) protocol conversion between WAP and HTTP; and (b) encoding/decoding. Content is encoded to make it compact for sending on a low bandwidth mobile network. The separation between the WAP gateway and Web server is only logical. The functionality of the WAP gateway also can be built into the Web server. The WAP content can be developed in WML and can reside in the Web server itself or it can reside in a separate machine—the WAP gateway. Because the WAP gateway acts as a proxy server to the Web server, it is also referred to as WAP proxy.

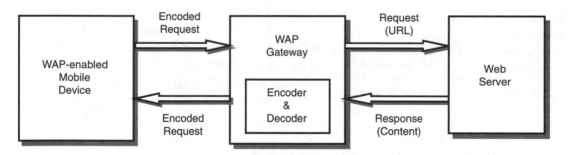

FIGURE 33.3 WAP programming model.

Most of the present Web content is written in HTML, but WAP requires that the content be in WML. For WAP applications, content needs to be rewritten.

Wireless Markup Language (WML) has been specially designed to develop content for access by mobile devices. WML is similar to HTML except that it has very few tags and does not support full-fledged graphics and animation.

Though HTML to WML conversion utilities are available, they are not efficient because WML does not have equivalents for all the tags supported by HTML.

WAP supports both pull and push models. In the pull model, the client sends a URL request and receives the content. In the push model, the content is sent automatically by the server. For obtaining the push service, the user has to register with the push server—this involves giving information on the push service to be requested (for example, the names of the companies for which stock quotes are to be obtained), the mobile number, the frequency of the push service, and the time of the day when the content has to be pushed.

 WAP did not become a great success mainly because rewriting the content in WML is a gigantic task, and very few Internet sites have WML content.

33.4.2 Wireless Application Environment

The wireless application environment (WAE) consists of:

- Wireless Markup Language (WML), a page description language that describes the content presented to the user. WML is similar to HTML and is based on XML.
- WMLScript, a scripting language similar to JavaScript.
- Content formats to describe the data, images, and other content.
- A microbrowser that runs on the mobile devices. The microbrowser occupies little system resources and provides functionality similar to a normal browser such as Internet Explorer or Netscape Navigator, though the (toolkit) functionality is limited.

The wireless application environment consists of Wireless Markup Language, WMLScript, content formats to describe the data and images, and micro-browser that runs on the mobile devices.

To create WAP applications—to develop the content using WML and WMLScript and test it in a laboratory environment before actually deploying the applications on the network—WAP toolkits are available from major vendors. These tool kits provide an environment to create, test, and demonstrate WAP applications. Both service providers and developers can use the toolkit to evaluate the usability of the wireless applications. The toolkit contains the applications to create, edit, and debug the WML and WMLScript programs, a microbrowser, simulators for WAP phones, and the WAP protocol stack. Some WAP tool kits are freely downloadable (a list of URLs is given in the References at the end of this chapter).

The content is divided into cards and decks. The basic unit of WML is a card. A card specifies the single interaction between the user agent and the user. Cards are grouped into decks. The mobile device receives a deck, and the first card is presented to the user. The user can navigate to the other cards or to a card in another deck.

WMLScript, which is similar to JavaScript, can be used for calculations, user input validation, to generate error messages locally, and for passing parameters to the server. Using the servlet APIs available on the server, complex applications can be developed using database access and processing.

A number of WAP toolkits are available that can be used to develop content using WML and WMLScript, and the content can be tested on the desktop before deploying on the wireless network.

33.4.3 WAP Architecture

The protocol stack of WAP (along with TCP/IP protocol suite for comparison) is shown in Figure 33.4. Below the wireless application environment (WAE), will be wireless session protocol (WSP), wireless transaction protocol (WTP), an optional wireless transport layer security (WTLS), and wireless datagram protocol (WDP). WSP and WTP together provide the functionality of HTTP; the optional WTLS is equivalent to TLS, and WDP provides the functionality of the transport layer.

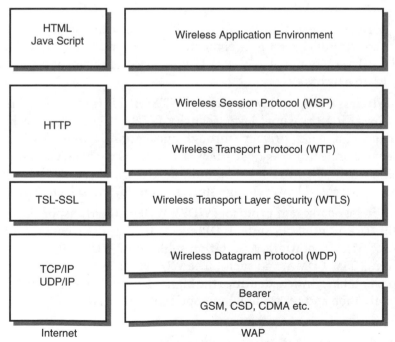

FIGURE 33.4 Internet and WAP protocol stacks.

The WAP protocol stack consists of wireless application environment, wireless session protocol, wireless transaction protocol, wireless transport layer security, and wireless datagram protocol. WTLS is an optional layer.

Wireless session protocol (WSP): This is the application layer of WAP. It supports two types of services—connection-oriented service over WTP and connectionless service (secure or nonsecure) over WDP. The WSP provides the complete HTTP Version 1.1 functionality with provision for session suspension and resumption and facility for unreliable and reliable data push. WSP in its present form is suited for browsing applications only.

The procedure for data transfer is:

- Establish a session
- Negotiate the capabilities of the mobile device
- Exchange content
- Suspend/resume session

Connectionless service is used to send/receive data without establishing a session. The core of WSP is a binary form of HTTP; all methods defined in HTTP are supported. Instead of sending the content in ASCII text format, the content is binary encoded.

Wireless transaction protocol (WTP): WTP runs on top of the datagram protocol. It provides a lightweight transaction-oriented protocol. Because the security layer is optional in WAP, WTP operates over secure or nonsecure wireless datagram networks.

Wireless transport layer security (WTLS): This is an optional layer in WAP. It is based on transport layer security (TLS). It ensures data integrity (no corruption of data), privacy (through encryption), authentication, and denial of service protection (against attacks). It is an optional layer, so it is possible to enable or disable WTLS. WTLS parameters are source address, destination address, proposed key exchange suite, session Id, and selected cipher suite.

Wireless datagram protocol (WDP): WDP is the transport layer equivalent in the TCP/IP protocol suite. WDP makes the upper layers transparent to the type of wireless network. Adaptation sublayer of WDP is different for different bearers (GSM, CDMA, and such). The WDP services are application addressing by port number, which identifies the higher layer entity above the WDP, optional segmentation and reassembly, and optional error detection.

WAP was designed initially only to support access to Web services. Hence, only HTTP-equivalent protocol is supported at the application layer.

The functions of a WAP gateway are protocol conversion between WAP and TCP/IP, obtaining the content from the origin server and converting it into WML if required, converting the WML code into binary form to compress the data, and subscriber data management.

Since the WAP gateway has to bridge the wireless network and the Internet, it has to do the protocol conversion—it has to run both protocol suites as shown in Figure 33.5. The functions of the WAP gateway are:

- Protocol conversion.
- Subscriber data management.
- Compiling the WML code into compact form and corresponding decoding.
- Content management. If the origin server gives HTML code to the WAP gateway, it has to be converted into WML format using HTML-to-WML conversion software. If the origin server gives the WML code, it has to be encoded and sent over the wireless network.

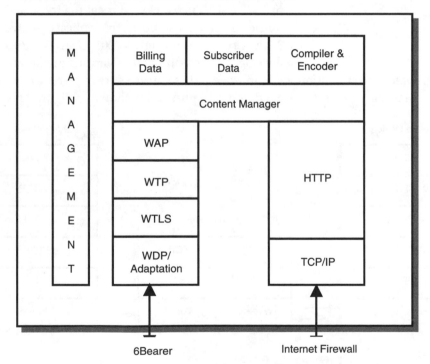

FIGURE 33.5 WAP gateway.

33.4.4 Evolution of WAP

WAP gave us a look at the possibilities of wireless Internet access. However, the initial versions (1.x) provided limited capability, mainly to access textual content. Due to lack of support for graphics and animation, WAP services did not become very popular. Moreover, Internet content is written mostly in HTML; to provide wireless access, it needs to be rewritten in WML. Conversion of the content from HTML to WML is a gigantic task, and very few organizations cared to write WML content for their Web sites.

WAP Version 2.0, released in August 2001, supports TCP/IP suitably modified to work in a wireless network environment. High-speed data services such as multimedia messaging service (MMS) can be offered using WAP 2.0.

In recent years, the capabilities of mobile devices increased significantly—powerful processors, higher memory capacities, increased battery life, and so on. Also, wireless network speeds are increasing with the advent of 2.5G and 3G networks. To meet the increased demand for richer content and powerful applications, WAP Version 2.0 was released in August 2001. Using WAP 2.0, we can develop services supporting various media such as text, graphics and animation, voice, and video. These services are referred to as multimedia messaging services (MMS). The WAP 2.0 protocol is shown in Figure 33.6.

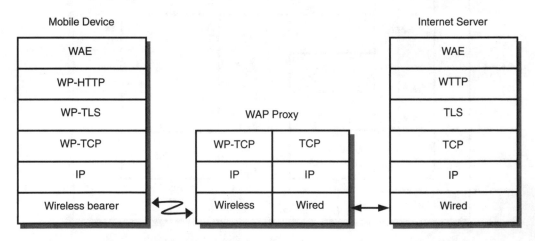

FIGURE 33.6 WAP 2.0 Protocol Architecture.

The protocol stack running on the mobile device consists of:

- Wireless Profiled HTTP (WP-HTTP), which supports all the functionality of HTTP and compresses the message body of the response from the server.

- Wireless Profiled WTLS (WP-TLS), which is an optional layer that defines the cipher suites, signing algorithms, and certificate formats.
- Wireless Profiled TCP (WP-TCP), which provides connection-oriented TCP service and provides less overhead as compared to normal TCP.
- IP, which provides standard IP service.
- Wireless bearer, which can be any wireless network, including 2.5G and 3G networks.

WAP 2.0 uses WML2 for developing content. WML2 is based on eXtensible HyperText Markup Language (XHTML), which in turn is derived from XML.

The WAP proxy runs the protocol stack as shown in Figure 33.6—it does the necessary protocol translation between the mobile client and the Internet server. In WAP 2.0, there is no real need for a gateway because the mobile device can communicate directly with the server using HTTP. However, if push functionality is required and additional services such as location-based services are to be provided, a WAP gateway is required.

To overcome the problem of developing new content for wireless environments, WAP2.0 uses WML2, which is based on eXtensible Hypertext Markup Language (XHTML), also derived from XML. The WAP 2.0 browser interprets the XHTML content and presents it to the user.

WAP 2.0 supports the following additional features:

- User agent profile (UAProfile) allows the mobile device to send its capabilities to the server. Based on the capabilities of the mobile device, such as display size, support for multimedia, and so on, the server can send the content to the mobile device.
- Wireless telephony application (WTA) facilitates accessing the voice application from the data application. For instance, the mobile device can receive the information about a telephone number of a hotel through the URL request, and then the user can call that number without cutting and pasting that number to make a call.
- Data synchronization synchronizes the data between two devices. For instance, the mobile device may contain an address book, if the address book has to be updated on to the desktop, it can be done easily using data synchronization. To achieve this, Synchronization Markup Language (SyncML) is defined, using which the data can be transferred between two devices.

- An external functionality interface provides support for external interfaces such as a GPS receiver, smart cards, digital cameras, and medical devices such as blood pressure measuring devices.

To facilitate accessing services from servers that support both WAP1.x and WAP2.x, the mobile device can support protocol stacks WAP1.x and WAP2.x. These devices are known as dual-WAP stack devices.

 When two devices need to synchronize data (for example, address books in the mobile phone and the PDA), the data transfer is done using a standard markup language called Synchronization Markup Language (SyncML), which is an industry standard.

33.5 2.5G WIRELESS NETWORKS

The 2.5 generation wireless networks evolve from 2G wireless networks to support data rates in the range 64–144 kbps. HSCSD and GPRS networks are 2.5G networks that evolve from GSM networks to support high data rates.

High-speed circuit switched data (HSCSD): In a GSM network, when a call is established using a circuit switched operation, one time slot is allocated to the handset. In one time slot, we can push 14.4kbps data using efficient coding techniques. If four consecutive time slots are allocated for a user, it gives 57.6kbps end-to-end connection. As shown in Figure 33.7, the MSC of the GSM network has to be interfaced with the PSTN or Internet through an inter working function (IWF) to access Internet services. The IWF carries out the necessary protocol conversion. Using this approach, download speeds up to 43.2kbps have been achieved. This data rate is sufficient to support services such as audio streaming and high-resolution graphics and animation.

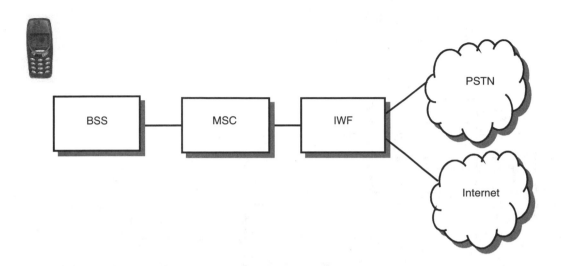

FIGURE 33.7 GSM circuit-switched data service.

High-speed circuit switched data (HSCSD) service is offered on the existing GSM networks by allocating multiple time slots to a user. Data rates up to 57.6kbps can be achieved using this approach.

General packet radio service (GPRS): GPRS is a packet architecture that overlays existing GSM infrastructure. Using eight time slots simultaneously, data rates up to 172kbps can be achieved with GPRS. GPRS employs variable rate coding schemes and multislot operation to increase data rates. Four possible coding systems corresponding to data rates of 9.05, 13.4, 15.6 and 21.4kbps in each slot have been recommended in the GPRS standard. GPRS is more efficient than HSCSD in spectrum utilization and is seen as the best route to 3G. Video transmission using the MPEG-4 standard has been demonstrated over the GPRS network.

The GPRS system offers packet switching services. The elements of GPRS network are Gateway GPRS support node (GGSN) and serving GPRS support node (SGSN), in addition to the standard GSM network elements.

The block diagram of a GPRS system is shown in Figure 33.8. Gateway GPRS support node (GGSN) is the gateway between a GPRS wireless network and the IP-based data network. GGSN carries out the necessary protocol conversion and address conversion. Serving GPRS support node (SGSN) routes and delivers the packet data between GGSN and the appropriate BSS of the GSM network. The functions of SGSN are:

- Packet routing and transfer

- Mobility management
- Logical link management
- Authentication
- Billing

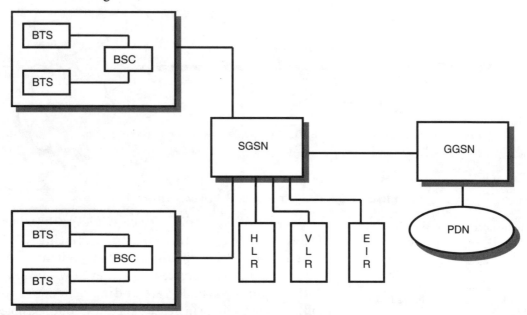

FIGURE 33.8 GPRS architecture.

Many versions of wireless networks based on CDMA are operational in the U.S., China, and Korea. Qualcomm is the pioneer in CDMA technology.

The location register of SGSN stores location information (current cell, current VLR) and the user profile of the subscriber registered with the SGSN.

CDMA-based wireless Internet: The early cellular systems were based on analog technology. Subsequently, digital technologies were used: In the U.S., the IS-136 standards-based systems and in Europe, GSM systems. However, both these technologies are based on TDMA as the multiple access technique. In the 1990s, the commercial viability of CDMA technology was demonstrated. CDMA was extensively used in defense systems but not for civilian communications. Led by Qualcomm Corporation, a pioneer in CDMA technology, many companies in China, Korea and elsewhere also developed CDMA-based systems for wireless communications. CDMA 2000-1x is the standard for 2.5G systems based on CDMA.

33.6 THIRD GENERATION (3G) WIRELESS NETWORKS

3G networks support 2Mbps for handheld devices, 384 kbps for walking mobile devices, and 144kbps for car-borne mobile devices.

To develop high-speed wireless networks, ITU initiated the standardization process in 1986—the Future Public Land Mobile Telecommunication Systems (FPLMTS) working group developed the initial set of standards. FPLMTS is now known as IMT 2000 (International Mobile Telecommunications in the year 2000). UMTS (Universal Mobile Telecommunications Systems) is the European version of IMT 2000. The work was challenging, keeping in mind the different technologies that were in vogue, the huge infrastructures put in place by different operators, the lack of availability of spectrum in the same bands in different countries and increasing demand for higher and higher data rates by the end users. Even today, no common spectrum is available worldwide for 3G networks, and no single technology could be standardized.

The broad objectives of 3G systems are:

- Should support 2Mbps for handheld devices, 384kbps for walking mobile devices, and 144kbps for car-borne mobile devices.
- Should support global roaming.
- Should work in all types of radio environments: urban, suburban, and mountainous regions. In these regions, depending on the terrain, the cell size may vary considerably. In suburban or rural areas, the cell size may be large; in indoor environments, the cell size may be very small (picocells).
- Should support asymmetric and symmetric services. In asymmetric services, the uplink (from handset to base station) data rates can be lower, and downlink (base station to handset) data rates can be higher. In symmetric services, both uplink and downlink speeds are the same.
- Should support the following services:
  - Data services such as mail, file transfer, mobile computing.
  - Telecom services such as telephony, video telephony, video and audio conferencing, audio/video on-demand, teleshopping, TV, and radio broadcast.

The objective of 3G networks is to support international roaming. Moreover, they should operate in all types of radio environments such as urban, suburban, and hilly and mountainous regions.

33.6.1 3G Standardization Activities

ETSI Special Mobile Group (SMG) in Europe, Research Institute of Telecommunications Transmission (RITT) in China, Association of Radio Industry and Business (ARIB) and Telecommunications Technology Committee in Japan, Telecommunication Technologies Association (TTA) in Korea, and Telecommunications Industries Association (TIA) and T1P1 in the U.S. submitted proposals for 3G standardization based on the call for proposals by ITU Radio Communications Standardization Sector (ITU-R) Task Group 8/1. ETSI SMG submitted UMTS Terrestrial Radio Access (UTRA), China proposed TD-SCDMA, ARIB proposed W-CDMA, and TTA Korea submitted two proposals—one close to ARIB's W-CDMA and one similar to TIA cdma2000. The U.S. proposals were UWC-136, which is an evolution of IS-136; cdma2000, which is an evolution of IS-95; and W-CDMA. Subsequently, two international bodies were established: 3GPP (3G Partnership Programme) and 3GPP2: 3GPP to harmonize and standardize the similar proposal proposals from ETSI, ARIB, TTC, TIA and T1 WCDMA, and related proposals and 3GPP2 for cdma2000-based proposals from TIA and TTA.

A summary of some of the proposed 3G standard systems is as follows:

A number of proposals have been submitted for standardization of 3G networks. These proposals are based on wideband CDMA and TDMA technologies.

W-CDMA (FDD): Wideband CDMA with frequency division duplex systems are likely to be deployed in the U.S., Europe, Japan, and other Asian countries. These systems operate in the following frequency bands: IMT 2000 band (1920–1980 MHz, 2110–2170 MHz), PCS 1900 band (1850–1910 MHz, 1930–1990 MHz), and DCS 1800 band (1710–1785 MHz, 1805–1880 MHz). Channel spacing is 5MHz.

W-CDMA (TDD): Wideband CDMA with time division duplex systems are likely to be deployed in U.S., Europe, Japan, and other Asian countries. The multiple access used is TDMA/CDMA. Channel spacing is 5MHz.

W-CDMA (low chip rate TDD): These systems are similar to W-CDMA (TDD) except that channel spacing is 1.6MHz and data support is limited to 384kbps.

TD-SCDMA: Time division synchronous CDMA systems will be deployed in China and other countries that deploy W-CDMA (TDD). The multiple access technique used is TDMA/CDMA. The channel spacing is 1.6MHz. In phase I, the system supports 384kbps, and in phase II, 2Mbps.

The architecture of Universal Mobile Telecommunications System (UMTS) being standardized by ETSI is shown in Figure 33.9. The radio base station does the functions of a base transceiver subsystem, and the radio network controller does the functions of a base station controller. The RNCs are interconnected through a backbone network called the core network. The core network can be based on ATM technology.

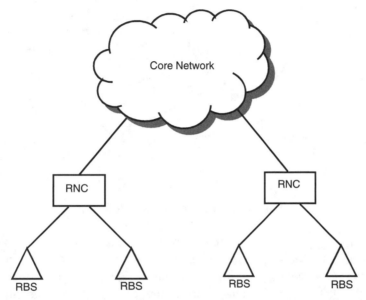

RNC: Radio Network Controller

RBS: Radio Base Station

FIGURE 33.9 UMTS.

Though a lot of activity is going on to standardize the 3G wireless networks, the progress is slow because every country already has a huge installation base of cellular networks operating in different frequency bands with different protocols. Achieving uniformity is a difficult and time-consuming task.

Japan, which has its own set of proprietary protocols for its 2G and 2.5G wireless networks, however, has made headway in deploying a 3G network and offering multimedia services.

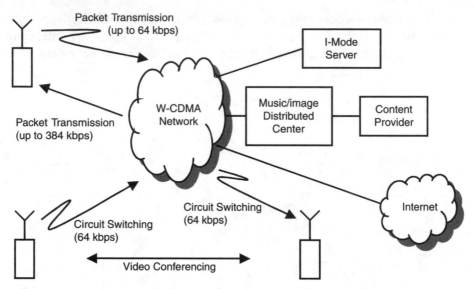

FIGURE 33.10 NTT DoCoMo's FOMA.

NTT DoCoMo is the first operator to provide 3G services in Japan. The network is based on W-CDMA and supports 64kbps uplink speed and 384kbps downlink speed. This service is known as Freedom of Mobile Multimedia Access (FOMA).

In October 2001, Japan's NTT DoCoMo became the first operator to introduce a 3G network on a commercial basis. The network is based on W-CDMA. The architecture of the network is shown in Figure 33.10. The network supports both packet switching and circuit switching operation. Data services use packet switching. The packet switching speeds are 64kbps in uplink and 384kbps in downlink. For applications like video conferencing, circuit switching is used at a data rate of 64kbps. In addition to all the data services, applications such as music and image distribution are also supported. This service is known as FOMA (Freedom of Mobile Multimedia Access). The protocols used are known as i-mode, proprietary to NTT DoCoMo.

Wireless Internet holds great promise—wireless subscribers are likely to grow from 400 million in 2000 to 1.8 billion by 2010. By 2010, 60% of the traffic is expected to be multimedia. Hence, operators and equipment manufacturers can reap rich benefits. More than that, users will be greatly benefited—a wide variety of services will be available to them anywhere, anytime and at a very low cost.

Summary

In this chapter, we studied the technologies for accessing the Internet through wireless devices. The second generation wireless networks support data rates up to 28.8kbps only. Even mobile devices to access the data services have limited capabilities, such as low processing power, small display, and little RAM. To access Internet services using such devices over wireless networks with low data rate support, the wireless application protocol (WAP) has been developed. The WAP protocol stack facilitates accessing Web service for applications such as obtaining stock quotes, weather information, and so forth that are basically text oriented. WAP has very limited capability to support high-resolution graphics or animation.

With the advent of 2.5G wireless networks, data rates up to 144kbps can be supported using high-speed circuit switched data (HSCSD) and general packet radio service (GPRS), which evolved from the GSM system. As the capabilities of the mobile devices also increased, services such as audio streaming multimedia services can be supported over these networks.

Third generation wireless networks based on CDMA technology will support data rates up to 2Mbps, and full-fledged multimedia applicatio0ns can be accessed over these networks. However, standardization of technologies for 3G is still in progress. 3G promises multimedia communications anytime, anywhere in the near future.

References

Dreamtech Software Team. *WAP, Bluetooth and 3G Programming.* Wiley-Dreamtech India Pvt. Ltd., 2002. This book gives the source code listings for a number of wireless applications.

Dreamtech Software Team. *Programming for Embedded Systems,* Wiley-Dreamtech India Pvt. Ltd., 2002. This book also contains source code listings for wireless applications using J2ME, BREW, and WAP.

J. Korhonen. *Introduction to 3G Mobile Communications,* Artech House, 2001.

T. Ojanpera, R. Prasad (Ed.): *WCDMA: Towards IP Mobility and Mobile Internet,* Artech House, 2001.

www.3gpp.org Official site of the 3G Partnership Program.

www.cdg.org Web site of CDMA development group.

www.java.sun.com/products /j2metoolkit/download.html You can download the J2ME toolkit from this link. Using J2ME, you can develop wireless mobile applications.

www.microsoft.com/mobile/pocketpc Microsoft's Pocket PC is another popular platform for mobile applications.

www.nttdocomo.com Gives information on i-mode and FOMA.

www.openwap.org Web site that provides links to WAP toolkits.

www.palmos.com Web site of Palm Corporation. Palm OS is a popular operating system for mobile devices.

www.qualcomm.com/Brew You can download the BREW (binary runtime environment for wireless) development kit from Qualcomm's Web site. BREW facilitates development of wireless data applications. Qualcomm is a pioneer in CDMA technology, and the site provides information on CDMA-based 3G systems.

www.sss-mag.com/w-cdma1.html Provides resources on CDMA.

www.the3gportal.com Provides rich information on 3G wireless networks.

www.tiaonline.org/standards IMT 2000 standards and other resources from Telecommunications Industries Association.

www.uwcc.org Web site of Universal Wireless Communications Consortium.

www.wapforum.org Provides information on technology and standards of WAP.

Questions

1. What are the problems involved in accessing the Internet through wireless networks?
2. Describe the various proposals/plans to evolve 3G wireless networks from the present 2G wireless networks.
3. Describe the WAP protocol architecture.
4. Differentiate between the push model and the pull model. What are the pros and cons of the push model?
5. Explain how location-based services can be provided over mobile phones. Discuss the pros and cons of such services. Do you think that the privacy of individuals is at stake because of such services?

Exercises

1. Download a WAP toolkit and experiment with it.

2. Download J2ME toolkit and experiment with it.

3. Download BREW development kit and experiment with it.

4. Write a technical paper on mobile operating systems (Palm OS, Windows CE, Symbian OS, embedded Linux, Embedded XP, and others).

5. Write a technical paper on Mobile IP.

6. Write a technical paper on location-based services.

Projects

1. Develop a mobile commerce application using WML and WMLScript. The program has to access a WAP server that sells pizza, burger and, soda. You should be able to order any item of your choice and also specify the quantity. Test the application using a WAP toolkit (for instance, that of Nokia).

2. Create a WAP push application for mobile advertising. The user has to log on to a server and give his preferences for receiving advertisements (for example, books, jewelry, cosmetics). The arrival of a new item has to be pushed to the user. Write the code using WML, WMLScript, and servlets. Use a WAP toolkit.

3. Develop a mobile portal for tourism. The user has to log on to a server to obtain information about tourist spots such as historical places, museums, and parks in a city. He also should be able to obtain information about the important hotels and restaurants in the city. You can use a WAP toolkit or a BREW toolkit.

4. Using a WAP toolkit, develop a push application that sends the latest football scores every 15 minutes to a mobile phone.

5. Develop software that converts an HTML file into a WML file. Tags supported by HTML but not by WML should be ignored.

34 Multimedia Communication over IP Networks

In This Chapter

- The Importance of Multimedia Communication over IP Networks
- The Protocols Used for Multimedia Communication over IP Networks
- Voice/Fax/Video Communication over IP Networks
- The Java Media Framework

Until a few years ago, the Internet was used mainly for text communication (e-mail, file transfer, chat, etc.). Though multimedia is supported in the form of graphics, voice, and video clips through the Web service, it is mainly one way—the server downloads the graphics or voice/video clips onto the client for the user to hear or view.

During the last few years, we have witnessed a revolution—using the Internet for two-way (and multiparty) voice and video communication, transmiting fax messages over the Internet, and broadcasting voice and video over the Internet. Similarly, multimedia communication over local area networks and corporate intranets spread over a large geographical area is also possible. This mode of communication results in tremendous cost savings for users. Multimedia communication over IP networks presents immense opportunities but many technical challenges as well. We will study the intricacies of this technology in this chapter.

34.1 IP NETWORKS AND MULTIMEDIA COMMUNICATION

Packet networks are now becoming widespread. Every corporate office now has a LAN and a WAN based on IP (such as IP over ATM or IP over Frame Relay). If this network infrastructure can be used not just for data services, but for voice, fax, and video services as well, there will be tremendous cost reduction.

Voice/fax/video communication over IP networks results in tremendous cost savings for organizations because billing is based on flat rates and not based on the distance.

If the multimedia PCs in an organization can be used for voice communication, we can do away with expensive PBX systems. With desktop video cameras, video conferencing and intranet-based learning are possible.

The Internet can be used for telephone conversations, fax transmission, and video conferencing. This will create tremendous cost savings because we pay flat rates for Internet connectivity, unlike PSTN where the billing is done based on the distance.

However, packet-based networks provide technical challenges for real-time transmission of multimedia content.

34.1.1 Limitations of IP Networks

Networks running TCP/IP protocol stack are not well suited for real-time multimedia communication because the desired quality of service is not guaranteed in these networks.

The TCP/IP protocol suite was developed mainly for data applications in the initial days of the Internet. To use the Internet for audio, video, and fax applications, real-time transmission of data is required. The TCP/IP architecture has inherent limitations for supporting real-time transmission. In a TCP/IP network, there is no guarantee that all packets will be received at the destination with constant delay and in sequence. This lack of guarantee of the quality of service is not acceptable for applications such as voice, fax, and video where real-time transmission is a must. Loss of packets can be minimized through the acknowledgements in the TCP layer, but the delay will be very high if a few packets are lost and the destination has to wait until these packets are received. The variable delay also causes problems in two-way conversations. Hence, to support real-time applications using multimedia, a new set of protocols is required.

Another problem with voice and video communication over IP networks is in terms of the data rates. The access speed for most Internet users is generally limited to 28.8 or 33.6 or at most 56kbps. This data rate is low as compared to the PCM coding rate of 64kbps. To overcome this problem, low bit rate coding of voice and video signals is done. Obviously when we reduce the coding rate, we compromise on the quality.

To achieve an acceptable quality of service for multimedia communication over IP networks, a new protocol architecture is defined with which these problems are overcome.

For voice and video transmission over IP networks, low bit rate coding techniques are used to overcome the limitations of low access speeds.

34.2 ARCHITECTURE FOR MULTIMEDIA COMMUNICATIONS OVER IP NETWORKS

The architecture for multimedia communications over the IP network is shown in Figure 34.1. Corporate LANs can be connected to an IP-based intranet or the Internet through routers/gateways. The corporate LAN will consist of nodes with multimedia capability. The multimedia PC can make voice calls, carry out video conferencing, or exchange fax messages with users on the Internet/intranet or with users on the PSTN. If a multimedia PC has to make a call to a telephone on the PSTN, there is a need for a special gateway (also called interworking function or IWF), which does protocol conversion between the IP-based protocols and PSTN protocols for call processing. Note the special IP Phone device connected to the corporate LAN—this is a telephone with an Ethernet interface that can be connected to the hub/switch of the LAN. Persons who do not need a PC can be provided with an IP phone for making voice calls. Like any PC on the LAN, an IP phone will have an IP address. This IP phone can be used to make calls to a local PC, an external IP phone, or a fixed telephone on the PSTN. Of course, for all this to happen, we need special protocols to overcome the deficiencies of TCP/IP protocols for real-time transmission of voice. Similarly, a fax machine on the LAN or a PC-based fax system can interchange fax messages with fax machines connected over the PSTN.

An IP network can be interworked with the PSTN through a gateway or interworking function (IWF), which does the necessary protocol conversions. Note that the signaling used in the PSTN and IP networks is different to establish voice calls.

For transmission of voice over IP networks, PCM coding will result in very high data rate—the number of packets to be transmitted will be very high, resulting in high delays. To reduce the delay and also the processing requirements, low bit rate speech coding techniques are used. The same is the case for video communication.

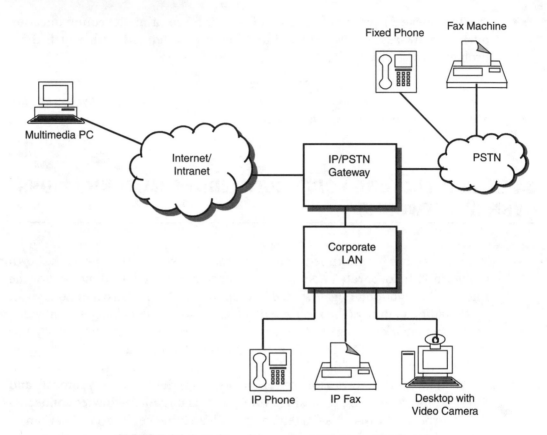

FIGURE 34.1 Multimedia communication over IP network.

 The IP phone is a gadget that functions as a normal telephone. However, it can be connected to an IP network through an Ethernet interface. An IP phone will have an IP address that can be mapped to a telephone number.

34.3 PROTOCOL STANDARDS FOR MULTIMEDIA COMMUNICATIONS OVER IP NETWORKS

For multimedia communication over IP networks, signaling is the most important aspect. During the initial days of voice and video conferencing over the Internet, proprietary signaling protocols were used for communication between multimedia PCs, because of which the products of different vendors did not interoperate. The H.323 set of protocol standards was released by ITU-T to bring in standardization to make interoperability feasible.

Until a few years ago, for voice over IP, proprietary protocols were used. Hence, software supplied by the same vendor had to run on both ends for two users to communicate with each other.

The H.323 series of standards specifies the protocols for multimedia communication over networks that do not offer guaranteed quality of service. Such networks are Ethernet-based local area networks and the Internet.

H.323 is the standard set of protocols for multimedia communication over networks that do not offer guaranteed quality of service (QoS), such as LANs and the Internet. Now, a large number of vendors offer products that support H.323.

An H.323 network is shown in Figure 34.2. The various components of H.323 network are:

- Terminal equipment
- Multipoint control unit
- Gateway
- Gatekeeper

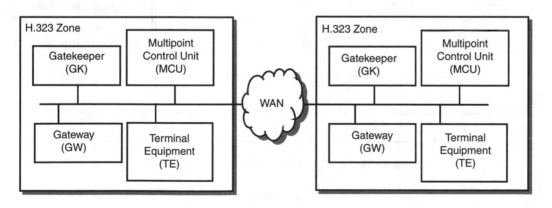

FIGURE 34.2 H.323 network.

Terminal equipment (TE): The TE provides a user with data/voice/video services over an H.323 network. The TE can be a multimedia PC, an IP phone, a fax machine, and so forth.

Multipoint controller unit (MCU): This is a conference server that allows three or more H.323 TEs to connect and participate in a conference. MCU contains a multipoint controller to manage the functions and capabilities of the TEs for conference calls and processors to process data and audio/video streams.

Gateway: Gateway makes H.323 TE on a LAN available to a WAN or another gateway. It provides the necessary protocol conversion, for example, to enable an H.323 TE to talk to a telephone connected to the PSTN.

Gatekeeper: Gatekeeper provides the network services to the H.323 terminals, MCUs and gateways. Its functions are (a) control the number and type of connections; (b) help route a call to the correct destination; and (c) determine the network addresses for incoming calls. Every H.323 gatekeeper controls one H.323 zone.

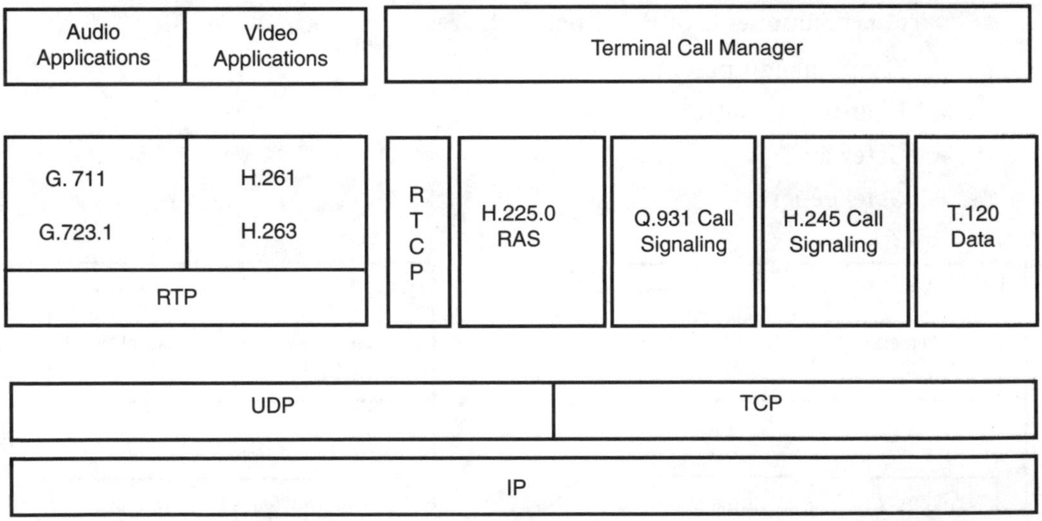

FIGURE 34.3 H.323 terminal-side protocol stack.

The four network elements in a H.323 network are (a) terminal equipment; (b) multipoint control unit; (c) gateway; and (d) gatekeeper.

The protocol stack that runs on an H.323 terminal is shown in Figure 34.3. The same stack also runs on the MCU, gateway, and gatekeeper. The various protocol standards used in H.323 are as follows:

H.225.0: This standard is to format data, audio, video, and control streams for outputting to network and retrieve from the network using packet format of RTP (Realtime Transport Protocol) and RTCP (Real Time Control Protocol) for:

- RAS (registration, admission, and status) control to communicate with the gatekeeper
- Error detection
- Logical framing to convert audio/video to bit streams
- Sequence numbering

Q.931: This is a datalink layer protocol for establishing connections and framing data. It provides logical channels within a physical channel.

H.245: This standard defines the call control procedures to enable two H.323 terminals to talk to each other. In addition to establishing audio/video connections, it also handles signaling, flow control, capability negotiation, and codec selection. Bit rate, frame rate, picture format, and algorithms to be used for coding are negotiated using this protocol.

The H.323 series of standards specifies the various protocols for call control such as call establishment and disconnection, formats for video coding, low bit rate coding techniques to be used for audio and video, and two special protocols to achieve real-time audio/video communication.

G.711: This standard defines audio coding at 48, 56, and 64 kbps data rates.

G.723: This standard defines audio coding at 5.3 and 6.3 kbps data rates.

H.261: This standard defines video coding at 64 kbps data rate.

H.263: This standard defines the format and algorithms to be used for video coding. The formats supported are Common Interchange Format (CIF), Quarter CIF (QCIF), Semi-Quarter CIF (SQCIF) picture formats.

T.120: This standard is defined for data applications.

RTP, RTCP: These are special protocols defined for real-time audio/video streaming applications.

The protocols RTP and RTCP run above the UDP, and not TCP. As compared to TCP, UDP is a lightweight protocol, and hence processing overhead is reduced if UDP is used.

34.3.1 Call Models

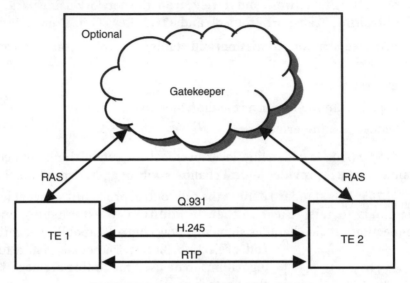

FIGURE 34.4 Direct call model of H.323.

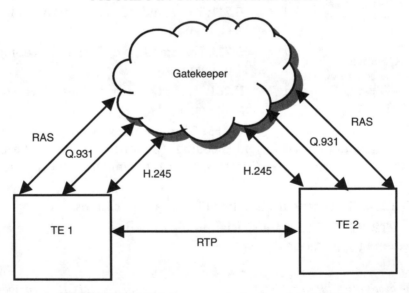

FIGURE 34.5 Gatekeeper-routed call model of H.323.

In an H.323 network, calls can be established in two ways. In direct call model shown in Figure 34.4, two TEs can communicate with each other directly. The

In an H.323 network, two types of call models are used. In direct call model, two terminals can talk directly with each other. In gatekeeper-routed call model, signaling is done through the gatekeeper, and the audio/video data is exchanged between the terminals directly.

gatekeepers are optional; if present, they provide RAS functionality only. The signaling is carried out directly between the two TEs. In Figure 34.5, the gatekeeper-routed call model is shown. In this model, the signaling is carried out through the gatekeepers, and the data streams (audio/video) are exchanged between the TEs.

34.4 VOICE AND VIDEO OVER IP

Organizations having their own IP-based intranets or virtual private networks will be greatly benefited by providing voice/video services over these networks.

Using the Internet, intranets and virtual private networks for voice communication offers great potential because all corporations can drastically reduce their communication costs. The initial euphoria created by VoIP was so dramatic that some journalists even started writing the obituaries of major telecom service providers. Certainly VoIP has the potential, and all these service providers followed the dictum "if you cannot beat them, follow them," and are getting into VoIP. It is still a long time before the traditional telecom operators close their shops, simply because the Internet infrastructure has not penetrated the market the way the traditional telephone infrastructure has. But then, VoIP is the technology of the future.

34.4.1 Applications of Voice over IP

An organization can use the intranet infrastructure to provide voice over IP service. However, to interact with the normal PSTN subscribers, a gateway needs to be installed.

Compared to transmitting voice over a circuit-switched network, transmitting voice over a packet-switched network has the main advantage of savings in cost—it is because of the fact that in packet-switched networks, the billing is not based on the distance but on the data sent. For example, on the Internet, irrespective of the location of the Web server we access, we pay only for the local calls for accessing the ISP, based on the duration of the connection. The same approach for billing can be followed for voice communication over IP networks, including the Internet.

Consider an organization with multiple branches in different parts of a country or in different countries. These branches can be connected through a virtual private network (VPN) using the Internet infrastructure. The gateway or IWF provides the interface between the subscriber equipment (such as PBX LAN) and the packet network. People at different branches can do voice or data communication through the packet network without paying for long distance calls.

Through this technology, the Internet can be used for voice communication between any two persons located anywhere on Earth. A person can communicate with another directly through the multimedia PCs, or one person can use a multimedia PC and another a normal telephone connected to the PSTN. In the second case, a VoIP gateway is required to do do the protocol conversion between the packet network and the PSTN.

VoIP is the first and foremost step in achieving convergence wherein the Internet infrastructure can be used for both voice and data communication. Also, the PSTN infrastructure and the Internet infrastructure will converge to provide a single interface to the subscriber for providing voice communication.

34.4.2 Issues in Voice and Video over IP

When we make voice calls over the PSTN, we get good quality speech. This is because the PSTN provides low transmission delay (except when the call uses satellite channel), and generally the network is reliable. So, the quality of service is guaranteed.

When voice is transmitted over an IP network, the quality of service is a major issue mainly because of the delays and unreliable service provided by the IP network. The major quality of service issues in IP networks for voice communication are delay, jitter, and packet loss.

Delay: Delay causes two problems in voice communication: echo and talker overlap. Echo is caused by signal reflections of the speaker's voice from the far end telephone equipment back to the ear of the speaker. If the round trip delay (sender to receiver and back) is more than 50 msec, echo is heard, and echo cancellation equipment has to be incorporated. ITU standard G.165 specifies the performance requirements of the echo cancellation equipment. Talker overlap occurs when the one-way delay is greater than 250 msec. The speaker talks into his telephone and, because of the delay, assumes that the other person is not responding and says a hello again—the conversation is very difficult to continue. This must be avoided by reducing the delay to the maximum extent possible.

Delay in IP networks is an accumulation of algorithmic delay, processing delay, and network delay.

- **Algorithmic delay:** To reduce the bandwidth requirements, voice is coded at low bit rates for transmission over the IP networks. The coding algorithm introduces algorithmic delay. Typical algorithmic delays for various coders are given below. It needs to be mentioned that the coding delays are decreasing because of the availability of digital signal processors with higher processing power.

Coding technique	ITU Standard	Coding Rate (Kbps)	Delay
ADPCM	G.726	16, 24, 32, 40	125 microseconds
CELP (Code Excited Linear Prediction)	G.728	16	2.5 milliseconds
Multirate coder	G.723.1	5.3, 6.3	30 milliseconds

- **Processing delay:** After coding the voice signals, packetization has to be done, which again contributes to delay. This delay is dependent on the performance of the processor used and also the algorithm.

- **Network delay:** Depending on the traffic on the network, there may be congestion in the network, and hence there will be delay in the arrival of the packets at the receiver. In IP networks, a network delay up to 100 milliseconds is not uncommon.

The delay in IP networks is the sum of algorithmic delay, processing delay, and network delay. Algorithmic delay is caused by the low bit rate coding algorithm, processing delay by packetization of the voice data, and network delay due to the congestion in the network.

The three important factors that need special attention for real-time communication over IP networks are delay, jitter, and packet loss.

In VoIP systems, the effect of the delay has to be compensated using different strategies—fast packet switching, use of low-delay codes, and so on.

Jitter: In IP networks, the delay is variable because the delay is dependent on the traffic. If there is congestion in the network, the delay varies from packet to packet. For voice, this is a major problem because if the packets are not received with constant delay, the voice replay is not proper, and there will be gaps while replaying.

In real-time voice communication, even if a voice packet is lost, the replay at the receiving end should be continuous. To achieve this, one approach is to replay the previous packet. The other approach is to send redundant information in each packet so that the lost packet can be reconstructed.

One way of solving this problem is to hold the packets in a buffer to take care of the highest delay anticipated for a packet. This calls for a high buffer storage at the receiver.

Lost packets: When congestion develops in the IP networks, some packets may be dropped. Loss of packets means loss of information. IP networks do not guarantee quality of service, and the problem is overcome using one of the following methods:

- **Interpolation for lost packets:** When a packet is lost, the previous packet is replayed in the place of the lost packet to fill the time. This approach is ok if the number of packets lost is small and if consecutive packets are not lost.

- **Send redundant information:** The information about the nth packet is sent along with $(n+1)^{th}$ packet. When one packet is lost, the next packet contains information about the previous packet, and this information is used to reconstruct the lost packet. Of course, this approach is at the expense of bandwidth utilization, but speech quality improves.

Use a hybrid approach: A combination of the previous two approaches can be used as a trade-off between bandwidth utilization and quality of speech.

34.4.3 Video over IP

Realtime Transport Protocol (RTP) identifies the type of data (voice/video) being transmitted, determines the order of the packets to be presented at the receiving end and synchronizes the media streams for different sources.

Video communication over IP networks follows the same principles as voice over IP networks. However, video requires a much higher data rate than voice. If we transmit 30 video frames per second, then the video appears normal. The effect of reducing the number of frames per second will be seen in the form of jerks. Because of the lack of high-speed Internet access and good streaming technologies, we encounter problems in video communication.

With higher access speeds to access the Internet and better video streaming technologies, in the future, video conferencing and video messaging will become popular.

Because the present IP networks do not support high bandwidths (particularly the Internet and the corporate LANs operating at 10Mbps), low bit rate video codecs are used to provide two-way and multiparty video conferencing. Video coding at 64kbps data rate would provide reasonably good quality for business applications.

For video communication over IP, the video signal is encoded at 64kbps, which provides a reasonably good quality video for business meetings.

To overcome the problems of real-time transmission over the TCP/IP networks, special protocols RTP and RTCP are defined for voice and video communication over IP networks. The protocol suite for voice/video communication in IP networks is shown in Figure 34.6. The RTP runs above the UDP, which runs above the IP (Version 4 or Version 6 or mobile IP).

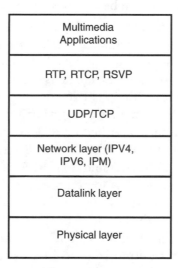

FIGURE 34.6 Protocol suite for voice and video over IP.

34.4.4 Realtime Transport Protocol (RTP)

RTP provides end-to-end delivery service for real-time data transmission. RTP runs over the UDP. RTP supports both unicast and multicast service. In unicast, separate copies of data are sent from source to each destination. In multicast service, the data is sent from the source only once, and the network is responsible for transmitting the data to multiple applications (such as in video conferencing over IP networks).

RTP provides the following services:

- Identify the type of data being transmitted (voice or video).
- Determine the order of the packets to be presented at the receiving end.
- Synchronize the media streams for different sources.

RTP does not guarantee delivery of packets. Based on the information in the packet header, the receiver has to find out whether the packets are error free or out of sequence. Since RTP does not guarantee quality of service (QoS), it is augmented by Real Time Control Protocol (RTCP), which enables monitoring the quality. RTCP also provides control and identification mechanisms for RTP transmission. If QoS is a must for an application, RTP can be used over a Resource Reservation Setup Protocol (RSVP) that provides connection-oriented service. Using RSVP, the resources can be reserved so that QoS can be assured.

RTP uses ports—one port is used for the media data and the other is used for RTCP. Sessions are established between hosts (called participants) to exchange data. One session is used for each media type. For example, in a video conferencing application, one session is used for audio and one session is used for video. So, for example, the participants can choose, to hear only audio or only video (if a beautiful woman is giving a boring lecture). The data for a session is sent as a stream of packets called an RTP stream. In a session, there can be multiple sources (audio source, video source). Each source is called a contributing source (CSRC). Each CSRC is a source of RTP packets. In a conferencing call, each source of RTP packets has to be uniquely identified, and this is achieved through a unique ID known as the synchronization source (SSRC) identifier, which is 32 bits in length. The RTP packets from a number of sources are combined through a mixer, and the combined output RTP stream is sent over the network. The mixer output consists of a list of SSRC identifiers known as the CSRC list. At the receiving host, the RTP packets are sent to different media players based on the identifiers—audio packets are sent to the sound card, video to the display unit and so on. These output devices are called the renderers.

Each RTP packet consists of a header and data (payload). The RTP data packet header format is shown in Figure 34.7.

Bits

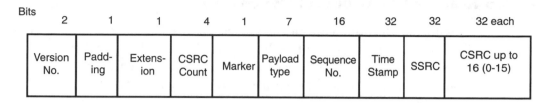

FIGURE 34.7 RTP data packet header format.

Version number (2 bits): the current version number is 2.

Padding (1 bit): If the bit is set, it indicates that there are more bytes at the end of the packet that are not part of the payload. This bit is used by some encryption algorithms.

Extension (1 bit): If the bit is set, the fixed header is followed by one header extension. This mechanism allows additional information to be sent in the header.

CSRC count (4 bits): Indicates the number of CSRC identifiers that follow the fixed header. If the count is zero, the synchronization source is the source of the payload.

Marker (1 bit): A marker bit defined by the particular media profile.

Payload type (7 bits): An index to a media profile table that describes the payload format. RFC 1890 specifies the payload mappings for audio and video.

Sequence number (16 bits): The packet sequence number. This value is incremented by one for each packet.

Timestamp (32 bits): The sampling instant of the first byte of the payload. Several consecutive packets can have the same timestamp—for example, when all the packets belong to the same video frame.

SSRC (32 bits): This field identifies the synchronization source. If the CSRC count is zero, the payload source is the synchronization source. If the CSRC count is not zero, the SSRC identifies the mixer.

The RTP header consists of the following fields: version number, padding, extension, CSRC count, marker, payload type, sequence number, timestamp, SSRC, and CSRC.

CSRC (32 bits each): Identifies the contributing sources for the payload. There can be a maximum of 16 contributing sources.

34.4.5 Real Time Control Protocol (RTCP)

In a session, in addition to the media data, control packets are sent. These RTCP packets are sent periodically to all the participants. These packets contain information on the quality of service, source of media being transmitted on the data port, and the statistics of the data so far transmitted. The types of packets are:

Sender report (SR) packets: A participant that recently sent data packets sends an SR packet that contains the number of packets and bytes sent and information that can be used to synchronize media streams from different sessions.

Real Time Control Protocol is used to send control packets to all participants. These control packets contain information about the quality of service and the statistics of the data so far transmitted.

Receiver report (RR) packets: Session participants periodically send RR packets. An RR packet contains information about the number of packets lost, highest sequence number received, and the timestamp that can be used to estimate the round-trip delay between the sender and the receiver.

Source description packets: This packet gives the source description element (SDES) containing the canonical name (CNAME) that identifies the source.

BYE packets: A BYE packet is sent by a source when it is leaving the session. The packet may include a reason as to why it is leaving.

Application-specific packets: An APP packet is sent to define and send any application-specific information.

Since the PSTN and the Internet use different protocols for voice communication, if we need to make calls from a H.323 network to the PSTN or vice versa, we need a gateway. The block diagram of the gateway is shown in Figure 34.8.

The gateway between the IP network and the PSTN handles two functions: (a) conversion of the signaling protocols between the PSTN and the IP network and (b) media transformation such as conversion of the PCM-coded speech to low bit rate coded speech.

The gateway has two modules. One module does the conversion of the signaling protocols between the PSTN and the Internet. The second module does the necessary transformation of the media—since the PSTN uses 64kbps PCM coding whereas the VoIP network uses low bit rate codes, this coding transformation (voice transcoding) is done by this module.

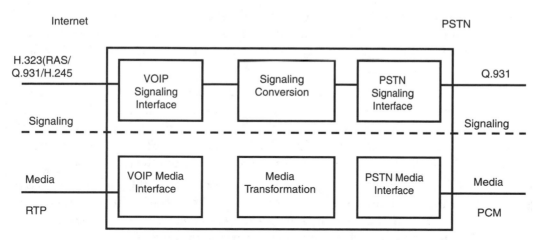

FIGURE 34.8 Interconnecting Internet and PSTN through gateway.

34.5 FAX OVER IP

Fax is used extensively because of its reliability and ease of use. Fax messages are transmitted using the PSTN and hence using the circuit switching operation. As in voice, we need to pay for long distance calls for sending messages outside the local area. If fax transmission is done using the IP networks, we can save a lot of money on fax calls. Original fax data is in digital form, and hence if packet transmission is used, there is no need to convert into analog form as is done for transmission over PSTN.

To transmit fax messages over IP networks, there are two approaches: the store-and-forward method and the real-time method.

In the store-and-forward method, the fax message is stored at intermediate servers and finally reaches the destination. It is similar to the store-and-forward mechanism used in e-mail except that the message is in fax format. It is easy to implement this method over the IP networks. ITU standard T.37 defines the store-and-forward method used for fax over IP.

In the store-and-forward method of fax transmission over an IP network, the fax message is stored at a server and then relayed to the destination. ITU standard T.37 defines the protocols used for this method.

ITU standard T.38 defines the real-time transmission of fax over IP; this standard is a part of the H.323 series standards. Delivering a fax message in real time is a challenging task because of the inherent problems in the IP networks. The problems are:

- Delay: This can be network delay or processing delay. Both are likely to create problems for fax transmissions—if the calling fax machine does not receive a response in a fixed time, it may time out and disconnect the call. Spoofing is used to overcome the problem of delay. In spoofing, a local response is generated without waiting for the response from the called machine.

- Jitter: In IP networks, we cannot guarantee that all the packets will arrive at the destination with the same delay. The variation in the delay causes problems in reassembling the packets at the receiving end. Generally, packets are time stamped to overcome this problem.

- Packet loss: In voice over IP, the packet loss can be compensated for by replaying the earlier packet or sometimes by ignoring that packet. However, we cannot use the same trick here—information in the fax message is lost if packets are ignored. The packet loss can be overcome using one of the following methods: (a) repeating the information in subsequent packets so that errors can be corrected at the receiving end or (b) using an error detection and correction protocol such as TCP, though it introduces additional delay.

In real-time transmission of fax over IP networks, the issues to be addressed are delay, jitter, and packet loss. ITU standard T.38 defines the protocol for real-time transmission of fax over IP networks. This standard is a part of the H.323 series of recommendations.

We need to devise methods to overcome these problems. Before we study the fax over IP software architecture, an understanding of fax transmission over the PSTN is required.

34.5.1 Fax Transmission over PSTN

For fax transmission, there are two standards defined by ITU. T.30 defines the mechanism for fax transmission, and T.4 defines the format for image data. For transmission over PSTN, the calling fax has to dial the called fax, transmit the data and then disconnect the call, just like a voice call. The fax call is divided into five stages:

- Establishment of a fax call
- Negotiation of capabilities

- Page transfer
- End of page and multipage signaling
- Disconnection of the call

Establishment of fax a call: A fax call can be established either manually or automatically. In manual call establishment, the user dials the destination fax number and then switches to the fax mode. In automatic calling, the calling fax machine dials automatically. The fax call is indicated with a calling tone (CNG), a short periodic tone we can hear. The called fax machine sends an answer tone called station identification (CID), a high-pitched tone that can be heard (an audible indication that things are going fine).

Negotiation of capabilities: In this stage, the capabilities, particularly the speed of transmission, are negotiated. This negotiation takes place at 300bps data rate because the called fax machine's capabilities are not known. The called machine starts the negotiation procedure by sending a digital identification signal (DIS) containing its capabilities (such as the data rate it supports). Also, called subscriber information (CSI) and non-standard facilities (NSF) messages are sent (which are optional). The calling machine responds with a digital command signal (DCS) to define the conditions for the call, after examining its own capabilities table. Then, the calling fax machine sends a training check field (TCF) to verify that the channel is suitable for the transmission at the accepted data rate. The called machine responds with confirmation to receive (CTR) to confirm the capabilities and speed. This process is known as high speed modem training.

Fax transmission over the PSTN is done in five stages: establishment of fax call, negotiation of capabilities of the fax machines, page transfer, end-of-page and multipage signaling, and disconnection of the call.

Page transfer: In this stage, the scanned page data is compressed and sent over the channel. T.4 protocol standard is used to format the page data.

End-of-page and multipage signaling: At the end of the page transmission, if there are no further pages, an end of procedure (EOP) message is sent by the calling machine to indicate that the call is complete. If additional pages are to be sent, a multipage signal (MPS) is sent. The called machine responds with a message confirmation (MCF) to indicate its readiness to receive the next page.

Disconnection of the call: The calling machine sends the disconnect message (DCN) to indicate the end of the call. However, the calling fax machine may disconnect without sending this message.

Essentially, the same procedure is followed for fax over IP as well. However, the problems inherent in the IP networks such as, delay, jitter, and packet loss need to be taken care. Let us see how we can overcome these problems.

34.5.2 Fax over IP

To transmit fax over IP networks (and in general, over the packet networks), an inter working function (IWF) or gateway is required. The function of this IWF is to carry out protocol conversion between the fax machine protocols and the IP-based network protocols. The IWF is implemented as a software solution with necessary hardware interfaces using a fax interface unit (FIU). FIU acts as the interface between the IP network and the fax machine. The block diagram of FIU is shown in Figure 34.9.

For fax over IP, an inter working function or gateway is required which carries out the protocol conversion between the PSTN protocols and the IP network protocols.

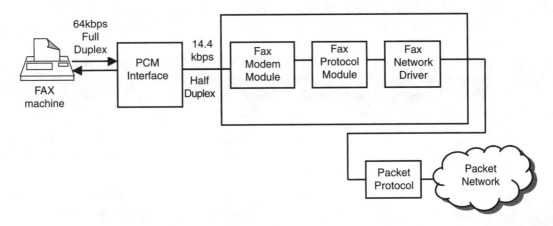

FIGURE 34.9 Fax interface unit.

The functions of the FIU are:

- Demodulate the analog voice band signal from the fax machine and convert it into digital format suitable for an IP-based packet network.
- Modulate data received from the packet network to be sent over the analog interface to the fax machine.

- Convert between analog fax protocols and digital fax protocols used in the packet networks.

As shown in Figure 34.9, the FIU contains three modules: fax modem module, fax protocol converter module, and fax network driver.

Fax modem module: This module has to support the functionality of the analog fax protocols, which include the following:

- CED detection and generation

- CNG detection and generation

- 300bps V.21 channel 2 binary signaling modulation and demodulation

- HDLC framing

- High speed modulation and demodulation per the following standards: V.27 (2400, 4800 bps), V.29 (7200, 9600 bps), V.17 (7200, 9600, 12000, 14400 bps) and V.33 (12000, 14400 bps).

Fax protocol converter module: This module compensates for the effects of delay and lost packets. It prevents the local fax machine from timing out if the response is not received from the called fax machine due to network delay. This is done by generating the HDLC flags. If the response is not received even after time out, it sends a command repeat (CRP) to resend the frame. The other function of this module is to carry out protocol processing.

Fax network driver: This module assembles and disassembles fax packets for transmission over packet networks. The functions of this module are to format the control information, format the fax data, remove the jitter by delaying the data through elastic buffering, and compensate for lost packets by repeating information of a packet in subsequent packets so that errors can be corrected using the information in the subsequent packets.

34.6 CISCO'S ARCHITECTURE FOR VOICE, VIDEO, AND INTEGRATED DATA (AVVID)

Cisco's Architecture for Voice, Video, and Integrated Data (AVVID) provides voice/data/fax/video services over IP networks.

Cisco, the pioneer in communications technology, developed the Architecture for Voice, Video, and Integrated Data (AVVID) to provide voice/video/fax services over IP networks. Using AVVID, an enterprise can provide efficient customer service through Internet-enabled call centers as well as reduce its communication costs drastically using voice over IP, video over IP, and fax over IP. The block diagram of AVVID is shown in Figure 34.10.

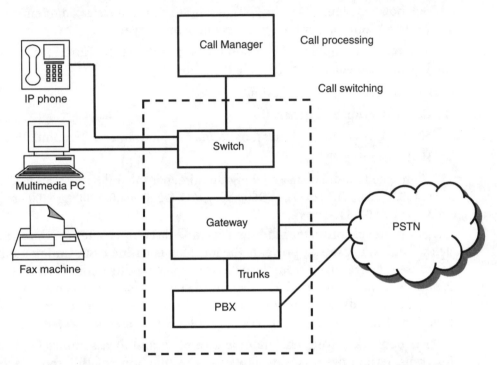

FIGURE 34.10 Architecture for Voice, Video and Integrated Data (AVVID).

The various components of this architecture are described in the following:

IP phone: The IP phone is a standalone phone running the H.323 protocols. It can be connected to the switch using an Ethernet interface. A user can make calls to any PSTN phone or any other IP phone. Voice will be packetized and sent over the IP network.

Multimedia PC: The PC can also act as an IP phone. The PC will be running the H.323 protocol stack to provide audio and video services. This will also be connected to the switch using an Ethernet interface.

Fax machine: The fax machine will be connected to the gateway to facilitate fax over IP as discussed in the previous section.

PBX: The legacy PBX can be connected to make telephone calls from the PBX extensions. The PBX will be connected to the gateway through trunks. The existing connectivity between the PSTN and the PBX can continue.

Switch: The switch provides the necessary functionality to switch the calls. In conjunction with the call manager, it obtains the necessary information for routing a call and switches the call to the appropriate destination either on the PSTN or on the PBX.

Gateway: The gateway does the necessary protocol translation. Since the call processing functions and voice-coding formats are different in PSTN and IP networks, the necessary translations are done here. For example, the PCM-coded speech is converted into H.323 standards–based speech coding. Similarly, the protocol differences in fax messaging are handled by the gateway.

Call manager: The call manager is an IP-based PBX running on Windows 2000 Server. Through software, all call processing software is implemented. The call manager also contains the directory of users (equivalent to a telephone directory, but with IP addresses of the phones). Using dedicated hardware, it carries out the necessary voice coding per H.323 standards.

> The call manager is an IP-based PBX that runs on Windows 2000 Server. The call manager carries out call processing and low bit rate coding of voice. It also keeps a directory of users.

The system shown in Figure 34.10 can be enhanced by integrating other products such as these:

- Customer relations management software for call center applications
- Interactive voice response systems
- Automatic call distribution
- Unified messaging by integration of voice mail, text-to-speech conversion and so on.

The architecture of a call center using these products is given in the next chapter.

34.7 MERGER OF DOTCOM AND TELECOM OR THE DEATH OF PSTN?

In the past, the PSTN was used exclusively for voice and fax communications and the Internet for data communications. Today, the Internet (in general, any IP network) is used for voice, fax, and video as well. This is an important paradigm shift, giving rise to the convergence of networks and the convergence of services

Multimedia communication over IP networks is paving the way for the convergence of various services. The Internet infrastructure can be effectively used for two-way communication, multiparty conferencing, and audio/video broadcasting.

to end users. Though the Internet in the present form is constrained by lack of bandwidth to support high-quality voice and video traffic, in the future, bandwidth will not be a constraint. That will lead to a lot of restructuring in the business strategies of the basic telecom operators and the Internet service providers.

If the Internet is going to be used for voice, fax, and video transmission, what will happen to the good old PSTN? During the initial days of introduction of VoIP, a few futurologists wrote the obituaries of major long distance telecom providers. In countries where long distance communication is controlled by governments, VoIP is considered illegal, but the use of the technology cannot be stopped, and sooner or later every country will have to allow voice and video communication over the Internet. Does it lead to long distance carriers going out of business? Does it lead to the death of PSTN? Well, the old adage came into action: "If you cannot beat them, join them." The long distance carriers joined hands with the ISPs to offer multimedia services over the Internet by offering their bandwidths.

Multimedia over IP is the integrating force for convergence. The PSTN, PLMN, Internet, and cable TV networks can all merge together to offer a unified service to the subscriber. The subscriber will be able to choose a single terminal (or multiple terminals) to access data, voice, fax, and video services without bothering about the underlying network or the protocols. Each subscriber can also be identified by a single personal telecommunications number (PTN), and the network can be programmed to route calls to the PTN that is identified by the terminal of choice. And, the subscriber can get a single bill for all the services.

Whether the PSTN as we know it will survive or die is a moot point. The subscriber has to be provided quality service for all the communication needs, and this service should be easy to use, of low cost, and reliable.

34.8 JAVA MEDIA FRAMEWORK

Java Media Framework (JMF) is an application programming interface (API) developed by Sun Microsystems to facilitate development of multimedia applications over IP networks. Java applications or applets can be developed incorporating multimedia content. JMF 2.0 allows storing/playing multimedia files, processing the media streams, and customizing the JMF functionality through plug-ins.

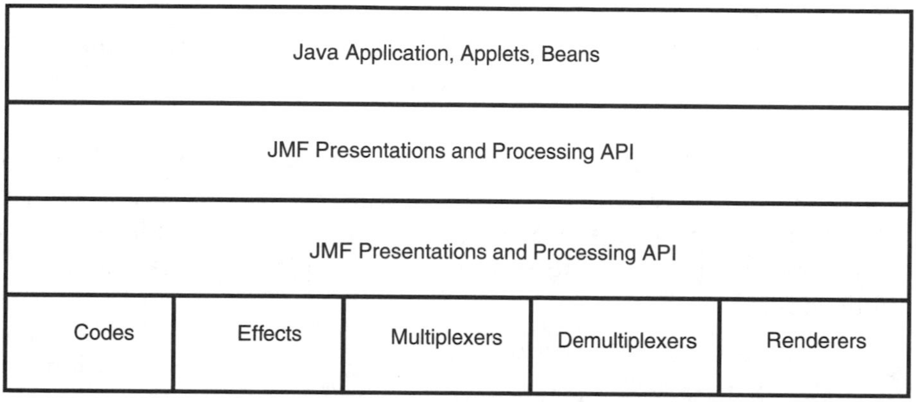

Java Application, Applets, Beans				
JMF Presentations and Processing API				
JMF Presentations and Processing API				
Codes	Effects	Multiplexers	Demultiplexers	Renderers

FIGURE 34.11 Architecture of JMF.

Java Media Framework (JMF) is an application programming Interface (API) that facilitates development of multimedia applications over IP networks using Java programming language.

The architecture of JMF is shown in Figure 34.11. Its various components are described in the following:

Codecs: Based on the type of application to be developed (keeping in view the bandwidth availability, CPU requirements, quality required, and so forth), for voice and video transmission, an appropriate coding technique has to be chosen. JMF supports the following audio codecs:

- PCM A-law and m law. (AVI, Quicktime, WAV RTP formats)
- ADPCM (AVI, QuickTime, WAV RTP formats)
- MPEG1 (MPEG format)
- MP3 (MPEG layer 3 format)
- GSM (content type: WAV RTP)
- G.723.1

Multiplexers: Multiplexers allow combination of various streaming media such as voice and video.

Demultiplexers: Demultiplexers allow the separation of the streams obtained from the multiplexers.

Effects: Effect filters allow modifying the data to create special effects such as echo blurring. Generally, effect filters are applied to the raw data before compression.

Renderers: Renderers are presentation devices. For audio, the renderer is the hardware that puts sound to the speakers; for video, the renderer is the monitor.

JMF APIs: JMF APIs consist of interfaces that define the behavior and interaction of objects that are used to capture, process, and present media. Through plug-in APIs and presentation and processing APIs, one can create Java applications using streaming media. Typical applications include the following:

- Intranet-based voice chatting
- Intranet-based voice mail
- Interactive voice response system
- Talking Web pages
- Internet-enabled call centers
- H.323 network elements
- IP phones
- Desktop video conferencing
- Fax transmission over IP networks

JMF provides the facility to develop multimedia applications over IP networks. Since Java provides platform independence, the applications developed can be run on various types of devices—desktops, palmtops, mobile phones, and so on.

Using JMF, applications such as intranet-based voice/video conferencing, interactive voice response systems, IP phones, and fax communication over IP networks can be developed.

Multimedia communication over packet networks will revolutionize the telecommunications industry in the coming years.

Summary

This chapter presented the details of multimedia communication over IP networks. IP networks were not originally designed for carrying multimedia traffic. Particularly for real-time communication such as audio/video conferencing, IP networks are not suitable. This is due to the fact that IP does not guarantee a desired quality of service—packets may be lost, delay is variable, and so on. To overcome these problems, special protocols have been designed.

The H.323 standards specify the protocols to run on IP networks for real-time communication of voice, video, and fax. To achieve real-time performance, instead of TCP, UDP is used as the transport layer. Above UDP, two protocols— Real Time Transport Protocol (RTP) and Real Time Control Protocol (RTCP)— are defined for real-time communication.

Applications using multimedia communication over IP networks can be developed using the Java Media Framework APIs developed by Sun Microsystems. We also discussed Cisco's Architecture for Voice, Video, and Integrated Data (AVVID), which facilitates developing advanced corporate networks for multimedia communication over IP networks.

References

M. Hasan et al. "Internet Telephony: Services, Technical Challenges and Products," *IEEE Communications Magazine*, Vol. 38, No. 4, April 2000.

K. McGrew. *Optimizing and Securing Cisco AVVID Applications*, Wiley Publishing Inc., 2003. This book gives the complete details of AVVID and how to build IP-based call centers by integration of Cisco's products such as IP phones, CRM software, IVR, ACD, etc. It also gives details of storage area networks and content distribution networks.

Dreamtech Software Team. *WAP, Bluetooth and 3G Applications*, Wiley Dreamtech India Pvt. Ltd. 2001. This book gives the source code for developing multimedia applications using JMF.

Internet telephony software can be obtained from these sites:

www.microsoft.com/netmeeting NetMeeting

www.vocaltech.com InternetPhone

www.freetel.com FreeTel

www.upine.com CU-SeeMe

www.sun.com Sun Microsystems' Web site, a repository of information and resources for developers of multimedia applications.

Questions

1. What are the limitations of IP networks for multimedia communication?
2. Describe the H.323 protocol architecture for achieving multimedia communication over IP networks.

3. Describe the fax over IP operation.

4. Explain how real-time performance can be achieved using RTP and RTCP.

5. Explain JMF architecture.

6. What is AVVID? Describe its components.

Exercises

1. Study the use of Microsoft's NetMeeting software for desktop video conferencing.

2. Which standard is followed for fax communication over the PSTN?

3. Study the details of Cisco's call manager and prepare a technical report.

4. Download the freely available H.323 protocol stack and study the source code.

5. Study the various API calls provided in the JMF.

Projects

1. Using Java Media Framework API, create an application for audio conferencing between users on a LAN.

2. Using Java Media Framework API, create a voice mail application. Every user should have a voice mailbox on the server. Necessary security features have to be built into the software so that a user can access his voice mailbox only through a password.

3. Develop a video conferencing application using JMF.

4. Develop an intranet messaging system. In a LAN, the user should be able to log on to a server (Windows or Linux) for accessing the following services: e-mail, chat, voice mail, and voice chat.

5. Obtain the freely available H.323 protocol stack and port it onto a system. Study how the software can be ported onto an embedded system to develop an IP phone.

35 Computer Telephony Integration and Unified Messaging

In This Chapter

- Computer Telephony Integration
- The Technology Ingredients of CTI
- Text-to-Speech Conversion
- Interactive Voice Response Systems
- Automatic Speech Recognition
- Unified Messaging

The telephone is the most widely used gadget for communication all over the world, and telephone use is very high compared to computers. A huge amount of information is available on computers, but the number of people who have access to that information through a computer is still far behind the number of those who use a telephone. If we can make information on computers accessible by telephones, information dissemination will be much more efficient. Computer telephony integration (CTI) does precisely this.

CTI also gave rise to interesting developments in messaging through unified messaging. Unified messaging enables people to use voice as the medium to access information available in text form using text-to-speech conversion technology. It facilitates accessing the information using voice recognition instead of the keyboard and mouse, and it also helps in using one device (such as a mobile phone) to access messages (text or voice) available on servers at different locations.

In this chapter, we will study the technology ingredients of CTI and unified messaging.

35.1 NEED FOR CTI

CTI technology is finding applications in every business sector. Why? The following are the reasons:

- Throughout the world, the use of telephones is very high. In most developed countries, the telephone density (number of telephones per 100 population) is anywhere between 50 and 90. In developing countries, it is lower—between 2 and 20, but still higher than the number of computers.

- To access the information using a telephone is much easier than using PC, mostly because voice is the most convenient mode of communication, and operating a telephone is much easier than operating a PC.

- With the advent of automation, much of information is available on computers—whether from banks, libraries, transport agencies, service organizations, and others—some of this information needs to be provided to users/customers.

- Throughout the world, manpower costs are rising. If the monotonous work of distributing routine information can be automated, organizations can save money and use their human resources for more productive work.

- CTI provides the platform for unifying the various messaging systems. It benefits the user because it provides a single device for accessing different messaging systems such as e-mail, voice mail, and databases.

The purpose of computer telephony integration (CTI) is to access information stored in computers through telephones. CTI is gaining practical importance due to the high number of telephones and availability of large amounts of information in computers.

For these reasons, CTI is emerging as one of the highest growth areas in IT. However, CTI is not a new technology; it is a combination of many existing technologies to provide innovative applications to users.

CTI finds applications in every business sector—banking, education, transportation, service organizations, entertainment, and others.

35.2 TECHNOLOGY INGREDIENTS OF CTI

CTI uses various technologies to provide value-added services to users. The three important technologies on which CTI is based are:

The three technology ingredients of CTI are: (a) text-to-speech conversion; (b) interactive voice response systems; and (c) automatic speech recognition.

- Text-to-speech conversion
- Interactive voice response systems
- Automatic speech recognition

We will study these technologies in the following sections.

35.3 TEXT-TO-SPEECH CONVERSION

Speech is the most natural and convenient mode of communication among human beings, and it enriches interaction between computers and humans as well. In the past few decades, significant progress has been made in achieving interaction between humans and computers through speech, though unrestricted speech communication between humans and computers still remains an elusive goal. Text-to-speech conversion and speech recognition are the two pillars on which human interaction with computers through speech rests.

Text-to-speech conversion is gaining importance in many information processing systems, such as information retrieval from databases, computer-aided instruction, conversion of e-mail into speech form so that e-mails can be accessed from telephones, reading machines for the blind, and accessing Web pages from telephones. As shown in Figure 35.1, text-to-speech conversion involves mainly three steps: machine representation of text, transliteration through which text is converted into its corresponding sound symbols, and synthesizing the speech signal from the sound symbols.

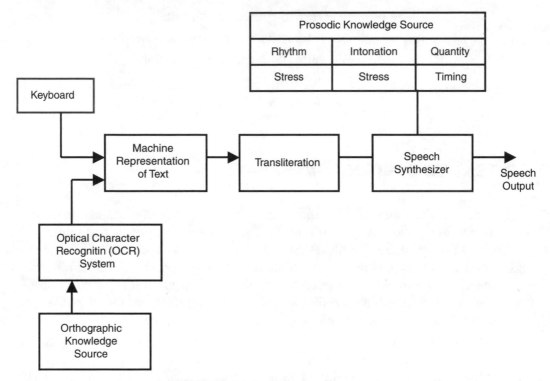

FIGURE 35.1 Text-to-speech conversion.

 Text-to-speech conversion has many practical applications such as retrieval of information stored in databases through telephones, reading machines for the blind, and accessing Internet content using telephones.

35.3.1 Machine Representation of Text

Typed text is represented in the computer using ASCII or Unicode. Alternatively, optical character recognition software is used to convert the written or typed text into machine code.

Text can be input into the computer either through a keyboard or through an optical character recognition (OCR) system. When the text is input through the keyboard, the characters are encoded using American Standard Code for Information Interchange (ASCII) format, in which 7 bits represent each character. For Indian language representation, ISCII (Indian Standard Code for Information Interchange) is used—both 7-bit and 8-bit ISCII formats are available. If the OCR system is used to input the text, an orthographic knowledge source that contains the knowledge of the written symbols is required. The OCR software interprets the scanned text and converts it into machine code.

Unicode is now used extensively for machine representation of text in all world languages. Unicode is a 16-bit code. Programming languages such as Java as well as markup languages such as XML support Unicode.

Optical character recognition (OCR) will be highly accurate for typed text but not for handwritten text. Many pattern recognition algorithms are used in OCR.

35.3.2 Transliteration

The process of converting text into its equivalent pronunciation is called transliteration. To convert English text into its equivalent pronunciation, nearly 350 pronunciation rules are required.

Conversion of text into corresponding pronunciation is called *transliteration*. This process is complicated for languages such as English. This is because in English, 26 letters are mapped onto 42 phonemes—a phoneme is the smallest speech sound in a language. These phonemes are represented by special symbols (such as a, aa, i., ii, u, uu). For English, 350 pronunciation rules are required to convert text into the corresponding pronunciation. Even with these 350 rules, all the words are not pronounced properly. Hence, a dictionary of exceptions is required. This dictionary contains the words and their corresponding pronunciation in the form [are] = aa r.

When a word is given as input to the transliteration algorithm, first the word is checked in the dictionary of exceptions. If the word is found in this dictionary, the corresponding pronunciation is assigned to it. If the word is not found in the dictionary, the pronunciation rules are applied and the pronunciation is obtained.

Unlike English, most Indian languages are phonetic languages, meaning that there is a one-to-one correspondence between the written form and the spoken form. Hence, for Indian languages, there is no need for a transliteration algorithm. An exception to this rule is Tamil, in which the pronunciation of a letter may depend on the following letter; so one symbol look-ahead algorithm is required for transliteration of Tamil.

Transliteration of most Indian languages is easy because there is a one-to-one correspondence between the written form and spoken form. An exception is that Tamil language.

35.3.3 Speech Synthesis

From the pronunciation of the text obtained from the transliteration algorithm, speech has to be generated by the computer. Conceptually, for each phoneme, the speech data can be stored in the computer and, combining the data for all the phonemes in a given word, speech can be generated. For example, if the speech data for the phonemes aa and r are combined, we can produce the sound for the word "are." But unfortunately, it does not work well, and the quality of speech produced this way will be very poor.

To produce good quality speech, various techniques have been tried. These techniques are based on using different basic units of speech—words, syllables, diphones, and phonemes.

Words: A simple mechanism is to use the word as the basic unit: each word in the language is spoken and recorded in the computer. When a sentence is to be spoken, the speech data corresponding to all the words in the sentence is concatenated and played. This approach gives very good quality speech. The only problem is that the number of words in any language is very high.

Assume that we store about 100,000 words of English and each word on an average takes about 0.4 seconds. We need to store 40,000 seconds of speech, which is 320Mbytes of data if 64kbps PCM coding is used. This used to be a very high storage space requirement in earlier days, but not any longer. One CD-ROM can hold nearly 200,000 words of speech data. Now the only requirement is to be able to pick up the data corresponding to the required words quickly from the database, and so a fast searching mechanism is all that is required. Nowdays, many text-to-speech conversion systems follow this approach.

Syllables: A syllable is a combination of two phonemes. The symbol for ka in any language is a syllable consisting of two phonemes—k and a. Similarly, kaa, ki, and kii are syllables. Each language has about 10,000 to 30,000 syllables. If a syllable is taken as the basic unit of speech, we can store the speech data corresponding to these syllables. From the transliteration algorithm, we obtain the pronunciation, and from the pronunciation and the syllable speech data, we can synthesize the speech. This approach gives good quality speech and is recommended if there is a constraint on the storage space.

Using the syllable as the basic unit is the best approach to obtain good quality speech in text-to-speech conversion systems. The number of syllables in a language will be between 10,000 and 30,000.

NOTE

For synthesizing speech, the waveforms corresponding to basic units of speech need to be stored in the computer. These basic units can be words, syllables, diphones, or phonemes. Depending on the storage space and quality of speech required, the best basic unit can be chosen.

Diphones: The sound from the middle of one phoneme to the middle of the next phoneme is called diphone. For instance, in the word put there are four diphones: #p, pu, ut, and t#, where # stands for blank. Diphone as the basic unit is considered an attractive choice because the transition from one phoneme to another phoneme is important for obtaining good quality speech. The number of diphones in any language is limited to about 1,500, and so the storage requirement is very small.

Phonemes: The number of phonemes in any language is very small—fewer than 63 in any language. Hence, if the phoneme is used as the basic unit, the storage requirement will be very small. However, it is difficult to get good quality speech because there will be subtle variation in the sound of a phoneme when it occurs in the context of other phonemes—for example, the phoneme b has different sounds in the two words "bat" and "but". How to obtain this effect in speech synthesis is still an active research area. The advantage of using phonemes as the basic unit is that we can manipulate the sound through software, for example to put stress or vary the pitch. This is required to obtain natural sounding speech.

To summarize, text-to-speech conversion using words and syllables as the basic units gives good quality speech. If the vocabulary is limited, as in the case of most systems (discussed in the next section), it is better to use words; if the vocabulary is very large, it is better to use syllables.

To make computers generate natural sounding speech is still very difficult. Speech produced by most of the text-to-speech conversion systems sounds rather artificial. This is because when people speak, they vary the pitch, they vary the stress on different sounds, and they vary the timing (some sounds are elongated). She is a beautiful woman has different meanings depending on how one pronounces the word "beautiful." Currently, the machines cannot do that kind of job.

To develop text-to-speech conversion systems that produce natural-sounding speech, we need to introduce three special effects (refer to Figure 35.1): intonation (variation of pitch with time), rhythm (variation of stress with time), and quantity (variation of duration of the phonemes). This calls for various knowledge components as shown in Figure 35.2. The knowledge of written symbols is represented by orthographic components, and the knowledge of phonemes is represented by phonological components. Using phonemes, words are formed with the lexical knowledge component. Using the syntactic component (grammar), words are combined to form sentences. Semantic components represent the meaning associated with the words. A prosodic component represents the context-dependent meaning of the sentences. Generation/understanding of speech using all these knowledge components is still complicated and it will require thousands of people their Ph.D. in this area to make computers talk like human beings. To produce natural-sounding speech by developing these knowledge sources is an active research topic in artificial intelligence.

> To produce natural-sounding speech, intonation, rhythm, and quantity are important. Intonation is variation of pitch with time. Rhythm is variation of stress with time. Quantity is variations in the duration of the phonemes.

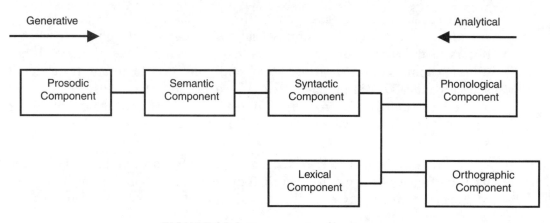

FIGURE 35.2 Components of language.

To produce natural-sounding speech is still an active research area because many artificial intelligence concepts need to be introduced for generating speech.

35.3.4 Issues in Text-to-Speech Conversion System Design

> To develop commercial text-to-speech conversion systems, the design issues are vocabulary size, basic speech units, number of languages, and the low bit rate coding technique to be used for storing the basic speech units.

To develop practical text-to-speech conversion systems that give very high quality speech is still a challenging task. The following issues need to be considered while designing commercial text-to-speech conversion systems:

Vocabulary size: If the vocabulary is limited (as in most of the IVR systems), the speech data corresponding to the words and phrases can be recorded directly. Text-to-speech conversion is basically concatenation of the speech data files and replaying the concatenated file to the user.

If the vocabulary size is not limited, conversion to the phonetic alphabet using a transliteration algorithm has to be done, and then the speech synthesis has to be done.

Basic speech units: Depending on the quality required and the storage space available, the basic speech unit has to be chosen—it can be word, syllable, diphone, phoneme. As discussed earlier, depending on the unit chosen, the quality of the speech will vary.

Number of languages: In some applications, multilingual support is required. Each language has to be considered separately, and the text-to-speech conversion mechanism has to be worked out.

Low bit rate coding of speech: Storing the speech data in PCM format (64 kilobits per second) requires lots of storage space. Though storage is not a major issue for many applications, some applications may demand conserving storage space, as in the case of talking toys. In such a case, low bit rate coding schemes such as ADPCM or LPC can be used. Of course, quality and storage space are trade-off parameters.

If you can store the speech waveforms of about 200,000 words of any language, you can get good quality speech by concatenating the words to form sentences. Though the storage requirement is high, storage is not a constraint on desktops.

35.4 INTERACTIVE VOICE RESPONSE SYSTEMS

The interactive voice response (IVR) system is a simple yet efficient system for retrieving the information from computers in speech form through telephones. As shown in Figure 35.3, the IVR hardware consists of a PC add-on card. This card has a telephone interface to connect to the telephone in parallel. Any telephone subscriber can call the IVR number (the number of the telephone to which the IVR card is connected) through the PSTN to access the information available on the computer in which the add-on card is located.

In an interactive voice response (IVR) system, a PC add-on card has telephone interface circuitry. A subscriber can call an IVR system and access the information stored in the PC. The information is converted into speech form and sent to the subscriber.

We will study the functioning of an IVR system, its architecture, applications, and design issues in this section.

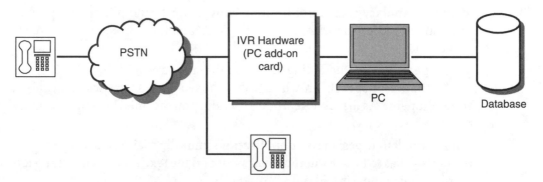

FIGURE 35.3 Interactive voice response system.

35.4.1 How an IVR System Works

Consider the example of obtaining train reservation status information through an IVR system. You hold a ticket for a particular train—the ticket contains the ticket number, the train number, and the reservation status when you purchased the ticket. The IVR system of the railway authorities contains the complete database with the latest information on the reservation status. To find out the status of the reservation at any time, you can call up the IVR system. You dial the IVR number and hear a response: "Welcome to the IVR system for reservation status enquiry. Please dial the train number." You dial the train number by pressing the digits on the keypad. The IVR system will prompt again: "Please dial your ticket number." You dial the digits 765391. The system will open the database corresponding to the reservation for the given train and check the status for the given ticket number. The database contains three fields: confirmed, S7, and 64, indicating that the ticket has a confirmed berth in coach number S7 and berth number 64. The system will convert this information into a sentence: "Your ticket is confirmed, coach number is S7, and berth number is 64. Note that coach number and berth number are obtained from the database and put at the appropriate places in the sentence; these fields will vary for different users. The next step is to convert this sentence to speech form. Speech data corresponding to all these words is stored in the computer along with data corresponding to all digits. This data is concatenated (joined together) to form the sentence, and the data is converted into analog form and sent over the telephone line.

In an IVR system, the user is presented a menu of items. The user can select a menu item by pressing the corresponding DTMF digit on the telephone keypad. The information corresponding to that menu item will be sent to the user.

We can now see the great advantage of an IVR system—the data available on computers can be retrieved without the need for a computer at the user's end, just through telephones. There is no need for an operator for providing reservation status—our IVR system is user friendly, does not take coffee breaks, and above all, gives polite replies. It is not surprising, that IVR systems have gained wide acceptance all over the world.

We can now see that the IVR system should have the following functional blocks for the hardware:

- Telephone interface circuitry, to pick up a call and to recognize the various tones (the user may put back the telephone after dialing a few digits, and the circuitry has to take care of such situations).

- PC interface circuitry to interact with the PC and obtain the data (the speech files).

- Circuitry to recognize the digits dialed by the user. Two types of dialing are possible: pulse dialing in which each digit is represented by a set of pulses and DTMF signals in which each digit is a combination of two sine waves.

- A PCM codec that converts the digital speech stored in the computer to analog form to transmit on the telephone line.

- Control circuitry to control all functional blocks.

An IVR system consists of telephone interface circuitry, PC interface circuitry, circuitry to recognize the digits dialed by the user, a PCM codec, and control circuitry, in addition to the software to generate the desired responses.

In addition, the IVR system contains the software that takes the recognized digits from the hardware, accesses the database, generates the required response, converts the response into speech form by concatenation of different speech data files, and passes the speech data to the IVR hardware. Before getting into the architecture, let us see the various applications of IVR.

IVR Application Panorama

IVR systems find use in almost every business sector. Some representative applications are as follows:

Automated attendant: Information required by clients/customers can be provided even after office hours and during holidays.

IVR systems have applications in a number of areas such as telemarketing, telebanking, tele-surveys, value-added service provisioning by service industries, entertainment, and fax-on-demand.

Telemarketing: Marketing agencies can provide the information required by customers (such as prices of items, stock availability, features of items) automatically to their customers.

Teleshopping: Customers can dial into the IVR system and place an order for the required item (by dialing the required item number, quantity, and so on). If payment is by credit card, the credit card number also can be input by the customer.

Banking: Banks can provide services whereby a customer can dial into the IVR system to find out the bank balance, currency exchange rates, information about the various deposit schemes, and so forth.

Talking Business Directory: An IVR system can contain category-wise information about various products. The user will be provided with a voice menu from which he can select the required product and hear a beautiful voice explaining the salient features of the product. These talking business directories will be user friendly and can be updated frequently.

Telesurveys: Marketing agencies, government organizations, broadcast agencies, and such can conduct surveys through IVR systems. Users will be asked to dial a specific number (telephone number of the IVR system) and then dial various codes to convey their opinions, which will be recorded and stored for subsequent processing.

Value-added services by telecom operators: Telecom operators can provide services such as morning alarm service, changed number announcement service, telephone bill reminder service, fault booking service, bill enquiry service, and so on using an IVR system without any human intervention.

Transport agencies: Airlines/railways/bus transport authorities can use an IVR system for automatic announcement of arrivals/departures, reservation status enquiry, cargo status enquiry, and others.

Entertainment: One popular IVR application is to obtain astrological predictions—the user has to dial his date and time of birth, and the IVR system will give the astrological predictions for the year or the month or the day. Another popular application is audio-on-demand—one can dial an IVR system and listen to a music recital, a Shakespearean drama, or a novel by Pearl S. Buck. To use an old cliché, for IVR in entertainment, imagination is the limit.

Fax-on-demand: One can dial an IVR system and listen to a menu that gives details of fax messages that can be received by a user (for example, product literature). The user can select an item and then input the fax number to which the fax message has to be sent. The selected message will be faxed to the user. Of course, an additional module that invokes fax transmitting software (such as Microsoft Fax, BitWare fax, or WinFax pro) is also required along with an IVR system.

Every market segment—government, corporate, health care, transport, telecommunication, education, entertainment, banking, social service organizations—can apply IVR technology to increase productivity and to reduce human resources for carrying out mundane activities. That is the reason IVR is a multibillion dollar industry in both developed and developing countries.

35.4.2 IVR System Architecture

As shown in Figure 35.3, the IVR system consists of IVR hardware (generally a PC add-on card), a PC, and application software running on the PC. The IVR hardware is connected in parallel to the telephone line. The requirements of the IVR hardware are:

- To go off-hook (automatic lifting of telephone) when a user calls.
- To recognize the digits dialed by the user.
- To interact with the PC to obtain the necessary speech files and also to send the recognized digits to the PC for further processing.
- To convert the digital speech into analog form and send it to the telephone line.
- To convert the analog speech from the telephone line (if recording of incoming speech is required) into digital form and send it to the PC for storage.

The functionality of the IVR hardware is to lift the telephone automatically when the user calls, recognize the digits dialed by the user, interact with the PC to obtain the necessary information, convert the information into speech form and convert the incoming speech into digital form and store it in the PC.

The entire circuitry corresponding to these requirements is generally in the form of a PC add-on card that sits on the motherboard of the PC.

This circuitry is for a one port IVR system because the IVR system has one telephone line interface. If the application demands multiple ports (such as 2, 4, 8), additional circuitry is required to handle that many ports. Applications such as a railway status enquiry require multiple ports because many users will access the IVR system simultaneously.

To reduce the storage requirement, instead of using PCM, low-bit rate coding techniques such as ADPCM LPC can be used to store the speech waveforms.

NOTE

35.4.3 IVR Application Design Issues

In designing an IVR system, the human factors are the most important—after all, the user has only a telephone keypad for interaction with the computer. Designing a cost-effective and user-friendly IVR system involves the following issues:

- **Selection of application:** The application should not be too general or too narrow. If it is very general, the user has to be presented with a long menu that will be difficult to remember. Too narrow an application will not make the best use of the IVR system's capabilities and will not be cost effective.

- **Design of user interface:** The menu to be presented to the user should have typically four or five choices. For each choice, there can be a submenu. Provision should be made to go back to the previous menu item.

- **Design of database:** The database can be a short text file or it can be a popular RDBMS package. From the user's dialed responses, the information has to be accessed by generating the appropriate query to the database.

- **Text-to-speech conversion:** If the response to the query is of limited vocabulary, speech corresponding to the words can be stored directly. If the vocabulary is large (or unlimited), text-to-speech conversion software needs to be used. It is important to ensure that proper names are pronounced correctly.

- **Number of ports:** The traffic that has to be handled by the IVR system decides the number of ports. Based on the number of simultaneous calls that have to be handled by the IVR system, the number of IVR ports can be anywhere between 1 and 64. The system can also be upgraded as the traffic goes up.

- **Pulse-to-tone conversion:** A good number of telephone users still use rotary (pulse) telephones, whereas most of the commercially available IVR systems are based on DTMF (tone) detection. A pulse-to-tone converter may be necessary if access is to be provided to all the telephone subscribers without restricting access to only those who have DTMF telephones.

- **Accuracy of digit recognition:** Accuracy of recognition of the DTMF digits dialed by the user is extremely important for an IVR system to respond correctly. Most of the DTMF recognition chips achieve an accuracy up to 99%, which is sufficient for most applications, but not for applications such as banking and credit card verification. In such sensitive applications, the IVR system can speak the dialed digits and ask for confirmation by the user

The various issues to be considered while designing practical IVR systems are selection of a focused application, user interface design, database design, text-to-speech conversion algorithm, number of ports, need for pulse-to-tone conversion, and the accuracy of digits recognition.

(for example, the user can dial one to confirm, zero to indicate that the digits recognized are not correct).

Nowdays, more sophisticated IVR systems are becoming available wherein voice recognition is employed. The user need not dial the digits; instead, he can speak the digits. Even for such systems, the above issues need to be considered in designing a user-friendly IVR system.

35.5 AUTOMATIC SPEECH RECOGNITION

To make computers recognize our speech is an extremely difficult task because of the variations in our speech—pronunciation and speech signal characteristics differ according to our mood and health.In addition, a person's accent indicates things such as his educational, geographical, and cultural background. The iceberg model of speech, shown in Figure 35.4, represents the information contained in the speech signal. The information in the lower portion of the iceberg is difficult to capture during the analysis of a speech signal for speech recognition applications.

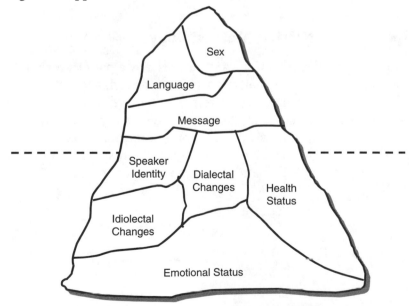

FIGURE 35.4 Iceberg model of speech.

Automatic speech recognition is a very challenging task mainly because speech characteristics vary widely from person to person. Practical speech recognition systems have many limitations and are capable of recognizing only limited vocabulary or a limited number of speakers.

Research in speech recognition technology is limited mainly to recognizing the message and speaker recognition. Recognizing the other features has not met with much success.

In the present speech recognition systems, the speech signal is analyzed and compared with prestored templates of the words to be recognized. Using this approach, a good recognition rate is obtained if the number of speakers is limited or if the computer is trained to recognize the speakers.

Future speech recognition systems need to incorporate various knowledge sources such as syntax (grammar) and semantics (meaning) knowledge sources to achieve high recognition rates. Though many such research systems are available, commercial systems are still a few years away. The iceberg model indicates the various knowledge components that need to be developed and incorporated for achieving a complete speech recognition system.

The iceberg model of speech depicts the various characteristics that need to be recognized from speech waveforms. Only those characteristics that are at the top of the iceberg can be recognized at present.

35.5.1 Speech Recognition Methodology

Automatic speech recognition (ASR) is basically a pattern recognition task, as shown in Figure 35.5. The input test signal is analyzed, and features are extracted. This test pattern is compared with reference patterns using a distance measure. Then a decision is made based on the closest match. Because speaking rates

Automatic speech recognition involves pattern recognition. It consists of three phases: feature extraction, time normalization, and distance calculation.

vary considerably, time normalization has to be done before calculating the distance between a reference pattern and the test pattern. The different stages of ASR are described in the following paragraphs.

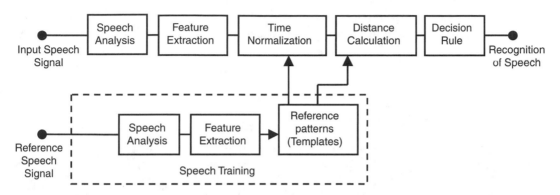

FIGURE 35.5 Speech recognition methodology.

In the feature extraction stage, the speech waveform is divided into frames of about 10 msec duration, and the parameters such as LPC coefficients are calculated for each frame. The feature set for each word to be recognized is stored in the computer and is called the reference pattern.

Feature extraction: The feature set extracted from the speech signal should be such that the salient properties of the signal are reflected in these features. A number of feature sets have been proposed, such as energy, zero crossing rate in selected frequency bands, and LPC coefficients. The most commonly used feature set is LPC coefficients. The speech signal is divided into small units called frames; each frame is of about 10 msec duration. For each frame, LPC parameters are obtained and stored. For example, if a word is of 0.3 second duration, the speech signal is divided into 30 frames and for each frame, the LPC parameters are calculated and stored as the feature set. If the speech recognition system is designed to recognize say 10 words, for each word spoken by the user, the feature set is extracted and stored. This is known as the training phase, in which the user trains the computer to recognize his voice. The feature set stored during the training phase is known as the *reference pattern*. Later, when the user speaks a word, the features are extracted again, and the feature set obtained is known as the *test pattern*. The test pattern is compared with all the reference patterns, and the closest match is found to recognize the word.

Time normalization is a process wherein the reference pattern is mapped on to the test pattern by normalizing in the time scale. This is required because the duration of the reference pattern and the test pattern may not be the same.

Time normalization: Speaking rates vary considerably, so time normalization has to be done to match the test pattern with the reference pattern. For instance, during the training phase, the word spoken by the user might be of 0.350 second duration, but during the recognition phase, the duration may be 0.352 second. To compare the reference pattern with the test pattern, the two patterns have to be normalized in the time scale. This is achieved using dynamic time warping, which can be done either by mapping the test pattern onto the reference pattern or bringing both patterns to a common time scale.

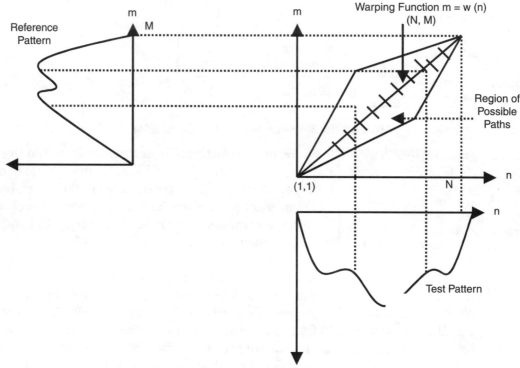

FIGURE 35.6 Time normalization of test pattern and reference pattern: the warping function.

Let T(n) be the feature vector obtained from nth frame of the test utterance and R(m) be the feature vector obtained from mth frame of the reference utterance. So,

T = { T(1), T(2),....T(N) } is the test pattern.

R = { R(1), R(2), ...R(M) } is the reference pattern.

For time normalization, a warping function defined by m = w (n) has to be found, as illustrated in Figure 35.6. The dynamic time warping procedure is used to find the optimal path m = w(n) that minimizes the accumulated distance between the test pattern and the reference pattern, subject to the end point and path constraints. For finding the path, the following recursive solution, proposed by L.R. Rabiner of Bell Laboratories, is used:

$$D_a (T(n), R(m)) = d (T(n), R(m)) + \min [D (T(n-1), R(q))]$$
$$q <= m$$

where $D_a(T(n), R(m))$ is the accumulated distance from $(T(1), R(1))$ to $(T(n), R(m))$ and $d(T(n), R(m))$ is the distance between nth test frame and mth reference frame.

Time normalization is done through a dynamic time warping function. This involves minimizing the accumulated distance between the test pattern and the reference pattern.

Distance calculation: Calculation of distance between test and reference patterns is done along with time normalization. Distance calculation is a computationally burdensome step in speech recognition. From a computational effort point of view, two distance measures, Euclidian and the loglikelihood, have been found to be reasonable choices.

The two distance measures used for calculating the distance between the reference pattern and the test pattern are Euclidean distance measure and loglikelihood distance measure.

A simple Euclidean distance is defined by

$$d(T,R) = \sum_i [T(i) - R(i)]^2$$

Another distance measure is the loglikelihood measure proposed by Itakura and is defined as

$$d(T,R) = \log [a_R v_T (a_R)^t / a_T v_T (aT)^t]$$

where a_R and a_T are the LPC coefficient vectors of the reference and test patterns, and v_T is the matrix of autocorrelation coefficients of the test frame.

Decision rule: Using the distance scores obtained form the dynamic time warping algorithm for the V reference patterns (V being the vocabulary size), a decision is made using either the nearest-neighbor (NN) rule or the K nearest-neighbor (KNN) rule. In NN rule, the pattern that gives the minimum distance is chosen as the recognized pattern. The KNN rule is applied when the number of reference patterns for each word in the vocabulary is P, with P > 2. In such a case, the distance scores between the test pattern and each of the reference patterns are calculated and arranged in ascending order. The average of the first K (= 2 or 3) distance score is taken as the final distance score between the test word and the reference word. Minimum distance criterion is applied on such final distance scores obtained for the entire vocabulary.

To develop a speech recognition system, the computer has to be trained by the user. Each word is spoken by the user, and the templates are created. The training methods are divided into (a) casual training; (b) averaging; and (c) statistical clustering.

Training procedure: For template creation, the training procedure is the most crucial step that reflects on the performance of the system. The training methods can be divided into three classes: (i) casual training; (ii) averaging; and (iii) statistical clustering. In the casual training method, a reference pattern is created for each word spoken during the training mode. In averaging method, a word is spoken n times, and the features are averaged to obtain a single reference pattern. In the statistical clustering method, each word in the vocabulary is spoken n times; these n patterns are grouped into clusters using pattern similarity measures, and from each cluster, templates are obtained.

In the statistical clustering method, each word is spoken a number of times and the patterns are grouped into clusters using pattern similarity measures. Templates are created from each cluster. This method increases the recognition accuracy.

35.5.2 Categories of Speech Recognition Systems

Automatic speech recognition is a complex task for the following reasons:

- The characteristics of speakers vary widely, and to store templates for all speakers requires huge memory. Processing is also time consuming and hence speech recognition without considerable delay is not possible, at least not yet.
- The characteristics of the speech signal vary even for the same individual. For instance, if the speaker is suffering from a cold, the speech recognition system will fail to recognize the speech even if the system has been trained for that user.
- Pronunciation varies from region to region (known as dialectal changes) and from individual to individual (known as idiolectal changes). As it is said, there is no English—there is American English, British English, Indian English, and so on. Even in a country, every region has peculiarities. To capture all these variations is still an active research area.
- Every country has many languages, and multilingual speech recognition systems are required to provide practical speech recognition systems.

Because of all these difficulties, practical speech recognition systems, to be commercially useful, can be categorized as follows:

Isolated word recognition systems: Isolated word recognition systems accept each word in isolation: between two successive words, there should be a gap, a silence of about 100 msec. In such systems, the end points of the word can be found easily, making the recognition highly accurate.

Continuous speech recognition systems: These systems accept continuous speech, speech without any gap between words. However, since recognizing word boundaries is difficult in continuous speech, accuracy may not be very high, particularly if the vocabulary is very large.

The main difficulty in continuous speech recognition is to identify the boundaries of words. Due to this difficulty, recognition accuracy is not very high; only up to 80% accuracy can be obtained.

Keyword extraction systems: In some applications, the system needs to recognize a certain number of predefined keywords. Continuous speech is given as input to the system, and the system will recognize the presence or absence of any of the keywords in the sentence.

Speaker recognition systems are divided into speaker verification systems and speaker identification systems. Speaker verification involves checking whether a speaker belongs to a set of known speakers. Speaker identification involves finding out the identity of the speaker.

Limited vocabulary speaker-dependent systems: For applications such as data entry and voice dialing, only a limited number of words are required. Such systems can be trained by the speaker for his voice to achieve 100% recognition accuracy. These systems are known as limited vocabulary speaker-dependent systems.

Limited vocabulary speaker-independent systems: For applications such as voice-based interactive voice response systems, the vocabulary will be limited but the speech of any speaker should be accepted. Instead of storing templates for only one individual, representative multiple templates are stored for each word, to take care of variations due to speakers. The processing and storage requirements will be high in this case, but reasonably good accuracy (about 95%) can be achieved for such limited vocabulary speaker-independent systems.

Unlimited vocabulary speaker-dependent systems: For applications such as dictation machines, any word spoken should be accepted, but the system can be trained with about 100 words that capture the phonetic data corresponding to that speaker. Such unlimited vocabulary speaker-dependent systems achieve an accuracy of about 60% to 80%.

Unlimited vocabulary speaker-independent systems: To develop speech recognition systems that will understand any word spoken by any person would lead to a world wherein we can talk to computers the way we talk to fellow human beings. This has remained an elusive goal.

Multilingual speech recognition systems: If we succeed in developing unlimited vocabulary speaker-independent systems for one language, we can extend the system to recognize multiple languages.

Language recognition systems: For a computer to recognize multiple languages, the first module required is the language recognition module. After recognizing which language is spoken, the corresponding language's knowledge source has to be invoked to recognize the words in that language.

Speaker recognition systems: These systems are used to find out who the speaker is. Speaker recognition systems are categorized into (i) speaker verification systems; and (ii) speaker identification systems. Speaker verification is to find out whether a test speaker belongs to a set of known speakers. Speaker identification involves finding out the identity of the speaker after making a binary decision whether the speaker is known or not. Speaker recognition can be one of two types: text-dependent or text-independent. In text-dependent speaker recognition systems, reference patterns are generated from the features extracted from a predetermined utterance spoken by the speaker, and the same utterance is used in the test pattern. For applications in which the restriction that a predetermined text has to be spoken cannot be imposed, text-independent speaker recognition is required. Speaker recognition has two types of errors: false rejection of a truthful speaker and false acceptance of an imposter.

> Speech recognition systems are divided into the following categories: isolated word recognition systems, continuous speech recognition systems, keyword extraction systems, limited vocabulary speaker-dependent systems, unlimited vocabulary speaker-dependent systems, limited vocabulary speaker-dependent systems, limited vocabulary speaker-independent systems, language recognition systems, and multilingual speech recognition systems.

Language recognition and multilingual speech recognition systems are still active research areas. It will be a few more years before such systems are successfully demonstrated.

35.5.3 Applications of Speech Recognition Technology

In spite of the present limitations on the type of speech input, speech interaction with computers is of immense practical application. Reaching the goal of completely natural speech communication with computers is still far off, but application-oriented systems in which the task is controlled are being developed. These applications include:

- Computer data entry and editing with oral commands using isolated word recognition systems.

- Voice response systems with limited vocabulary and syntax.
- Retrieval of information from textual databases in speech form through telephones.
- Aids to the physically handicapped. Reading machines for the blind and speech synthesizers for the speech impaired are of immense use.
- Computer-aided instruction for teaching foreign languages and pronunciation.
- Entertainment on personal computers.
- Military applications such as (a) human-machine interaction in command and control applications; (b) automatic message sorting from radio broadcasts using keyword extraction systems; and (c) secure restricted access using speaker recognition.

Some of the applications of speech recognition are computer data entry through speech, voice response systems, aids to the physically handicapped, and military applications for human-machine interaction through speech.

Hopefully, in the not too distant future, we will be able to communicate with computers the way we communicate with each other, in speech. The graphical user interface (GUI) will be slowly replaced by voice user interface (VUI). Subsequently, automatic language translation may be possible.

35.6 CALL CENTERS

Every service organization—telephone service provider, hotel, bank, credit card company, and trading company—is now making efforts to provide efficient and timely service to its customers. To achieve this objective, call center technology is being widely used.

Nowdays, many developed countries are setting up call centers in developing countries such as India to gain the advantage of lower manpower costs and infrastructure costs. People at these call centers provide customer support to the overseas clients.

Call centers use the CTI technologies extensively. These include IVR, fax-on-demand, voice and video over IP, and video conferencing.

A call center is not a new technology, it is just a combination of various systems such as IVR, customer relations management software, fax-on-demand, voice over IP, and video conferencing. In this section, we will study call center applications and the technology ingredients.

35.6.1 Call Center Applications

In call centers, the CTI technologies, in conjunction with customer relations management software, are used to provide efficient customer support by service organizations.

Consider the typical operation of a customer support division of a credit card company. Prospective credit card customers call the support staff to obtain details of the credit card, and current customers call the support staff to make inquiries regarding their credit limits, to check whether payments are received, to make requests for monthly statement, and so on. Invariably, the customer calls the customer support telephone number and listens to music (while on hold) until one of the customer support persons is free to take the call. It is a waste of time to the customer. It may also happen that the support staff will not be able to provide the information requested by the customer immediately.

In this competitive world, any service organization cannot survive unless efficient customer support is provided. Call centers achieve this objective by integrating of various technologies and software, such as customer relations management (CRM) software, IVR, fax-on-demand, and such.

Typically, a customer will call up a call center and hear a prerecorded message indicating the various options available such as "dial 1 to hear about credit card facilities," "dial 2 to talk to a customer support specialist," "dial 3 to receive a brochure of the services provided." If the customer dials 1, the IVR system will be invoked. If she dials 2, the call will be routed to the customer support staff and if she dials 3, the system will prompt the user to key in the fax number, and the fax message will be sent to that fax machine. When the user dials 2, the call will be picked up by a call center agent who will invoke the CRM software to get the information related to the user and answer the queries.

This is a typical application of a call center. Call centers can be deployed by all service organizations.

35.6.2 Call Center Architecture

The various components that make up a call center are CRM software, IVR, fax-on-demand, automatic call distribution software, and PC-based PBX.

The typical architecture of a call center is shown in Figure 35.7. It consists of a LAN with one node for each call center agent. The server will have the CRM software, which can be invoked by any of the agents. In addition, there will be a CTI server which has the necessary hardware and software for the IVR application. A fax server will also be connected on the LAN to provide fax-on-demand service. For voice communication, a separate PBX can be installed or a PC-based PBX can be integrated that will have advanced features such as call routing and call hold, and direct inward dialing.

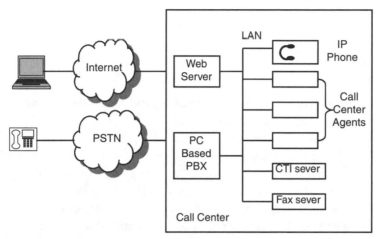

FIGURE 35.7 Call center architecture.

A customer can reach the call center through the PSTN by dialing the specified number of the customer support division. The IVR system will pick up the call, give the options to the user and, based on the user response, take action such as routing the call to a call center agent. This routing can be done intelligently by checking which agent is free and routing the call to that particular agent (thus avoiding playing music to the caller). This is known as automatic call distribution (ACD).

35.6.3 Internet-Enabled Call Centers

With the widespread use of voice over IP networks, there is no need to reach the call center through the PSTN. Internet-enabled call centers integrate voice over IP, fax over IP, and video over IP technologies to provide low-cost access to the call center.

In Internet-enabled call centers, voice and video over IP are used extensively to eliminate the need for long distance calls.

In a typical scenario, a customer can invoke the Web page of a credit card company and click on customer support. The customer can then click on a telephone icon that automatically connects to the customer support staff so that the customer can interact through voice using

VoIP protocols. As shown in Figure 35.7, an IP phone can be connected to the LAN directly. Using VoIP protocols, the call center agent can interact with the customer. Similarly, it is also possible to have video conferencing between the customer and the customer support manager.

35.7 UNIFIED MESSAGING

Technology has made it possible for us to communicate with anyone, anywhere, and anytime using different media—voice, data, fax, or video. However, the user has to use different devices to access different networks and call different numbers, depending on the location of the person called, and receives multiple bills for the different services. Unified messaging aims at solving this problem by providing an ability to access different networks using a device of one's choice. Also, it provides a single mailbox to access messages of different types such as voice, data or fax. The driving factor for this unified messaging is the demand of subscribers for simple, easy-to-use interfaces for meeting their communication needs.

Globalization of communication services is driving the telecom service providers to make the lives of their subscribers easier. To survive and grow in this immense competitive market, service providers have no choice but to provide unified messaging services. Providing enhanced services will ensure subscriber loyalty, and nowdays all service providers are focusing on providing unified messaging services.

> Unified messaging makes it possible to use a single mailbox for e-mail, voice mail, fax mail, and video mail. In addition, different services can be accessed by a subscriber through a device of his choice.

35.7.1 Unification of Messaging Systems

With increased use of communication facilities, subscribers are demanding a number of services, such as:

- Multimedia services such as voice, data, fax, graphics, and video.
- One mailbox for all types of media, not different mailboxes for voice, e-mail, and so forth.
- Access to different services from one device of their choice—the device may be a normal telephone, a mobile phone, a multimedia PC, PDA or a palmtop.
- Simple, easy-to-use interface for accessing different services.

- Single, consolidated bill for all the services, not multiple bills for different networks.
- A single directory number to call a person, irrespective of the location of the person.

  To meet the above demands of the subscribers, the important requirements are:

- **Personal telecommunications number (PTN):** A single number to reach a person, irrespective of the location. If the subscriber does not want to be reached, the calling person will be able to leave a message in the mailbox. The called person will be notified of the waiting message.
- **Single interface to access different networks:** The subscriber must be able to access the PSTN, the Internet, a mobile network or a paging network using a single interface of his choice. The device can be a PC, PDA, palmtop, fixed telephone, mobile telephone, or a fax machine. Of course, depending on the device, the capabilities may vary.
- **Integration of different networks:** Service providers or operators need to work out arrangements so that the various networks—PSTN, Internet, mobile network (PLMN), of paging network—are integrated together. This ensures that a single mailbox can be used by a subscriber for accessing messages.

The convergence of various networks and various services is paving the way for unified access to services.

As shown in Figure 35.8, a unified network provides unification of different messaging systems through a personal telecommunications terminal (PTT).

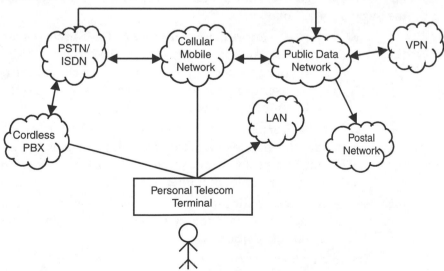

FIGURE 35.8 Unified messaging: a futuristic scenario.

35.7.2 Applications of Unified Messaging

Based on the architecture shown in Figure 35.8, some typical applications of unified messaging are described below. As you can see, this calls for many technologies—IVR, speech recognition, text-to-speech conversion, SMS, and fax-to-text conversion.

Voice messaging: When a called party does not want to be reached or is not available, a voice mail message is left in the voice mailbox. However, presently the voice mailboxes are many—at the PSTN service provider, at the mobile service provider, or at the subscriber premises. Instead, a single voice mailbox can be provided that can be used for voice mails from PSTN or a mobile phone. The voice mailbox can be accessed from any telephone—fixed or mobile or through a PC.

E-mail: To access text mail, there is a separate mailbox (or multiple mailboxes if one has multiple mail addresses). The voice mailbox can also be used for storing e-mail. In addition, text mail can be retrieved through a telephone (fixed or mobile). When one accesses the mailbox, the text is converted into speech form (through text-to-speech conversion software) and played to the user.

Fax mail: Fax messages also can be stored in the mailbox. The fax messages can be retrieved from the mailbox using a normal fax machine, or they can be read through a telephone (of course with the limitation that the pictures cannot be read). This calls for special software that converts the fax text into normal text and then converts the text into speech form. Another possibility is to reroute a fax message to a mailbox.

The various services that can be supported in the unified messaging framework are e-mail, voice mail, fax mail, and video mail, IVR, SMS, call forwarding etc.

Short messaging service (SMS): Whenever a mail (e-mail, voice mail, or fax mail) arrives in his mailbox, the user can be alerted through a short message. The user can program to receive the short message on his mobile phone or pager or on his PC.

Call forwarding: Call forwarding is nowdays supported on many networks. A person can program his mobile device for forwarding all calls to a fixed line or vice versa. This allows a person to be in touch with office/home all the time, and the calling party is saved of lot of bother trying different numbers.

Voice dialing: To access mailboxes or to dial telephones, voice dialing is a boon. However, with the present technology, the user has to train his device for his voice to obtain good recognition accuracy.

Interactive voice response systems: Users can access information available in databases through IVR systems. Advanced IVR systems also facilitate receiving fax messages. The fax message can be routed to a normal fax machine or a message box.

Video messaging: Presently, video messaging has not caught on very well because the video occupies large bandwidth and, if low bit rates are used, quality is not good. However, with good video streaming technologies coming up, desktop video conferencing is catching up. Once the Internet backbone can support higher data rates, video messaging will be used extensively and will be an integral part of unified messaging.

With unified messaging, exciting times are ahead. Communicate with anyone, anywhere, anytime using just one number and with any device of your choice—life cannot be better!

35.8 VOICEXML

The VoiceXML forum (*http://www.voicexmlforum.org*) was founded by AT&T, IBM, Lucent Technologies, and Motorola to promote Voice eXtensible Markup Language (VoiceXML). VoiceXML was designed to make Internet content available through voice from telephones. VoiceXML, in short, makes it possible to achieve voice-enabled Web. VoiceXML Version 1.0 was released in March 2000 and a working draft of Version 2.0 in October 2001.

Web access is normally through desktop PCs. The information obtained will be rich in content with graphics. But PC penetration is very low in developing countries, and computer literacy is a must to access Web services. Accessing the Web through mobile phones using WAP protocols is another alternative, but WAP-enabled mobile phones are costly and are not within the reach of many. However, because of the limited display on mobile phones, WAP services are not user friendly.

If Web services are accessible through normal telephones or mobile phones, with the output in voice form, Web reach can be much greater, and services will be very user friendly because speech is a very natural way of communicating among people. VoiceXML provides this possibility.

VoiceXML provides the capability of accessing Internet content through telephones. VoiceXML is derived from XML.

Consider a simple example of obtaining weather information from an Internet Web server. The dialogues between the computer (C) and the Human (H) can take one of two forms: (a) directed dialogue and (b) mixed initiative dialogue.

(a) Directed dialogue: In this approach, the interaction between C and H can be as follows:

C: Please say the state for which you want the weather information

H: Indiana

C: Please say the city

H: Fort Wayne

C: The maximum temperature in Fort Wayne is 63 degrees Fahrenheit

(b) Mixed initiative dialogue: In this approach, the interaction between C and H can be as follows:

C: Please say the state and the city for which you want the weather information.

H: Fort Wayne Indiana.

C: The maximum temperature in Fort Wayne Indiana is 63 degrees Fahrenheit.

> Accessing Internet content through telephones is done either through directed dialogue or mixed initiative dialogue.

This kind of interaction is possible (completely through speech) for obtaining information available on the Web. This calls for interfacing text-to-speech conversion system, a speech recognition system, and if required a IVR system to the Web servers. It is possible to provide voice-enabled Web service without VoiceXML, but the problem is that because all these components are built around proprietary hardware and software, it is difficult to port the application for different platforms.

NOTE

VoiceXML separates the service logic from the user interaction code. Hence, it is possible to port the application from one platform to another because VoiceXML is an industry standard for content development.

VoiceXML has been designed with the following goals:

- To integrate data services and voice services.
- To separate the service logic (CGI scripts) to access databases and interface with legacy databases from the user interaction code (VoiceXML).
- To facilitate portability of applications from one platform to another; VoiceXML is based on an industry standard for content development.

- To shield application developers from the low-level platform-dependent details such as hardware and software for text-to-speech conversion, IVR digit recognition, and speech recognition.

The operation of voice-enabled Web is shown in Figure 35.9. The VoiceXML server contains the necessary hardware and software for: (a) telephone interface; (b) speech recognition; (c) text-to-speech conversion; and (d) audio play/record. The Web server contains the information required for the specific application in the form of VoiceXML documents, along with the service logic in the form of CGI scripts and necessary database interfaces. When a user calls an assigned telephone number to access weather information, for instance, through PSTN or PLMN, the call reaches the VoiceXML server and, this server converts the telephone number to a URL. The weather information corresponding to the URL is obtained by this server from the Web Server, which is in the format of VoiceXML. The VoiceXML server converts the content into speech form and plays it to the user. When the user utters some words (the state and city names for obtaining weather information), the VoiceXML server recognizes those words and, based on the information available in the database, plays the information to the user.

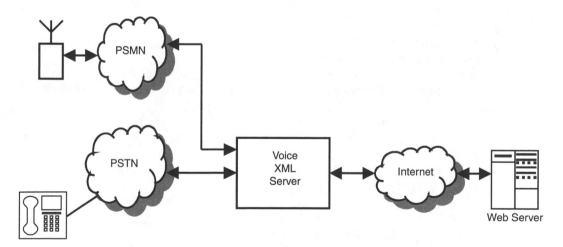

FIGURE 35.9 Operation of voice-enabled Web.

The human-machine interaction is carried out using the following:

- DTMF digits dialed by the user from the telephone.
- Text-to-speech conversion.

- Speech recognition.
- Recording of speech input of the user.
- Playing of already recorded speech to the user.

A VoiceXML server has the necessary hardware and software to facilitate this human-computer interaction. It gets the user inputs in the form of DTMF digits or voice commands and gives the output in speech format. The dialogues are of two types: menus and forms. Menus provide the user with a list of choices, and forms collect values for a set of variables (such as an account number). When a user does not respond or requests help, events are thrown.

The implementation platform of the VoiceXML server generates events in response to user actions (such as a spoken word or a pressed key) and system events (such as a timeout, in case the user does not respond). The implementation platform is different from the content, and the content is independent of the hardware used for developing voice-enabled Web.

A VoiceXML document contains tags to generate prompts, and obtain user responses and grammar to indicate the service logic.

A typical VoiceXML document is shown in the following. It contains the tags to generate prompts and obtain user responses and the grammar to indicate the service logic.

```
<?xml version="1.0"?>
<vxml version="1.0">
<form id = "weather">
      <block> Welcome to the weather information service </
block>
        <field name="state">
        <prompt> Please tell for which state you want weather
information </prompt>
            <grammar src="state.gram" type="application/x-jsgf"/>
             </field>

                <field name="city">
                <prompt> What city </prompt>
                 <grammar scr="city.gram" type="application/x-jsgf"/
>
                </field>
<submit next="/servlet/weather" namelist=" city state"/>
</block>
</form>
</vxml>
```

Using VoiceXML, content can be developed without bothering about the implementation details of the various components, such as text-to-speech conversion, speech recognition, or an interactive voice response system.

As can be seen from this code, VoiceXML provides a simple and efficient method of providing the content for developing voice-enabled Web applications. In the next decade, these services will catch up for very user-friendly web Browsing through telephones.

Summary

Computer telephony integration facilitates accessing the information available in computers through telephones. CTI technology has become very popular in recent years, particularly in developing countries because telephone density is very high compared to computer density. The three technology components of CTI are text-to-speech conversion, speech recognition, and interactive voice response (IVR). The details of all these technology components are presented in this chapter.

Text-to-speech conversion involves converting the text into its phonetic equivalent and then applying speech synthesis techniques. For English, a set of pronunciation rules is required to convert the text into its phonetic equivalent. For Indian languages, this step is very easy because there is a one-to-one correspondence between the written form and spoken form. For generating speech, the basic units can be words, syllables, diphones, or phonemes.

Speech recognition is a very complex task because speaker characteristics vary widely. Present commercial speech recognition systems can recognize limited vocabulary of limited speakers very accurately. Unlimited vocabulary speaker-independent speech recognition is still an active research area. Speech recognition is pattern recognition wherein prestored templates obtained during the training phase are compared with the test patterns.

Interactive voice response systems are now being widely deployed to provide information to consumers such as in railway/airline reservation systems, banking and so on. An IVR system consists of a hardware module to take care of the telephony functions and software to access the database and convert the text into speech.

CTI technology is now being used effectively in call centers by service organizations to provide efficient customer service. CTI also is very useful for accessing the Web services of the Internet through voice-enabled Web.

References

Many universities all over the world are working on text-to-speech conversion and speech recognition as research topics. You can visit the Web sites of the leading universities to obtain the latest information. *IEEE Transactions on Acoustics, Speech and Signal Processing (ASSP)* and *Bell System Technical Journal* (BSTJ) are two excellent references that publish research papers in this area. Also, a number of vendors such as Cisco, Nortel, Microsoft, and Sun Microsystems supply CTI products. Their Web sites also give the latest information on the state of the art in these areas. Selected references and Web resources are given below.

W.A. Ainsworth. "A System for Converting English Text into Speech," *IEEE Transactions on Audio Electroacoustics*, Vol. AU 21, No. 3, June 1974. This paper gives a complete list of pronunciation rules required for converting British English into speech.

Dreamtech Software Team. *Instant Messaging Systems*, Wiley Publishing Inc., 2002. This book gives excellent coverage of instant messaging. Using the source code listings given in the book, you can develop full-fledged instant messaging software.

Elaine Rich and Kevin Knight. *Artificial Intelligence*. McGraw Hill Inc., 1991. Speech synthesis and speech recognition fall under the realm of Artificial Intelligence. This book gives an excellent introduction to artificial intelligence concepts and systems.

J.L. Flanagan. *Speech Analysis, Synthesis and Perception*, Springer Verlag, New York, 1972. This book is considered the bible for researchers in speech. A must-read for everyone interested in research in speech processing.

Special issue on speech synthesis, *IEEE Transactions on Speech and Audio Processing*, Vol. 9, No. 1, January 2001. This special issue contains articles that describe research efforts in text-to-speech conversion in different languages.

www.call-center.net Resources about call centers.

www.cdacindia.com Web site of Centre for Development of Advanced Computing (CDAC), which carries out research in computing in Indian languages. CDAC also has products for development of content in Indian languages.

www.cisco.com Cisco's Web site. Cisco is one of the leading vendors of CTI products.

www.ibm.com IBM's Web site gives a wealth of information on CTI. You can get the details of the IBM's WebSphere voice server also from this site.

www.linuxtelephony.org/ Web site that provides resources for computer telephony on Linux platform.

www.philips.com Philips' FreeSpeech 2000 is software that facilitates development of CTI applications.

www.voicexmlforum.org The Web site of VoiceXML Forum. You can obtain VoiceXML standards documents from this site.

Questions

1. List the various technology components used in computer telephony integration.

2. One way of developing a text-to-speech conversion system in English is to record and store about 200,000 words of English and then concatenate the required words to generate speech. For example, if the input sentence in text is "she is a beautiful woman," the software has to pick up the speech data corresponding to all the words and concatenate these words and play through the sound card. Which search algorithm is good to search for the words in the database? Study the algorithmic complexity and the storage requirement for the search algorithm.

3. For text-to-speech conversion, is it easier to handle English or the Indian languages (such as Hindi, Telugu, Kannada, Bengali, Marathi, Gujarati, and Malayalam)? Why?

4. If you have to develop an automatic speech recognition system that recognizes any word spoken by a person, what are the issues to be addressed? If this system has to recognize anybody's speech, what are the issues?

5. What are the different categories of speech recognition systems? List the potential applications for each category.

6. Is it possible to communicate with computers the way we communicate with each other in a natural language such as English or Hindi? If not, why not?

7. What is an interactive voice response system? What are its potential applications?

8. Describe the architecture of a call center.

9. Call centers are now being set up in major Indian cities for foreign clients. Study the various market segments for which such call centers are being set up.

10. What is unified messaging? What are its advantages?

11. What is the need for a new standard for a markup language (VoiceXML) for CTI applications? What are the salient features of VoiceXML?

12. Study the Unicode representation of different Indian languages.

Exercises

1. Using the sound card of your PC, record about 100 words in English and store them in different files. Write a program that takes an English sentence as input and speaks out the sentence by concatenating the words. If the gap between two successive words is high, the speech does not sound good. Try to edit the voice files to reduce the silence at the beginning and end of each word and then try to do the text-to-speech conversion.

2. Design an IVR system for a telebanking application. Create a database that contains bank account number, password, type of account, and balance amount. Simulate a telephone keypad on the monitor using a Java applet. Design the conversation between the IVR system and the user.

3. Search for freely available text-to-speech conversion and speech recognition software packages available on the Internet. Experiment with these packages.

4. For any Indian language, find out the number of words, syllables, diphones, and syllables required to achieve unlimited vocabulary text-to-speech conversion.

5. List the various components required for developing a call center.

6. Study the commercial equipment available for setting up a cell center.

Projects

1. Develop a full-fledged text-to-speech conversion system for your native language. You can store the speech data for a reasonably large number of words, say 500. Using this database of words, create a database of syllables. Write the software that takes the text as input and converts it into speech using concatenation of words and syllables. If you develop a good database of syllables, you will achieve very good quality text-to-speech conversion.

2. Using Microsoft's Speech SDK, create a voice browsing application. Microsoft's Speech SDK can be used to recognize words. When a particular word is recognized, the system has to jump to a specific link. You can use VoiceXML to create the content.

3. Using a voice/data/fax modem connected to a PC, develop an IVR system. You can use Microsoft's Telephony API (TAPI) to control the modem and generate the responses based on the user's input of DTMF digits.

4. Using IBM's WebSphere Voice Server Software Developers Kit, develop a telebanking application that facilitates banking through voice commands.

5. Develop fax-on-demand software. A set of five documents (MS Word files) should be stored in the PC. These files can be brochures of five products. When a user dials a telephone number to which the voice/data/fax modem is connected, the user should get the response "Please dial 1 to get the brochure of TV, dial 2 to get the brochure of refrigerator, dial 3 to get the brochure of microwave oven, dial 4 to get the brochure of DVD player and dial 5 to get the brochure of washing machine." When the user dials a number, the user should hear the message "Please enter your fax number." When the user dials the fax message, the corresponding brochure should be faxed to the user's fax machine.

36 Wireless Personal/Home Area Networks

In This Chapter

- Personal Area Network (PAN)
- Bluetooth Technology
- PAN Technologies

With the widespread use of computers and other electronic gadgets, every office and every home is now a myriad of wires. These wires interconnect computers, peripherals, and office/home appliances. These wires create lots of maintenance problems. If we can interconnect these devices without wires through radio, there will be a tremendous advantage in terms of less maintenance, more reliability, and of course better looks. A number of technologies, such as Bluetooth, HomeRF, IrDA, and IEEE 802.11 provide this solution—we can interconnect devices through low-cost reliable radio. Out of these technologies, Bluetooth gained lot of popularity and market hype. During the next few years, every office and every home will have lots of Bluetooth-enabled devices. A large number of manufacturers have come out with Bluetooth hardware and software, but at present the cost of making a device Bluetooth enabled is high. In the years to come, every electronic device may be Bluetooth enabled. In this chapter, we will study the Bluetooth technology in detail and also briefly review competing technologies such as HomeRF and IrDA.

36.1 INTRODUCTION TO PERSONAL AREA NETWORKS (PANS)

A typical office or home or even a car is equipped with a number of electronic gadgets such as desktop, laptop, printer, modem, and mobile phone: These devices are interconnected through wires for using a service or for sharing information (such as transferring a file from desktop to laptop). These devices form a personal area network (PAN). When we bring two devices, say a laptop and a mobile phone, close to each other, these two can automatically form a network and exchange data. For example, we can transfer the address book from the mobile phone to the laptop. When two or more devices come close to one another, they form a network spontaneously; such networks are called ad hoc networks. In an ad hoc network, the topology and the number of nodes at any time are not fixed—the topology may change dynamically with time, and the number of nodes in the network also may change with time. All the headaches associated with administering such networks can be avoided if these devices are made to communicate through radio links and also if one device can find out the presence of other devices and their capabilities. The need for such PANs is everywhere—in offices at home, and also in cars.

A personal area network (PAN) is a network that interconnects various personal devices such as the desktop, laptop, or mobile phone within a radius of about 10 meters. A PAN can be formed in an office, at home, or in a car.

A number of technologies have been proposed for PANs. Notable among them are Bluetooth, HomeRF, IrDA, and IEEE 802.11.

 Bluetooth, HomeRF, IrDA, and IEEE 802.11 are the important technologies used for personal area networking.

36.2 OVERVIEW OF BLUETOOTH

Bluetooth technology enables various devices to be interconnected without the need for wires. The salient features of this technology are:

- It is a low-cost technology—its cost will soon be as low as a cable connection. Since most Bluetooth-enabled devices operate through a battery, power consumption is also very low.

- It is based on radio in the ISM band. ISM (Industrial, Scientific and Medical) band is not controlled by any government authority, and hence no special approval is required to use Bluetooth radio systems.

- It caters to short ranges—the range of a Bluetooth device is typically 10 meters, though with higher power, it can be increased to 100 meters.

- It is based on open standards formulated by a consortium of industries, and a large number of equipment vendors are committed to this technology.

Bluetooth is the most popular technology for PANs. Devices within a radius of 10 meters can form a network automatically when they come near to each other.

Bluetooth Special Interest Group (SIG), founded in February 1998 by Ericsson, Intel, IBM, Toshiba, and Nokia, released Version 1.0 of Bluetooth specifications in July 1999. Version 1.1 of Bluetooth specifications was released in February 2001.

Any electronic device, whether it is a PC, laptop, PDA, digital camera, mobile phone, pager, MP3 player, or headset, peripherals such as printer, keyboard, mouse, or LCD projector, domestic appliances such as TV, microwave oven, and music players can be Bluetooth enabled by attaching a module that contains the hardware to that device and running a piece of software on the device. A Bluetooth-enabled device communicates with another Bluetooth-enabled device over the radio medium to exchange information or transfer data from one to the other.

A device can be Bluetooth enabled by attaching a piece of hardware to it and running the Bluetooth protocol stack on it.

NOTE

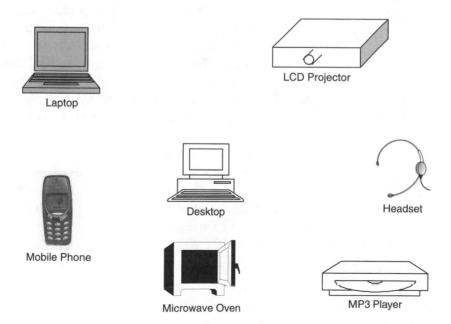

FIGURE 36.1 Wireless personal area network.

Bluetooth facilitates forming of an ad hoc network. When two Bluetooth-enabled devices come near to each other, they can automatically form a network and exchange data.

As shown in Figure 36.1, a set of devices can form a personal area network if they are in the radio vicinity of each other (typically 10 meters radius). When a device comes in the vicinity of another device, the Bluetooth protocols facilitate their forming a network. A device can find out what services are offered by the other device and then obtain that service. For example, a laptop can discover the printer automatically and then obtain the print service. Such networks are called ad hoc networks because the network is formed on-the-fly and, once the device gets out of sight, the network is no longer there. Such networks can be formed in the office, at home, in cars and also in public places such as shopping malls and airports.

36.2.1 Bluetooth System Specifications

The specifications of the Bluetooth system are as follows:

Bluetooth operates in the ISM band in the frequency band 2400–2483.5 MHz. This band is divided into 79 channels each of 1MHz bandwidth. Frequency hopping at the rate of 1600 hops per second is used to transmit the data.

Frequency of operation: Bluetooth devices operate in the ISM band in the frequency range 2400–2483.5 MHz. This band consists of 79 channels each of 1MHz bandwidth, with a lower guard band of 2MHz and upper guard band of 3.5MHz. When a device transmits its data, it uses frequency hopping: the device transmits each packet in a different channel. The receiving device has to switch to that channel to receive that packet. Though the radio design becomes complex when frequency hopping is used, the advantage is that it provides secure communication. Nominal frequency hop rate is 1600 hops per second.

Modulation: Gaussian frequency shift keying (GFSK) is used as the modulation technique. Binary 1 is represented by a positive frequency deviation and 0 by negative frequency deviation. The radio receiver has to be designed in such a way that the Bit Error Rate (BER) of minimum 0.1% is ensured; that is, the radio should provide a link that ensures that there will not be more than one error for every 1000 bits transmitted.

Operating range: Three classes of devices are defined in Bluetooth specifications:

- Class 1 devices transmit maximum of 100 mW. The range of such devices is 100 meters.

- Class 2 devices transmit 10 mW. The range is 50 meters.
- Class 3 devices transmit 1 mW. The range is 10 meters.

Most of the commercially available devices have a transmitting power of 1 milliwatt and hence a range of 10 meters.

The normal range of Bluetooth device is 10 meters. However, with increased transmit power, a range of 100 meters can be achieved.

Bluetooth supports both voice and data services. synchronous connection-oriented (SCO) links carry voice. asynchronous connectionless (ACL) links carry data.

Services supported: Both data and voice services are supported by Bluetooth devices. For voice communication, synchronous connection-oriented (SCO) links are used that support circuit switching operation. For data communication, asynchronous connectionless (ACL) links are used that use packet switching. The SCO links carry voice. Two types of voice coding are defined in the specifications: PCM based on G.711 standard at 64kbps and continuously variable slope delta (CVSD) modulation technique also at 64 kbps. There is no retransmission of voice packets if they are lost or received in error.

For data services, devices exchange data in the form of packets. The receiving device acknowledges the packets or reports that the packet is received in error. If a packet is received with errors, the packet is retransmitted. It is also possible to broadcast packets by one device to all the other devices in the network. However, in broadcast mode there is no acknowledgement or indication that the packet is received with errors. The broadcasting device indicates to the receiving devices how many times a broadcast packet will be transmitted so that at least once every device will receive the packet without errors.

In Bluetooth, two standard voice coding techniques are used for voice communication. These are 64kbps PCM and 64kbps CVSD (continuously variable slope delta) modulation.

Data rates: A Bluetooth device can support three synchronous voice channels and one asynchronous data channel. For voice communication, 64kbps data rate is used in both directions. For asynchronous links, two types of channels are defined with different data rates. In asymmetric channel, data rates are 723.2kbps in one direction and 57.6kbps in the other direction. In symmetric channel, data rate is 433.9kbps in both directions.

Network topology: In a PAN, a set of devices form a small network called a *piconet*. In a piconet, there will be one master and one or more slaves. All the slaves tune to the master. The master decides the hop frequency sequence and all the slaves tune to these frequencies to establish communication links. Any device can be a master or slave. The master/slave terminology is only for the protocols, the device capabilities are not defined by this terminology. It is also possible for a master and slave to switch roles—a slave can become a master. A piconet can have maximum number of seven slaves that can actively communicate with the master. In addition to these active slaves, a piconet can contain many slaves in parked mode. These parked devices are synchronized with the master, but they are not active on the channel. The communication between the master and the slave uses time division duplex (TDD).

The network formed by a set of devices is called a piconet. In a piconet, a device acts as a master and all others are slaves. A piconet can have a maximum of seven active slaves.

Figure 36.2 shows the various topologies of a Bluetooth piconet. In Figure 36.2(a), a piconet is shown with one master and one slave. It is a point-to-point communication mode. In Figure 36.2(b), the piconet consists of a master and a number of slaves. It is a point-to-multipoint communication mode. Figure 36.2(c) shows a scatternet, which is formed by a number of piconets with overlapping coverage areas. In this scatternet, each piconet will have a master and a number of slaves. The master of a piconet can be a slave in another piconet. Each piconet in the scatternet will have its own frequency hopping sequence, and hence there will be no interference between two piconets.

A scatternet is formed by a number of piconets with overlapping coverage areas. Within a scatternet, a master in one piconet can be a slave in another piconet.

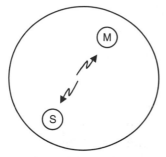

(a) Point-to-Print Communication
Between Master and Slave

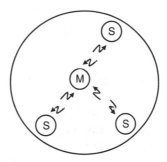

(b) Pont-to-Multipoint Communication
Between Master and Multiple Slaves

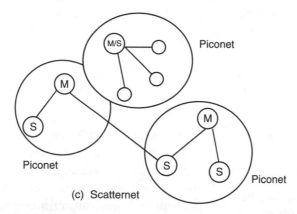

(c) Scatternet

FIGURE 36.2 **Bluetooth piconet** (a) Point-to-Print Communication Between Master and Slave
(b) Pont-to-Multipoint Communication Between Master and Multiple Slaves (c) Scatternet.

Security: To provide security of data over the radio medium, the
specifications contain the following features:

- Each Bluetooth device is given a 48-bit address. This address uniquely
identifies the device.

- When two devices have to communicate with each other, an authentication
procedure is used. Every Bluetooth device has a random number generator
that generates random numbers that are used for authentication.

- Data on the channel is encrypted so that only the intended recipients can
receive it.

- The frequency hopping scheme provides built-in security because only those
devices that know the hopping sequence can decode the data sent by the
master.

Security is a major issue in all radio systems. Bluetooth provides security through a frequency hopping scheme, encryption of the data, and an authentication procedure.

Communication between master and slave: The master and slave communicate in the form of packets. Each packet is transmitted in a time slot. Each time slot is of 625 microseconds duration. These slots are numbered from 0 to $2^{27} - 1$. The master starts the transmission in even slots by sending a packet addressed to a slave, and the slave sends the packets in odd-numbered slots. A packet generally occupies one time slot but can extend up to five slots. If a packet extends to more than one slot, the hop frequency will be the same for the entire packet. If the master starts the transmission in slot 0 using frequency f1, the slave transmits in slot 1 using frequency f2, the master transmits in slot 2 using frequency f3, and so on.

Within a piconet, the master always transmits in the even-numbered slots and the slave sends in the odd-numbered slots. The slots are continuously numbered from 0 to $2^{27} - 1$.

States of a Bluetooth device: A Bluetooth device can be in one of two major states: connection state and standby state. In connection state, the device is communicating with another device by exchanging packets. In standby state, the device is not communicating with another device and will be in low-power mode to save battery power, which is the default state. There can be seven substates: page, page scan, inquiry, inquiry scan, master response, slave response, and inquiry response. The state transition diagram is shown in Figure 36.3.

A device can be in one of the two major states: connection state and standby state. The various substates are page, page scan, inquiry, inquiry scan, master response, slave response, and inquiry response.

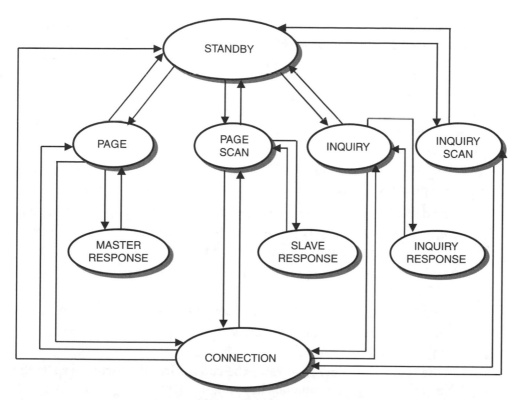

FIGURE 36.3 State transition diagram.

To start with, an application program in a Bluetooth device can enter the inquiry state to inquire about other devices in the vicinity. To respond to an inquiry, the devices should periodically enter into inquiry scan state and, when the inquiry is successfully completed, it enters the inquiry response state. When a device wants to connect to another device, it enters the page state. In this state, the device will become the master and page for other devices. The command for this paging has to come from an application program running on this Bluetooth device. When the device pages for the other device, the other device may respond, and the master enters the master response state. Devices should enter the page scan state periodically to check whether other devices are paging for it. When the device receives the page scan packet, it enters the slave response state.

Once paging of devices is completed, the master and the slave establish a connection, and the connection is in active state, during which the packet transmission takes place. The connection can also be put in one of the three modes: hold, sniff, or park mode. In hold mode, the device will stop receiving the data traffic for a specific amount of time so that other devices in the piconet can use the bandwidth. After the expiration of the specific time, the device will start listening to traffic again. In sniff mode, a slave will be given an instruction such as "listen starting with slot number S every T slots for a period of N slots." So, the device need not listen to all the packets, but only as specified through the sniff parameters. The connection can be in park mode, wherein the device listens to a beacon signal from the master only occasionally; it synchronizes with the master but does not do any data transmission.

A typical procedure for setting up a Bluetooth link is as follows:

- The device sends an inquiry using a special inquiry hopping sequence.
- Inquiry-scanning devices respond to the inquiry by sending a packet. This packet contains the information needed to connect to it.
- The inquiring device requests a connection to the device that responded to the inquiry.
- Paging is used to initiate the connection with the selected device.
- The selected device that has entered the page scan state responds to the page.
- If the responding device accesses the connection, it synchronizes with the master's timing and frequency hopping sequence.

Packet format: The packet format is shown in Figure 36.4. The packet consists of access code (68 or 72 bits), header (54 bits), and payload (0 to 2745 bits). Packets can contain only access code (shortened access code with 68 bits only), access code and header, or access code, header, and payload.

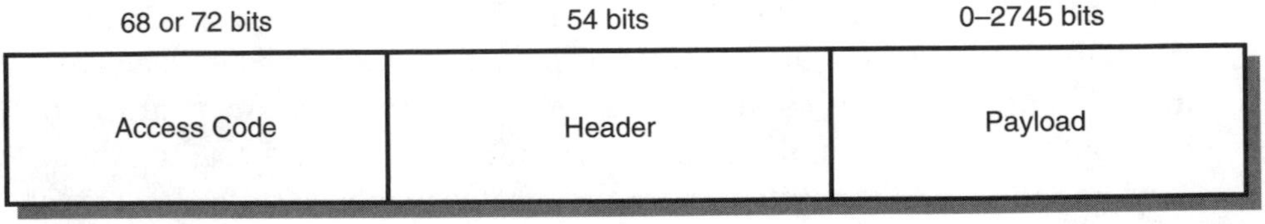

68 or 72 bits	54 bits	0–2745 bits
Access Code	Header	Payload

FIGURE 36.4 Bluetooth packet format.

Access code: All packets in a piconet will have the same access code. Access code is used for synchronization and identification of devices in a piconet. Access code is used for paging and inquiry procedures, and in such cases no header or payload is required because only signaling information is carried.

Access code can be of three types:

(i) **Channel access code (CAC):** Identifies a piconet; all packets in a piconet contain this code.

(ii) **Device access code (DAC):** This code is used for paging and response to paging.

(iii) **Inquiry access code (IAC), which is of two types:** General IAC is used to discover which Bluetooth devices are in the radio range. Dedicated IAC is common for devices with a common characteristic. Only those devices can be discovered.

Packet header: Packet header is 54 bits long.

- Three bits for active member address (all zeros for broadcast)
- Four bits for type code (SCO link or ACL link, how many slots the packet will occupy and such)
- One bit for flow control (if buffer is full, 0 for stop and 1 to go)
- One bit for acknowledgement indication (1 indicates that packet is OK, 0 indicates packet error)
- One bit for sequence number (for each packet, this bit is reversed)
- Eight bits for header error control for error checking

These total 18 bits. Rate 1/3 FEC is used to make it 54 bits by repeating each bit three times to help in error correction at the receiving end if there are transmission errors.

Bluetooth profiles specify the precise characteristics and protocols to be implemented for specific applications such as file transfer, serial communication, and cordless telephony.

Note that three bits are allocated for the address of the active device, and so the number of addresses in a piconet to eight. Out of these, one address (all zeros) is for broadcasting the packets in a piconet. So, we are left with seven addresses and hence only seven active devices can be in a piconet.

Payload: This field contains the user information, which can be either data or voice.

Bluetooth addressing: Each Bluetooth module (the radio transceiver) is given a 48-bit address containing three fields; LAP (lower address part) with 24 bits, upper address part (UAP) with 8 bits, and non-significant address part with 16 bits. This address is assigned by the manufacturer of the Bluetooth module and consists of company ID and a company-assigned number. This address is unique to every Bluetooth device. In Bluetooth specifications, this address is referred to as BD_ADDR.

Each active member in a piconet will have a 3-bit address. In addition to the maximum of seven active members, many more devices can be in parked mode. The parked members also need to have addresses so that the master can make them active for exchange of packets. A parked member address is either the BD_ADDR of 48 bits or an 8-bit parked member address denoted by PM_ADDR.

Bluetooth profiles: To ensure interoperability between devices manufactured by different vendors, Bluetooth SIG released the Bluetooth profiles, which define the precise characteristics and protocols supported by these devices. The Blueooth profiles are defined for headset, cordless phone, fax machine, LAN access point, serial communication, dial-up networking, file transfer, synchronization of data between two devices, and others.

36.2.2 Bluetooth Protocol Architecture

The complete protocol stack of a Bluetooth system is seen in Figure 36.5.

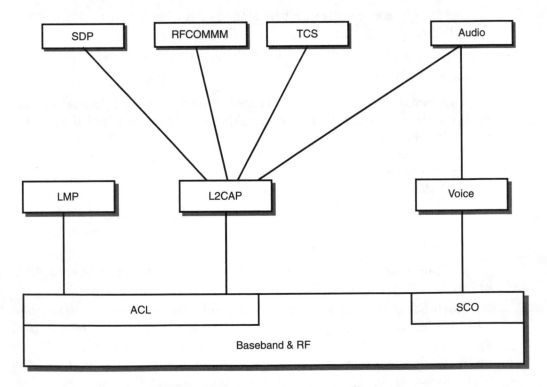

FIGURE 36.5 Bluetooth protocol architecture.

Baseband and RF

The baseband layer is for establishing the links between devices based on the type of service required—ACL for data services and SCO for voice services. This layer also takes care of addressing and managing the different states of the Bluetooth device. The RF portion provides the radio interface.

Link Manager Protocol

The Link manager protocol (LMP) is used to set up and control links. The three layers RF, link controller and the link manager will be on the Bluetooth module attached to the device. The link manager on one device exchanges messages with the link manager on the other device. These messages, known as LMP messages, are not propagated to higher layers. Link messages have higher priority than data. LMP messages are sent as single slot packets, with a header of 1 byte. The functions of the LMP are as follows:

- **Authentication:** When two devices have to communicate with each other, one has to verify the other device. So, one device is called the verifier and the other is called the claimant. The verifier sends a packet containing a random number, which is called a challenge. The claimant calculates the response, which is a function of the challenge, and sends the response along with its Bluetooth address (48-bit address) and secret key. This is known as a challenge-response scheme—you throw a challenge and check whether the other device can correctly respond to that challenge.

- **Encryption:** To maintain confidentiality of data over the radio link, the data is encrypted. The master sends a key with which the data is encrypted to all the slaves, through an LMP message.

- **Clock offset request:** To synchronize the clocks between the master and the slaves is a must for proper exchange of data. If the clock has to be offset, the LMP exchanges messages to ensure clock synchronization.

- **Timing accuracy information request:** To ensure synchronization, the master can ask the slaves for timing accuracy information.

- **LMP version:** It needs to be ensured that both devices use the same version of LMP. To achieve this, the version number of the LMP protocol is exchanged.

- **Type of packets supported:** Different Bluetooth-enabled devices may support different features, so an LMP features request and response are exchanged between the devices.

- **Switching master/slave role:** In a piconet, a device will act as a master and other devices will act as slaves. The master and a slave in a piconet can switch roles using the LMP messages. The master or the slave can initiate the switching operation.

- **Name request:** Each device can be given a user-friendly name having a maximum of 248 bits in ASCII format. A device can ask for the name through an LMP message and obtain the response.

- **Detach:** Messages exchanged to close a connection.

- **Hold mode:** Places an ACL link in hold for a specified time when there is no data to send. This feature is mainly to save power.

- **Park mode:** In synchronization with the master but not participating in data exchange.

- **Power control:** Asks to transmit less power. This is useful particularly for class 1 devices, which are capable of transmitting 100 mW power.

- **Quality of service (QoS) parameters exchange:** In applications that require a good quality transmission link, quality of service parameters can be specified. These parameters include the number of repetitions for broadcast packets, delay, and bandwidth allocation.

- **Request SCO link:** Request an SCO link after the ACL link is established.

- **Multislot packet control:** Controls the procedure when data is sent in consecutive packets.

- **Link supervision:** Monitors link when device goes out of range (through a timeout mechanism).

- **Connection establishment:** After paging is successfully completed, establishes the connection.

The functions of link manager protocol are to set up and control the radio links. Authentication, switching between master and slave roles, and link supervision to monitor the link are done at this layer.

The Bluetooth device will implement these three layers in a hardware/firmware combination. These three layers ensure establishment of a connection and managing the connection for transfer of voice or data. To ensure that the whole application runs per user requirements, we need lots of other protocols.

Logical Link Control and Adaptation Protocol (L2CAP)

L2CAP runs above the baseband and carries out the datalink layer functionality. L2CAP layer is only for ACL links. L2CAP data packets can be up to 64 kilobytes long. L2CAP protocol runs on the host such as laptop, cellular phone, or other wireless devices.

L2CAP does not do any checksum calculation. When L2CAP messages are exchanged between two devices, it assumes that an ACL link is already established between two devices. It also assumes that packets are delivered in sequence. Note that L2CAP does not support SCO links for voice communication. L2CAP does not support multicasting.

The functions of L2CAP layer are:

- **Protocol multiplexing:** In the protocol stack seen in Figure 36.5, above L2CAP, a number of other protocols can be running. A packet received by L2CAP has to be passed on to the correct higher layer. This is protocol multiplexing.

- **Segmentation and reassembly:** Baseband packets are limited in size as we saw in packet format. Large L2CAP packets are segmented into small baseband packets and sent to the baseband. Similarly, the small packets received from the baseband are reassembled and sent to higher layers.

- **Quality of service:** Quality of service (QoS) parameters such as delay can be specified, and this layers ensures that the QoS constraints are honored.

The functions of the L2CAP layer are protocol multiplexing, segmentation and reassembly, and ensuring that the quality of service parameters are honored. This layer is only for ACL links, in other words, for data applications.

The L2CAP layer sends connection request and QoS request messages from the application programs through the higher layers. It receives from the lower layers the responses for these requests. The responses can be connection indication, connection confirmation, connect confirmation negative, connect confirmation pending, disconnection indication (from remote), disconnect confirmation, timeout indication, and quality of service violation indication.

Service Discovery Protocol

The service discovery protocol (SDP) provides the Bluetooth environment the capability to create ad hoc networks. This protocol is used for discovering the services offered by a device. SDP offers the following services:

- A device can search for the service it needs in the piconet.

- A device can discover a service based on a class of services (for example, if a laptop wants a print service, it can find out the different printers available in the piconet—dot matrix printer, laser printer, etc., and then subsequently select the desired print service).

- Browsing of services.
- Discovery of new services when devices enter RF proximity of other devices.
- Mechanism to find out when a service becomes unavailable when the device goes out of RF range (when there is no RF proximity).
- The details of services such as classes of services and the attributes of services.
- It can discover services on another device without consulting the third device.

When a device wants to discover a service, the application software initiates the request (the client), and the SDP client sends SDP request to the server (the device that can provide the required service). SDP client and server exchange SDP messages. Note that the server and client can be any two devices—the server is the device that can provide the service being requested by the client.

The server maintains a list of service records. Each record is identified by a 32-bit number for unique identification. The service record will have a number of attributes. The attributes can be service class ID list (type of service), service ID, protocol description list (protocol used for using the service), provider name, icon URL (an iconic representation of the service), service name, and service description. Each attribute will have two components: attribute ID and attribute value.

The service discovery protocol (SDP) provides the capability to form ad hoc networks in a Bluetooth environment. When two devices come near to each other, using SDP, a device can obtain the list of services offered by the other device and then access the desired service.

For instance, consider a laptop that requires a print service. The laptop is a client looking for a print service in a Bluetooth environment. The procedure for obtaining this service is as follows:

1. Client sends a service search request specifying the print service class ID to the server.
2. Server sends a service search response to the client indicating that two print services are provided.
3. Client sends a service attribute request and a protocol descriptor list to the server, asking for the details of the service.
4. Server sends the response to the client indicating that PostScript print service is provided.

The SDP is the heart of the Bluetooth system because it provides the capability to discover availability of services and the details of the services, along with the necessary information such as protocols to access the service.

RFCOMM

RFCOMM is a transport protocol to emulate serial communication (RS232 serial ports) over L2CAP. Through RFCOMM, two devices can communicate using serial communication protocols over Bluetooth radio. To achieve this, RFCOMM emulates the nine signals of RS232. These signals are

102 for signal common

103 transmit data (TD)

104 received data (RD)

105 request to send (RTS)

106 clear to send (CTS)

107 data set ready (DSR)

108 data terminal ready (DTR)

109 data carrier detect (DTR)

125 ring indicator (RI)

RFCOMM is a transport layer protocol to emulate serial communication over L2CAP. A Bluetooth-enabled PC can communicate with a Bluetooth-enabled modem using this protocol.

RFCOMM is derived from the GSM specification TS 07.10 for serial emulation. It supports two types of devices. Type 1 devices are communication end points such as computers and printers. Type 2 devices are part of the communication segment such as modems.

Telephony Control Protocol Specifications (TCS)

TCS protocol handles call control signaling protocols to establish voice and data calls between Bluetooth devices. This protocol is based on the Q.931 standard, which is used for signaling in Integrated Services Digital Network (ISDN).

To establish voice communication between two Bluetooth devices, we need the SCO links. SCO links are not handled by L2CAP protocol. However, L2CAP handles the signaling required for establishing voice connections through the Telephony Control Protocol Specification (TCS). Note that it is not abbreviated as TCP: TCP stands for Transmission Control Protocol used in the Internet protocol architecture. TCS defines call control signaling for establishing speech and data calls between Bluetooth devices and mobility management procedures. This protocol is based on the International Telecommunications Union (ITU) standard Q.931, which is the standard for ISDN signaling. TCS messages are exchanged between devices to establish and release calls and to provide supplementary services such as calling line identification (to identify the telephone number of the calling subscriber).

Host Control Interface

To Bluetooth-enable a laptop computer, we can connect a small Bluetooth module to the USB port of the laptop and run the protocol stack on the laptop (called the host). The Bluetooth device will have two parts: a module implementing the lower layers (LMP and below) and a software module implementing higher layers (L2CAP and above). The software module runs on the laptop (the host). The host controller interface (HCI) provides a standard interface so that we can buy the hardware module from one vendor and the software module from another vendor. HCI uses three types of packets:

- Commands, which are sent from the host to the module
- Events, which are sent from the module to the host
- Data packets, which are exchanged between the host and the module

  The functions of HCI are:

- Setup and disconnection of the links and configuring the links.
- Control of baseband features such as timeouts.
- Retrieving of status information of the module.
- Invoking the test modes to test the module for local testing of Bluetooth devices.

  HCI provides command interface to the baseband controller and link manager as well as access to hardware status and control registers. HCI has to reside in the Bluetooth module connected to the laptop as well as the host. In the Bluetooth module firmware, HCI commands are implemented so that the host can access the baseband commands, link manager commands, hardware status registers, control registers, and event registers. The Bluetooth module is connected to the USB port. The physical bus is the USB port. Three transport layers are defined to get HCI packets from host to the Bluetooth module: (a) USB, (b) RS232 and (c) UART (universal asynchronous receive transmit), a serial interface without error correction.

 HCI software has to reside on both the Bluetooth module attached to a device (such as the PC) and the host (the PC).

In the host, the bus driver is implemented as software above which the HCI driver software and other higher layer protocol software are implemented. The HCI commands can be categorized as:

HCI commands are used to establish piconets and scatternets, to get information about the local Blutooth hardware, and to test the local Bluetooth hardware.

- Link control commands to establish piconets and scatternets
- Link policy commands to put devices in hold mode/sniff mode
- Commands to get information about the local hardware
- Commands to get the status parameters
- Commands to test the local Bluetooth module

In summary, Bluetooth provides an efficient protocol stack to support voice and data services. The wireless application protocol (WAP) stack can run over the Bluetooth to provide many interesting wireless applications. The reader can refer to the book by Dreamtech Software Team (given in the References section) for details of WAP over Bluetooth.

36.3 HOMERF

In many homes, there will be multiple PCs and a number of peripherals such as printers, modems, variety of telephones such as fixed telephones, cordless telephones and mobile telephones. In such homes, a low-cost solution to network PCs and peripherals is required. With such a network, we can develop applications such as:

- Operating home appliances (air conditioner, music player, etc.) through a cordless phone.
- Sharing the same modem and telephone line to access the Internet from two or more computers.
- Exchanging data between two PCs or a PC and a laptop.
- Forwarding calls to different cordless handsets.
- Backing up of data of a PC without wiring up two PCs.

Home networks are now gaining popularity to support such applications. Though wired networks based on standards such as IEEE 1394 are available, wireless home networks are more attractive because of ease of use, fast installation, and easier maintenance. The two standards that have been proposed for wireless home networks are IEEE 802.11b and HomeRF.

Shared wireless access protocol (SWAP) sponsored by the HomeRF working group *(http://www.homerf.org)*, provides a low-cost solution to home networking requirements. The broad specifications of this standard are:

- **Range:** Up to 150 feet (covering a home, backyard and garage).
- **No. of devices:** Up to 127 per network.
- **Frequency band:** 2.4GHz (ISM band).
- **Transmit power:** 100 mW.
- **Speed:** 10Mbps peak data rate with fallback mode to 5Mbps, 1.6Mbps, 0.8Mbps. For 10Mbps data rate, 15 5MHz channels are used and for 1.6Mbps data rate, 75 1MHz channels are used.
- **Access:** Frequency hopping spread spectrum with 50 hops per second.

   The protocol stack for the HomeRF system is shown in Figure 36.6. It uses the TCP layer above the IP for regular data services, UDP for audio and video streaming applications and, above the MAC layer, DECT (digital enhanced cordless telecommunications) standards for toll quality voice applications.

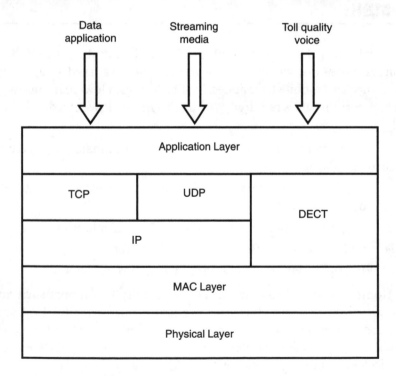

FIGURE 36.6 HomeRF protocol layers.

HomeRF is a radio technology to interconnect various home appliances. It operates in the 2.4GHz ISM band with speeds up to 10Mbps. A HomeRF network can support up to 127 devices.

Physical layer: The physical layer specifications are: 10Mbps peak data rate with fallback modes to 5Mbps, 1.6Mbps, 0.8Mbps. Data is modulated using constant envelope FSK modulation technique and transmitted over the 2.4GHz channel using frequency hopping.

MAC layer: For data and streaming audio/video applications, CSMA/CA protocol is used. For voice applications, reservation TDMA is used. The bulk of the MAC frame is used for data communication, but the streaming media services get priority. Based on the number of voice calls, time is reserved for voice slots. If voice packets fail, they can be resent at the start of the next frame.

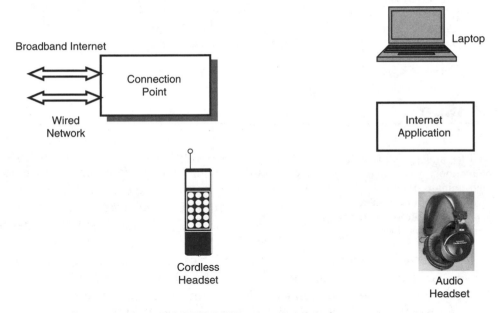

FIGURE 36.7 HomeRF network.

As shown in Figure 36.7, the HomeRF network consists of a connection point (CP) and different home devices. The CP is connected to the Internet or the fixed network on one side and communicates with the home devices (such as laptop, PC, audio headset, and cordless handset through the radio medium. A network can support a maximum of 127 nodes. The nodes of the network can be of four types:

1. Connection point that supports both voice and data services.

2. Voice terminals, which use TDMA to communicate with the base station.

3. Data nodes, which use CSMA/CA to communicate with the base station or other devices.

4. Voice and data nodes, which can support both types of services.

Up to eight simultaneous streaming media sessions and up to eight simultaneous toll-quality two-way cordless device connections are supported.

Each network is given a 48-bit ID, so the concurrent operation of multiple collocated networks is possible. The network ID, frequency hopping, and 128-bit data encryption provide the necessary security for the network and the data.

HomeRF supports data and voice services. For voice services, the TDMA access mechanism is used, and for data services, CSMA/CA access mechanism is used.

HomeRF technology holds lot of promise but, perhaps because of the competition from other technologies, the standardization committee was disbanded in January 2003.

36.4 IRDA

Infrared technology (IR) is used to interconnect two devices through point-to-point links. IR has a range of about 1 meter. Data rates up to 4Mbps can be supported by these links.

For many mobile phones and palmtops, the infrared (IR) port is available. When two palmtops are close to each other, they can exchange data. IR port is a very low-cost solution. The Infrared Data Association (*http://www.irda.org*) was founded in 1993 to develop standards for low-cost solutions for point-to-point infrared communication. IR communication can be implemented in devices such as PDA, digital camera, printer, overhead projector, bank ATM, fax machine, copying machine, and even a credit card. An IR device can communicate with another IR device within a range of 1 meter. Serial IR (SIR) supports data rates up to 115kbps, and Fast IR (FIR) supports data rates up to 4Mbps. The directed IR systems allow one-to-one communication in point-to-point mode and are not subject to regulations. However, the disadvantage with IR is that infrared rays cannot penetrate walls, and it supports only data. In spite of these limitations, IR-based systems are now being widely deployed on laptops, mobile phones, and computer peripherals (for instance to provide a wireless keyboard and wireless mouse) mainly because of the low cost.

36.5 IEEE 802.1X

IEEE 802.1x standards-based wireless LANs have become very popular in recent years. IEEE 802.11b and IEEE 802.15.3 are for personal/home area networks. The details of these standards are discussed in Chapter 17 "Local Area Networks" where all the IEEE Standards for LANs and PANs are covered. While deploying these networks, base stations need to be installed strategically to avoid interference because these networks are interference prone when installed in apartments.

IEEE 802.11b and IEEE 802.15.3 are the IEEE standards for Personal/Home Area Networks.

NOTE

Summary

Personal area networks and home area networks are receiving a lot of attention. The objective of PAN and HAN is to interconnect various devices used in offices, homes, and even in cars using wireless technology. The important technologies for achieving this objective are Bluetooth, HomeRF, IrDA, and IEEE 802.1x. These technologies are reviewed in this chapter.

Bluetooth enables creation of ad hoc spontaneous networks using wireless technology. A group of devices form a piconet. Each piconet will have a master and up to seven active slaves. Communication is in the 2.4GHz band using frequency hopping. Point-to-point and point-to-multipoint communication can be done between the master and the slaves. Both connection-oriented and connectionless services are supported for voice and data applications. Necessary security is built into this technology through encryption as well as frequency hopping. Bluetooth is gaining popularity, and devices such as PCs, laptops, palmtops, mobile phones, speakers, microphones, and headphones are being Bluetooth enabled. The range of Bluetooth is about 10 meters but can be extended to 100 meters with higher transmission power.

HomeRF is another wireless technology to interconnect up to 127 devices within a radius of about 150 feet. HomeRF systems also operate in the 2.4GHz band using frequency hopping with data rates support up to 10Mbps.

Infrared as the medium to interconnect devices has the main attraction of low cost. Devices such as palmtops, mobile phones, desktops, and laptops are being provided with infrared port for exchanging data within a range of about one meter.

In addition to providing wireless connectivity between devices, these technologies provide the capability for creating ad hoc networks wherein two or more devices spontaneously form a network when they come close to one another.

References

C. Bisdikian. "An Overview of the Bluetooth Wireless Technology." *IEEE Communications Magazine*, Vol. 39, No. 11, December 2001. This paper gives an excellent overview of Bluetooth.

B. Rose. "Home Networks: A Standards Perspective." *IEEE Communications Magazine,* Vol. 39, No. 11, December 2001. This paper gives the details of various standards for home networking.

J. Karaoguz. "High Rate Personal Area Networks." *IEEE Communications Magazine,* Vol. 39, No. 11, December 2001. This paper gives the details of the various personal area network technologies.

Dreamtech Software Team. *WAP, Bluetooth and 3G Programming.* Hungry Minds Inc., 2001. This book contains the theory of Bluetooth and also the source code for developing applications using Ericsson's Bluetooth development kit.

www.anywhereyougo.com Provides information on wireless technologies and resources for developers.

www.bluetooth.com Official site of Bluetooth special interest group.

www.comtec.sigma.se Web site of Sigma Comtec AB, Sweden, distributor of Ericsson's Bluetooth application toolkit.

www.cstack.com Bluetooth site by engineers for engineers.

www.homerf.org Web site for HomeRF standards and products.

www-124.ibm.com/developerworks/projects/ IBM's web resources for developers. Bluehoc 2.0, a network simulator, can be downloaded from this site.

standards.ieee.org You can obtain the IEEE standards on local area networks and personal area networks form this site.

www.irda.org IrDA Web site.

www.lucent.com/micro/bluetooth Bluetooth information area of Lucent Technologies.

www.motorola.com/bluetooth Motorola's Bluetooth resources.

www.nokia.com/bluetooth/index.html Nokia's Bluetooth resources.

www.semiconductors.philips.com/bluetooth Philips' Bluetooth resources.

Questions

1. Compared to wired LANs, what are the advantages of PANs with reference to ad hoc networking and spontaneous networking?
2. Explain the Blutooth protocol architecture.
3. Explain the salient features of HomeRF technology.
4. What are the important features and limitations of infrared communication systems?
5. Compare the various PAN technologies.
6. Explain the service discovery protocol with an example.

Exercises

1. Carry out a comparison of different technologies used for personal/home area networking. The comparison can be based on the data rates, security features, number of devices supported, and cost.
2. SyncML is a markup language developed to synchronize data between two devices (for example, between desktop and palmtop). Study SyncML and prepare a technical report.
3. Study how wireless networks can be simulated on a LAN. Study the Bluehoc development kit.
4. Write a technical paper on WAP over Bluetooth and its applications.

Projects

1. Using a Bluetooth development kit, develop software for a chat application between two PCs using Bluetooth as the wireless link.
2. Simulate a wireless personal area network on a LAN using IBM's Bluehoc.
3. Develop a WAP server that can be installed at a railway station. This server has to periodically broadcast information about the details of trains (train number, platform number, and the departure/arrival time) over a Bluetooth link. Whenever any mobile phone goes near this WAP server, automatically the information has to be sent to the mobile device by the WAP server. Use a WAP toolkit to test the application.

4. Home appliances such as air conditioner and VCD player can be controlled from a mobile phone if the appliance and mobile phone are Bluetooth enabled. Write the software for this type of remote control. You can simulate the control functions for the air conditioner (switch on/off, temperature, fan on/off) on a PC, and another PC can be used for simulating the mobile phone. The link between the two PCs can be either Bluetooth or an RS232 link.

37 Telecommunications Management Network

In This Chapter

- The Need for a Network to Manage Telecommunication Networks
- The Architecture of a Telecommunications Management Network
- Standard Interfaces Used in a TMN
- Legacy Network Elements in a TMN

Managing a telecommunications network is a complex task. Telecom service providers procure telecom equipment from different vendors, and each piece of equipment uses proprietary architecture. Even for managing the equipment, proprietary protocols are used. Multivendor, multiplatform network management is very important to provide reliable services as well as to reduce overhead in telecom network management.

A telecommunications management network (TMN) provides a framework for achieving this goal—to efficiently manage multivendor complex networks. ITU-T (International Telecommunications Union Telecommunications Services Sector) defined TMN in its M.3000 series of recommendations. When telecom equipment manufacturers implement these standard protocols in their network elements, it will be very easy to manage network elements. In this chapter, we will study the architecture of TMN.

37.1 OVERVIEW OF TMN

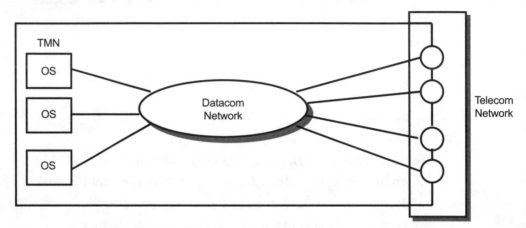

FIGURE 37.1 A telecommunications management network.

Telecommunications network management involves ensuring reliable functioning of various network elements, collecting traffic information, and optimal utilization of network resources.

Telecommunication network management is very important to provide reliable and efficient service to customers. The telecom service provider needs to have the necessary information on the functioning of various network elements such as subscriber terminals, switches, routers, multiplexers, radio base stations, digital cross connect equipment, and so on. Presently, the functioning of these network elements is generally known only locally, where the network element is installed. Ideally, all network elements should be monitored from a central location.

Furthermore, network management involves obtaining statistical information about the traffic. If the traffic is heavy on a particular route, the operator has to increase the bandwidth for that route to get more revenue. Also, billing information has to be collected at a central location for easy administration—generation and distribution of bills. All this calls for efficient centralized network management. TMN provides the solution.

The objective of a telecommunications management network is to achieve multivendor multiplatform network management using standard protocols.

As shown in Figure 37.1, TMN achieves interoperability by defining a data communication network that interacts with the various network elements for obtaining management information and presenting it in a user-friendly manner.

The operations systems (OSs) carry out network management operations such as monitoring and controlling the telecom network elements. However, the challenge lies in developing a framework that will ensure that legacy systems can also be brought into the TMN. Network elements (NEs), such as switching systems, multiplexers, and radio equipment, are managed through standard interfaces defined in TMN. If the NE is TMN compliant, it is fine; otherwise it communicates through a Q-adapter (QA). Defining the standard interfaces is the important feature of TMN, As shown in Figure 37.2, public networks such as PSTN and PLMN, consisting of various types of equipment such as satellite systems, digital terrestrial radio systems, optical fiber systems, multiplexers, and all types of switching and transmission systems can be controlled through TMN. The power of TMN is evident from this figure: every network element based on cable, radio, or fiber technology can be brought under the purview of TMN. In this figure, the switch, add-drop multiplexer (ADM), digital cross connect (DXC), digital radio terminal, satellite radio terminal, and mobile switching center of a mobile communication system are the network elements managed by the TMN.

A telecommunications management network (TMN) is a data communication network exclusively for managing telecommunications networks. Standard interfaces and protocols have been developed to manage the network elements.

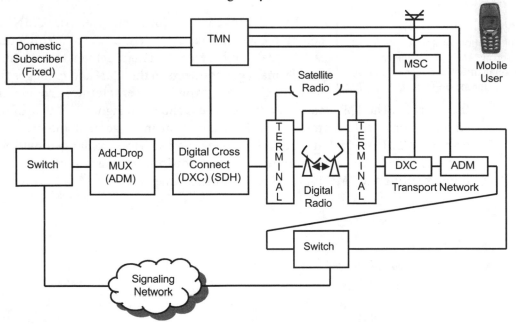

FIGURE 37.2 Public networks with TMN.

Communication equipment such as switches, multiplexers, repeaters, and radio equipment are called network elements. All these network elements can be controlled by TMN.

In line with the developments in programming languages, TMN uses object-oriented technology—each managed resource is viewed as an object. TMN defines guidelines for definition of managed objects (GDMO), which provide templates for classifying and describing the managed objects. The network-managed information and the rules by which the information is managed and presented are referred to as the management information base (MIB). This feature of TMN enables real-life modeling of telecom systems, so TMN implementation is pretty easy. The MIB framework is similar to the MIB discussed in SNMP.

TMN uses object-oriented technology. Each managed resource is viewed as an object. Hence, TMN allows real-life modeling of telecom systems.

In TMN, network management information is exchanged using the management information base (MIB). The MIB framework is similar to that of Simple Network Management Protocol (SNMP).

Common Management information protocol (CMIP) is used in TMN to define the management services exchanged by different objects.

To enable the various entities in TMN to communicate with each other, a protocol architecture is defined that is based on OSI management architecture. A TMN manager connects to the TMN and manages the network elements through a user interface provided to the operator. The TMN manager architecture is shown in Figure 37.3. One of the standard protocols defined in TMN is common management information protocol (CMIP), which defines the management services exchanged between peer entities. All telecommunications equipment manufacturers are implementing the TMN for efficient network management.

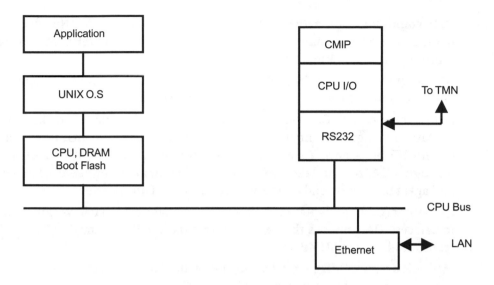

FIGURE 37.3 TMN manager architecture.

37.2 TMN BUILDING BLOCKS

TMN consists of the following functional blocks (See Figure 37.4):

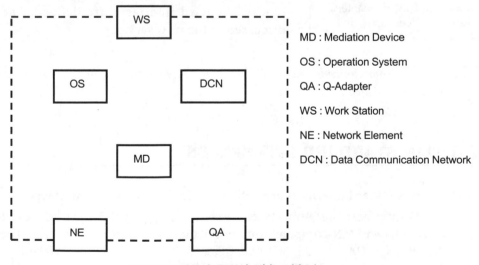

MD : Mediation Device

OS : Operation System

QA : Q-Adapter

WS : Work Station

NE : Network Element

DCN : Data Communication Network

FIGURE 37.4 TMN building blocks.

- **Data communication network (DCN):** The DCN within the TMN is used to manage the telecommunication network. It is based on OSI architecture representing layers 1 to 3.
- **Operations systems (OS):** Operations system carries out functions such as operations monitoring, and controlling network elements.
- **Network element (NE):** Every network element to be managed by the TMN contains manageable information that is monitored and controlled by an OS. An NE must have a standard TMN interface, otherwise it is managed through a QA. NE can take two possible roles: manager or agent. Manager and agent processes send and receive requests and notifications using CMIP.
- **Q-adapter (QA):** Q-adapter enables TMN to manage NEs that have non-TMN interfaces. QA provides the necessary translation. For example, SNMP QA translates between SNMP and CMIP.
- **Mediation device (MD):** A mediation device mediates between the local TMN interface and the OS information model. MD ensures that the information, scope and functionality are presented to the OS in the way they are expected.
- **Work station (WS):** A work station translates information between the TMN format and the format of a user-friendly interface. It provides a graphical user interface to give commands to obtain the management information from an NE. The command is sent using the CMIP protocol to the NE, and the information is obtained and presented to the user.

The building blocks of TMN are data communication network, network elements, Q-adapters, mediation devices, and work stations.

For communication among these functional blocks, standard interfaces are defined. These are discussed in the next section.

The data communication network (DCN) of TMN is based on the first three layers of the OSI protocol architecture.

37.3 TMN STANDARD INTERFACES

Three standard interfaces are defined in TMN: Q, F, and X interfaces.

Q-interface: The Q-interfaces are shown in Figure 37.5. Q-interface is defined between two TMN-compliant functional blocks. Q3 interface is defined between NE and OS, QA and OS, MD and OS, and OS and OS. Qx interface is defined between MD and QA, MD and NE.

F-interface: The F-interface is defined between WS and OS, WS and MD.

X-interface: X-interface is defined between two TMN-compliant OSs in two separate TMN domains. This interface is for two TMNs to exchange management information.

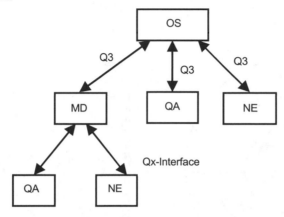

FIGURE 37.5 TMN Q-interface.

The three standard interfaces defined in TMN are Q-interface, F-interface, and X-interface. The heart of TMN is Q-interface or Q-adapter.

The heart of TMN is the Q-interface or the Q-adapter. As shown in Figure 37.6, a legacy network element that is presently running, such as SNMP for network management, can be brought into the purview of TMN through the Q-interface. The NE can run the SNMP software and through the Q-adapter, the OS can manage the NE. Between the OS and the Q-adapter, the CMIP/Q3 protocol is running and the Q-adapter interfaces between the legacy NE and the TMN-compliant OS.

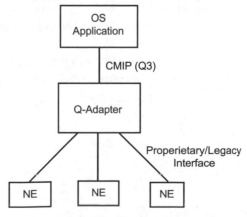

FIGURE 37.6 TMN Q-adapter.

The architecture of a Q-agent is shown in Figure. 37.7. The Q-agent runs a real-time operating system above which the application is running. The Q-agent is connected to the TMN through CMIP and to the NE through a data communication link that can run, for instance, HDLC protocol. It also can be connected to the LAN through a normal Ethernet interface. Note that the Q-agent is basically an embedded system running a real-time operating system.

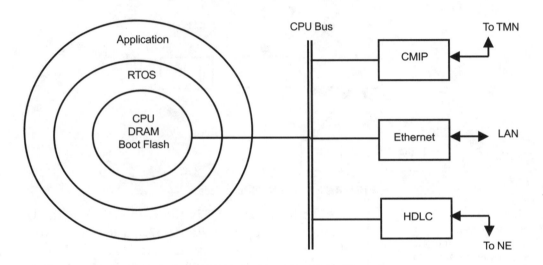

FIGURE 37.7 Q-agent architecture.

The complete TMN architecture and the building blocks and protocols are shown in Figure 37.8. A network element (such as a switch, multiplexer, or repeater) will be running the Q-agent software. The Operations System manages these network elements through the Q3 protocol. The OS is connected to the operation, administration, and maintenance (OAM) center through the work station. The TMN consists of a number of OSs networked together to get a global view of the whole telecommunication network and to manage it effectively.

A network element such as a switch, multiplexer, or repeater will be running the Q-agent software. The operations system manages the network element using Q3 protocol.

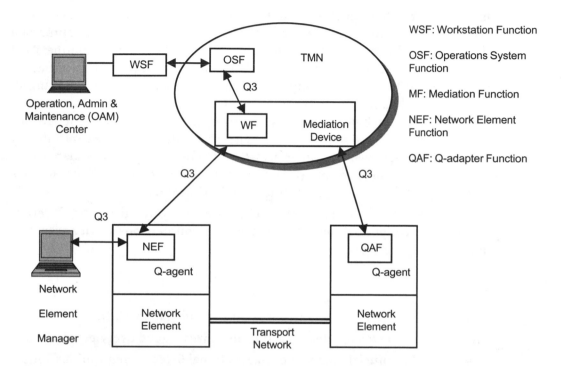

FIGURE 37.8 Typical TMN architecture.

Though many of the network elements presently do not have TMN compliance, all telecom equipment suppliers are now integrating TMN protocols in the network elements. The telecom service providers also are insisting on TMN interface for equipment because it will make network management extremely easy. TMN provides the capability to manage complex telecommunication networks of the future.

 Many of the present network elements do not have TMN compliance. However, using a Q-adapter, these network elements still can be managed by the TMN.

Summary

As the complexity of telecommunications networks increases managing the networks has become an important issue. A telecommunications management network (TMN) is a network based on standard protocols for efficient management of telecommunications networks. Even legacy network elements such as switches, and routers can be managed through the TMN using appropriate interfaces. In this chapter, the architecture of TMN is presented. In the TMN framework, a special protocol named common management information protocol (CMIP) is defined through which management services are exchanged. The TMN is a data communication network (DCN) that carries the management information through operations systems (OSs) that monitor different network elements (NEs). The NEs will have standard TMN interfaces to be controlled by OSs. If NEs do not have TMN interfaces, they can be managed by Q-adapters. Q-adapters are required for old (legacy) NEs. TMN will also have work stations (WSs) to provide the user interface for monitoring and control.

TMN provides an excellent mechanism for controlling the different telecommunications equipment, and all telecommunications service providers are deploying the TMN.

References

www.iec.org/online/tutorials/tmn This link gives a tutorial on TMN.

www.simpleweb.org/tutorials/tmn This link also gives a good overview of TMN.

www.itu.int The official Web site of International Telecommunications Union. An excellent repository of ITU standards.

Questions

1. Explain the architecture of a telecommunications management network.
2. What are the building blocks of TMN? Explain the functionality of each block.
3. What are the standard TMN interfaces?
4. Explain how a network element that does not have a standard TMN interface can be managed through the TMN.

Exercises

1. Study the various TMN products offered by vendors such as Cisco, Nortel, Ericsson, and Lucent Technologies.
2. Study the details of Q-interface and explore how non–TMN-compliant network elements can be brought under the purview of TMN.

Projects

1. Simulate a TMN on a local area network. Each node on the LAN can be considered a network element of the TMN. The LAN cable provides the connectivity between the network elements. Simulate the CMIP protocol.

2. Implement Q-agent architecture on an embedded system. You can use ARM/Motorola/Intel processor-based hardware, port a real-time operating system such as QNX or RTLinux, and then implement the software as shown in Figure 37.7.

38 ∷ Information Security

In This Chapter

- Strategies to Protect Information
- The Architecture of Surveillance Equipment
- Cryptography Techniques
- Steganography Techniques

With the advent of computerization, information security has gained importance because, after all, "information is power." Organizations have to ensure that their information is not stolen or tampered with, either by employees or outsiders. This has become a very challenging task, particularly with the convergence technologies. On one hand, antisocial elements use the technology to hack computers, spread viruses, and pose many security threats to the organizations. On the other hand, the organizations need to devise methods to contain the security threats by employing the latest technologies. Cryptography is used extensively for providing secure applications. Steganography is now being used in conjunction with cryptography to provide highly secure information systems. In this chapter, we will study the various security threats and how to overcome them using cryptography and steganography. We have discussed many security protocols in various chapters of this book, and this chapter gives an overview of security issues and solutions, with special emphasis on multimedia-based information security products.

38.1 INFORMATION LEAKS

Every organization has to protect its information. Depending on the type of organization, this information is of different categories—organizations involved

To protect valuable information, every organization needs to make use of surveillance equipment to contain information leaks.

in design have to protect their design documents as the intellectual property developed is very precious; organizations involved in marketing have to protect their market research organizations involved in services (for example, hospitals) have to protect the information of their clients (for example, patient information), and so on. Surveys show that in 70% of cases, the employees of an organization leak information. Sometimes employees leak information out of ignorance, and some employees intentionally commit a crime. Many organizations are now resorting to surveillance—monitoring the activities of employees to check information leaks.

As shown in Figure 38.1, information can be leaked by an employee through an office telephone, a fax machine, a mobile phone, or a computer. Many organizations listen in on the telephone conversations of employees. Similarly, all e-mails can be checked to find out whether any information is being leaked.

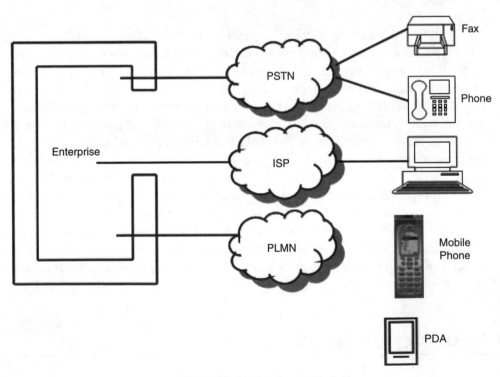

FIGURE 38.1 Information leaks.

38.2 SURVEILLANCE

Surveillance is the first step to gather the necessary intelligence regarding possible security attacks. Organizations concerned with public safety need to use surveillance equipment to monitor voice/fax/data communication.

Obtaining the necessary information regarding possible attacks is the most fundamental aspect of information security management. Organizations concerned with public safety, such as police and defense forces need to keep track of the activities of antisocial elements by monitoring their voice conversations and fax messages. Individual organizations need to monitor the voice conversations of suspicious employees. Surveillance equipment helps in monitoring voice/fax/data communications.

38.2.1 Surveillance of Voice/Fax Communication

Security agencies such as government intelligence agencies and police obtain the necessary permission from the appropriate authorities and monitor the calls of selected persons. The scheme for this surveillance is shown in Figure 38.2. The telecom service provider will give parallel telephone lines to the security agency. These parallel lines are extended from the main distribution frame (MDF) of the telephone switch. The security agency will have a digital recording system (DRS) that records the complete conversation. The recording system will be capable of recording conversations from either fixed telephones or mobile telephones.

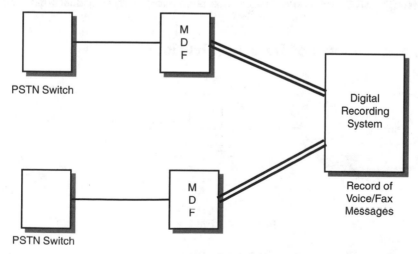

FIGURE 38.2 Surveillance of Voice/Fax Communication by Security Agencies.

Surveillance of voice/fax communication is done by a digital recording system that automatically records all voice/fax calls for later analysis. This type of surveillance provides the necessary intelligence regarding possible security threats.

Organizations can also monitor their telephone lines using the DRS as shown in Figure 38.3. In this system, incoming telephone lines will be connected to the PBX. Parallel lines will be connected to the DRS. All incoming and outgoing calls will be recorded in the DRS for later processing.

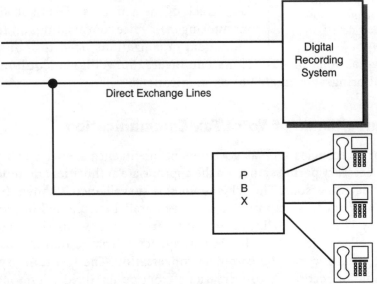

FIGURE 38.3 Surveillance of voice/fax communication by individual organizations.

38.2.2 Surveillance of Data Applications

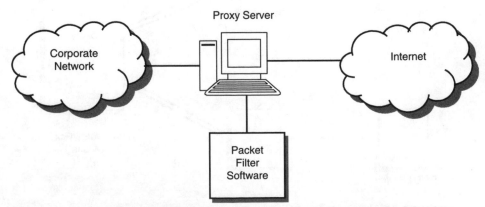

FIGURE 38.4 Surveillance of data applications.

Surveillance of data applications such as e-mail and file transfer is done through firewalls that monitor and if necessary filter the messages. A firewall can prevent access to specific URLs, or it can filter e-mail messages containing specific keywords.

To ensure that employees do not pass on secret information through e-mail, file transfer, and so on, surveillance is done from a proxy server through which the employees access the Internet. Packet-filtering software is used to monitor every incoming and outgoing packet. All the packets corresponding to a particular source/destination IP address can be collected to check the message content of the data application such as e-mail or file transfer. The packet filter can also filter out the packets—not allow the packets to pass out of the server.

38.3 THE DANGERS OF MOBILE PHONES

The mobile phone is an excellent communication gadget that can be used anywhere, anytime. However, it also poses a great security threat, as indicated in Figure 38.5.

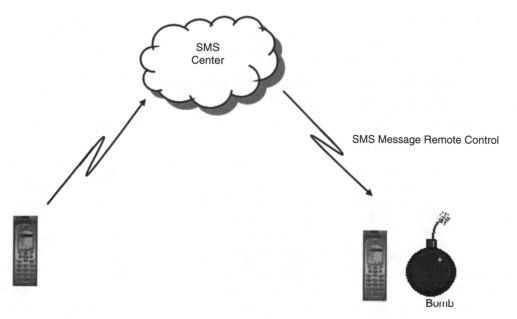

FIGURE 38.5 The SMS bomb.

Consider a scenario in which a bomb is fitted with a Bluetooth device. The bomb is programmed to explode if it receives a small text message. A few meters away from the bomb, a Bluetooth-enabled mobile phone is placed. From anywhere on the earth, another mobile phone can send a short message using the short messaging service (SMS) to the mobile phone located near the bomb. Over the Bluetooth link, this message is passed on to the electronics in the bomb, and the bomb is detonated. This example illustrates the dangers of technology and the need for surveillance to gather the necessary intelligence to stop such dangerous activities.

> The mobile phone is a security threat due to its potential in making a remote controlled detonator through the short messaging service.

38.4 CRYPTOGRAPHY

In any communication system, the three important security services to be provided are:

- Privacy
- Authentication
- Message integrity

These are achieved by cryptography.

Suppose you want to send a message to a friend. The message is

let us meet at sundowner

Instead of sending this message directly, you can send the following message:

mfu vt nffu bu tvoepxfs

> In cryptography, the message bit stream is modified using an encryption key. Only the person who has the encryption key can decode the message.
> Cryptographic algorithms are divided into two categories: secret key algorithms and public key algorithms.

You have encrypted the message. The encryption algorithm is to replace each letter with the next letter in the alphabet (replace a with b, b with c, c with d... and z with a). You have to tell your friend in advance that this is the algorithm and, whenever you send a message, your friend will be able to decode it. You can perhaps write a program to take the normal text message and convert it into the encrypted message. If your friend also uses this software, he can easily decode the message. Conceptually this is the mechanism used in cryptography. But then, if this message is seen by your foe, he may not find it difficult to decode the message—

it is not very difficult to find out the algorithm used. Many experts, called cryptoanalysts, can easily decode messages even if very complicated algorithms are used.

We can modify the above algorithm slightly. The new algorithm will shift each letter by two letters (replace a with c, b with d, etc.) or three letters (replace a with d, b with e, etc.), and so on. Now you have two things: an algorithm to shift the letters and a key that tells the number by which each letter is shifted (two, three, and so on). The algorithm can be known to anybody, but only you and your friend know the key, and you can use this mechanism to exchange secret information.

Cryptographic algorithms are broadly divided into two categories:

- Secret key algorithms
- Public key algorithms

38.4.1 Secret Key Algorithms

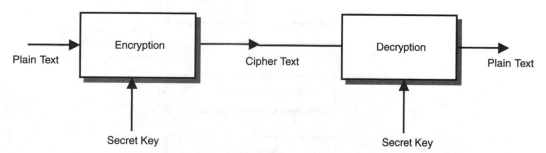

FIGURE 38.6 Encryption using secret key.

In secret key algorithms, both the sender and receiver share the same algorithm and the encryption key. The encrypted message is called ciphertext. Note that in this type of encryption, the algorithm can be made public, but the key should be kept confidential.

In a secret key algorithm, the sender and receiver (called participants) share the same algorithm and key. At the sending side, the original message (called plain text) is encrypted using the secret key. This encrypted message (called ciphertext) is transmitted to the receiver, and the receiver uses the same key to decrypt the message. The sender has to reveal the secret key beforehand to the intended recipient. If the secret key is leaked to someone else, he will be able to decode the message. The Data Encryption Standard (DES) developed by the U.S. Department of Defense, is the most widely known secret key algorithm.

Data Encryption Standard (DES) developed by the U.S. Department of Defense, has been used extensively as a secret key algorithm. In DES, each block of 64 bits of information is modified using a 56-bit encryption key to obtain 64-bit ciphertext.

In DES, 64 bits of information is manipulated using 56 bits of encryption key. The process of encryption using DES is shown in Figure 38.7. In the first step, the 64 bits of information are shuffled. Then the modified bit stream is modified using a 56-bit key in 16 iterations. The next step is to exchange the rightmost 32 bits with the leftmost 32 bits. The last step is to carry out the reverse operation of the first step. The output is a 64 bit encrypted data. At the receiving end, the operations are carried out in reverse order to get back the original plain text, of course using the same key.

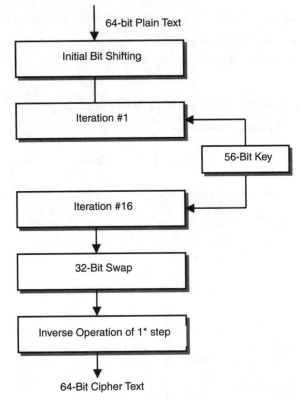

FIGURE 38.7 Data Encryption Standard.

DES is not based on any mathematical foundation. If the data is received by an unauthorized person, it is very difficult to decode the information because of the heavy computation involved—he has to try all possible combinations of ones and zeros of the 56-bit key to get the plain text. Through jugglery of bits, DES is a complicated algorithm.

To make it more secure, triple DES has been developed, which encrypts the data three times. For each iteration, a separate key can be used, so three keys are used for the encryption.

The intended recipient should know the key to decrypt the data. The key can be exchanged in person, or sent through the mail or separate secure communication links can be used.

38.4.2 Public Key Algorithms

In public key encryption, everyone will have two keys. One is called the public key, and the other is called the private key. The public key is known to everyone. You can keep the public keys in a database and make it accessible to everyone. The private key is kept confidential by everyone. As shown in Figure 38.8, if A wants to send a secret message to B, A encrypts the message using B's public key. B decrypts the message using his private key.

> In public key encryption, there will be two keys—public key and private key. To send a secret message, the message is encrypted using the recipient's public key, and the recipient decodes the message using his private key.

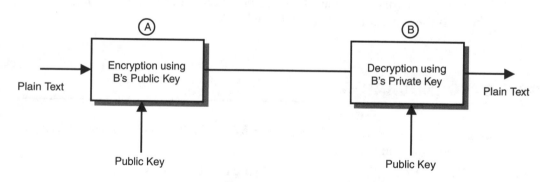

FIGURE 38.8 Public key encryption.

RSA algorithm is the most widely used public key algorithm. RSA gets its name from the initials of the three inventors: Rivest, Shamir, and Adleman. RSA algorithm is based on number theory. To understand this algorithm, you need to recall the following definitions:

> The RSA algorithm is the most widely used public key algorithm. It is based on the fact that factoring large numbers is computationally intensive. Two large prime numbers are used to generate the public key and private key.

A number that has no factors except 1 and itself is known as a prime number. Examples of prime numbers are 2, 3, 5, 7, 11, 13 and 17.

If two numbers do not have any common factors other than 1, they are called relatively prime numbers or co-primes. Examples are 7 and 60, 8 and 15, 24 and 77, 105 and 143, and 108 and 77.

In the RSA algorithm, we need to generate a public key and a private key. For this, choose two large prime numbers p and q such that both are at least 256 bits.

Let n = a * b

Choose a number e such that e and (p–1) * (q–1) are relatively prime.

Calculate the value of d using the formula $d = e^{-1} \mod((p-1)(q-1))$.

The public key is the pair (e,n) and the private key the pair (d, n).

Encryption is done using the formula

Ciphertext = (message)^e mod n

Decryption is done using the formula

Message = (ciphertext)^d mod n

The RSA algorithm is based on number theory and on the fact that factoring large numbers is computationally intensive.

38.5 STEGANOGRAPHY

Steganography is defined as "the art of information hiding" or "covered writing." Suppose you send a message:

HOW   ARE   YOU? ICAN   MEETYOU   SOON.

The spaces between the successive words are 3, 7, 5, 0, 8, 0, 4 (between HOW and ARE there are four spaces, between ARE and YOU seven spaces and so on). The number 3750804 can convey a telephone number. This is covered writing.

A more sophisticated steganography technique is to embed secret messages in image files, voice files or video clippings. For instance, the photograph shown in Figure 38.9 has a secret message embedded in it (but you will not know what the secret message is!). The secret message can be text or another image.

FIGURE 38.9 Secret message embedded in an image.

Steganography is the art of information hiding. Secret messages can be sent by hiding them within text, image, voice, or video.

The process of steganography is shown in Figure 38.10. The secret message is encrypted using an encryption key and is embedded in an image (or voice or video file). The image containing the secret message is transmitted over the network. The modified image is called a *stegano*. At the receiving end, the secret message is extracted from the stegano, and the message is decrypted. The important requirement in steganography is that the image modified with the secret message should look normal. A number of techniques are used to meet this requirement by applying many image analysis algorithms. For instance, the least significant bit of each pixel may contain one bit of the secret message. By modifying the least significant bit, the image quality is not going to degrade much. Hence, the modified image looks almost normal.

Another technique used is to modify only certain portions of the image (for instance, the background portion of the image in Figure 38.9). Similarly, a secret message can be embedded in a voice file by replacing the least significant bit of each PCM-encoded voice sample with one bit of the secret message.

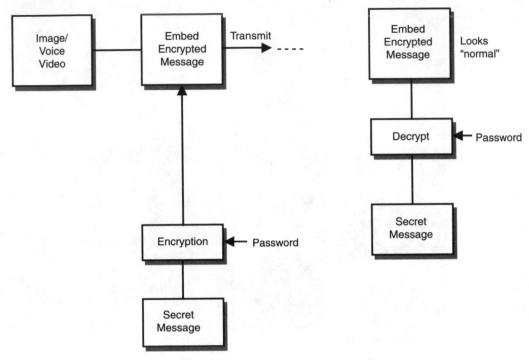

FIGURE 38.10 Process of steganography.

Steganography is a very powerful technique for sending secret information. Unfortunately, it has been used extensively by terrorists and other antisocial elements, but not by organizations to provide a public service.

38.5.1 Digital Watermarking

Digital watermarking is a mechanism wherein information can be hidden in multimedia content using digital techniques. Secret messages (text, image, voice, or video) can be hidden in a cover document, which can also be text, image, voice, or video.

Steganographic technique, when applied to digital images, digital voice, or digital video files, is known as *digital watermarking*. Hence, hiding a secret message in multimedia content using digital techniques is digital watermarking. The secret message is known as a watermark. The watermark can be text, image, voice, or video. This secret message is embedded in another document known as the cover document. The cover document can also be text, image, voice, or video.

Software product development organizations use digital watermarking techniques to embed copyright information, product serial number, and so on in an image such as the company logo. If you want to hide copyright information and serial number of your software product, you can create a JPEG image of your company logo and embed the copyright message and serial number in this logo. The copyright message is not visible to the eye. If the software is pirated and installed on another system, your company can analyze the logo and find out from whom the software is copied.

 Digital watermarking is nowdays used extensively for hiding copyright information in software products as well as music files to detect piracy.

38.5.2 Secure Multimedia Communication

The layered architecture for secure multimedia communications is shown in Figure 38.11. In addition to cryptography and key management, a new layer that uses steganography needs to be introduced. Any data, voice, fax, or video application can use steganography to transmit secret information. Images can be used as passwords using this architecture.

In the future, attacks based on steganography are likely to increase because of the widespread use of convergence technologies, particularly voice and video communication over IP networks. The surveillance equipment discussed earlier in this chapter needs to take care of these aspects as well. Organizations involved in public service such as banks, insurance agencies, and governments also need to use steganography to provide more secure communications.

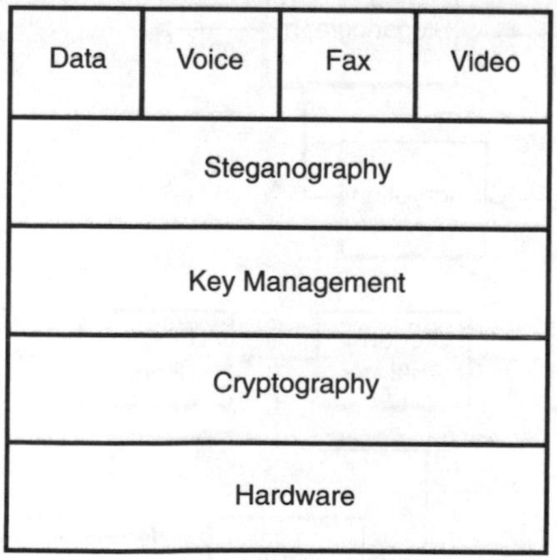

Data	Voice	Fax	Video
Steganography			
Key Management			
Cryptography			
Hardware			

FIGURE 38.11 Layered architecture for secure multimedia communication.

Combining steganography and encryption provides a highly secure communication system. With the advent of convergence, more security threats are also likely, due to steganography.

Presently, encryption devices are used for data applications. Devices that can encrypt data, voice, and fax messages are used at the two ends of the communication link, as shown in Figure 38.12.

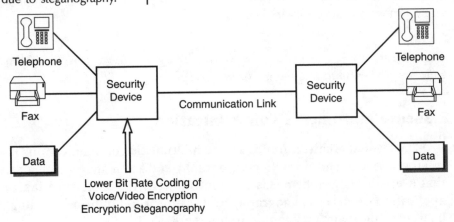

FIGURE 38.12 Multimedia security products.

The security device must be capable of low bit rate coding of messages and also support encryption and steganography. Such products can provide real secure information transfer. A representative application that uses biometrics and steganography is shown in Figure 38.13.

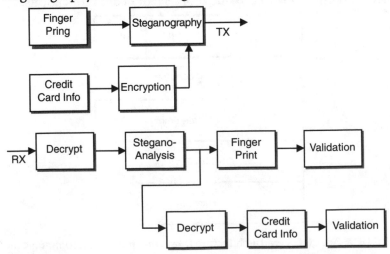

FIGURE 38.13 An example of biometrics and steganography.

By combining biometric security measures and steganography, highly secure banking applications can be developed. Biometric techniques such as fingerprint recognition can be done in conjunction with credit card verification.

Consider a bank that provides Internet banking. At present, the user logs into a Web server, types his account number, credit card number, and so forth for a bank transaction. The information typed is encrypted and sent to the server. Sometimes the credit card information is stolen. This is due to the fact that encryption alone does not provide full security. As shown in Figure 38.13, encryption and steganography can be combined. The fingerprint (biometric information) is converted into an image, and the encrypted credit card information is embedded into the image. At the receiving end, the stegano is analyzed, and the fingerprint and credit card information are extracted. The fingerprint is validated against the stored database of fingerprints. Similarly, credit card information is also validated. If the bank uses a proprietary algorithm for steganography, then a very highly secure application can be built.

Summary

In this chapter, the various issues involved in information security are discussed. Every organization needs to take measures to protect its information. Surveillance of voice, fax, and data is required to do this.

To provide secure communication, the two important cryptographic techniques are private key algorithm and public key algorithms. In private key algorithms, a private key is shared by the participants to encrypt and decrypt the data. The most widely used private key encryption algorithm is Data Encryption Standard (DES) developed by the U.S. Department of Defense. In public key encryption, two keys are used—a public key and a private key.

Steganography is a technique in which secret information is embedded in an image, a voice file, or a video file. In embedding the secret message, it has to be ensured that the original image does not change much. The simple technique to achieve this is by replacing the least significant bit of each pixel by one bit of the secret message. Steganography is a powerful technique and, by combining steganography and encryption, highly secure communication systems can be developed.

References

M. Barni et al. "Watermark Embedding: Hiding a Signal Within a Cover Image." *IEEE Communications Magazine,* Vol. 39, No. 8, August 2001

L.L. Peterson and B.S. Davie. *Computer Networks: A Systems Approach*. Morgan Kaufman Publishers Inc., 2000.

A.S. Tanenbaum. *Computer Networks*. Prentice Hall Inc., 1996.

Questions

1. Describe the various surveillance equipment used for surveillance of voice, fax, and data.
2. What is steganography? Explain the process of steganography.
3. What is digital watermarking? What are its uses?
4. Explain the various mechanisms for encryption.
5. Distinguish between secret key algorithms and public key algorithms.
6. Explain the RSA algorithm.
7. What is a firewall? Explain the various options for development of firewalls.

Exercises

1. In a voice surveillance equipment, 100 hours of telephone conversations have to be stored. Calculate the secondary storage requirement if PCM coding scheme is used for storing the voice.
2. Write a C program for the RSA algorithm.
3. Survey the commercially available audio and video surveillance equipment and prepare a technical report.
4. Download freely available steganography software and experiment with it.
5. Mathematicians claim that encryption algorithms cannot be broken for many, many years even if very powerful computers are used. Still, hackers are able to break the algorithms. How is that possible?
6. Work out the design of a telephone tapping system. This system has to be connected in parallel to your telephone, and it should record all telephone conversations.

Projects

1. Develop software to tap the telephone conversations in your office. You can connect a voice/data modem in parallel to the telephone. The modem is connected to a PC. When an incoming call is received and the telephone is picked up, the conversation has to be recorded in the PC automatically.

2. Develop a microcontroller (for example, 8051 or 8085) based embedded system that can be connected in parallel to the telephone for tapping telephone calls. The embedded system should have enough memory to store speech up to 15 minutes. You can use low bit rate coding chips to reduce the memory requirement.

3. Develop steganography software to embed secret text messages in a voice file. To ensure that there is no degradation in the voice quality, you can replace the least significant bit of each voice sample value with one bit of the secret message. You also need to write the software to retrieve the secret message from the stegano.

4. Develop steganography software to embed secret text messages in an image file. You can use the same strategy as in Project #3 or use a more sophisticated image processing technique.

5. Develop software that implements the Data Encryption Standard (DES) algorithm.

6. Develop a firewall that can be run on the proxy server of your LAN whose job is to filter all packets from a given IP address. This IP address corresponds to the address of a node on the LAN.

39 Futuristic Technologies and Applications

In This Chapter

- The Evolution of Telecommunication Networks
- How the Different Networks Are Converging
- Future Applications That Will Be Supported by the Networks
- The Challenges Involved in Developing Future Networks

I n the last decade of the twentieth century, many exciting technologies have become popular. Notable among them are the World Wide Web and wireless Internet access. Commercialization of services using computer telephony integration and multimedia communication over IP networks also happened on a large scale in recent years. In the first decade of the twenty-first century, we will witness more exciting developments. These developments will be catalyzed by the availability of high bandwidth transmission media to our homes and offices as well as the availability of high bandwidth wireless networks. Anywhere, anytime high-speed communication for anybody will be the theme of the first decade of the twenty-first century. In this chapter, we will trace past developments based on which the present architecture of the telecommunication networks evolved. We will also peep into some futuristic technologies and applications.

39.1 EVOLUTION OF THE PRESENT TELECOMMUNICATION NETWORK ARCHITECTURE

Ever since Alexander Graham Bell invented the telephone, there have been tremendous improvements in telecommunication technology. The Pubic Switched Telephone Network (PSTN) initially consisted of mechanical and

643

electromechanical switches and copper cables. These switches are being replaced by digital switches. Today the telephone networks in the developed countries are mostly digital and driven by software power. Though copper cables are still widely used as the transmission medium, trunk routes are now using optical fiber that can support millions of voice calls. Today, making a telephone call to anyone, anywhere on the Earth is child's play!

In the early 1970s, the Internet was developed to provide data services. In a span of about 30 years, the Internet has spread all over the world, connecting millions of computers and providing data services, and communicating has become much easier. Though the Internet was mainly intended for data services, it is now becoming the platform for voice and video services as well.

Though voice over IP is slowly gaining popularity, voice service over the Internet is not very high quality compared to the PSTN.

In the 1960s, mobile communication systems were deployed, paving the way for communicating while on the move. The mobile networks of earlier days supported only voice communication, but slowly data services, though at low speeds, became available. In the 1990s, wireless Internet took off, to access Internet content through mobile phones. Multimedia messaging on wireless networks is now in the works.

Due to the advances in low bit rate coding of voice and video signals coupled, with advances in efficient modulation and multiple access techniques, mobile communication systems can now support data rates up to 50Mbps.

The broadcasting systems, for radio and TV broadcasting, are the main entertainment systems. TV in particular has brought the entire world to our living rooms. The news of any corner of the world is made known to the entire world almost instantaneously. TV program distribution is done through the cable TV network or through satellites direct to homes.

Though current TV transmission is mostly analog, digital TV transmission will be predominant in the future. This will give much better quality video, called high-definition TV (HDTV).

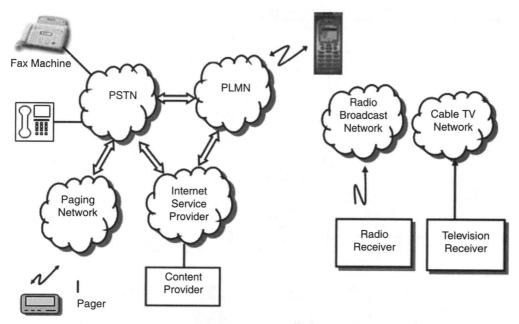

FIGURE 39.1 Architecture of telecommunications networks.

All these developments have taken place at a breath-taking pace, and today the telecommunication systems provide us the means of communicating very effectively. However, as shown in Figure 39.1, today we use different networks for doing different things—we use the Public Switched Telephone Network (PSTN) for making telephone calls and sending fax messages, we use the public land mobile network (PLMN) to make calls from mobile phones, we use the paging network for paging services, we access data services from the Internet through an Internet service provider, the cable TV network (in conjunction with satellite TV network) provides us with TV programs, and the radio broadcasting network provides us the audio programs. We have to use different gadgets to use different networks to get different services. These gadgets include a fixed telephone, a cordless phone at home and office, a fax machine, a mobile phone, a pager, a TV set, a desktop computer, a laptop computer, a palmtop computer.

Using multiple networks for different services is very inconvenient for users because we need different devices, we get multiple bills, and need to access different mailboxes.

Using so many gadgets is a problem. We also subscribe to different services and pay bills separately—one bill for Internet access, one bill to the fixed telephone, one bill to the mobile phone, and so on. For end users of telecommunication services, all this causes a lot of inconvenience and so on.

39.2 CONVERGENCE OF NETWORKS

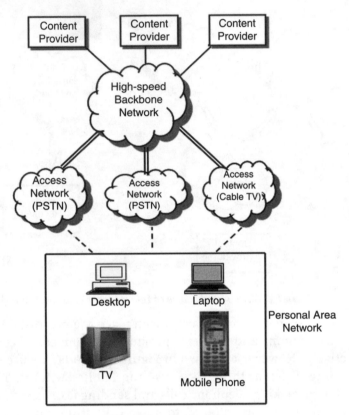

FIGURE 39.2 Convergence of Networks.

Convergence is paving the way for unification of different communication networks and services. Users can access any service using a device of his choice.

We are now witnessing a revolution called convergence. This convergence is leading to unification of different networks and services, paving the way for a telecommunication network architecture as shown in Figure 39.2. There will be a backbone network that is a very high-speed optical fiber network. The various content providers will be connected to this backbone network. Users can access the content through access networks which can be PSTN, PLMN, cable TV network, or others. The access can be through any device of the user's choice—for example, he can obtain e-mail messages on his desktop through the PSTN, on the mobile device through PLMN, or on WebTV through the cable TV network. The various gadgets used by a person at office or home are networked as a personal area network. This convergence of networks and services in turn is the basis for unified messaging. A user can use any terminal (a telephone, laptop, or mobile phone) and obtain any service without bothering about the underlying telecommunication network. For instance, a person can be notified on his mobile phone about the arrival of new mail in the mailbox, and if he wants the server can read out the message. He can access the Internet through a mobile phone. Multiparty video conferencing between several people (perhaps all of them traveling) can be done cost effectively. In a nutshell, the convergence technologies provide us the benefit of a wide variety of telecommunication services—anywhere, anytime communication at a very low cost.

39.3 EMERGING APPLICATIONS

Optical fiber provides high-capacity transmission bandwidths. Though optical fiber is used extensively in the backbone network, fiber to the home will become predominant in the coming years. When gigabits of bandwidth is available on the local loop, and the backbone network consists of high-speed switches and trunks, what kind of services can be envisaged? The number of television channels can increase drastically. Though the present TV transmission systems are still analog, digital TV is making inroads. Up to 1000 TV channels can be provided through the optical fiber using the present state of the art. Even if we surf the TV channels at the rate of 15 seconds per channel, it will take four hours to see them all.

Today, TV manufacturers advertise home theatres. Home theatres facilitate watching high-quality TV programs and listening to high fidelity music. With the availability of large bandwidth, there is a distinct possibility of having home studios. As shown in Figure 39.3, every home can become a studio with a full-fledged video camera and video reception equipment. People at two or more homes can participate in full-fledged video conferencing.

Convergence is paving the way for unification of different communication networks and services. Users can access any service using a device of his choice.

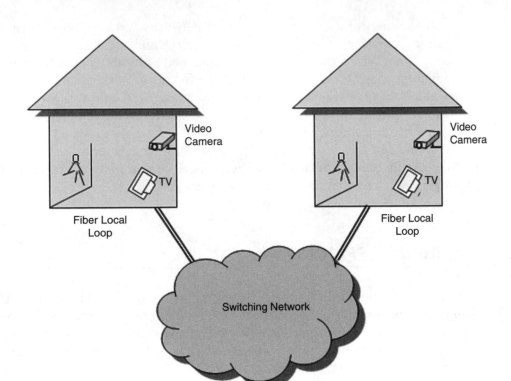

FIGURE 39.3 Home studio.

Recent advances in wireless technologies to provide data rates up to 50Mbps will facilitate high-bandwidth services to be supported even while we are on the move. Because mobile devices will have the power of today's desktop computers, mobile applications will support integrated data, voice, and video services.

 Wireless local loops are also being used in urban areas because this technology obviates the need for burying lines, and also maintenance is easy.

Future telecommunications networks will consist of high-speed backbone networks using optical fiber as the transmission medium. The access to these networks will be through high-speed wireless networks that support multimedia services.

Wireless local loop technologies will be a great boon to remote and rural areas. Particularly in developing countries, the majority of the people live in rural areas, and they can be provided with low-cost communication services to improve their quality of life.

The high bandwidths also facilitate a number of value-added services to be offered to all the people. Some important value-added services are:

- Telemedicine, or medicine at a distance: In areas where there is no access to medical doctors, patients can use video conferencing to interact with a doctor at a far-off place and get the necessary medical advice.

- Web-based learning: Though Web-based learning is becoming popular, because of lack of good Internet infrastructure in most parts of the world, accessing multimedia content with high-quality video and animation is a problem. Web-based learning will attain prominence in future.

- Collaborative working: Groups of researchers in different parts of the world can carry out combined research using the communication infrastructure. Collaborative work is also likely to be used effectively in software development.

- Virtual tours: Sitting in one's home, one can take a virtual tour of any of the historical places located anywhere in the world. Though as a concept it has been there for a few years, the communication infrastructure makes it a distinct commercial possibility.

Telemedicine is used extensively in developed countries for providing health care for remote and rural areas. This technology is of great benefit to developing countries, where there is a dearth of medical professionals.

A number of universities offer formal education programs (M.S. and M.B.A. degrees) through e-learning mode.

High-bandwidth wireless access will make anytime, anywhere communication a reality. New, exciting services such as telemedicine, e-learning, collaborative working, and virtual tours will be available to users.

If we let our imaginations fly, we can think of many more exciting applications. The communication infrastructure is only a platform; how we use it for our basic needs, our entertainment, our education, our social interactions, and our business activities depends on us.

39.4 THE CHALLENGES

Though the roadmap for communications technology is very clear, there are many challenges for communications engineers.

- Providing the necessary quality of service (QoS) is a major problem. For multimedia services that require high bandwidths, if the user requires a specific throughput and a minimum delay, achieving these QoS parameters is presently a research challenge, particularly in wireless networks.

- Though reliability of the communication networks and network elements improved substantially with the introduction of digital systems, achieving 100% availability continues to be a challenge. Fault-tolerant systems and fault-tolerant networking are areas of active research.

- The complexity of the communications networks is increasing day by day, and managing the networks continues to be a great problem. Though the telecommunications management network will be a big leap forward in efficient network management, legacy equipment continues to be used. Network management will be another area with research potential.

- Information security and network security are very important issues to be addressed. Attacks on information systems and networks are growing, and elimination of security threats is a high-priority research area.

- As the security of people and nations come under attack, governments resort to more and more surveillance. Telephone calls will be monitored, e-mails will be monitored, and so on. The debate on individual privacy versus public safety will heat up in coming years. As specialists in the communication technology, the communication engineers need to address this issue as well.

As the complexity of the communication networks, increases, reliability, security, and network management will be the major design issues.

As it is said, even in communications technology, "the greatest songs are still unsung." In the coming years, we will see many more exciting developments, many of which we never even imagined.

Summary

In this chapter, we discussed how the architecture of the telecommunications networks evolved. The PSTN, the PLMN, the Internet, cable and satellite broadcasting networks, and so on provide us many communication facilities. However, we need to access different networks through different gadgets for obtaining different services. With the emergence of convergence, we will have the option of accessing any service through a device of our choice. Instant messaging and unified messaging will facilitate accessing communication services easier, faster, and cheaper.

In the future, with the availability of high-bandwidth local loops and high-speed backbone networks, many value-added services will be provided. Even remote/rural areas can be brought into the mainstream through wireless local loops. The global village will no longer be a utopia, it will be a reality.

References

The landscape of communications technology is changing very fast. It is important to keep pace with these developments. Journals published by international professional societies such as the Institute of Electrical and Electronics Engineers (IEEE) and the *Association for Computing Machinery* (ACM) give the latest information on the emerging technologies. In particular, the *IEEE Communications Magazine*, the *IEEE Computer Magazine*, and the *Communications of ACM* are excellent journals that can keep us "future-proof."

A   Glossary

1G: First generation wireless networks. Analog cellular mobile communication systems are referred as 1G systems.

2G: Second generation wireless networks. The digital wireless mobile communication systems that can support data services up to 28.8kbps data rates are known as 2G systems.

2.5G: Two and a half generation wireless networks. The digital wireless mobile communication systems that can support data services up to 144kbps data rates are known as 2.5G systems.

3G: Third generation wireless networks. The digital wireless mobile communication systems that can support data services up to 2.048 Mbps data rates are known as 3G systems.

3GPP: 3G Partnership Program. The committee set up to arrive at common standards for third generation wireless networks.

4G: Fourth generation wireless networks. The 4G networks being planned can support very high data rates. 4G systems are still in the concept stage.

AAL: ATM adaptation layer. This layer runs above the ATM layer to support multimedia services over ATM networks. The main function of this layer is segmentation and reassembly.

ACK: Abbreviation for acknowledgement. ACK is sent by the receiver to inform the sender that the packet was received successfully.

ACL: A asynchronous communication link. In Bluetooth, data services are supported on an ACL link which provides connectionless service.

ACM: Association for Computing Machinery. An international professional body for computer engineers.

ADPCM: Adaptive differential pulse code modulation. A technique used to code voice at 32kbps data rate. The quality will be same as that of 64kbps PCM. Used in DECT.

AMs: Amplitude modulation. Analog modulation technique in which the carrier amplitude is varied in proportion to the amplitude of the modulating signal.

AMPS: Advanced mobile phone system. First generation analog cellular system that was in use in North America.

API: Application programming interface. A set of software routines provided by software tool vendors to develop applications. For example, Java Media Framework (JMF) APIs provided by Sun Microsystems are used to develop multimedia applications over IP networks.

ARP: Address resolution protocol. ARP is used to map IP address to the physical address such as an Ethernet address.

ARPA: Advanced Research Projects Agency. This agency of the U.S. Department of Defense, funded the ARPANet project, which evolved into the Internet.

ARPANet: The experimental network setup in the early 1970s to interconnect a few sites using packet switching and TCP/IP protocols. This network expanded rapidly to many other sites. ARPANet evolved into the present Internet.

ARQ: Automatic Repeat Request. When the transmission link is not reliable, ARQ protocol is used for retransmission of data. The sender sends a packet and, if the acknowledgement is not received within a specified time, the packet is retransmitted. A number of protocols such as stop-and-wait protocol and sliding window protocol belong to the ARQ family of protocols.

ASK: Amplitude shift keying. A digital modulation technique in which binary 1 is represented by presence of carrier and binary 0 by absence of carrier. Also called on-off keying.

ASR: Automatic speech recognition. The input speech is compared with already stored templates to make the computer recognize spoken words.

ATM: Asynchronous Transfer Mode. A fast packet-switching network technology that supports broadband services. The packets, called cells each of 53 bytes, are used to carry the data.

Audio: Sounds that can be heard. Voice and music are examples of audio signals. In audio broadcasting, the bandwidth of audio signals is limited to 15kHz.

Authentication: Authentication is a process to verify the identity of parties who exchange information.

Autonomous system: An internet, or network of networks, within the administrative control of an organization. An autonomous system can use its own protocols for routing its packets.

Bandwidth: A measure of the highest frequency component present in a signal. For example, voice signal has a bandwidth of 4kHz. The capacity of a communication channel is also referred to as bandwidth. The capacity can be measured either in Hz or in bits per second. Example: The radio channel has a bandwidth of 25kHz. The optical fiber has a bandwidth of 1Gbps.

BER: Bit Error Rate. A measure of the performance of a communication system. The number of bits received in error divided by the total number of bits.

BGP: Border gateway protocol. On the Internet, BGP is used by routers to exchange routing information. When the routers belong to different autonomous systems, this protocol is used.

Bluetooth: Standard for wireless personal area networks. Devices such as a desktop, palmtop, or laptop computer, mobile phone, fax machine, printer, headset and microphone can be Bluetooth enabled by attaching a small Bluetooth module.

BPF: Band pass filter. A filter that passes only a band of frequencies. BPF will have a lower cutoff frequency and an upper cutoff frequency.

BREW: Binary runtime environment for wireless. A tool to develop wireless applications on CDMA-based wireless networks. Developed by Qualcomm Corporation.

Bridge: A bridge interconnects two LANs. If the two LANs use different protocols, the bridge does the necessary protocol conversion.

Broadcast: To deliver information to a large number of receivers. Audio and video broadcasting facilitate distribution of audio/video programs to a large number of receivers. In a data network, it means to send the data to all the nodes on the network. A broadcast address is used for this operation.

BSC: Base station controller. In a mobile communication system, a number of base transceiver systems (BTSs) are controlled by a BSC. The BSC assigns the radio channels and time slots to the mobile devices.

BTS: Base transceiver subsystem. In a mobile communication system, each cell is controlled by a BTS. BTS consists of radio transmitters and receivers.

Carrier frequency: The frequency of the carrier used to modulate a signal.

CCITT: Comite Consultif International de Telegraphique et Telephnique, the French name of the International Consultative Committee for Telegraphy and Telephony. This organization is now called ITU-T (International Telecommunications Union-Telecommunications Sector).

CDMA: Code division multiple access. A spread spectrum technology that provides secure communication. Frequency hopping and direct sequence are examples of CDMA.

Cell: In cellular mobile communication systems, service area is divided into small regions called cells. Each cell contains a base station transceiver. In ATM networks, the packet is called a cell, the size of the cell being 53 bytes.

Certificate: A digitally signed document that contains the name and public key of the other party. Public keys are distributed using certificates.

Channel: The transmission medium (twisted pair, radio frequency band, optical fiber) is referred to as a channel.

Channel capacity: The highest data rate that can be supported by a communication medium.

Channel coding: A coding mechanism used to detect and correct errors.

Checksum: The additional bits added by the sender for a packet. The one's complement sum of the bytes of the packet is calculated and added to the packet. At the receiver, the checksum is again calculated and compared with the received checksum. If the two match, the receiver is assured that the packet is received without errors.

Chipping rate: A pseudo-random sequence of bits XORed with the data bits to implement direct sequence code division multiple access.

CIDR: Classless inter-domain routing. A mechanism of routing in which a group of networks with contiguous class C IP addresses are considered as a single network.

Circuit switching: Switching mechanism used in telephone networks. Circuit switching involves setting up a circuit, data transfer, and disconnection of the circuit.

CLNP: Connectionless network protocol. Equivalent ISO protocol of IP.

Clock recovery: The process of deriving the clock from the received serial data.

Congestion: A state of the packet switch or router caused by buffer or memory limitation. When the switch cannot process the incoming packets, it may discard the packets. Congestion control mechanisms are used to reduce the congestion. The switch/router may discard the packet, but it informs the other switches/hosts to reduce the packet transmission rate.

CRC: Cyclic redundancy check. A code appended to the packet at the transmitting side so that the receiver can detect if there is any error in the packet. CRC-16, CCITT-16, and CRC-32 are the standards for CRC calculation.

CSMA/CD: Carrier sense multiple access/collision detection. Protocol used for medium access in Ethernet LAN. A node on the LAN checks whether the medium is free. If the medium is free, the packet is immediately transmitted. If it is busy, the node will wait until it becomes free and then transmit the packet. If two nodes transmit their packets simultaneously, it results in collision. When there is collision, the node will back off and retransmit the packet.

Data compression: A technique used to reduce the data rate/storage requirement. Data compression techniques are classified as lossless compression techniques and lossy compression techniques.

Datagram: A chunk of data. The packet in the Internet Protocol (IP) layer is referred as an IP datagram in IP Version 4. The packet of user datagram protocol (UDP) is also referred to as a datagram.

Data rate: The rate at which data is transmitted. Measured in bits per second (bps), kilobits per second (Kbps), megabits per second (Mbps), gigabits per second (Gpbs), or terabits per second (Tbps).

Decryption: The reverse process to encryption. Restoration of the encrypted data.

Delay-bandwidth product: The product of round-trip delay and bandwidth. Indicates the amount of data in transit on the medium.

Demultiplexing: Reverse process of multiplexing. The data that is combined during multiplexing is separated by demultiplexing.

DES: Data Encryption Standard. An algorithm for encryption of data using a 56-bit key. Developed by the U.S. Department of Defense.

DHCP: Dynamic host configuration protocol. A protocol used to configure the IP address of the nodes in a network.

Digital watermark: The technique of embedding a secret message in another message (text, voice, image, or video) is called digital watermarking. The secret message is called the watermark, and the message carrying the watermark is called the host message. Digital watermarking is used to embed copyright information, the company logo, or the serial number of the software package in another image.

DS-CDMA: Direct sequence code division multiple access. A spread spectrum multiple access technique in which ones and zeros in a bit stream are replaced by special codes unique for each transmitter.

DSL: Digital subscriber line. DSL is used to access the Internet service provider over twisted copper pair at high data rates. A family of DSL standards known as xDSL has been developed, that include HDSL, ADSL, and VDSL.

EGP: Exterior gateway protocol. A routing protocol used by the IP layer to route packets. EGP was replaced by BGP on the Internet.

Encapsulation: Packet of one protocol layer embedded in the data field of the packet of the layer below it. For example, the protocol data unit (PDU) of the TCP layer is encapsulated in the IP datagram.

Encryption: A technique wherein the bit stream is modified using a key. Only the destination that knows the key can decode the bit stream.

Entropy: Entropy is a measure of the information content produced by an information source.

Error correction: A mechanism by which the receiver corrects the errors introduced by the channel. Additional bits are added by the sender to facilitate error correction by the receiver.

Error detection: A mechanism by which the receiver detects the errors introduced by the channel. The errors can be detected using parity, checksum, or CRC.

Ethernet: The most popular local area network standard. An equivalent LAN standard is IEEE 802.3.

Fax: Short form for facsimile. Fax facilitates transmission of documents.

FDDI: Fiber distributed data interface. A local area network standard that uses optical fiber as the transmission medium.

FEC: Forward error correction. Redundancy is introduced in the data bits at the transmitting side so that the receiver can correct the errors.

Filter: A module that removes the unwanted frequency components. Low pass filters, band pass filters, and band reject filters are examples.

Firewall: Software that runs on a server or router that provides the necessary security features to the network.

Flow control: A mechanism used to control the flow of packets in a network. If a host/router cannot receive the packets as fast as the source is transmitting, the host/router sends a message to reduce flow.

Fragmentation: Division of a packet into smaller units by a host/router. These smaller units have to be reassembled at the other end to form the original packet.

Frame: The packet of data at the datalink layer is called a frame.

Frame Relay: A network technology that provides fast packet switching by reducing the overhead at the datalink layer. Frame Relay networks are used extensively in wide area networks.

Frequency hopping: One form of spread spectrum multiple access. In frequency hopping, a device uses different radio channels to transmit its packets. For example, in Bluetooth, a device transmits each packet in a different radio channel. Hopping is done in a pseudo-random sequence. Only the intended receiver that knows the sequence can decode the data.

FM: Frequency modulation. Analog modulation technique in which the frequency deviation of the carrier is proportional to the amplitude of the modulating signal.

FSK: Frequency shift keying. Digital modulation technique in which binary $\emptyset$ is represented by one frequency and binary 1 by another frequency. The difference between these two frequencies is the frequency deviation.

ETSI: European Telecommunications Standards Institute, which formulates the telecommunications standards for Europe. In India, most of the telecommunications standards are based on European standards.

FDMA: Frequency division multiple access. An access mechanism in which different users share a pool of radio channels. A radio channel is assigned to a user when the user has data to send.

FTP: File Transfer Protocol. Application layer protocol in TCP/IP networks for file transfer.

GSM: Global System for Mobile Communications. The standard specified by ETSI for cellular mobile communications. This standard has been adapted by many African and Asian countries and in Europe.

H.323: The series of protocols for multimedia communication over IP networks.

HDLC: High level data link control. Datalink layer protocol that is the basis for many other link layer protocols such as LABP, LAPD, and LAPF.

HomeRF: The wireless technology developed to network various devices at home.

Host: An end system (or node) connected to a computer network.

HTML: Hypertext Markup Language. A markup language to develop Web content. The markup specifies how the document is organized (through anchor tags) and how to present the content (through tags for bold, underline and so on).

HTTP: Hypertext Transfer Protocol. The application layer protocol in TCP/IP networks for World Wide Web service.

ICMP: Internet Control Message Protocol. A protocol at the same level as IP, that uses IP service. ICMP is used to send error and status messages by routers.

IEEE: Institute of Electrical and Electronics Engineers. The largest professional body of electrical and electronics engineers.

IEEE 802 Committee: The committee formed by IEEE to develop standards for local area networks.

IETF: Internet Engineering Task Force is the body that coordinates and approves the technical specifications of protocols used on the Internet.

Internet: The global network of computer networks.

IP: Internet Protocol. The protocol responsible for addressing and routing.

IP address: Address assigned to each host and router on the Internet. In IP Version 4, the address length is 32 bits, in IP Version 6, the address length is 128 bits.

IPng: IP next generation. This protocol is now known as IP Version 6.

IPv6: The new version of IP that will supersede IP Version 4.

IPN: Interplanetary Internet. Vincent Cerf proposed the project which aims to develop Internets on different planets and spacecraft and to interconnect them.

IPSEC: The protocol that provides the security at IP level. Used in virtual private networks.

IrDA: Infrared Data Association. IrDA specifies the standards for low-cost infrared communication systems.

IRTF: Internet Research Task Force. The group that coordinates the research activities of the Internet.

ISDN: Integrated Services Digital Network. A network architecture that provides multimedia services through standard interfaces. Narrowband ISDN and broadband ISDN standards have been developed by ITU-T.

ISO: International Organization for Standardization. An international body that develops standards for many things, from paper sizes to telecommunications.

ITU-T: International Telecommunications Union Telecommunications Sector. The international body that develops standards for communications. The standards are referred as recommendations.

JMF: Java Media Framework. The APIs for the Java programming language facilitate development of multimedia applications over IP networks.

JPEG: Joint Photography Experts Group. The group that standardized the image compression technique. Using JPEG compression technique, a compression ratio of 30:1 can be achieved.

LAN: Local area network. A network that spans an area of about 10 km radius.

Latency: Delay.

LDAP: Lightweight Directory Access Protocol. A protocol for accessing, updating, searching, and retrieving information from a directory. Derived from X.500 standard.

Link: A communication path interconnecting two end points.

LLC: Logical link control. A sublayer in the datalink layer of LANs. LLC layer runs above the MAC layer. LLC layer provides a logical link between two end points and takes care of error control and flow control.

Lossless coding: A compression technique in which the original information is not lost. File compression utilities such as WinZip are lossless coding schemes.

Lossy coding: A compression technique that causes information loss. The JPEG compression technique is an example of lossy coding.

LPC: Linear prediction coding. A coding technique used for coding voice signals at low bit rates. This technique is based on the fact that a voice sample can be predicted from the previous samples. Variations of LPC coding are used in wireless communication systems.

LPF: Low pass filter. A filter that does not allow frequency components above a certain value. An LPF with a cutoff frequency of 4kHz allows only frequency components below 4kHz to pass through it.

MAC: Media access protocol. The sublayer of the datalink layer in LANs. MAC layer runs above the physical layer and below the LLC sublayer. The medium access by different nodes is governed by this protocol. CSMA/CD is an example of MAC.

MAN: Metropolitan area network. A network that covers an entire city.

MIB: Management information base. The information pertaining to a network element (host, router, switch, etc.) is stored in the MIB and can be accessed by a network manager using a network management protocol such as SNMP.

Microbrowser: A small browser that runs in a wireless device such as a mobile phone. The microbrowser interprets the content obtained from a server. A microbrowser running on a mobile device that supports WAP can interpret the content written in WML.

MIME: Multipurpose Internet Mail Extension. The protocol that runs above SMTP to support multimedia email.

MP3: MPEG layer 3. A standard for coding audio signals. Defined as part of the MPEG standard for video coding.

MPEG: Moving Picture Experts Group. The group that defined the standards for video coding. MPEG2 and MPEG7 are used for digital video transmission.

MSC: Mobile switching center. The entity that carries out the switching operations in a cellular mobile communication system. MSC has a number of databases that store subscriber information. These databases are home location register (HLR), visitor location register (VLR), equipment identity register (EIR), and authentication center (AuC). MSC is connected to the BSCs.

Multicast address: Address corresponding to multiple recipients.

Multiplexing: The process of combining a number of channels for transmission over a communication link.

NSF: National Science Foundation. The agency of U.S. government that supports advanced research.

OC: Optical carrier. In SONET hierarchy, the hierarchy levels are represented as OC, such as OC-3, OC-12, and OC-768.

Octet: The standards documents (ITU-T recommendations and IETF RFCs) refer to a byte (8 bits) as an octet.

OSI: Open Systems Interconnection. The OSI model is a 7-layer protocol architecture for computer communication. Developed by International Organization for Standardization.

OSPF: Open shortest path first. A routing protocol used within autonomous systems.

Packet: A block of data. The information to be transmitted is divided into packets and sent over the network in a packet-switching network.

Packet switching: Switching technology in which the information is divided into packets and transmitted from the source to the destination via packet switches. The packet switch takes the packet from an incoming channel, analyzes it, and based on the destination address of the packet, routes the packet to the appropriate outgoing channel.

PAN: Personal area network. A PAN interconnects devices within a range of about 10 meters. PANs can be established in homes or offices.

PCM: Pulse code modulation. In PCM, the signal is sampled at Nyquist rate, and the samples are quantized and converted into bit stream. For instance, voice signal has a bandwidth of 4kHz. The voice signal is sampled at 8000 samples per second, and each sample is coded with 8 bits (with 256 quantization levels). Using PCM, a voice signal can be digitized at 64kbps.

PCS: Personal communication system. The cellular mobile communication system in North America.

Phoneme: The smallest voice sound in a language. Most languages have about 40 to 60 phonemes.

Piconet: When two or more Bluetooth-enabled devices come in the vicinity of one another, a piconet is formed. A piconet can support seven active devices.

POTS: Plain old telephone service. The traditional fixed telephone service is referred to as POTS.

Protocol: An established procedure for entities to communicate with each other.

PSK: Phase shift keying. The phase of the carrier is changed based on the input bit stream. In binary PSK, binary 1 and binary 0 are represented by two phases of the carrier. In quadrature PSK, 00, 01, 10, and 11 are represented by four phases of the carrier.

Public key encryption: An encryption mechanism in which there will be a public key and a private key. Public key is used for encryption and private key for decryption.

QoS parameters: Quality of service parameters. The performance characteristics that define the service quality. Parameters such as delay, throughput, and packet loss can be specifed as QoS parameters.

Recommendations: The standards of ITU-T are called recommendations. Examples: X.25 for packet switching network protocols, X.400 for message handling systems.

Repeater: An entity that boosts the signal levels. For example, a GSM repeater boosts the radio signals transmitted by the BTS.

RFC: Request for Comments. IETF specifications on various protocols are referred to as Request for Comments.

Router: A network element in a packet-switching network that routes the packets toward the destination.

Routing: The process of forwarding packets toward the destination.

RTCP: Real Time Control Protocol. The protocol used above RTP for real-time communication in IP networks. Used for real-time voice and video communication over IP networks.

RTP: Real-time Transport Protocol. The protocol used for real-time communication in IP networks. RTP is used for voice and video communication over IP networks.

Segment: The packet at TCP layer is called a TCP segment.

SCO: Synchronous connection oriented link. SCO links are used for connection-oriented services such as voice communication over Bluetooth links.

SDH: Synchronous digital hierarchy. The standard proposed by ITU-T for optical transmission.

Signaling: The information sent by the subscriber/switch for call setup, disconnection, and management, such as billing.

SMTP: Simple Mail Transfer Protocol. Application layer for sending e-mail on TCP/IP networks.

SNA: System network architecture. Network architecture developed by IBM.

SNMP: Simple Network Management Protocol. Protocol used for network management in TCP/IP networks.

Source coding: The process of coding the signals produced by a source. Text, voice, image, and video are encoded to reduce the data rate required for transmission.

SONET: Synchronous optical network. The optical network standard developed by American National Standards Institute (ANSI).

Source routing: A routing mechanism in which the source decides the route to be taken by each packet. The complete route will be embedded in the packet, and each router has to forward the packet based on this routing information.

SSMA: Spread spectrum multiple access. A channel access mechanism in which high bandwidth is used to access the channel simultaneously. Frequency hopping and direct sequence are two examples of SSMA.

SSL: Secure Socket Layer. A layer that runs above the TCP layer to provide secure communication over TCP/IP networks.

Steganography: A technique used to send secret information embedded in text, voice, image, or video files.

Stop-and-wait protocol: Protocol used for flow control and request for retransmissions. The sender sends a packet and waits for the acknowledgement. Only after the acknowledgement is received the next packet sent.

Switch: A network element that takes data from incoming lines and puts it in the outgoing lines. A packet switch takes packets from incoming ports and puts them in outgoing ports based on the address contained in the packet.

TCP: Transmission Control Protocol. The transport protocol used in TCP/IP networks. This layer provides end-to-end reliable data transfer between end systems. It takes care of flow control and error control and provides a connection-oriented service.

TDMA: Time division multiple access. A channel access technique in which multiple users share the channel in different time slots. Each user is allowed to transmit data in a given time slot. The slots can be assigned permanently or dynamically only when traffic is there.

TDMA-FDD: Time division multiple access-frequency division duplex. In TDMA-FDD, a pair of frequencies is used—one frequency for uplink and one frequency for downlink.

TDMA-TDD: Time division multiple access-time division duplex. In TDMA-TDD, only one frequency is used for both uplink and downlink. The TDMA frame is divided into uplink time slots and downlink time slots.

TE: Terminal equipment. The equipment used by end users.

Telnet: The protocol for remote login in TCP/IP networks.

Throughput: The effective data rate in a communication channel.

TLS: Transport layer security. The protocol layer that runs above the TCP layer to provide secure communication in TCP/IP networks.

Transport layer protocol: The protocol that provides end-to-end data transfer. TCP and UDP are examples of transport layer protocols.

TTL: Time-to-live. A field in an IP datagram. This field decides the number of routers the packet can pass through before it is discarded. The default value is 64. At each router, this value is decremented by 1, and when it reaches a router with the value 0, that router discards the packet.

TTS: Text-to-speech conversion. The process by which the text input to the computer is converted into speech form.

UDP: User datagram protocol. The transport layer protocol on the Internet that provides connectionless service.

URL: Uniform Resource Locator. The resource on a server that can be accessed over the Internet. The URL *http://www.iseeyes.com* contains the protocol and the resource.

Virtual circuit: A logical connection established between two end points.

VPI: Virtual path identifier. The identification assigned to a virtual path in an ATM network. The virtual path contains a number of virtual circuits.

VPN: Virtual private network. A logical network of an enterprise that uses the Internet infrastructure.

WAN: Wide area network. A network that spans a large geographic area. A WAN may interconnect network elements spread over a country or the world.

WAP: Wireless application protocol. The WAP protocol stack is used to access Internet content by mobile devices.

Watermark: See digital watermark.

WiFi: Wireless Fidelity. The popular name given to an IEEE 802.11b wireless local area network.

WML: Wireless Markup Language. A markup language for developing content that can be accessed by mobile devices using wireless application protocol.

X.25: The ITU-T recommendation for wide area packet networks. X.25 provides a connection-oriented service.

X.28: The ITU-T recommendation for an asynchronous terminal/PC to communicate with a packet assembler disassmbler in an X.25 packet-switching network.

X.121: ITU-T recommendation for addressing in X.25-based packet networks. The X.121 address will have 14 digits.

X.400: The ITU-T recommendation for message handling systems. This is the application layer protocol for e-mail in the ISO/OSI protocol architecture.

X.500: ITU-T standard for directory services.

B | Acronyms and Abbreviations

2G	Second Generation (wireless networks)
2.5G	Two and a half generation (wireless networks)
3G	Third Generation (wireless networks)
3GPP	3G Partnership Program
AAL	ATM Adaptation Layer
ACL	Asynchronous Connectionless Link
ACM	Association for Computing Machinery
ADPCM	Adaptive differential pulse code modulation
ADSL	Asymmetric Digital Subscriber Line
AI	Artificial Intelligence
AIN	Advanced Intelligent Network
AM	Amplitude Modulation
AMPS	Advanced Mobile Phone System
ANSI	American National Standards Institute
API	Application Program Interface
ARPA	Advanced Research Projects Agency
ARQ	Automatic Repeat Request
ASCII	American Standard Code for Information Interchange
ASK	Amplitude Shift Keying
ASP	Application Service Provider
ATM	Asynchronous Transfer Mode
AuC	Authentication Center
AVVID	Architecture for Voice, Video, and Integrated Data
BER	Bit Error Rate

BGP	Border Gateway Protocol
BPSK	Binary Phase Shift Keying
BSC	Base Station Controller
BSU	Base Station Unit
BTS	Base Transceiver Subsystem
CAS	Channel Associated Signaling
CCS	Common Channel Signaling
CDMA	Code Division Multiple Access
CDN	Content Distribution Network
CDR	Call Details Record
CDROM	Compact Disk Read-Only Memory
CHAP	Challenge Handshake Authentication Protocol
CMIP	Common Management Interface Protocol
CRC	Cyclic Redundancy Check
CRM	Customer Relations Management
CSMA/CA	Carrier Sense Multiple Access/Collision Avoidance
CSMA/CD	Carrier Sense Multiple Access/Collision Detection
CTI	Computer Telephony Integration
CUG	Closed User Group
CVSD	Continuously Variable Slope Delta (modulation)
DAMA	Demand Assigned Multiple Access
DAP	Directory Access Protocol
DARPANET	Defense Advanced Research Projects Agency NETwork
DCE	Data Circuit Terminating Equipment
DECT	Digital Enhanced Cordless Telecommunications
DES	Data Encryption Standard
DGPS	Differential Global Positioning System
DLC	Digital Loop Carrier
DM	Delta Modulation
DNS	Domain Name System
DSL	Digital Subscriber Line
DTE	Data Terminal Equipment
DTMF	Dual Tone Multi-Frequency

DVB	Digital Video Broadcast
DWDM	Dense Wave Division Multiplexing
EBCDIC	Extended Binary Coded Data Interchange Code
EGP	Exterior Gateway Protocol
EHF	Extra High Frequency
EIA	Electronic Industries Association
EIR	Equipment Identity Register
ETSI	European Telecommunications Standards Institute
FCC	Federal Communications Commission
FEC	Forward-acting Error Correction
FDDI	Fiber Distributed Data Interface
FDM	Frequency Division Multiplexing
FDMA	Frequency Division Multiple Access
FH	Frequency Hopping
FOMA	Freedom of Mobile Multimedia Access
FM	Frequency Modulation
FSK	Frequency Shift Keying
GEO	Geosynchronous Earth Orbiting (satellite)
GMSK	Guassian M-ary Shift Keying
GPRS	General Packet Radio Service
GPS	Global Positioning System
GSM	Global System for Mobile Communications
GUI	Graphical User Interface
HCI	Host Controller Interface
HDLC	High Level Datalink Control
HDSL	High Data Rate Digital Subscriber Line
HF	High Frequency
HiperLAN	High Performance Local Area Network
HLR	Home Location Register
HMI	Human Machine Interface
HTML	Hypertext Markup Language
HTTP	Hypertext Transfer Protocol
ICMP	Internet Control Message Protocol

IEEE	Institute of Electrical and Electronics Engineers
IETF	Internet Engineering Task Force
IGP	Interior Gateway Protocol
IMEI	International Mobile Equipment Identity
IMSI	International Mobile Subscriber Identity
IN	Intelligent Network
IP	Internet Protocol
IPN	Interplanetary Internet
IPsec	IP Security Protocol
IrDA	Infrared Data Association
IRP	Interior Router Protocol
ISCII	Indian Standard Code for Information Interchange
ISDN	Integrated Services Digital Network
ISM	Industrial, Scientific, and Medical (band)
ISO	International Organization for Standardization
ISP	Internet Service Provider
ITU	International Telecommunications Union
IVR	Interactive Voice Response
IWF	Interworking Function
JMF	Java Media Framework
JPEG	Joint Photography Experts Group
JVM	Java Virtual Machine
L2CAP	Logical Link Control and Adaptation Protocol
L2F	Layer 2 Forwarding (protocol)
L2TP	Layer 2 Tunneling Protocol
LAN	Local Area Network
LDAP	Lightweight Directory Access Protocol
LEO	Low Earth Orbiting (satellites)
LLC	Logical Link Control
LPC	Linear Prediction Coding
MAC	Media Access Control
MAN	Metropolitan Area Network
MD	Mediation Device

MEO	Medium Earth Orbiting (satellite)
MHS	Message Handling System
MIB	Management Information Base
MIDP	Mobile Information Device Profile
MIME	Multipurpose Internet Mail Extension
MPEG	Moving Picture Experts Group
MS	Mobile Station
MSC	Mobile Switching Center
NC	Network Computer
NE	Network Element
NMC	Network Management Center
NMT	Nordic Mobile Telephony
OCR	Optical Character Recognition
OFDM	Orthogonal Frequency Division Multiplexing
OMC	Operation and Maintenance Center
OS	Operating System
OSI	Open Systems Interconnection
OSPF	Open Shortest Path First
PAN	Personal Area Network
PBX	Private Branch Exchange
PC	Personal Computer
PCM	Pulse Code Modulation
PDA	Personal Digital Assistant
PDN	Public Data Network
PDU	Protocol Data Unit
PLMN	Public Land Mobile Network
POCSAG	Post Office Code Standardization Advisory Group
PoP	Point of Presence
PPG	Push Proxy Gateway
PPTP	Point-to-Point Tunneling Protocol
PSK	Phase Shift Keying
PSTN	Public Switched Telephone Network
PVC	Permanent Virtual Circuit

QoS	Quality of Service
QPSK	Quadrature Phase Shift Keying
RADIUS	Remote Authentication Dial In User Service
RFC	Request for Comments
RMI	Remote Method Invocation
RSU	Remote Station Unit
RTCP	Realtime Control Protocol
RTP	Realtime Transport Protocol
SAN	Storage Area Network
SCO	Synchronous Connection Oriented (link)
SCP	Signaling Control Point
SCPC	Single Channel Per Carrier
SDP	Service Discovery Protocol
SMS	Short Messaging Service
SMTP	Simple Mail Transfer Protocol
SNMP	Simple Network Management Protocol
SNR	Signal to Noise Ratio
SS7	Signaling System No. 7
SSMA	Spread Spectrum Multiple Access
SSP	Signaling Switching Point
STP	Signaling Transfer Point
SVC	Switched Virtual Circuit
TACS	Total Access Control System
TCP	Transmission Control Protocol
TDD	Time Division Duplex
TDM	Time Division Multiplexing
TDMA	Time Division Multiple Access
TDMA-FDD	Time Division Multiple Access-Frequency Division Duplex
TDMA-TDD	Time Division Multiple Access-Time Division Duplex
TLS	Transport Layer Security
TMN	Telecommunications Management Network
TMSI	Temporary Mobile Subscriber Identity
UART	Universal Asynchronous Receiver/Transmitter

UDP	User Datagram Protocol
UHF	Ultra High Frequency
UMTS	Universal Mobile Telecommunication System
URL	Uniform Resource Locator
UTP	Unshielded Twisted Pair
VC	Virtual Circuit
VLF	Very Low Frequency
VLR	Visitor Location Register
VoIP	Voice over IP
VSAT	Very Small Aperture Terminal
VPN	Virtual Private Network
WAN	Wide Area Network
WAP	Wireless Application Protocol
WARC	World Administrative Radio Conference
WDP	Wireless Datagram Protocol
WDM	Wave Division Multiplexing
WiFi	Wireless Fidelity
WLL	Wireless Local Loop
WML	Wireless Markup Language
WS	Work Station
WSP	Wireless Session Protocol
WTP	Wireless Transaction Protocol
WTLS	Wireless Transport Layer Security
XHTML	eXtensible Hypertext Markup Language
XML	eXtensible Markup Language

C █ Solutions to Selected Exercises

Solutions to selected exercises given in various chapters are provided in this Appendix. For exercises involving software development, the source code listings are given. For exploratory exercises, references to Internet resources are provided.

CHAPTER 1

1. To generate a bit stream of 1s and 0s, you can write a program that takes characters as input and produces the ASCII bit stream. A segment of the VC++ code is given in Listing C.1. The screenshot for this program is given in Figure C.1. Please note that you need to create your own project file in VC++ and add the code given in Listing C.1.

Listing C.1 To generate a bit stream of 1s and 0s.

```
/* Converts the text into ASCII */
CString CAsciiMy::TextToAscii(CString text)
{
    CString ret_str,temp_str;
    int length,temp_int=0;
    length=text.GetLength();
    for(int i=0;i<length;i++){
        temp_int=text.GetAt(i);
        temp_str=ConvertBase(temp_int,2);
        ret_str=ret_str+temp_str;
    }
    return ret_str;
}
```

```
CString CAsciiMy::ConvertBase(int val, int base)
{
    CString ret_str,temp_str;
    int ret_val=0;
    int temp=val;
    while(1){
        if(temp>0){
            temp_str.Format("%d",temp%base);
            ret_str=temp_str+ret_str;
            temp=temp/base;
        }
        else
            break;
    }
    while(ret_str.GetLength()<7){
        ret_str="0"+ret_str;
    }
    return ret_str;
}
```

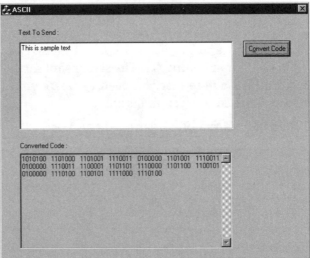

FIGURE C.1 Screenshot that displays the input text and equivalent ASCII code.

2. To generate noise, you need to write a program that generates random numbers. The random numbers can be between –1 and +1. Listing C.2 gives the VC++ code segment that does this. You need to create a project file in the VC++ environment and add this code. The waveform of the noise generated using this program is shown in Figure C.2.

Listing C.2 To generate random numbers between –1 and +1 and display the waveform.

```
/* To display the signal on the screen */
int CNoise_signalDlg::NoiseFunction()
{
    CWindowDC dc(GetDlgItem(IDC_SINE));
    CRect rcClient;
    LOGBRUSH logBrush;
    logBrush.lbStyle =BS_SOLID;
    logBrush.lbColor=RGB(0,255,0);
    CPen pen(PS_GEOMETRIC | PS_JOIN_ROUND,1, &logBrush);
    dc.SelectObject(&pen);
    dc.SetTextColor(RGB(255,255,255));

    while(continueThread){
        m_sine.GetClientRect(rcClient);
        dc.FillSolidRect(rcClient,RGB(0,0,0));
        dc.MoveTo(0,rcClient.bottom/2);
        int x,y;
        dc.MoveTo(0,rcClient.bottom/2);
        for   (x =0 ; x < (rcClient.right); x++)   // display Input
        {
                y      = rcClient.bottom/2 - Noise();
                dc.LineTo(x,y);
        }
        Sleep(200);
    }
    return 0;
}


/* To generate the noise signal */

int CNoise_signalDlg::Noise()
{
    int NISample;
    double NSample;
    double N2PI  = 2*TPI;
    double NWT;
    NoiseFreq = 300+rand()%4300;
    NoiseAMP  = 8+rand()%32;
    NWT = NoiseFreq*0.00125;
```

```
    NSampleNo++;
    NSample =NoiseAMP*sin(N2PI*NWTn);
    NWTn += NWT;
    if (NWTn > 1.0) NWTn -= 1.0;
    NISample = (int) NSample;
    return NISample;
}
```

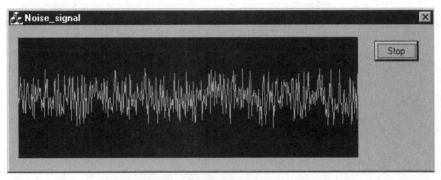

FIGURE C.2 Waveform of the noise signal.

3. To simulate a transmission medium, you need to modify the bit stream at random places by converting 1 to 0 and 0 to 1. Listing C.3 gives VC++ code segment that generates the bit stream and then introduces errors at random places. Figure C.3 gives the screenshot that displays the original bit stream and the bit stream with errors.

Listing C.3 To simulate a transmission medium.

```
/* To convert the text into bit stream and introduce errors in
the bit stream */
void CBitStreamDlg::OnDisplay()
{
    // TODO: Add your control notification handler code here
    CString strdata,binary,s,ss,no;
    int bit,i=0,count=0;
    char ch;
    m_text.GetWindowText(strdata);
    for(int j=0;j<strdata.GetLength();j++)
    {
        ch=strdata[j];
        for(int k=0;k<8;i++,count++,k++)
        {
            bit = ch%2;
```

```
                bin_val[i]=bit;
                ch=ch/2;
                s.Format("%d",bin_val[i]);
                binary = binary + s;
        }
    }

    m_bin_data.SetWindowText(binary);
    for(int  n=0;n<10;n++)
    {
        int  ran_no;
        srand((unsigned)time( NULL ) );
        ran_no =   rand() % 100;
        ss.Format("%d",ran_no);
        AfxMessageBox(ss);
        no = no + "," + ss;
        if(bin_val[ran_no]==0)
                bin_val[ran_no]=1;
        else
                bin_val[ran_no]=0;
    }
    CString bin1;
    for(i=0;i<104;i++)
    {
        s.Format("%d",bin_val[i]);
        bin1 = bin1 + s;
    }
    m_con_text.SetWindowText(bin1);
    m_Random_no.SetWindowText(no);
}
```

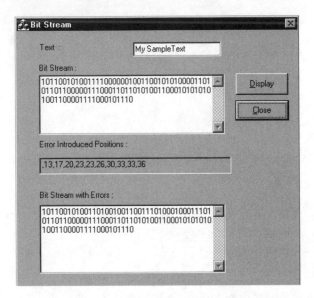

FIGURE C.3 Screenshot that displays the original bit stream and the bit stream with errors.

4. Many semiconductor vendors provide integrated circuits that generate noise in the audio frequency band. Suppliers of measurement equipment provide noise generators used for testing communication systems. The best way to generate noise is through digital signal processors.

CHAPTER 2

1. For a source that produces 42 symbols with equal probability, the entropy of the source is

   $H = \log_2 42$ bits/symbol

   $= 5.55$ bits/symbol

2. For a source that produces two symbols A and B with probabilities of 0.6 and 0.4, respectively, the entropy is

   $H = -\{0.6 \log_2 0.6 + 0.4 \log_2 0.4\} = 0.970$ bits/symbol

3. In ASCII, each character is represented by seven bits. The frequency of occurrence of the English letters is not taken into consideration at all. If the frequency of occurrence is taken into consideration, then the most frequently occurring letters have to be represented by small code words (such as 2

bits) and less frequently occurring letters have to be represented by long code words. According to Shannon's theory, ASCII is not an efficient coding technique.

However, note that if an efficient coding technique is followed, then a lot of additional processing is involved, which causes delay in decoding the text.

4. You can write a program that obtains the frequency of occurrence of the English letters. The program takes a text file as input and produces the frequency of occurrence for all the letters and spaces. You can ignore the punctuation marks. You need to convert all letters either into capital letters or small letters. Based on the frequencies, if you apply Shannon's formula for entropy, you will get a value close to 4.07 bits/symbol.

5. You can modify the above program to calculate the frequencies of two letter combinations (aa, ab, ac,...ba, bb, ... zy, zz). Again, if you apply the formula, you will get a value close to 3.36 bits/symbol.

CHAPTER 3

1. For a terrestrial radio link of 30 kilometers, assuming that speed of light is 3×10^8 meters/second, the propagation delay is $30,000 / (3 * 10^8)$ seconds = 100 microseconds.

2. When the cable length is 40 kilometers and the speed of transmission in the coaxial cable is 2.3×10^8 meters/second, the propagation delay is $40,000 / (2.3 * 10^8)$ seconds = 173.91 microseconds.

3. The propagation delay in an optical fiber of 100 kilometers if the speed is 2×10^8 meters/second, is $100,000 / (2 \times 10^8)$ seconds = 0.5 msec.

4. In satellite communication, broadcasting and voice communication are done in the C band and Ku band.

Band Designation	Frequency Range
L band	1-2 GHz
S band	2-4 GHz
C band	4-8 GHz
X-band	8-12 GHz
Ku band	12-18 GHz
K band	18-27 GHz
Ka band	27-40 GHz
Q band	33-50 GHz
U band	40-60 GHz
V band	40-75 GHz
W band	75-110GHz

5. In mesh configuration, the communication is from one Earth station to another Earth station. Hence, the propagation delay is 240 msec. In star configuration, the transmission is from VSAT to the satellite, satellite to the hub, hub to the satellite, and then satellite to the other VSAT. Hence, the propagation delay is 480 msec.

CHAPTER 4

1. To record your voice on your multimedia PC, you can use the sound recorder available on the Windows operating system. You can also use a more sophisticated software utility such as GoldWave (*www.goldwave.com*). You will have the options to select the sampling rate (8kHz, 16kHz, etc.) and the quantization levels (8 bits, 16 bits, etc.). GoldWave provides utilities to filter background noise, vary the pitch, and so on.

2. When you use a software package such as Microsoft's NetMeeting over a LAN, the video will be transmitted at very low bit rates, so the video appears jerky.

3. If the video is transmitted at the rate of 30 frames per second, with each frame divided into 640 × 480 pixels, and coding is done at 3 bits per pixel, the data rate is

$30 \times 640 \times 480 \times 3$ bits per second $= 3 \times 64 \times 48 \times 3$kbps $= 27,648$kbps
$= 27.648$Mbps

4. The Indian Standard Code for Information Interchange (ISCII) is a standard developed by Department of Electronics (Ministry of Information Technology), Government of India. ISCII code is used to represent Indian languages in computers. Center for Development of Advanced Computing (CDAC) supplies the hardware and software for Indian language processing based on ISCII. You can obtain the details from the Web site *www.cdacindia.com*.

5. MP3 software can be obtained from the following sites:

   www.dailymp3.com

   www.mp3machine.com

   www.mp3.com

6. To store 100 hours of voice, the memory requirement is given below if the coding is done using (a) PCM at 64kbps, (b) ADPCM at 32kbps, and (c) LPC at 2.4kbps.

   (a) To store 100 hours of voice using PCM at 64kbps data rate,

   Total duration of voice conversation $= 100$ hours $= 100 \times 3600$ seconds

   Memory requirement $= 100 \times 3600 \times 64$ kbps $= 100 \times 3600 \times 8$ Kbytes
   $= 360 \times 8$ Mbytes $= 2880$Mbytes

   (b) 1440Mbytes

   (c) $100 \times 3600 \times 2.4$kbps $= 100 \times 3600 \times 0.3$Kbytes $= 36 \times 3$ Mbytes $= 108$Mbytes

7. If the music signal is band-limited to 15kHz, the minimum sampling rate required is twice the bandwidth. Hence,

   Minimum sampling rate $= 2 \times 15$kHz $= 30$kHz

   If 12 bits are used to represent each sample, the data rate $= 30,000 \times 12$ bits/second $= 360,000$ bits per second $= 360$kbps.

8. The image is of the size 640×480 pixels. Each pixel is coded using 4 bits. To store the image,

   Memory requirement $= 640 \times 480 \times 4$ bits $= 153.6$Kbytes.

CHAPTER 5

1. You can experiment with various parameters for serial communication using the Hyper Terminal program on Windows. You need to connect two PCs using an RS232 cable. Figure C.4 shows the screenshot to set the serial communication parameters.

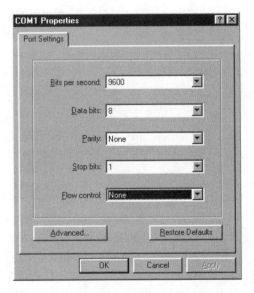

FIGURE C.4 Screenshot to set the serial communication parameters.

To get this screen, you need do the following:

- Click on Start
- Go to Programs
- Go to Accessories
- Go to Hyper Terminal
- Double click on HYPERTERM.EXE
- Enter a name for New Connection "xxxx" and Click OK
- You will get a screen with the title Connect To. Select COM1 or COM2 from the Connect Using listbox and click OK.

2. "Parity can detect errors only if there are odd number of errors". To prove this statement, consider the bit stream 1010101. If we add an even parity bit,

the bit stream would be 10101010. If there is one error, the bit stream may become 11101010. Now the number of ones is 5, but the parity bit is 0. Hence, you can detect that there is an error. If there are two errors, the corrupted bit stream is 11111010. However, in this case, the calculated parity bit is 0, the received parity bit is also 0, and the even number of errors cannot be detected.

3. If the information bits are 1101101, the even parity bit is 1, and the odd parity bit is 0.

4. If the information bits are 110110110, when Rate 1/3 FEC is used for error correction, the transmitted bit stream will be

    111111000111111000111111000

    Suppose the received bit stream is

    101011100011110001110110100

    There is one bit error for every three bits. Even then at the receiver, the correct bit stream can be obtained. However, suppose the first three bits received are 001, then the receiver will decode it as 0 though the actual information bit is 1.

CHAPTER 6

1. For the bit pattern 10101010100, the NRZ-I waveform is as follows:

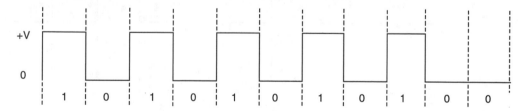

    The NRZ-L waveform is as follows:

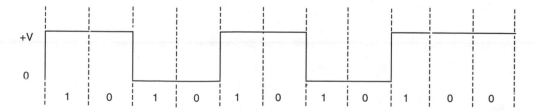

2. For the bit pattern 10101010110, the waveform using Manchester coding is as follows:

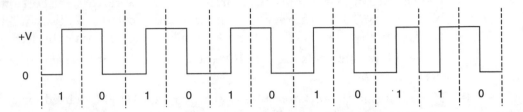

3. For the bit pattern 101010101110, the waveform appear on an RS232 port will appear as:

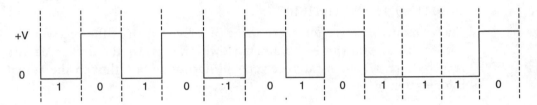

4. For the bit pattern 10101010100, bipolar AMI waveform is as follows:

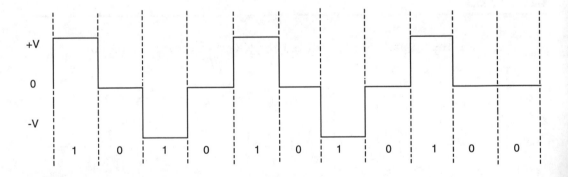

The HDB3 waveform is as follows:

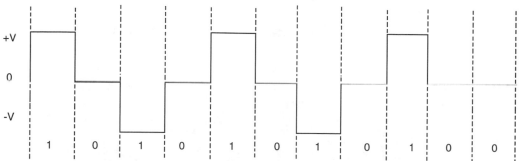

Note that the HDB3 encoding will be the same as AMI. When four consecutive zeros are in the bit stream, the waveforms will be different.

5. The UART (universal asynchronous receive-transmit) chip is used to control the serial communication port (COM port) of the PC. The functions of this chip are converting parallel data into serial data and vice versa, adding start and stop bits, creating a hardware interrupt when a character is received, and flow control. Different UART chips support different speeds: 8250 supports 9.6kbps, 16450 supports 19.2kbps, and 16550 supports 115.2kbps. 8250 and 16450 have one byte buffer, and 16550 has 16 bytes of buffer.

6. To generate 5 bit codes from the RS232 interface of your PC is not possible. In RS232 communication, there will be start bit and a parity bit. Even if there are five data bits, additional bits are added. To generate five bit codes, you need to have special hardware with UART chips.

CHAPTER 7

1. The T1 carrier supports 24 voice channels with an aggregate data rate of 1.544 Mbps out of which 8000 bps is for carrying signaling information. In T1 carrier, each frame consists of one bit for framing following by 24 slots. Each slot contains 7 bits of voice and one bit for signaling. Four T1 carriers are multiplexed to from T2 carrier. The data rate of T2 carrier is 6.312 Mbps whereas 4 x 1.544 Mbps = 6.176 Mbps. Hence the additional overhead is 0.136 Mbps.

The complete hierarchy is given in the following Table:

Level	Data Rate (Mbps)	No. of 64kbps Channels
T1	1.544	24
T1C	3.152	48
T2	6.312	96
T3	44.736	672
T4	274.176	4,032

2. In satellite communication, the frequency band is divided into small bands, and a number of users share it. Hence, FDMA is used. TDMA access scheme is also used for multiple users to share the same band. In low-earth orbiting satellite systems, CDMA technique is also used.

3. The multiplexing hierarchy used in optical fiber communication systems is given in Chapter 14.

4. C band and L band are used in optical fiber for WDM and DWDM. C band corresponds to the wavelengths 1530 – 1565 nm, and L band corresponds to 1565– 1625 nm.

CHAPTER 8

1. (a) Mobile communication systems based on GSM standards use FDMA/TDMA. (b) A Bluetooth radio system uses frequency hopping. (c) 802.11 wireless local area network uses frequency hopping. (d) HiperLAN uses TDMA-TDD. (e) A digital enhanced cordless telecommunications (DECT) system uses TDMA-TDD.

2. In satellite communication systems, TDM-TDM, TDMA, and CDMA technologies are used. The details are given in Chapter 13.

3. In GSM, the service area is divided into small regions called cells. Each cell is assigned a number of channels that are shared by the mobile phones in that cell. Hence, FDMA is used. In addition, each channel is time-shared by eight mobile phones. Hence, TDMA is also used. So, the multiple access technique is referred to as FDMA/TDMA. Each radio channel has a pair of frequencies—one for uplink and one for downlink. Hence, the TDMA scheme is TDMA-FDD.

4. In a TDMA system, each station will give a small time slot during which it has to pump its data. However, the source will generate the data continuously. Hence, the source has to buffer the data and send it fast over the channel when it gets its time slot. For instance, the user talks into his mobile phone continuously. But the mobile phone gets its time slot only at regular intervals. Hence, the mobile device has to buffer the digitized voice and send it in its time slot.

CHAPTER 9

1. The code segments for generation of ASK and FSK signals are given in Listing C.4. The screen shots are given in Figure C.5 and Figure C.6.

Listing C.4 Generation of ASK and FSK waveforms.

```
/* To draw ASK waveform for 1 kHz tone */
void CSendDialog::ASKPlotter()
{
    int n,j1,j2,var=0,mm=0;
    int i=0,xr[8],bit;
    char c;

    wn=8*0.01745 ;

    CWindowDC dc(GetDlgItem(IDC_SINE));
    LOGBRUSH logBrush;
    logBrush.lbStyle =BS_SOLID;
    logBrush.lbColor=RGB(0,255,0);
    CPen pen(PS_GEOMETRIC | PS_JOIN_ROUND,1, &logBrush);
    dc.SelectObject(&pen);
    dc.SetTextColor(RGB(255,255,255));

    CString strData;
    strData="My Sample Text";

    for(int ii=0;ii<strData.GetLength();ii++)
    {
        c = strData[ii];
        dc.TextOut(20,10,"Amplitude1 = ±1V");
        dc.TextOut(20,30,"Amplitude2 = ±2V");
        dc.TextOut(150,10,"Freq = 1200Hz ");
```

```
        var=0;
        xs=0;ys=0;
        for(j2=0,i=0;j2<8;j2++){
                bit=c%2;
                xr[i]=bit;
                i=i+1;
                c=c/2;
        }
        for(i1=0;i1<=360;i1++){
                ya[i1]=amp1*sin(wn*i1);
                yb[i1]=amp2*sin(wn*i1);
        }
        for(n=0;n<8;n++){
                if(xr[i-1]==1){
                        PlotSine1(&dc,var);i-;
                }
                else{
                        PlotSine2(&dc,var);i-;
                }
                var=var+45;
        }
        incr=0;
        _sleep(200);
        CRect rect;
        m_sine.GetClientRect(rect);
        dc.FillSolidRect(rect,RGB(0,0,0));
    }// End For str.Length
        m_start.SetWindowText("Send Code");
}

/* To draw the sin waveform when signal is present */
void CSendDialog::PlotSine1(CWindowDC *dc, int var)
{
    if(incr!=0)
        incr-=1;

    flagctrl=1;

    line(*&dc,25,110,385,110);
    for( i1=var; i1<=var+45; i1=i1+1 )
    {
        xe=i1; ye=ya[i1];
        spect[incr+=1]=ya[i1];
```

```
           line(*&dc, 25+xs, 110-ys, 25+xe, 110-ye );
           xs=xe;
           ys=ye;
        }
}


/* To draw the sin waveform when signal is not present */
void CSendDialog::PlotSine2(CWindowDC *dc, int var)
{
    if(incr!=0)
       incr-=1;
    line(*&dc,25,110,385,110);

    for( i1=var; i1<=var+45; i1=i1+1  )
      {
         xe=i1; ye=yb[i1];
         spect[incr+=1]=yb[i1];
         line(*&dc, 25+xs, 110-ys, 25+xe, 110-ye);
         xs=xe;
         ys=ye;
      }
       flagctrl=0;
}

/* To draw the horizontal line */
void CSendDialog::line(CWindowDC *dc, short x1, short y1, short
x2, short y2)
  {
     dc->BeginPath();
     dc->MoveTo(x1,y1);
     dc->LineTo(x2,y2);

     dc->CloseFigure();
     dc->EndPath();
     dc->StrokePath();
  }
/* To draw FSK waveform for 1 kHz tone */
void CSendDialog::FSKPlotter()
  {
     int n,j1,j2,var=0;
     int i=0,xr[1000],bit;
     char c;
```

```
wn1 = 8*0.01745 ;
wn2 = 32*0.01745;
CWindowDC dc(GetDlgItem(IDC_SINE));
LOGBRUSH logBrush;
logBrush.lbStyle = BS_SOLID;
logBrush.lbColor = RGB(0,255,0);

CPen pen(PS_GEOMETRIC | PS_JOIN_ROUND, 1, &logBrush);
dc.SelectObject(&pen);
dc.SetTextColor(RGB(255,255,255));

CString strData;
strData="My Sample Text";

for(int ii=0;ii<strData.GetLength();ii++)
{
   c=strData[ii];
   dc.TextOut(20,10,"Freq1 = 1200Hz");
   dc.TextOut(20,30,"Freq2 = 2000Hz");
   var=0;
   xs=0;
   ys=0;
   for(j2=0,i=0;j2<8;j2++){
           bit=c%2;
           xr[i]=bit;
           i++;
           c=c/2;
   }
   for(i1=0;i1<=360;i1++){
           ya[i1] = amp1*sin(wn1*i1);
           yb[i1] = amp1*sin(wn2*i1);

   }
   for(n=0;n<8;n++){
           if(xr[i-1]==1){
                   PlotSine1(&dc,var);i-;
           }else{
                   PlotSine2(&dc,var);i-;
           }
           var = var + 45;
   }
   incr=0;
   _sleep(200);
```

```
        CRect rect;
        m_sine.GetClientRect(rect);
        dc.FillSolidRect(rect,RGB(0,0,0));
    }
}
```

Figure C.5 shows a screenshot for ASK waveform for data.

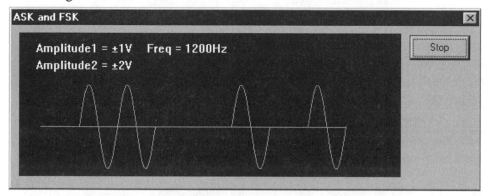

FIGURE C.5 ASK waveform.

Figure C.6 shows a screenshot of an EFSK waveform.

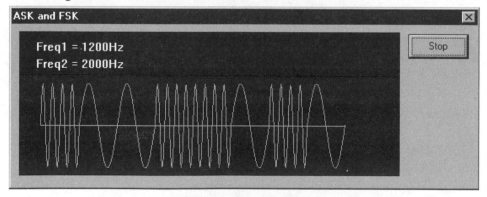

FIGURE C.6 FSK waveform.

2. The code segment for generation of a 1kHz sine wave is given in Listing C.5. You can use this as the modulating signal to generate the modulated signals. The waveform is shown in Figure C.7.

Listing C.5 To generate a 1KHz sine wave.

```
/* To Display the signal on the screen*/
void CSingleToneDlg::Tone(int i)
{
    CWindowDC dc(GetDlgItem(IDC_SINE));
    CRect rcClient;
    LOGBRUSH logBrush;
    logBrush.lbStyle =BS_SOLID;
    logBrush.lbColor=RGB(0,255,0);
    CPen pen(PS_GEOMETRIC | PS_JOIN_ROUND,1, &logBrush);
    dc.SelectObject(&pen);
    dc.SetTextColor(RGB(255,255,255));
    while(continueThread){
        m_sine.GetClientRect(rcClient);
        dc.FillSolidRect(rcClient,RGB(0,0,0));

        dc.MoveTo(0,rcClient.bottom/2);
        int x,y;
        dc.MoveTo(0,rcClient.bottom/2);
        for   (x =0 ; x < (rcClient.right); x++)   // display Input
        {
            y      = rcClient.bottom/2 - ToneNextSample();
            dc.LineTo(x,y);
        }
        Sleep(200);
    }
}


/* To initialize the frequency*/
void CSingleToneDlg::ToneInitSystem(double Freq)
{
    T2PI = 2*TPI;
    TSampleNo = 0;
    ToneFreq = Freq;
    TWT = ToneFreq*0.000125;
    TWTn = 0;
    TSampleNo = 0;
}


/* To calculate the next sample value */
int CSingleToneDlg::ToneNextSample()
{
```

```
        int TISample;
        double TSample;
        int c;
        TSampleNo++;
        TSample = KTONE_AMPL*sin(T2PI*TWTn);
        TWTn += TWT;
        if (TWTn > 1.0) TWTn -= 1.0;
        TISample = (int) TSample;
        return TISample;
}
```

Figure C.7 shows a 1kHz sine wave.

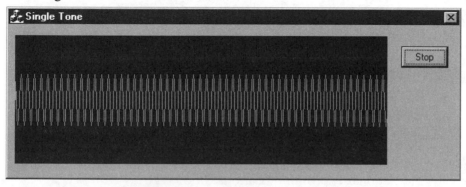

FIGURE C.7 1kHz sine wave.

3. Comparison of modulation techniques is done based on the noise immunity, bandwidth requirement, error performance, and implementation complexity of the modulator/demodulator. ASK is not immune to noise. QPSK occupies less bandwidth, but implementation is complex. The waterfall curve given in the Chapter 10 gives the performance of different modulation techniques.

4. The modems used in radio systems are called radio modems, and the modems used on wired systems are called line modems. GSM and Bluetooth are radio systems. The modulation used in GSM is GMSK. Bluetooth uses Gaussian FSK. There are various standards for line modems specified by ITU. These are V.24, V.32, V.90, and so on.

5. If the bandwidth of the modulating signal is 20kHz, the bandwidth of the amplitude-modulated signal is 40kHz.

6. If the bandwidth of a modulating signal is 20kHz and the frequency deviation used in frequency modulation is 75kHz, the bandwidth of the frequency modulated signal is 2(20 + 75) kHz = 190kHz.

CHAPTER 10

1. The various standardization bodies are
- American National Standards Institute (ANSI)
- Advanced Television Systems Committee (ATSC)
- Alliance for Telecommunications Industry Solutions (ATIS)
- Cable Television Laboratories, Inc. (CableLabs)
- European Committee for Standardization (CEN)
- European Committee for Electrotechnical Standards (CENELEC)
- European Computer Manufacturers Association (ECMA)
- Electronic Industries Alliance (EIA)
- European Telecommunications Standards Institute (ETSI)
- Federal Communications Commission (FCC)
- International Electrotechnical Commission (IEC)
- Institute of Electrical and Electronics Engineers (IEEE)
- International Organization for Standardization (ISO)
- Internet Society (ISOC)
- International Telecommunications Union (ITU)
- Telecommunications Industry Association (TIA)
- World Wide Web Consortium (W3C)

   A number of industry consortia and special interest groups (SIGs) formulate the telecommunications specifications. Some such consortia and groups are

- 10 Gigabit Ethernet Alliance
- ATM Forum
- Bluetooth SIG
- CDMA Development Group
- DSL Forum
- Enterprise Computer Technology Forum
- Frame Relay Forum
- HomeRF Working Group
- International Multimedia Telecommunications Consortium
- Infrared Data Association

- Object Management Group
- Personal Communications Industry Association
- Personal Computer Memory Card International Association
- Satellite Broadcasting and Communications Association
- SONET Interoperability Forum
- Society for Motion Picture & Television Engineers
- WAP Forum
- Wireless Ethernet Compatibility Alliance
- Wireless LAN Association

2. To plan and design a cellular mobile communication system, a radio survey needs to be carried out. A radio survey involves study of the terrain and deciding where to install the base stations. The radio propagation characteristics need to be studied for which computer models are available. Leading mobile communication equipment suppliers such as Motorola, Ericsson, and Nokia provide these tools.

3. BER is a design parameter. You need to find out the BER required for the communication system being designed. For this BER, the modulation technique, which requires the least energy per bit, is the best. However, other considerations such as the implementation complexity or cost of the modulator/demodulator also play an important role while choosing a modulation technique.

4. The standards for various types of LANs have been formulated by the professional body IEEE. IEEE 802.3 LAN is based on the Ethernet, FDDI is for optical fiber–based LANs, and IEEE 802.11 is for wireless LAN. You can design a backbone network based on fiber and then interconnect 802.3 LANs and wireless LANs to obtain a campus-wide network.

CHAPTER 11

1. Each T1 carrier supports 24 voice channels. To support 60 voice channels, three T1 carriers are required. Two T1 carriers are not sufficient as they can support only 48 voice channels. However, when 3 T1 carriers are used, 12 voice channels remain free. It is a good practice to keep some spare capacity for future expendability.

2. You can see the datasheet of the IC 8880 from MITEL. This is a DTMF chip, and you can use the reference circuit given in the data sheet.

3. The IC 44233 generates PCM-coded speech.

4. The PSTN is based on circuit-switching. The switches contain (a) the line cards that interface with the subscriber equipment; and (b) processor-based hardware/software to switch the calls and establish a circuit between the two subscribers. Now the trend is to use packet switching for voice calls using the computer networking protocols. Switches that handle packet-switched voice are called the soft switches.

CHAPTER 12

1. Digital TV transmission has many advantages: improved signal quality, bandwidth efficiency through use of compression techniques, and effective control and management. Two standards HDTV (High Definition TV) and SDTV (Standard Definition TV) have been developed for digital video broadcasting. In HDTV, the aspect ratio is 16:9. Twenty-four or 30 frames are sent per second. Each frame is divided into 1080 × 1920 pixels or 720 × 1280 pixels. In SDTV, the aspect ratio is 4:3, with 24 or 30 frames per second, and each frame with 480 × 640 pixels.

2. To provide wireless local loops, small base stations are installed at various locations. The subscriber terminal (telephone instrument) communicates with the base station using CDMA access mechanism. You can get the details of wireless local loop using CDMA from the Web site *www.qualcomm.com*.

3. Though most of the trunked radio systems are analog, the latest standard developed by European Standards Telecommunications Institute (ETSI) is based on digital technology—it is called TETRA (Terrestrial Trunked Radio). You can get the details from *www.etsi.org*. Motorola is a leading supplier of trunked radio systems. You can get product details from *www.motorola.com*.

4. Path loss calculation involves finding out the loss/gain of different network elements. The network elements are filters, amplifiers, antennas, and the cable connecting the RF equipment with the antenna. A major contributor to path loss is the propagation loss in the medium.

5. Underwater communication systems are used in sports, search and rescue operations, and military applications. As the attenuation of the electrical signal is very high in water, high-power transmitters are required. VLF band

is used in underwater communication. For instance, the phones used by divers operate in the frequency band 30 to 35 kHz. If the transmit power is ½ watt, the range is about 400 yards, and if the transmit power is 30 watts, the range is about 5 miles. You can get the details of some commercial underwater communication systems from the site *www.oceantechnologysystems.com*.

CHAPTER 13

1. Direct broadcast satellites transmit TV programs directly to homes. These systems operate in the frequency band 17/12GHz. However, due to the wide penetration of cable TV, direct broadcast satellite technology has not taken off well in India, though DBS is extensively used in North America and Europe.

2. Remote sensing satellites are used for a variety of applications: to find out the areas in which natural resources are available, to determine the water table under the earth's surface, to analyze the fertility of lands, and so on. Remote sensing satellites have sensors that operate in the infrared and near-infrared bands. The satellite imagery is sent to the ground stations. The imagery is analyzed based on the application. Indian Space Research Organization (ISRO) launched the IRS (Indian Remote Sensing Satellite) series satellites exclusively for remote sensing.

3. In satellite communication, efficient use of the bandwidth is very important, as bandwidth is much costlier than the terrestrial media bandwidth. Low bit rate coding of voice is done using ADPCM, LPC, and CELP techniques.

4. In satellite-based video surveillance, there should be a system that captures the video and transmits it to the ground continuously. The video camera captures the scene and digitizes it using a standard coding technique such as MPEG2 or MPEG4 and transmits to the Earth station. The minimum data rate to be supported by the link is 1Mbps to obtain reasonably good quality video.

5. In a satellite network, the round-trip delay is about 0.5 second. If stop-and-wait protocol is used on such a network for data communication, the satellite channel is not used effectively. After a packet is transmitted from one side, it takes 0.5 seconds to reach the destination (assuming that there is no other delay). Then the acknowledgement will be sent by the other station, and it will be received by the station after another 0.5 seconds. So, effectively one packet is transmitted every one second!

CHAPTER 14

1. You can get the details from *www.efiber.net* and *www.fiberalliance.net*.
2. Latest information on DWDM commercial products can be obtained from *www.cisco.com*, *www.nortelnetworks.com*, *www.ericsson.com*, and *www.siemens.de*.
3. A Raman amplifier is a device that amplifies the optical signals directly. Hence, there is no need to convert the optical signal to electrical signal, amplify it, and reconvert it into optical signal.
4. Presently C band and L band are used in optical fiber, mainly because of the availability of optical components in these bands.

CHAPTER 15

1. Every computer is given an IP address. The IP address is generally assigned by your system administrator.

The screen that displays the IP address of a PC is shown in Figure C.8.

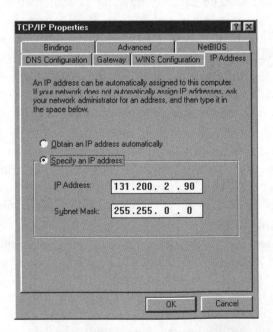

FIGURE C.8 IP address of a computer.

To obtain this screen, do the following:

- Right click on Network Neighborhood icon and select the Properties option from the pop-up menu. The following screen (Figure C.9) will appear:

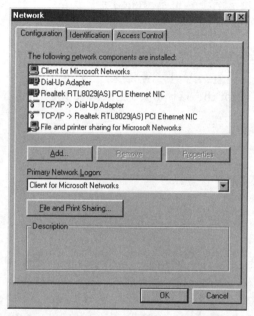

FIGURE C.9 Selection of the network interface card.

- In the screen seen in Figure C.9, select TCP/IP => Realtek RTL8029(AS) PCI Ethernet NIC and click Properties. Note that you need to select the correct Ethernet network interface card installed on your PC to obtain the properties. (Realtek is the name of the company that manufactures the Ethernet cards.)

2. When you access a Web site through a browser, the Domain Name Service of your Internet service provider gives the IP address of the origin server in which the resource is located. Then the TCP connection is established between the client and the origin server. When you see the message Connecting to followed by the IP address of the Web server, it is an indication that the DNS of your ISP has done its job. If the DNS is down, you will not get this message and will not be able to access the URL. Note that the DNS may be working, but still you may not be able to access the resource if the origin server is down.

3. The Ethernet LAN uses a maximum packet size of 1526 bytes. X.25 network uses a maximum packet size of 1024 bytes. In both cases, the packet size is

variable. Variable packet size leads to more processing by the switches. Also, the switches should have variable size buffers. However, if large size packet is negotiated, data transfer is fast, and protocol overhead is less. Fixed-size packets certainly can be switched much faster. In Asynchronous Transfer Mode (ATM) networks, fixed-size packets are used.

4. ATM uses a fixed packet size of 53 bytes. This is a small packet compared to Ethernet or X.25 packet sizes. The small size causes fast packet switching and fixed size buffers at the switches. The only disadvantage is slightly higher overhead. Out of the 53 bytes, 5 bytes are for header information.

CHAPTER 16

1. A six-layer architecture for computer communication can be just the elimination of the session layer in the ISO/OSI architecture. Session layer functionality is minimal and can be eliminated. An eight-layer architecture can have an additional layer to provide security features that runs above the transport layer.

2. Windows 9x/2000/XP operating systems provide the Hyper Terminal with which you can establish communication between two systems through an RS232 link.

3. In Linux operating system, you will have access to the complete source code for serial communication. You can interconnect a Linux system and a Windows system using RS232.

CHAPTER 17

1. You can obtain the LAN connection statistics using the procedure given below.

    The screen shot given in Figure C.10 shows the LAN activity. It displays the sent packets and received packets as well as connection statistics.

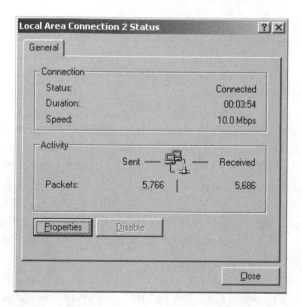

FIGURE C.10 Local area network connection status.

On Windows 2000 operating system, you need to do the following:

In the Start menu, select My Network Place and then right-click on it. In the pop-up menu that appears, select Properties option. A window with the title Network connections will appear. Right-click on the icon in the Network connections window and select the Status option from the pop-up menu.

2. Security is a major issue in wireless LANs. The 802.11 standard-based LANs do not provide complete security of information. The security is optional and in many installations, this feature is disabled. The encryption key is common to all the nodes and is stored as a file in the computers. If the computer is stolen, the security key is known to the person who stole the computer. Of course, work is going on to improve the security features of wireless LANs.

3. To plan a LAN, you need to obtain a map of your campus and find out the traffic requirements. If there are two buildings separated by say, more than 500 meters, you need to install two LAN segments and interconnect them. If laying the cable in a building is not feasible, you need to consider wireless LAN option.

4. You can get the details of IEEE 802.11 products from *www.palowireless.com*.

CHAPTER 18

1. You can obtain the details of X.25 products from the site *www.nationaldatamux.com*.

2. In X.25, there will be a number of packet transfers between the switches mainly for acknowledgements, flow control, and error control. Hence, the protocol overhead is very high as compared to Frame Relay, which is mainly used in optical fiber systems in which there are fewer errors.

3. The addressing based on X.121 for a nationwide network should be done in a hierarchical fashion. X.121 address will have 14 digits, out of which 3 digits are for country code. Out of the remaining 11 digits, 2 digits can be assigned to the state, 2 digits for each district, and the remaining 7 digits for the different end systems.

4. There are 18 PAD parameters per X.3 standard. The important parameters are baud rate, local echo mode, idle timer, line delete, line display, character editing, input flow control, and discard output.

CHAPTER 19

1. The type of network element used to interconnect two LANs depends on the protocols used in the two LANs. If both LANs run the same protocols, there is no need for any protocol conversion. If the LANs use different protocols, a network element needs to do the protocol conversion.

2. The earlier routers were only to handle the protocol conversion for data application. Nowdays, routers are capable of handling voice/fax/video services as well. Cisco's AVVID (Architecture for Voice, Video, and Integrated Data) supports multimedia application and the routers are capable of protocol conversion between the PSTN and the Internet. Routers now support IP Version 6.

CHAPTER 20

1. You can obtain the details of Interplanetary Internet at *www.ipnsig.org*.

2. The TCP/IP protocol stack does not perform well on a satellite network because of the large propagation delay. There will be timeouts before the acknowledgement is received, so packets are retransmitted by the sender though the packets are received at the other end. This causes unnecessary traffic on the network. To overcome these problems, spoofing and link accelerators are used.

3. When the communication is one-way only, the TCP protocol cannot be used at the transport layer because acknowledgements cannot be sent in the reverse direction. In such a case, the connectionless transport protocol UDP has to be used. To transfer a file, the file has to be divided into UDP datagrams and sent over the communication link. At the receiving end, the datagrams have to be assembled by the application layer protocol. It is possible that some of the datagrams are received with errors, but retransmission cannot be done because of lack of the reverse link. The receiver has to check every datagram for errors, and if there is an error even in a single packet, the whole file is discarded. The sender may send the file multiple times so that at least once all the datagrams are received without error.

4. If the transmission medium is very reliable, the packets will be received correctly. Also, if there is no congestion, the packets are likely to be received without variable delay and in sequence. Hence, UDP provides a fast data transfer, and the transmission medium is utilized effectively.

5. Stop-and-wait protocol is very inefficient because the communication channel bandwidth is not utilized well. After the first packet is sent, an acknowledgement has to be received, and then only the second packet can be sent. On the other hand, in sliding window protocol, a number of packets can be sent without waiting for acknowledgements. Hence, channel utilization is better if sliding window protocol is used.

CHAPTER 21

1. You can use the procedure described in Exercise #1 of Chapter 15 to obtain the IP address of your computer.

2. In class B IP address format, 14 bits are used for network ID and 16 bits for host ID. Hence, 2^{14} networks can be addressed, and each network can have 2^{16} hosts.

3. In class C IP address format, 24 bits are used for network ID and 8 bits for host ID. Hence, 2^{24} networks can be addressed and in each network 2^8 hosts.

4. The maximum number of addresses supported by IP Version 6 is 340,282,366,920,938,463,463,374,607,431,768,211,456.

5. You can obtain the RFC from the site *www.ietf.org*.

CHAPTER 22

1. The open source for TCP/IP protocol stack is available with the Linux operating system.

2. The Java code for implementation of UDP server and UDP client are given in Listing C.6 and Listing C.7, respectively. The server software is used to transfer a file to the client. The server divides the file into UDP datagrams and sends it. The client will receive each datagram and assemble the file. This code can be tested on a LAN environment.

Listing C.6 UDP server software.

```java
import java.net.*;

public class UDPServer
{
   public static DatagramSocket ds;
   public static int buffer_size=10;
   public static int serverport=555;
   public static int clientport=444;
   public static byte buffer[]=new byte[buffer_size];

   public static void Server() throws Exception
   {
      int pos=0;
      byte b[] = { 'H','e','l','l','o'};
                        ds.send(new             DatagramPacket
(b,b.length,InetAddress.getLocalHost(), clientport));
   }
   public static void main(String args[])
   {
     try{
```

```
      System.out.println("Server is ready");
        ds=new DatagramSocket(serverport);
        Server();
      }catch(Exception e){ }
    }
}
```

Listing C.7 UDP Client software.

```
import java.net.*;

public class UDPClient
{
   public static DatagramSocket ds;
   public static int buffer_size=5;
   public static int serverport=555;
   public static int clientport=444;
   public static byte buffer[]=new byte[buffer_size];

   public static void Client() throws Exception
   {
      while(true)
      {
   DatagramPacket dp = new DatagramPacket(buffer,buffer.length);
         ds.receive(dp);
         byte b[] = dp.getData();
         for(int i=0;i<=b.length;i++)
         System.out.print((char)b[i] + " ");
      }
   }
  public static void main(String args[])
  {
    try{
        System.out.println("Client is ready");
        ds=new DatagramSocket(clientport);
        Client();
    }catch(Exception e){ }
  }
}
```

3. In sliding window protocol used in TCP, the receiver must advertise its window size. Receiver may advertise a small window size due to various reasons such as buffer full. In such a case, the sender has to transmit small segments. This results in inefficient utilization of the bandwidth. To avoid this problem, the receiver may delay advertising a new window size, or the sender may delay sending the data when the window size is small.

4. In a satellite communication system, if the VSATs are receive only, it is not possible for VSAT to send an acknowledgement to the server located at the hub. In such a case, the server at the hub has to use the UDP as the transport layer to transmit the file. The server software will divide the file into UDP segments and broadcast each datagram. The datagram contains the VSAT address as the destination address. The VSAT will receive the datagarams and assemble the file.

5. In sliding window protocol used in the TCP layer, the receiver has to advertise the window size, and the sender must adhere to this size. This may cause the silly window syndrome.

CHAPTER 23

1. You can get the details of FTP from the RFC at *www.ietf.org*.

2. When you interconnect two PCs using an RS232 link, the data is transferred character by character. You need to write the software to implement SMTP and then send the data via the RS232 port.

3. When the information in the directory is stored using a proprietary mechanism, it is not possible to share the directory information among devices. For instance, the address directory stored in a digital diary cannot be transferred to the Outlook Express address book because the directory formats are different. LDAP solves this problem by defining the directory structure, the operations to be performed on the directories, and so on.

4. The standard protocols for file transfer are FTP, TFTP, and FTAM.

5. The network management software has to collect the status of various network elements to ensure the reliable operation of the network. In addition, traffic data has to be collected. The traffic data can be used to identify the busy routes and plan for expansion.

CHAPTER 24

1. When you start your own organization, you need to have a Web address. To obtain a unique URL for your organization, you can go to a site such as *www.register.com* or *www.networksolutions.com* and find out whether the required URL is already registered by someone or is available. If it is available, you can pay the registration charges (about US$70 for two years). Then you can develop your Web site and host it on a Web-hosting service provider's server by paying the hosting charges (about US$10 per month).

2. When you go to *www.register.com*, and give a URL, you will get information about the owner of the URL.

3. XHTML code for displaying text, playing audio and video clips, and displaying images in the background and foreground is given below. You need to have the necessary audio clip, video clips, or the image files in your directory before running this code. The names of these files are indicated in the code itself.

Listing C.8 XHTML code to display text.

```
<?xml version="1.0" encoding="UTF-8"?>
<!DOCTYPE html PUBLIC "-//W3C//DTD XHTML 1.0 Transitional//
EN"
            "http://www.w3.org/TR/xhtml1/DTD/xhtml1-
transitional.dtd">
   <html xmlns="http://www.w3.org/1999/xhtml1" xml:lang="en"
lang="en">

  <head>
  <title>Text Sample</title>
  </head>

<body>
   <h1> XHTML Document welcomes you</h1>
</body>
</html>
```

Listing C.9 XHTML code to play an audio file.

```
<?xml version="1.0" encoding="UTF-8"?>
<!DOCTYPE html PUBLIC "-//W3C//DTD XHTML 1.0 Transitional//EN"
            "http://www.w3.org/TR/xhtml1/DTD/xhtml1-
```

```
transitional.dtd">
   <html xmlns="http://www.w3.org/1999/xhtml1" xml:lang="en"
lang="en">

   <head>
   <title>Audio Sample</title>
   </head>

<body>
   <a href="sample.mid">Audio</a>
</body>
</html>

using embed tag:
   <embed src ="sample.mid" autostart="true" width="200"
height="100" hidden="true" />
using object tag:
<object data ="sample.mid" type="audio/midi" autostart="true"
width="400" height="400" hidden="true" />

Listing C.10 XHTML code to play a video file.
<?xml version="1.0" encoding="UTF-8"?>
<!DOCTYPE html PUBLIC "-//W3C//DTD XHTML 1.0 Transitional//
EN"
             "http://www.w3.org/TR/xhtml1/DTD/xhtml1-
transitional.dtd">
   <html xmlns="http://www.w3.org/1999/xhtml1" xml:lang="en"
lang="en">

   <head>
   <title>Video Sample</title>
   </head>

<body>
   <a href="search.avi">Video</a>
</body>
</html>

using embed tag:
<embed src ="search.avi" autostart="true" width="200"
height="200" loop="true" />
using object tag:
```

```
        <object data ="search.avi" type="video/x-msvideo" autostart="true"
width="400" height="400" hidden="true" />
```

Listing C.11 XHTML Code to display an image in the background.

```
<?xml version="1.0" encoding="UTF-8"?>
<!DOCTYPE html PUBLIC "-//W3C//DTD XHTML 1.0 Transitional//
EN"
              "http://www.w3.org/TR/xhtml1/DTD/xhtml1-
transitional.dtd">
   <html xmlns="http://www.w3.org/1999/xhtml1" xml:lang="en"
lang="en">

   <head>
   <title>Background Image Sample</title>
   </head>

<body background="TajMahal.jpg">
<br /><br /><br /><br />
<h1 align="center">Sample Text</h1>

</body>
</html>
```

Listing C.12 XHTML code to display an image in the foreground.

```
<?xml version="1.0" encoding="UTF-8"?>
<!DOCTYPE html PUBLIC "-//W3C//DTD XHTML 1.0 Transitional//
EN"
              "http://www.w3.org/TR/xhtml1/DTD/xhtml1-
transitional.dtd">
   <html xmlns="http://www.w3.org/1999/xhtml1" xml:lang="en"
lang="en">

   <head>
   <title>Image Sample</title>
   </head>

<body>
   Sample Text
   <img src="rose.gif" alt="Rose Blum" />
</body>
</html>
```

4. Search engines can be compared based on the number of Web pages searched, the amount of time it takes to present the results, and the relevance of information presented to the user. You can search for your name using different search engines the engine that gives the highest number of results is the best! Intelligent search engines are now being developed, and they will be able to present more relevant information.

5. Some of the Web-based learning portals are:

   www.techonline.com

   www.elearn.cdacindia.com

   www.gurukulonline.co.in

CHAPTER 25

1. An e-learning portal has to provide a user-friendly interface to make learning a pleasant experience. E-learning has many advantages. You can learn at your own pace at a time of your convenience. You need not always be at the same place and you can access the portal from anywhere on Earth. However, e-learning has some drawbacks, too. You cannot have face-to-face interactions with the instructors, and hence you need to use facilities such as e-mail, and chat to interact with the instructors and other students. E-learning is now catching up very fast, and M.S. and M.B.A. degrees are being offered by leading universities through e-learning mode.

2. An e-learning portal must have the following modules:

- An e-lecture module that delivers the lecture material. The lecture material has to be supplemented with lecture notes. The lecture notes can be read by inserting voice files in the presentation.

- E-mail module for interaction between the students and the instructors.

- Chat module for online chatting among the students and the instructors.

- Online examination module to test the students.

- An electronic notice board to make announcements periodically.

- A database module that stores complete information about the various students.

- A virtual laboratory module that facilitates experimentation.

3. To learn scientific and engineering concepts, laboratory work plays an important role. To facilitate hands-on lab work in e-learning, virtual laboratories are used. As an example, suppose you want to learn about digital signal processing. Through the e-learning mode, you learn a lot of theory on DSP. But unless you write some DSP code and run it, you will not get a good understanding of the theory. You can use a virtual lab—a DSP kit will be connected to the server in which the e-learning portal is hosted. You can write the code, run it on the DSP kit, and view the results on your screen. So, even if you do not have a DSP kit, you gain experience on using it—that is the power of virtual laboratories. You can use the virtual laboratory provided on *www.techonline.com*.

4. Web services use XML, SOAP (Simple Object Access Protocol), and UDDI for directory service. Sun ONE development tools and Web Sphere are used for developing the Web services. You can obtain the details from *www.sun.com, www.ibm.com, www.microsoft.com*.

5. What are J2EE, J2SE, and J2ME?

    J2EE stands for Java 2 Enterprise Edition and is used to develop enterprise applications. Using J2EE SDK, you can develop server components such as Java Server Pages, servlets, and Enterprise Java Beans. J2SE stands for Java 2 Standard Edition and is used to develop Java applications on desktops. J2ME stands for Java 2 Micro Edition and is used to develop applications for small devices such as mobile phones, pagers, and PDAs. Such devices have small memory and small display, and also do not have secondary storage such as hard disk. For such devices, a small Java Virtual Machine, known as Kilobytes Virtual Machine (KVM), is developed by Sun Microsystems.

6. XML is a markup language. It is a meta-language—you can develop new markup languages using XML. WML, XHTML, SyncML, and VoiceXML are derived from XML.

7. A networked information appliance is a device that is networked using TCP/IP protocols. Consider a Web camera. The Web camera can have the necessary hardware and software to run the TCP/IP protocol stack software and the HTTP server software. So, the Web camera is an embedded system that acts as a Web server. The Web camera can be connected to the Internet. Anyone can access the Web camera through the Internet. The Web camera can be installed in a floor of a factory to monitor the functioning of important equipment. Such devices are called networked information appliances. Networked information appliances can be of great use to obtain real-time weather information, monitoring of important systems in nuclear plants, and so forth.

8. For many governments throughout the world, e-governance is the buzzword. In all countries, decision-making by governments is very slow. E-governance is a concept that facilitates collection of the right information by governments, use of that information for better planning, execution of plans through electronic monitoring, and so forth. The citizens can be benefited immensely by e-governance. Payment of taxes can be done very easily. Even obtaining birth certificates, driving license, and so on, governments are providing Internet-based services. Using the Internet, the government departments can disseminate the information so that the governments becomes more transparent. In a truly democratic setup, the government can obtain feedback on proposed legislation using electronic means.

9. In Internet voting, citizens can vote sitting at their place of work or at their homes. However, it is a long way before Internet voting is used by governments due to security reasons.

CHAPTER 26

1. When you buy a prepaid telephone calling card, the card contains a number and a telephone number you need to call. The procedure for verification is as follows:

- You dial the telephone number given on the calling card.
- You hear a message to dial the calling card number.
- You dial the calling card number.
- The server checks the database and finds out the value of the calling card, how much has already been spent, and how much is still left.
- You will hear this information, and then you can dial the telephone number of your choice.
- You can converse with the person.
- The total cost of the call will be deducted from the remaining amount, and the new amount will be stored in the database.

The entire signaling is done through the SS7 network.

2. In a GSM network, a separate data communication network is used to carry the signaling information. This network is based on SS7. Even for communication between mobile phone and the base station, there will be a

separate time slot that carries the signaling information. The voice slots are only to carry the voice data.

3. You can obtain the information on SS7 over ATM from *www.cisco.com*.

4. SS7 protocol stack is available from *www.openss7.org*.

CHAPTER 27

1. Basic rate interface (BRI) provides two 64kbps B channels and one 16kbps D signaling channel. The services supported are voice, slow scan TV at 64kbps, data services such as e-mail, file transfer, and remote login.

2. Primary rate interface (PRI) provides thirty 64 kbps B channels and one 64kbps signaling channel. High-speed services such as video conferencing at 384kbps and video streaming at 2.048Mbps can be supported.

3. In Telemedicine, patient information has to be sent to the doctor located at a distant place. Ideally, the video of the patient is sent to the doctor. In addition, x-ray, ECG, EEG, and so forth can be sent. A full-fledged tele-medicine application requires 2Mbps link.

4. You can obtain the details from *www.cnet.com* and *www.askey.com*.

CHAPTER 28

1. Information about Frame Relay commercial products can be obtained from *www.lucent.com* and *www.cisco.com*.

2. You can study the FRAD interfaces provided by Cisco equipment at the Web site *www.cisco.com*.

3. In X.25, packets are exchanged for flow control and error control. In Frame Relay, there is no layer 3 protocol.

4. When optical fiber is the transmission medium, the error rate is very low. Hence, the Frame Relay protocols enable fast transmission of data without the need for flow control and error control.

5. Because low bit rate coding also gives good quality voice, the bandwidth can be better utilized. Using low bit rate coding allows fast packet transmission, so the delay will be less.

CHAPTER 29

1. Type 1 ATM adaptation layer is used for supporting real-time voice communication and video conferencing.

2. You can obtain the details at *www.cisco.com* and *www.nortelnetowrks.com*.

3. To support voice services with 64kbps PCM over ATM, Type 1 AAL protocol is used.

4. Type 2 AAL is used for variable rate video transmission. For fixed rate transmission, Type 1 AAL is used.

5. The ATM cell has 53 bytes, out of which 5 bytes contain the header information. Hence, the overhead is 5/53 or about 10%.

CHAPTER 30

1. In automatic paging systems, there is no need for an operator at the paging terminal interface. A user can dial into the paging terminal's computer and speak the paging message and the pager number. The automatic speech recognition software will convert the spoken message into text format and then send the message. However, with the present state-of-the-art, speech recognition accuracy is not very high. An alternative mechanism for automatic paging is to use a computer to send text directly to the paging terminal. The paging terminal can be Web-enabled. Any user can access the terminal's computer over the Internet and submit a message for paging.

2. Each message is of 30 seconds duration. Voice coding is done at 5.3kbps. To store 10 messages, the storage requirement is 300 seconds × 5.3 kbps = 1590kbps = 198.8KBytes.

3. For a voice paging system that uses 2.4kbps voice coding, the memory requirement on a voice pager that has to store 20 messages each of 30 seconds duration is 20 × 30 × 2.4 Kbits = 6 × 240 Kbits = 1440 Kbits = 180 KBytes.

4. You can obtain the POCSAG decoder software from *www.foni.net* or *www.qsl.net*.

CHAPTER 31

1. You can obtain the GSM codec software from the Web site *kbs.cs.tu-berlin.de/ ~jutta/toast.html*.

2. You can obtain the details of the GSM standard at *www.3gpp.org*.

3. You can obtain the details form *www.qualcomm.com*.

4. When a mobile phone is lost, the first thing to do is to call that number and check whether it is ringing. If the thief is using the mobile phone without removing the SIM card, the operator can trace approximately, but not precisely, where the thief is located. However, the operator can disable the phone so that both incoming and outgoing calls be barred. The GSM standard provides another feature using which lost mobile phones cannot be used. If the IMEI is registered with the operator, then the thief cannot use the mobile phone.

CHAPTER 32

1. The format of the location data received at the RS232 interface needs to be obtained from the data sheet of the GPS receiver vendor. The code given in Listing C.13 is for the Conexant (*www.conexant.com*) GPS receiver. As you can see from the code, the longitude and latitude positions are obtained as a series of characters. Figure C.11 gives a digitized map.

Listing C.13 Location data from the GPS receiver.

```
/* Receive the data from the GPS modem */
void CGPSDlg::GpsReceiver()
{
    CString temp;
    CPoint point;
    int iLat,iLon;
    unsigned char ch[1],Lat[12],Lon[13],Alt[11],string[60];
    unsigned int i;
    CSerial serObj;
    if(!serObj.Open(1,9600,'n',8,1))
    // Comm port opening using given settings
    {
        AfxMessageBox("failed to open Comm Port");
        return;
```

```
        }
        while(1)
        {
            if(serObj.ReadData(ch,1) > 0)
            {
                    if((ch[0] == 'g') || (ch[0] == 'G'))
                    {
                            serObj.ReadData(ch,1);
                            if((ch[0] == 'p') || (ch[0] == 'P'))
                            {
                                    serObj.ReadData(ch,1);
                                    if((ch[0] == 'g') || (ch[0] == 'G'))
                                    {
                                            serObj.ReadData(ch,1);
                                    if((ch[0] == 'g') || (ch[0] == 'G'))
                                            {
                                    serObj.ReadData(ch,1);

                                    if((ch[0] == 'a') || (ch[0] == 'A'))
                                                    {
                                            serObj.ReadData(ch,1);    // ','
                                                    if(ch[0] != ',')
                                                    continue;
                                    serObj.ReadData(ch,1); //first char of
time of ','
                                                        if(ch[0] == ',')
                                                            continue;

serObj.ReadData(string,55);
                                        //remaining char of time
                                                    string[55] = '\0';
                                        // Assigning Latitude value

                                                    Lat[0] = string[6];
                                                    Lat[1] = string[7];
                                                    Lat[2] = string[8];
                                    Lat[3] = string[9];
                                    Lat[4] = string[10];
                                    Lat[5] = string[11];
                                    Lat[6] = string[12];
                                    Lat[7] = string[13];
                                    Lat[8] = string[14];
```

```
                             Lat[9] = ' ';
                             Lat[10]  =  string[16];
                                 Lat[11]  =  '\0';
           //  Assigning  Longitude  value
                             Lon[0]  =  string[18];
                             Lon[1]  =  string[19];
                             Lon[2]  =  string[20];
                             Lon[3]  =  string[21];
                             Lon[4]  =  string[22];
                             Lon[5]  =  string[23];
                             Lon[6]  =  string[24];
                             Lon[7]  =  string[25];
                             Lon[8]  =  string[26];
                             Lon[9]  =  string[27];
                                 Lon[10]  =  ' ';
                             Lon[11]  =  string[29];
                                 Lon[12]  =  '\0';
       //  Assigning  Altitude  value
                             for(i=0;i<10;i++)
                                 {
                             Alt[i]  =  string[41+i];
                             if(string[41+i]  ==  'M')
                                 {
                                 Alt[41+i-1]  =  ' ';
                                 Alt[41+i+1]  =  '\0';
                                 break;
                                 }
                                 }

                   temp.Format("Lat  :  %s",Lat);

m_lat.SetWindowText(temp);

                   temp.Format("Lon  :  %s",Lon);

m_lon.SetWindowText(temp);
                   temp.Format("Alt  :  %s",Alt);

m_alt.SetWindowText(temp);

                             Lat[4]  =  '\0';
                             Lon[5]  =  '\0';
```

```
                                               temp.Format("%s",Lat);
                                                  iLat = atoi(temp);


  temp.Format("%s",Lon);
                                                  iLon = atoi(temp);

                                                  point.x = iLon;
                                                  point.y = iLat;
     // Plot the location on the map

  m_location.SetWindowPos(NULL,point.x,point.y,0,0,SWP_NOSIZE);
                                                  continue;


                                      }
                                  }
                              }
                          }
                      }
                  }
              }
          }
```

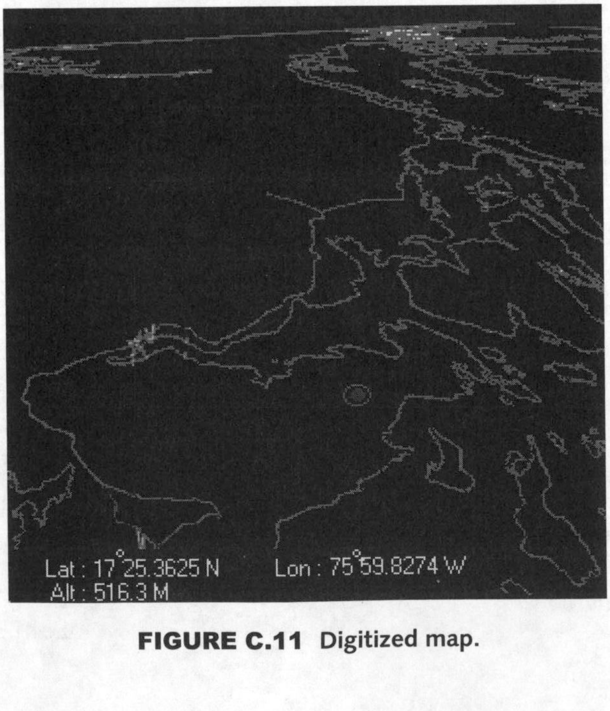

FIGURE C.11 Digitized map.

2. To develop a navigation system, you need to calculate the distance between two points when the longitude and latitude of the two points are known.

   At point P1, longitude = x1 and latitude = y1

   At point P2, longitude = x2 and latitude = y2

   Distance between P1 and P2 $= \cdot \ddot{O}((x2 - x1)^2 + (y2 - y1)^2)$

   Note that 1° = 60 minutes, 1° = 3600 seconds (1 minute = 60 seconds) and 1° corresponds to 108 KM (1 second corresponds to 30 Meters)

   The listing of the program to calculate the distance given the longitude and latitude of two points is given in Listing C.14.

Listing C.14 Calculation of distance between two points given their latitude and longitude.

```
#include "stdafx.h"
#include <cmath>

int main(int argc, char* argv[])
{
    float p1Longitude,p1Latitude,p2Longitude,p2Latitude;
    double res1,res2,fres;
    printf("\n \n Distance between two points when Longitude
        and Latitude are known \t\n\n");
    printf("Longitude of I point: ");
    scanf("%f",&p1Longitude);
    printf("Latitude of I point: ");
    scanf("%f",&p1Latitude);
    printf("\nLongitude of II point: ");
    scanf("%f",&p2Longitude);
    printf("Latitude of II point: ");
    scanf("%f",&p2Latitude);

    res1 = pow((double)(p2Longitude - p1Longitude),2.0);
    res2 = pow((double)(p2Latitude - p1Latitude),2.0);

    fres = pow((double)(res1 + res2),0.5) * 108;
    //(Since 1 degree = 108 KM)

    printf("\nDistance between two points: %lf KM \n",fres);

    return 0;
}
```

The output screen is shown in Figure C.12.

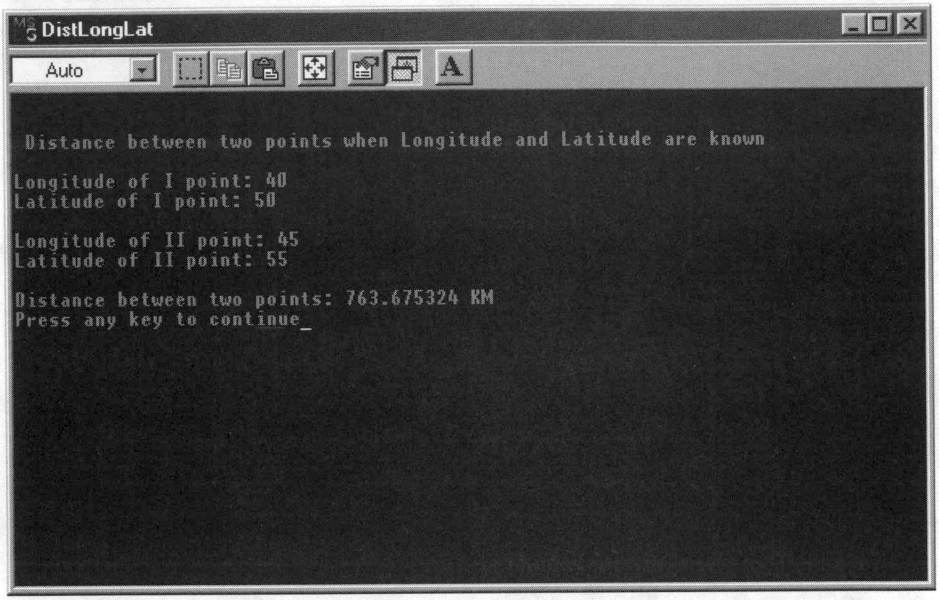

FIGURE C.12 Screenshot showing the output of the program to calculate distance.

3. Information about the GIS software can be obtained from the following sites: *www.mapinfo.com*, *www.freegis.org*, and *www.opensourcegis.org*.

4. In differential GPS, the mobile devices will have a GPS receiver and an FM receiver using which the location parameters can be found very accurately. If the mobile device is also fitted with another FM transmitter that continuously transmits its GPS data to a centralized location, then the mobile device can be continuously tracked. Without the owner's knowledge, such a gadget can be fixed to a car and the car movement can be continuously tracked. Such applications are required for organizations involved in public safety. But certainly, this is invading the privacy of individuals.

CHAPTER 33

1. You can download WAP toolkits from the Web sites of Nokia, Ericsson, Motorola, and others. These tookkits can be used to develop and test mobile

applications in the lab before deploying the applications in the actual mobile network environment. Using WML and WML Script, you can create good applications and test them. The framework for developing a mobile commerce application using the push model is given in the following.

Assume that there is a Web site that provides the facility to buy various items through your mobile phone. The Web site also sends the latest information about new arrivals in the items of your choice. To start with, you need to register yourself with the Web site and give details such as your mobile phone number, your preferences, and the frequency with which you would like to receive the advertisement information. A sample registration form is shown in Figure C.13.

Once you are a registered user of the Web site, you keep getting information from the Web site using the push model. You can also use the mobile cart to place an order for different items.

To achieve this functionality, you need to create the necessary HTML files, a database that stores the complete information about all the registered users, and Java programs that run on the server and send the information to the users. You also need to create the WML file that provides the interface to the user on the mobile device.

The screenshot with the first screen of the mobile commerce application is shown in Figure C.14.

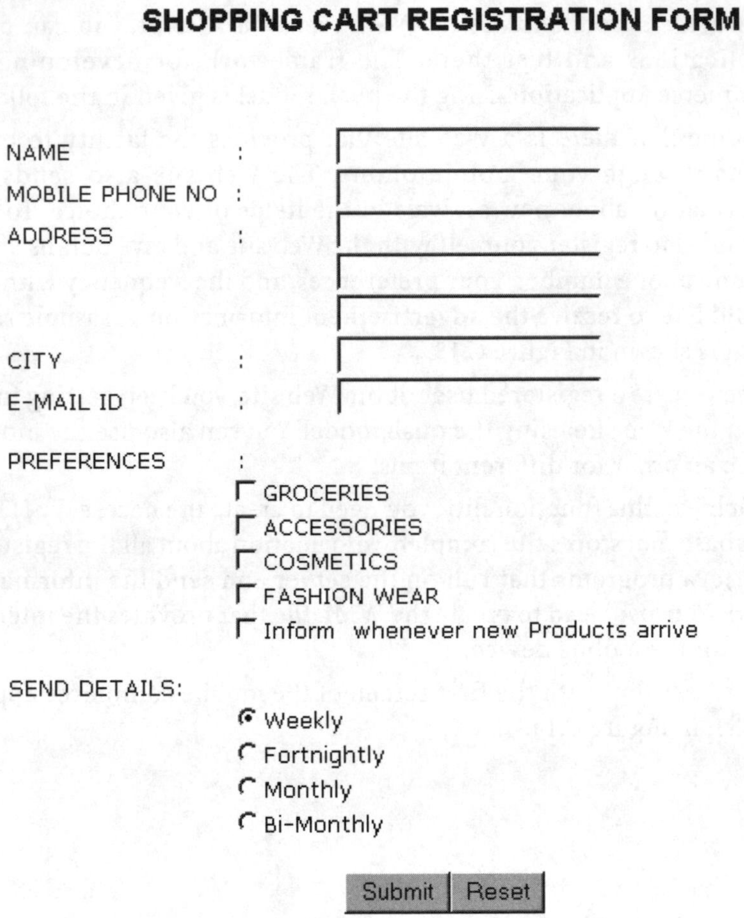

SHOPPING CART REGISTRATION FORM

NAME : _____

MOBILE PHONE NO : _____

ADDRESS : _____
_____
_____

CITY : _____

E-MAIL ID : _____

PREFERENCES :
☐ GROCERIES
☐ ACCESSORIES
☐ COSMETICS
☐ FASHION WEAR
☐ Inform whenever new Products arrive

SEND DETAILS:
◉ Weekly
○ Fortnightly
○ Monthly
○ Bi-Monthly

Submit | Reset

FIGURE C.13 Registration form for mobile push application.

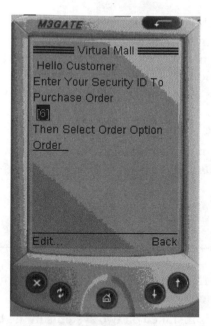

FIGURE C.14 M-commerce application on a mobile phone.

2. J2ME facilitates the development of mobile applications using the Java programming language. Keeping in mind the limitations of the mobile devices such as small display, limited functionality keypad, and so on. The J2ME provides limited GUI capabilities and is optimized to work on devices with less memory. You can download the J2ME kit from Sun Microsystems Web site.

3. Binary runtime environment for wireless (BREW), developed by Qualcomm, facilitates development of mobile applications for CDMA-based wireless networks. You can download the kit from Qualcomm's Web site *www.qualcomm.com*.

4. Mobile phones do not have an operating system. However, devices such as personal digital assistants and palmtops have an operating system. The operating systems that are ported on the mobile devices are known as mobile operating systems. Palm OS, developed by Palm Computing Inc., is the most famous mobile OS. Other operating systems that now have a large installation base are Symbian OS, Windows CE and Windows XP. Embedded Linux has also gained importance in recent years.

5. When a mobile device moves from one service area to another service area, to ensure that there is no disruption in the service, mobile IP is required. In mobile IP, the mobile device is given two addresses, called home address and care-of address. The home address is a permanent address, and the care-of address is a temporary address that is given when the mobile device is in a foreign area. The packets addressed to the mobile device will be received by the home network and forwarded to the care-of address.

6. You can obtain information from the following sites: *www.nokia.com*, *www.ericsson.com*, *www.microsoft.com*.

CHAPTER 34

1. Microsoft's NetMeeting software can be used for chat, audio conferencing, and desktop video conferencing.

   If NetMeeting is installed on two systems in the LAN, you can do video chat using the following steps:

   - Open NetMeeting application on two systems.
   - On the caller PC, type the receiver's system name or receiver's system IP address in the address box, as shown in the Figure C.15 and click the Place Call button.

FIGURE C.15 NetMeeting.

- The screen in Figure C.16 will appear on the receiver's system, indicating an incoming call.

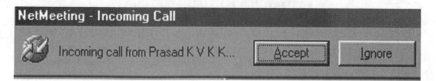

FIGURE C.16 NetMeeting incoming call indication.

- After the receiver accepts the caller's request, the screen in Figure C.17 will appear at the receiver's system.

FIGURE C.17 NetMeeting video.

- If one of the users wants to chat with another user, he can click on Chat mode button, then the screen in Figure C.18 will appear.

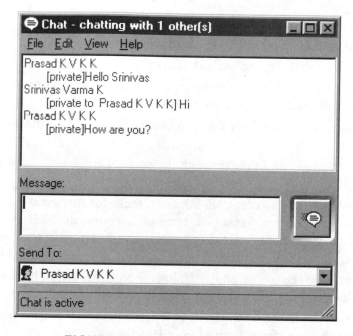

FIGURE C.18 Chat using NetMeeting.

2. The first and second generation fax machines were called Group I and Group II fax machines and used electrochemical processes. These are now obsolete, and current fax machines are called Group III fax machines. These devices transmit at speeds in the range 9.6kbps to 14.4kbps over analog telephone lines. Group IV fax machines operate at speeds up to 64kbps.

3. You can obtain the details from *www.cisco.com*.

4. You can obtain the source code for H.323 codecs from *www.openh323.org*.

5. You can obtain the details from *www.javasoft.com*.

CHAPTER 35

1. You can develop text-to-speech conversion software using words as the basic units. You can record a large number of words and store each word in a separate file. The program for text-to-speech conversion has to do the following:

  ■ Read the input sentence.

- Remove punctuation marks.
- Convert all capital letters to small letters.
- Expand abbreviations such as Mr., Prof., and Rs.
- Scan each word in the sentence and pick up the corresponding sound file from the database of spoken words.
- Create a new sound file that is a concatenation of all the sound files of the words.
- Play the new sound file through the sound card.

2. To design an IVR system for a telebanking application, you need to create a database (in MS Access, MS SQL, or Oracle, for instance) that contains bank account number, password, type of account, and balance amount. The information that needs to be stored in the database is: account holder name, address, account number, type of account, and present bank balance.

   You also need to design the dialogues for interaction between the account holder and the IVR system. A typical dialogue is as follows:

   IVR: Welcome to ABC Bank's IVR system. Please dial 1 for information in English, dial 2 for information in Hindi.

   User: Dials 1.

   IVR: Please dial your account number.

   User: Dials 2346.

   IVR: The account number you dialed is two three four six. Please dial your password.

   User: Dials 4567.

   IVR: You have a savings bank account. The present balance is Rupees Ten Thousand Four Hundred. Thank you for calling the IVR.

3. You can experiment with Microsoft's Speech SDK to develop text-to-speech and speech recognition–based applications. IBM's WebSphere can also be used for development of such applications.

4. For most Indian languages, the number of phonemes is about 60. The number of diphones is about 1500. The number of syllables is about 20,000. You need to store nearly 200,000 words if you want to use the word as the basic unit for text-to-speech conversion. It is better to use syllables.

5. A call center consists of the following components:

- A local area network in which one node (computer) is given to each agent.
- A server that runs the customer relations management software.

- A PBX with one extension to each agent.
- Automatic call distribution (ACD) software.
- Fax-on-demand software.
- A interactive voice response system.

6. Nortel and Cisco are the two major suppliers of call center equipment. You can get the details from their Web sites *www.cisco.com* and *www.nortelcommuncations.com*.

CHAPTER 36

1. For PANs, Bluetooth and IEEE 802.11 are the most popular technologies.

2. You can get information about SyncML from *www.syncml.org* and also from *www.symbian.com*, *www.nokia.com*, *www.ibm.com*, *www.motorola.com*, and *www.ericsson.com*.

3. You can obtain the Bluehoc Development Kit from *www-124.ibm.com/ developerworks/opensource/bluehoc*.

4. Bluetooth provides the radio link between various devices that form a network when they are close to one another. This radio link can be used to transfer WAP applications. Consider the following example:

   At a railway station, a WAP server can be installed that provides the latest information about the train arrivals/departures and the platform numbers. This WAP server can be Bluetooth enabled. When a Bluetooth-enabled mobile device comes near this WAP server, automatically the server can send the details of the train information to the mobile device over the Bluetooth link. This is known as WAP Kiosk.

CHAPTER 37

1. You can get the details of TMN products from the following sites: *www.cisco.com*, *www.nortelnetworks.com*, *www.ericsson.com*, *www.lucent.com*, and *www.sun.com*.

2. You can obtain the details from *www.netmansys.fr* and *www.eurescom.de*.

CHAPTER 38

1. To store one second of voice conversation using PCM, the storage requirement is 8 Kbytes. To store 100 hours of conversation,

   Storage requirement = 360,000 × 8 KB = 360 × 8MB = 2880MB = 2.8GB

2. Listing C.15 encrypts the given message using RSA algorithm in C.

Listing C.15 Encryption using RSA algorithm.

```c
#include "stdio.h"
#include "stdlib.h"
#include "math.h"

int main(void)
{
    int p=7,q=11;
    int n,e,d,res1,k;
    int flag=0;
    int i=0,ch;
    double fraction, integer;
    double ex,d1;
    char Message[] = {9,8,4,8,1,4,4,3,3,7,'\0'};
    char EncryptedMessage[20];
    char DecryptedMessage[20];
    n = p * q;
    k = (p -1) * (q - 1);
    e = p;
    while(flag != 1){
        i++;
        res1 = (k*i) + 1;
        if((res1%e) == 0){
            d = res1 / e;
            flag = 1;
        }
    }
    printf("\n\n\t\t\t\tRSA Algorithm\n\n");
    printf("\n Public key: <%d,%d> ",e,n);
    printf("\n Private key: <%d,%d> \n",d,n);
    printf("\n Original Message: ");
    for(i=0;i<strlen(Message);i++){
        printf("%d ",Message[i]);
        ch =(int)(Message[i]);
```

```
    ex = pow((double)ch,(double)e);
    d1 = (ex / n);
    fraction = modf(d1, &integer);
    EncryptedMessage[i] = (char)(fraction * n);
}
EncryptedMessage[i] = '\0';
printf("\n Encrypted Message: ");
for(i=0;i<strlen(EncryptedMessage);i++){
    printf("%d ",EncryptedMessage[i]);
}
printf("\n");
return 0;
}
```

When you execute the program, the public key, private key, original message and encrypted message will be displayed as shown in Figure C.19.

FIGURE C.19 RSA algorithm output screen.

3. You can obtain the details form *www.iseeyes.com*, *www.audiosoft.com*.

4. You can obtain Steganography software from *www.outguess.com* or *www.stegoarchive.com*.

5. The hackers use trial-and-error methods to break the algorithms, and hence statistically, they will succeed! Every credit card holder is given a PIN, which is a 4-digit number. If the hacker tries these numbers systematically, he has to try all the possible 9999 combinations to succeed. However, the credit card company generates the PIN using an algorithm, and if the algorithm is stolen, the job is very simple!

6. To tap a telephone, you need to have a PC into which the interactive voice response hardware (a PC add-on card) is plugged in. The IVR hardware has to be connected in parallel to the telephone. The software running on the PC should be automatically invoked when the telephone is lifted, and the entire conversation needs to be recorded on the PC.

Index

E

EBCDIC, 46
EGP, 282
EIA, 129, 222
Electronic Industries Association , 129, 222
Electronic mail, 359, 576
Email, 359, 576
Entropy of an information source, 19
Error correction, 69
Error detection, 63
Ethernet addresses, 237
Ethernet frame format, 238
Ethernet LAN, 233
ETSI, 129, 160, 165
European Telecommunications Standards
      Institute, 129, 160, 165
Exterior gateway protocol, 282

F

Fading (of terrestrial radio), 34
Fast Fourier transform, 55
Fax interface unit, 540
Fax mail, 576
Fax over IP, 537, 540
Fax transmission over PSTN, 538
FDDI LAN, 246
FDM, 82
FDMA/TDMA, 98
FECN, 425–426
FH, 100
File transfer protocol, 282
File transfer, 360
Firewalls, 368
Flooding protocol, 316
Forward explicit congestion notification, 425
FPLMTS, 513
FRAD, 430
Frame relay, 421
Frame relay frame format, 424
Frame relay network, 422
Frequency hopping, 100

Frequency modulation, 113
Frequency shift keying, 117
FSAN, 194
FTP, 282
Full services access networks, 194
Full-duplex communication, 8
Future Public Land Mobile
      Telecommunication Systems, 513

G

G.711 (network protocol), 527
G.723 (network protocol), 527
G.723.1 (network protocol), 545
Gatekeeper, 526
Gateway GPRS support node, 511
Gateway, 362, 526
Gaussian frequency shift keying, 590
General packet radio service, 511
Generic communication system, 16
GFSK, 590
GGSN, 511
Global positioning system, 483
Global system for mobile communications,
      457, 463–465
Globalstar, 478
GPRS, 511
GPS, 483
Ground segment (of satellite communication), 171
GSM, 457, 463–465
   architecture, 464
   services, 465

H

H.245 (network protocol), 527
H.261 (network protocol), 527
H.263 (network protocol), 527
H.323 (network protocol), 525, 542–543
Half-duplex communication, 8
HDLC, 224, 226
HDLC frame structure, 226

V

VC, 423
VCC, 436
VCI, 438
Very small aperture terminal, 171, 173
Video coding, 56
Video conferencing (QCIF format), 58
Video messaging, 577
Video over IP, 532
Virtual circuit, 206
Virtual connection, 423
Virtual private network, 370–371
Vocoding, 52
Voice dialing, 576
Voice/fax communication, surveillance, 627
Voice mail, 576
Voice paging, 454
Voice user interface, 571
VoiceXML, 577
VoIP, 529, 530
VPC, 436
VPI, 438
VPN, 370–371
VSAT, 171, 173
VUI, 571

W

WAE, 504
WAN, 260
WAP, 496
Wave division multiplexing, 87, 187
Waveform coding, 48
WDM, 87, 187

WDP, 505–506
Web-based learning, 362
Wide area networking, 259
WiFi, 251
Wireless access to the Internet, 370
Wireless application environment, 504
Wireless application protocol, 496
Wireless Datagram Protocol, 505–506
Wireless Fidelity, 251
Wireless LANs, 248
Wireless local loop, 135, 138, 152
Wireless Profiled HTTP, 508
Wireless Profiled TCP, 509
Wireless Profiled WTLS, 509
Wireless session protocol, 505–506
Wireless transaction protocol, 506
Wireless transport layer security, 505–506
WLAN, 248
WLL, 135, 138, 152
WML, 503
WMLScript, 504
World Wide Web, 347, 360
WP-HTTP, 508
WP-TCP, 509
WP-TLS, 509
WSP, 505–506
WTLS, 505–506
WTP, 506

X

X.121 (network protocol), 267
X.25 (network protocol), 260, 428
X.500 (network protocol), 351